# Recent Advances in Sensors for Chemical Detection Applications

# Recent Advances in Sensors for Chemical Detection Applications

Guest Editor

**Michele Penza**

MDPI

Basel • Beijing • Wuhan • Barcelona • Belgrade • Novi Sad • Cluj • Manchester

*Guest Editor*
Michele Penza
Department for Sustainability
ENEA, Italian National
Agency for New
Technologies, Energy and
Sustainable Economic
Development
Brindisi
Italy

*Editorial Office*
MDPI AG
Grosspeteranlage 5
4052 Basel, Switzerland

This is a reprint of the Special Issue, published open access by the journal *Sensors* (ISSN 1424-8220), freely accessible at: https://www.mdpi.com/journal/sensors/special_issues/FG5030JOZ2.

For citation purposes, cite each article independently as indicated on the article page online and as indicated below:

| Lastname, A.A.; Lastname, B.B. Article Title. *Journal Name* **Year**, *Volume Number*, Page Range. |
|---|

**ISBN 978-3-7258-5537-7 (Hbk)**
**ISBN 978-3-7258-5538-4 (PDF)**
**https://doi.org/10.3390/books978-3-7258-5538-4**

# Contents

# About the Editor

**Michele Penza**

Michele Penza earned a degree in Physics from the University of Bari, Italy. He worked with CNRSM SCpA (IT S&T Park) before joining ENEA in 2001. He currently manages regional, national, and international research projects and teams on material science, sensors, and solid-state devices at the Brindisi Research Center, Italy. His research interests include sensor materials, nanomaterials, gas sensors, portable sensor systems, functional applications, and environmental technologies. He received the ENEA E2 Excellence Special Mention in Environmental Research in 2008. He is the author of three national patents, 180 publications, and three book chapters, and has delivered over 40 invited or keynote talks. He has co-organized scientific meetings, chaired sessions, and served on international committees. He has acted as expert for FP7 and HORIZON projects, reviewer, SME research manager, and associate or guest editor of several Special Issues and Topics. He chaired COST Action TD1105 EuNetAir—European Network on New Sensing Technologies for Air-Pollution Control and Environmental Sustainability (2012–2016) and headed the Laboratory Functional Materials and Technologies for Sustainable Applications in Brindisi (2015–2024). He serves as ENEA Delegate in local ecosystems for education, training, research, and innovation, and coordinated the EIT Raw Materials Hub—Regional Center Southern Italy (2019–2022). He is Section Editor-in-Chief and Associate Editor of *Sensors* (MDPI), Associate Editor of *Chemosensors* (MDPI) and *Journal of Sensors and Sensor Systems* (Copernicus), and Academic Editor of *Journal of Sensors* (Wiley). He is Adjunct Professor at the University of Salento, Department of Engineering for Innovation. He holds the Italian National Scientific Qualification as a Full Professor in Experimental Physics of Matter (02/B1 - ASN 2021–23, valid 2023–2035) and ranks among the world's Top 2% most-cited scientists (2021–present) in Analytical Chemistry and Applied Physics.

# Preface

The Reprint focusses on low-cost sensors for chemical sensing applications. Multidisciplinary approach includes advanced materials, transducers, modeling, algorithms, proof of concepts, sensing solutions and applications. Recent advances have been reported for chemical detection and gas monitoring.

**Michele Penza**
*Guest Editor*

sensors

MDPI

*Editorial*

# Special Issue on Recent Advances in Sensors for Chemical Detection Applications

**Michele Penza**

Brindisi Research Center, Division of Technologies and Advanced Materials for Sustainable Manufacturing Industry, Department for Sustainability, ENEA—Italian National Agency for New Technologies, Energy and Sustainable Economic Development, I-72100 Brindisi, Italy; michele.penza@enea.it

**Abstract:** This Special Issue based on 15 articles/reviews focusses on low-cost sensor technology, gas sensors, chemical sensors, advanced active materials, sensing nanomaterials, sensor nodes, hardware innovation, data communication, system integration, sensor testing, functional characterization, sensor modeling, processing and correction algorithms, new sensing solutions, advanced proof of concepts, and chemical detection applications. Proper calibration techniques of chemical sensors have been explored, both in the laboratory and in field applications. Sensing solutions have been applied in the context of biochemical detection and gas monitoring.

**Keywords:** gas sensors; chemical sensors; sensor active materials; advanced functional nanomaterials; portable chemical sensor-systems; chemical sensor modeling; IoT devices; sensor engineering; chemical sensor applications; new concepts in chemical sensing

## 1. Introduction

Chemical detection based on low-cost sensor technologies [1–4] has become increasingly popular for several emerging applications, such as industrial process control, chemical threat monitoring, green chemistry, environmental sustainability, smart cities, hydrogen economy, energy saving, wearable devices, IoT applications, public health protection, sustainable mobility, autonomous vehicles, and community sensing.

Functional materials [5–9] are cross-cutting technologies for chemical detection to provide advanced gas sensors at the laboratory level and real-world testing in many industrial applications. Low-power consumption, high-quality data, and optimal performance are some important parameters for a new generation of low-cost chemical sensors. Portable sensor systems and wireless sensor networks are typical approaches to monitoring chemical threats in long-term operation [10–14].

Current low-cost sensor technologies include numerous types of transducers, such as chemiresistors, electrochemical, transistor, optical, mass-sensitive, catalytic, and other hybrid configurations, evolving quickly with different open questions and considerable challenges, such as sensitivity, selectivity, stability, limit of detection, calibration, accuracy, and so on. Understanding the limitations and capabilities of current low-cost sensor technologies for chemical detection is a key issue for future applications.

This Special Issue of the *Sensors* (MDPI journal), titled "Recent Advances in Sensors for Chemical Detection Applications", brings together fifteen articles presenting the recent developments in these areas. The publications range from advanced sensing materials, sensor devices, chemical detection, biosensing, calibration algorithms, new sensing solutions, proof of concepts, practical applications for environmental monitoring, and bio-sensing measurements. The featured research highlights significant advancements in biochemical sensing applied as smart technology for a sustainable future based on environmental, social, and governance methodology.

## 2. Overview of Published Papers

José Carlos Santos-Ceballos et al. (Contribution 1—Article) present a study devoted to the electrochemical modification of laser-induced graphene (LIG) with polyaniline (PANI), which led to the development of a chemo-resistive nanocomposite (PANI@LIG) for detecting ammonia levels at room temperature. The composite is characterized by Field Emission Scanning Electron Microscopy (FESEM), Fourier Transform InfraRed (FTIR), and Raman and X-ray Photoelectron Spectroscopy (XPS). Gas sensing mechanisms and functional tests have been discussed to fix sensing performance. The proposed sensor may offer higher response to ammonia, processing convenience, low-cost scalability, and a low limit of detection (LOD) of 2.38 ppb, as well as the sensor's performance in real-world conditions, making this sensor a candidate for applications such as environmental monitoring and industrial safety. This work marks the first utilization of PANI@LIG for gas sensing and introduces a simple but effective approach for fabricating low-cost wearable gas sensors with high sensitivity and flexibility.

Rongqing Dong et al. (Contribution 2—Review) present a review devoted to conducting polymer-based gas sensors. Conducting Polymers (CPs) are promising materials for gas sensors due to their organic nature coupled with unique and versatile optical, electrical, chemical, and electrochemical properties. The fundamental gas sensing mechanisms in CPs-based sensors are elucidated, covering diverse transduction modes including electrochemical, chemo-resistive, optical, piezoelectric, and field-effect transistor-based sensing. Various types of conducting polymers employed in gas sensors, such as polypyrrole, polyaniline, polythiophene, and their composites, are introduced, with emphasis on their synthesis methods, structural characteristics, and gas-sensing response properties. Finally, the wide range of applications of these sensors is discussed, spanning industrial process control, environmental monitoring, food safety, biomedical diagnosis, and other fields, as well as existing issues such as long-term stability and humidity interference. The review has presented a comprehensive understanding of the CP-based sensors by examining their sensing mechanisms, sensitive materials, and device components. It also highlights future research directions, including device miniaturization, AI-assisted gas identification, multifunctional integrated sensing systems, and wearable and flexible sensor platforms.

Mahmoud Torkamani Cheriani et al. (Contribution 3—Review) present a review devoted to the plasma-treated nanostructured resistive gas sensors. Resistive gas sensors are among the most widely used sensors for the detection of various gases. In this type of gas sensor, the gas-sensing capability is linked to the surface properties of the sensing layer, and accordingly, modification of the sensing surface is of importance to improve the sensing output. Plasma treatment is a promising way to modify the surface properties of gas sensors, mainly by changing the amounts of oxygen ions, which have a central role in gas sensing reactions. After an introduction to air pollution, toxic gases, and resistive gas sensors, the main concepts regarding plasma are presented. Then, the impact of plasma treatment on the sensing characteristics of various sensing materials is also discussed. Generally, oxygen plasma causes the addition of surface oxygen functional groups on the sensor surface, and hence, the reactions between adsorbed gases with oxygen increase, leading to a higher sensing performance relative to pristine sensors. Also, exposure to other plasma atmospheres such as Ar or He causes the generation of oxygen defects, which act as favorable sites for oxygen adsorption and accordingly contribute to enhanced sensing performance. Different sensing materials such as metal oxides, TMDs, MXenes, CNTs, graphene, and CPs have been subjected to plasma treatment. In this regard, the combination of plasma exposure with other high irradiation techniques such as ion beams, electron beams, and gamma rays can lead to interesting sensing results.

Sanket Naresh Nagdeve et al. (Contribution 4—Review) propose a review devoted to the perspectives on the application of biosensors for the early detection of oral cancer. This study evaluates the significance of biomarkers and recent advancements in oral cancer detection, emphasizing cutting-edge electrochemical methods. The paper provides an epidemiological and etiological overview, outlining its clinical importance and reviewing the current state of the art in detection methods. Despite considerable progress, conventional methods exhibit limitations such as invasiveness, long wait times, and a lack of accuracy, creating a critical need for more robust technologies. This review emphasizes the significance of oral cancer biomarkers, which are considered promising cues for early detection, facilitating the development of innovative biosensing technologies. The review seeks to illuminate the recent advances in early detection and precision diagnostics, along with the usage of artificial intelligence strategies, ultimately contributing to significant progress in the battle against oral cancer. Integrating biomarkers and biofluids into the development and application of biosensors enhances the potential for accurate, reliable, and non-invasive oral cancer detection methods. Further research and development are essential to address challenges such as optimization for clinical settings, the validation of real-world applications, and integration into established diagnostic pathways. As the field of electrochemical biosensing continues to advance, the strategic integration of these biosensors into clinical practice holds promise to transform the landscape of oral cancer detection and management.

Chiheb Walleni et al. (Contribution 5—Article) propose a report on the synergistic effect of decorating nitrogen-doped reduced graphene oxide (N-rGO) with nickel oxide (NiO) nanoparticles for developing highly selective and sensitive chemi-resistive $NO_2$ gas sensors. The N-rGO/NiO sensor was synthesized straightforwardly, ensuring uniform decoration of NiO nanoparticles on the N-rGO surface. Comprehensive characterization using SEM, TEM, XRD, and Raman spectroscopy confirmed the successful integration of NiO nanoparticles with N-rGO and revealed key structural and morphological features contributing to its enhanced sensing performance. As a result, the NiO/N-rGO nanohybrids demonstrate a significantly enhanced response five orders of magnitude higher than that of N-rGO toward low $NO_2$ concentrations (<1 ppm) at 100 °C. In the first tests, the sensor showed a very high selectivity toward $NO_2$ (the other gaseous species tested were $CO_2$, ethanol, and $NH_3$). Consequently, NiO NPs proved their potential for boosting the sensitivity of N-rGO toward $NO_2$ gas, thanks to the p-p junctions created that facilitate carrier conduction, as explained by the underlying sensing mechanisms. Moreover, the present device has an outstanding performance, high sensitivity, and very low limit of detection (<1 ppb). The findings pave the way for integrating these sensors into advanced applications, including environmental monitoring and IoT-enabled air quality management systems. The nanomaterial presented robust performances, such as a high sensitivity and very low limit of detection, showing high prospects for being integrated in the next generation of advanced chemo-resistive sensors.

Bogdan-Catalin Serban et al. (Contribution 6—Article) propose a study on the ethanol vapor sensing performance of a resistive sensor that utilizes a quaternary nanohybrid sensing layer composed of holey carbon nanohorns (CNHox), graphene oxide (GO), $SnO_2$, and polyvinylpyrrolidone (PVP) in an equal mass ratio of 1:1:1:1 ($w/w/w/w$). The sensing device includes a flexible polyimide substrate and interdigital transducer (IDT)-like electrodes. The sensing film is deposited by drop casting on the sensing structure. The morphology and composition of the sensitive film are analyzed using Scanning Electron Microscopy (SEM), Energy Dispersive X-ray (EDX) Spectroscopy, and Raman spectroscopy. The manufactured resistive device presents good sensitivity to concentrations of alcohol vapors varying in the range of 0.008–0.16 mg/cm$^3$. The resistance of the proposed sensing structure increases over the entire range of measured ethanol concentration. Different types of sensing mechanisms are recognized. The decrease in the hole concentration in CNHox, GO, and CNHox due to interaction with ethanol vapors, which act as electron donors, and the swelling of the PVP are plausible and seem to be the prevalent sensing pathway. The hard-soft acid-base (HSAB) principle strengthens the proposed chemical analysis. Unlike conventional ethanol sensors, which primarily rely on metal oxides and rare elements, this innovative approach combines the synergistic properties of its components to enhance performance. CNHox provides high conductivity and porosity, improving electron transport and gas diffusion, while GO increases surface area and introduces functional groups that enhance ethanol interaction. $SnO_2$ further strengthens ethanol adsorption and sensing response, and PVP ensures structural integrity and dispersion stability. Beyond performance, this sensor offers a cost-effective and environmentally friendly alternative to traditional designs.

Paniz Vafaei et al. (Contribution 7—Article) propose a study on low-power gas sensors that can be used in IoT (Internet of Things) systems, consumer devices, and point-of-care devices that will enable new applications in environmental monitoring and health protection. We fabricated a monolithic chemi-resistive gas sensor by integrating a micro lightplate with a 2D sensing material composed of single-layer graphene and monolayer-thick $TiO_2$. Applying ultraviolet (380 nm) light with quantum energy above the $TiO_2$ bandgap effectively enhanced the sensor responses. Low (<1 μW optical) power operation of the device was demonstrated by measuring $NO_2$ gas at low concentrations, which is typical in air quality monitoring, with an estimated limit of detection less than 0.1 ppb. The gas response amplitudes remained nearly constant over the studied light intensity range (1–150 mW/cm$^2$) owing to the balance between the photoinduced adsorption and desorption processes of the gas molecules. The rates of both processes followed an approximately square root dependence on light intensity, plausibly because the electron-hole recombination of photoinduced charge carriers is the primary rate-limiting factor. These results pave the way for integrating 2D materials with micro-LED arrays as a feasible path to advanced electronic noses. Finally, the authors developed a monolithic gas microsensor by integrating a UV microlight plate with a 2D sensing material made by CVD graphene and a less than a nanometer thick layer of $TiO_2$ for advanced electronic noses with large sensor arrays.

Hsuan-Yu Chen et al. (Contribution 8—Article) propose a study on chemical analysis adopting a calibration curve to establish the relationship between the measuring technique's response and the target analyte's standard concentration. The calibration equation is established using regression analysis to verify the response of a chemical instrument to the known properties of materials that served as standard values. An adequate calibration equation ensures the performance of these instruments. There are two kinds of calibration equations: classical equations and inverse equations. For the classical equation, the standard values are independent, and the instrument's response is dependent. The inverse equation is the opposite: the instrument's response is the independent value. This study used measurement data sets from two kinds of humidity sensors and nine data sets from the literature to evaluate the predictive performance of two calibration equations. Four criteria were proposed to evaluate the predictive ability of two calibration equations. The study found that the inverse calibration equation could be an effective tool for complex calibration equations in chemical analysis. The precision of the instrument's response is essential to ensure predictive performance. The inverse calibration equation could be embedded into the measurement device, and then intelligent instruments could be enhanced. The results of this study show that the inverse equation has excellent predictive performance for the calibration equation of the capacitive humidity sensor. The classical equation has better accuracy, and the inverse equation has better precision for the predictive performance of resistive humidity sensors. If the instrument response has good repeatability, the inverse equation performs excellently for the nine data sets collected in the literature. If the repeatability of the instrument response is poor, two calibration equations have similar predictive performance.

Trine Juul-Kristensen et al. (Contribution 9—Article) demonstrate a study devoted to the detection of the malaria-causing Plasmodium parasite in non-invasive saliva samples (N = 61) from infected individuals by combining a DNA-based Rolling-circle-Enhanced-Enzyme-Activity-Detection (REEAD) sensor system with a chemiluminescence readout that could be detected with an in-house-developed affordable and battery-powered portable reader. The authors' team successfully transferred the technology to sub-Saharan Africa, where the malaria burden is high, and demonstrated a proof of concept in a small study (N = 40) showing significant differences ($p < 0.00001$) between malaria-positive individuals (N = 33) and presumed asymptomatic negative individuals (N = 7), all collected in Gabon. This is the first successful application of the REEAD sensor system for the detection of malaria in saliva in a high-epidemic area and holds promise for the potential future use of REEAD for malaria diagnosis or surveillance based on non-invasive specimens in sub-Saharan Africa. This study was conducted as a pilot investigation; more comprehensive field trials will be necessary to validate the diagnostic accuracy of the method, assessing its sensitivity and specificity.

Murugaiya Sridar Ilango et al. (Contribution 10—Article) present a study devoted to membrane permeability monitoring of the antipsychotic olanzapine using platinum black-modified electrodes. The blood–brain barrier (BBB) is key to the regular functioning of the central nervous system. The dysfunction of the BBB has been described in various neurological disorders, including schizophrenia. Schizophrenia (SCZ) is a chronic psychiatric disorder characterized by hallucinations, delusions, and negative symptoms. The Olanzapine (OLZ) drug is an electroactive species, and its levels can be monitored using electrochemical sensors. The detection of OLZ was demonstrated previously by using electrochemical sensors, and this technique can be used to monitor the levels of OLZ in real time. The challenge is to identify the permeability of OLZ through the BBB, so a replica model was designed with the BBB based on a Transwell membrane seeded with endothelial cells. A microfabricated electrode consisting of a 3 mm Au disk was modified with platinum black; this enables higher selectivity of electrochemical signals from OLZ. The dose–response of OLZ was characterized in a phosphate-buffered saline solution (10 mM, pH 7.4) by adding 20–200 nM (in steps of 20) of OLZ stock solution. The observed chronoamperometric electrochemical signals showed an increasing current at 0.45 V vs. Ag/AgCl with an increasing OLZ concentration. The controls for the experiments were performed in phosphate-buffered saline solution (10 mM, pH 7.4). The detection limit was calculated as $9.96 \pm 7.35 \times 10^{-6}$ nM from the calibration curve. The membrane permeability of the OLZ drug tested with five SCZ patients was monitored by studying the TEER measurements and permeability rate constant data. This study highlights the potential of electrochemical sensors for predicting human responsiveness to antipsychotic drugs. Platinum-black-modified electrodes were employed to detect the concentration of OLZ after their penetration through the BBB in various cell line models. The effective surface area of the platinum-black-modified electrodes is $5.23 \times 10^{-2} \pm 2.3 \times 10^{-3}$ cm$^2$, which is four times higher than the bare gold electrode, $3.81 \times 10^{-2} \pm 1.2 \times 10^{-3}$ cm$^2$. The dose–response of OLZ with platinum-black-modified electrodes was characterized using chronoamperometric electrochemical signals, which showed an increasing current at 0.45 V vs. Ag/AgCl with an increasing OLZ concentration.

Haixia Mei et al. (Contribution 11—Article) present a study devoted to research on the binary mixed VOCs gas identification method based on multi-task learning. Traditional volatile organic compound (VOC) detection models separate component identification and concentration prediction, leading to low feature utilization and limited learning in small-sample scenarios. Here, the authors realize a residual fusion network based on multi-task learning (MTL-RCANet) to implement component identification and concentration prediction of VOCs. The model integrates channel attention mechanisms and cross-fusion modules to enhance feature extraction capabilities and task synergy. To further balance the tasks, a dynamic weighted loss function is incorporated to adjust weights dynamically according to the training progress of each task, thereby enhancing the overall performance of the model. The proposed network achieves an accuracy of 94.86% and an $R^2$ score of 0.95. Comparative experiments reveal that using only 35% of the total data length as input data yields excellent identification performance. Moreover, multi-task learning effectively integrates feature information across tasks, significantly improving model efficiency compared to single-task learning. In summary, the proposed method offers a new solution for gas detection tasks in fast detection and low-resource consumption scenarios, which shows great application potential. Future work can further optimize the network structure to enhance task collaboration, particularly in more complex gas mixtures or dynamic response scenarios.

Vadim Platonov et al. (Contribution 12—Article) present a report on the synthesis of perovskite-type Ba-doped $LaFeO_3$ ($La_{1-x}Ba_xFeO_3$, x = 0.00, 0.02, 0.04, and 0.06) nanofibers (NFs) using the electrospinning method. The synthesized $La_{1-x}Ba_xFeO_3$ materials have a fibrous structure with an average fiber diameter of 250 nm. The fibers, in turn, consist of smaller crystalline particles of 20–50 nm in size. The sensor properties of $La_{1-x}Ba_xFeO_3$ nanofibers were studied when detecting 20 ppm CO, $CH_4$, methanol, and acetone in dry air in the temperature range of 50–350 °C. Doping with barium leads to a significant increase in sensor response and a decrease in operating temperature when detecting volatile organic compounds (VOCs). The process of acetone oxidation on the surface of the most sensitive $La_{0.98}Ba_{0.02}FeO_3$ material was studied using in situ Diffuse Reflectance Infrared Fourier Transform Spectroscopy (DRIFTS) and Temperature-Programmed Desorption in combination with Mass Spectrometry (TPDMS). A mechanism for the formation of the sensor signal is proposed. The obtained materials were single phase, had an orthorhombic structure, and consisted of nanocrystallites with a size of about 14–16 nm. The introduction of barium led to the inhibition of crystallite growth during isothermal annealing and promoted an increase in the sensor response of the $LaFeO_3$ nanofiber-based sensors toward VOCs. The $La_{0.98}Ba_{0.02}FeO_3$ sample demonstrated the highest sensor response and a decrease in the operating temperature. The improvement in the gas-sensitive properties of the doped materials can be explained by the high catalytic activity of the surface of synthesized materials associated with the formation of oxygen vacancies, highly active iron cations ($Fe^{4+}$), and coordinatively unsaturated cations ($Fe^{3+}$). The mechanism of acetone oxidation on the sensor surface, studied using DRIFTS and TPD-MS methods, is assumed to have a multi-stage nature.

Alexey Vasiliev et al. (Contribution 13—Article) analyze the influence of micro hotplate size on the convective heat exchange of gas sensors. Usually, the role of convection in the heat exchange of gas sensors is not considered in thermal simulation models because of the complexity of the convection process. As a result, the contribution of this process to the overall heat loss of sensors remains without detailed analysis. The authors' team analyzed convection issues in two groups of gas sensors: semiconductor and thermos-catalytic (calorimetric) sensors and, on the other hand, in the oxygen sensors of the thermomagnetic type. It is demonstrated that there is a critical size leading to the formation of convective heat exchange flow. Below this critical value, only thermal conductivity of ambient air, IR (infrared) radiation from the heated micro hotplate surface, and thermal conductivity of the micro hotplate supporting elements should be considered as channels for heat dissipation by the micro hotplate, and the contribution of free convection can be neglected. Similar results were obtained in the analysis of the behavior of thermal magnetic sensors of oxygen, which use paramagnetic properties of molecular oxygen for the determination of $O_2$ concentration. In this case, the critical size of the sensor is also of significance; if the size of the magnetic sensor is much below this value, the oxygen concentration value measured with such a device is independent of the orientation of the sensor element. The results of the simulation were compared with the measurement of heat loss in micromachined gas sensors. The optimal dimensions of the sensor micro hotplate are given as a result of these simulations and measurements. The authors' team investigated the influence of convection on the heat exchange processes of micro hotplates used in the fabrication of semiconductor and thermos-catalytic (calorimetric) gas sensors, as well as the heat exchange of thermal magnetic sensors of oxygen. The analysis was based on the consideration of the competition of convection flow and back diffusion. It was shown that there is a certain critical size of the micro hotplate. If the size of the micro hotplate $d << d_{cr}$, the influence of convection heat exchange can be neglected, and only the thermal conductivity of air and of the elements of the sensor, together with the IR (infrared) radiation (if the sensor is heated up to a very high temperature), should be taken into account as channels of heat losses.

O. L. Gribkova et al. (Contribution 14—Article) report on optical ammonia sensors based on spray-coated polyaniline complexes with polysulfonic acids. The optical ammonia-sensing properties of water-dispersible polyaniline (PANI) complexes chemically synthesized in the presence of polysulfonic acids of different structures and chain flexibility were compared for the first time. Flexible-chain poly(styrene-4-sulfonic acid) and poly-(2-acrylamido-2-methyl-1-propanesulfonic acid), as well as semi-rigid-chain poly-4,4′-(2,2′-disulfonic acid)diphenylene-iso-phthalamide and rigid-chain poly-4,4′-(2,2′-disulfonic acid) diphenylene-tere-phthalamide (t-PASA) were used. The sensor films were prepared by a convenient and scalable method: spray coating of aqueous solutions on glass substrates. The optical response time and amplitude of the sensor films in the range of ammonia concentrations from 5 to 200 ppm were investigated. To overcome the influence of humidity and the presence of over-stoichiometric protons of the polyacid on the accuracy of ammonia determination, treatments of the films in aqueous solutions of NaCl, $CaCl_2$, and $BaCl_2$ were tested. The treatment in 1 M $CaCl_2$ solution for all the PANI complexes results in a significant improvement in the response time, amplitude, and reproducibility. The films of PANI complexes with the flexible-chain polyacids have the highest response amplitude in the range of ammonia concentrations 5–25 ppm. PANI-t-PASA film demonstrated the best sensory properties at ammonia concentrations more than 50 ppm. FTIR spectroscopy showed that $CaCl_2$ treatment results in cross-linking of sulfoacid groups from adjacent polyacid chains by $Ca^{2+}$ ions. Thus, such a treatment results both in the neutralization of excessive protons and a significant reduction in the films' swelling at high humidity. Among the films of the PANI complexes treated with $CaCl_2$, the best sensory properties were demonstrated by the PANI-t-PASA film at ammonia concentrations more than 50 ppm. The films of PANI complexes give reproducible results at 2–4 reuses, the reversibility decreased at high ammonia concentration. Therefore, at this stage of the investigations, these films are preferably to be used as alarm detectors.

Anas Mohd Noor et al. (Contribution 15—Article) report on a wearable device for continuous and real-time monitoring of human sweat sodium. Wearable sweat-sensing devices hold significant potential for non-invasive, continuous health monitoring. However, challenges such as ensuring data accuracy, sensor reliability, and measurement stability persist. This study presents the development of a wearable system for the real-time monitoring of human sweat sodium levels, addressing these challenges through the integration of a novel microfluidic chip and a compact potentiostat. The microfluidic chip, fabricated using hydrophilic materials and designed with vertical channels, optimizes sweat flow, prevents backflow, and minimizes sample contamination. The developed wearable potentiostat, as a measurement device, precisely measures electrical currents across a wide dynamic range, from nanoamperes to milliamperes. Validation results demonstrated accurate sodium concentration measurements ranging from 10 mM to 200 mM, with a coefficient of variation below 4% and excellent agreement with laboratory instruments (intraclass correlation = 0.998). During physical exercise, the device measured a decrease in sweat sodium levels, from 101 mM to 67 mM over 30 min, reflecting typical physiological responses to sweating. These findings confirm the system's reliability in providing continuous, real-time sweat sodium monitoring. This work advances wearable health-monitoring technologies and lays the groundwork for applications in fitness optimization and personalized hydration strategies. Future work will explore multi-biomarker integration and broader clinical trials to further validate the system's potential. The microfluidic chip, fabricated using a water-washable resin and 3D printing technology, offers high resolution, rapid fabrication, and excellent hydrophilicity. Additionally, investigating the performance of the proposed device in larger-scale clinical trials could further validate its potential for real-world applications.

Finally, the research presented is pivotal in shaping the future of chemical sensing by means of advanced sensor systems and devices integrating innovative functional materials for advancements in biochemical applications.

## 3. Statistics and Trend Analysis

Trend analysis is reported in Table 1. The Special Issue offers an outstanding overview of recent advancements in biochemical sensors for detection applications. The statistics consisted of 15 published papers, including 3 reviews and 12 articles co-authored by 89 international scientists from 15 countries located in Europe, Africa, Asia, and North America. The total number of the rejected manuscripts is two.

**Table 1.** Main features of the papers published in the Special Issue.

| Paper | Type | Nr. of Authors | Corresponding Author Country | Other Authors' Countries | Keywords |
|---|---|---|---|---|---|
| Contribution 1 | Article | 6 | Spain | Spain | laser-induced graphene; polyaniline; gas sensing; RT ammonia sensing |
| Contribution 2 | Review | 7 | China | China | conducting polymer; gas sensor; sensing mechanisms; environmental monitoring |
| Contribution 3 | Review | 2 | Iran | Iran | plasma treatment; toxic gas; gas sensor; sensing mechanisms |

**Table 1.** *Cont.*

| Paper | Type | Nr. of Authors | Corresponding Author Country | Other Authors' Countries | Keywords |
|---|---|---|---|---|---|
| Contribution 4 | Review | 3 | USA | USA | biomarkers; biofluids; electrochemical sensors; molecular analytical techniques; diagnostic tools; commercial test kits |
| Contribution 5 | Article | 4 | Spain | Spain; Tunisia | N-doped reduced graphene oxide; nickel oxide nanoparticles; sub-ppb $NO_2$ sensing; gas sensing |
| Contribution 6 | Article | 9 | Romania | Romania | ethanol sensor; holey carbon nanohorns; graphene oxide; swelling; HSAB (hard soft acid base) principle |
| Contribution 7 | Article | 7 | Estonia | Estonia; Spain; Germany | gas sensor; $NO_2$ gas sensing; micro-lightplate; graphene/$TiO_2$ heterostructure |
| Contribution 8 | Article | 2 | Taiwan | Taiwan | calibration; classical equation; inverse equation; predictive performance |
| Contribution 9 | Article | 13 | Denmark | Denmark; Gabon; Germany | malaria; diagnosis; rolling circle amplification; saliva; topoisomerase 1 |
| Contribution 10 | Article | 6 | Israel | Israel | olanzapine; platinum black; blood–brain barrier; electrochemical sensors; schizophrenia |
| Contribution 11 | Article | 6 | China | China | gas sensor; multi-task learning; mixed gases; feature fusion |
| Contribution 12 | Article | 5 | Russia | Russia | Ba-doped $LaFeO_3$; perovskites; semiconductor gas sensor; VOCs; DRIFTS (diffuse reflectance infrared Fourier transform spectroscopy); TPD-MS (temperature-programmed desorption in combination with mass spectrometry) |
| Contribution 13 | Article | 4 | Russia | Russia | microheater; convective heat losses; thermal conductivity; Grashof number |
| Contribution 14 | Article | 6 | Russia | Russia | polyaniline complexes; polyacid; spray coating; ammonia sensors; optical gas sensors |

**Table 1.** *Cont.*

| Paper | Type | Nr. of Authors | Corresponding Author Country | Other Authors' Countries | Keywords |
|---|---|---|---|---|---|
| Contribution 15 | Article | 9 | Malaysia | Malaysia; Japan | microfluidic chip; sweat sodium measurement; wearable device |
| Total Authors involved | | 89 | | | |
| Total Author Countries | | 15 | | | |
| Total Rejected Papers | | 2 | | | |
| Total Published Papers | | 15 | | | |

The challenges of the Special Issue deal with advanced sensing materials (graphene-based materials, carbon nanomaterials and composites, metal oxide nanoparticles, heterostructures, perovskites, hybrid functional materials, and polymeric complexes) and related material processing (laser and plasma) for optimal biochemical sensing properties. Furthermore, the findings push the knowledge in the advanced transduction of biochemical sensing (electrochemical, chemo-resistive, optical, analytical, spectroscopic, microfluidic, micro-hotplate, thermal, biomarker, and test kit). The calibration of sensors, predictive performance, multi-task learning, and feature fusion have been explored as well.

Challenging applications have been addressed, such as sub-ppb detection of toxic gases and volatile organic compounds for environmental monitoring, malaria detection by fast test kits for in-field biosensing, schizophrenia by the biomarker of olanzapine using platinum electrodes, and sweat sodium measurement by microfluidic chip for wearable applications.

## 4. Summary and Conclusions

The work presented in this Special Issue reflects the drive to increase the knowledge in chemical sensors by applied research in advanced materials, sensing devices, new transducers, and practical applications. In this Special Issue, front-line scientists have been kindly invited to submit original research and review articles on exploring recent advances in sensors for chemical detection applications.

Potential topics included, but were not limited to, gas sensors, chemical detection, advanced materials for chemical sensing, novel gas sensor materials, sensor calibration, sensor systems, machine learning algorithms, wireless sensor networks, chemical threat monitoring, environmental measurements, sensors for smart city applications, sensors for environmental sustainability, sensors for energy applications, sensors for IoT applications, sensors for industrial applications, sensors for sustainable mobility, case studies of chemical detection campaigns, and new concepts and trends in chemical sensing.

The impactful achievements presented in this Special Issue are a valuable contribution to the theory and practice of chemical sensing. Additional development of these trends can be expected in the future to consolidate the development of a new generation of chemical sensors applied for practical applications.

**Acknowledgments:** The Guest Editor (M.P.) would like to acknowledge all the authors, peer-reviewers and other Academic Editors for their valuable time and contributions to this Special Issue "Recent Advances in Sensors for Chemical Detection Applications". Furthermore, M.P. would like to thank all editorial staff from MDPI and Sensors for kind and professional assistance in the production of the open access Special Issue and related reprint (Volume I).

**Conflicts of Interest:** The author declares no conflicts of interest.

**List of Contributions:**

1. Santos-Ceballos, J.C.; Salehnia, F.; Güell, F.; Romero, A.; Vilanova, X.; Llobet, E. Room-Temperature Ammonia Sensing Using Polyaniline-Coated Laser-Induced Graphene. *Sensors* **2024**, *24*, 7832. https://doi.org/10.3390/s24237832
2. Dong, R.; Yang, M.; Zuo, Y.; Liang, L.; Xing, H.; Duan, X.; Chen, S. Conducting Polymers-Based Gas Sensors: Principles, Materials, and Applications. *Sensors* **2025**, *25*, 2724. https://doi.org/10.3390/s25092724
3. Cheriani, M.T.; Mirzaei, A. Plasma-Treated Nanostructured Resistive Gas Sensors: A Review. *Sensors* **2025**, *25*, 2307. https://doi.org/10.3390/s25072307
4. Nagdeve, S.N.; Suganthan, B.; Ramasamy, R.P. Perspectives on the Application of Biosensors for the Early Detection of Oral Cancer. *Sensors* **2025**, *25*, 1459. https://doi.org/10.3390/s25051459
5. Walleni, C.; Ali, M.B.; Ncib, M.F.; Llobet, E. Synergistic Enhancement of Chemiresistive $NO_2$ Gas Sensors Using Nitrogen-Doped Reduced Graphene Oxide (N-rGO) Decorated with Nickel Oxide (NiO) Nanoparticles: Achieving sub-ppb Detection Limit. *Sensors* **2025**, *25*, 1631. https://doi.org/10.3390/s25051631
6. Serban, B.-C.; Dumbravescu, N.; Buiu, O.; Bumbac, M.; Brezeanu, M.; Pachiu, C.; Nicolescu, C.-M.; Brancoveanu, O.; Cobianu, C. Holey Carbon Nanohorns-Based Nanohybrid as Sensing Layer for Resistive Ethanol Sensor. *Sensors* **2025**, *25*, 1299. https://doi.org/10.3390/s25051299
7. Vafaei, P.; Kodu, M.; Alles, H.; Kiisk, V.; Casals, O.; Prades, J.D.; Jaaniso, R. Graphene/$TiO_2$ Heterostructure Integrated with a Micro-Lightplate for Low-Power $NO_2$ Gas Detection. *Sensors* **2025**, *25*, 382. https://doi.org/10.3390/s25020382
8. Chen, H.-Y.; Chen, C. Comparison of Classical and Inverse Calibration Equations in Chemical Analysis. *Sensors* **2024**, *24*, 7038. https://doi.org/10.3390/s24217038
9. Juul-Kristensen, T.; Thiesen, C.; Haurum, L.W.; Keller, J.G.; Lendamba, R.W.; Manego, R.Z.; Ongwe, M.E.B.; Knudsen, B.R.; Pareja, E.; Pareja-Tobes, E.; et al. Non-Invasive Malaria Detection in Sub-Saharan Africa Using a DNA-Based Sensor System. *Sensors* **2024**, *24*, 7947. https://doi.org/10.3390/s24247947
10. Ilango, M.S.; Desagani, D.; Jagadeesan, S.; Snezhko, A.; Vatine, G.; Ben-Yoav, H. Membrane Permeability Monitoring to Antipsychotic Olanzapine Using Platinum Black-Modified Electrodes. *Sensors* **2025**, *25*, 2266. https://doi.org/10.3390/s25072266
11. Mei, H.; Yang, R.; Peng, J.; Meng, K.; Wang, T.; Wang, L. Research on Binary Mixed VOCs Gas Identification Method Based on Multi-Task Learning. *Sensors* **2025**, *25*, 2355. https://doi.org/10.3390/s25082355
12. Platonov, V.; Malinin, N.; Filatova, D.; Sapkov, I.; Rumyantseva, M. Electrospun (La,Ba)$FeO_3$ Nanofibers as Materials for Highly Sensitive VOC Gas Sensors. *Sensors* **2025**, *25*, 2790. https://doi.org/10.3390/s25092790
13. Vasiliev, A.; Shaposhnik, A.; Kul, O.; Mokrushin, A. The Role of Convection and Size Effects in Microhotplate Heat Exchange: Semiconductor and Thermomagnetic Gas Sensors. *Sensors* **2025**, *25*, 2830. https://doi.org/10.3390/s25092830
14. Gribkova, O.L.; Kabanova, V.A.; Rodina, E.I.; Teplonogova, M.A.; Demina, L.I.; Nekrasov, A.A. Optical Ammonia Sensors Based on Spray-Coated Polyaniline Complexes with Polysulfonic Acids. *Sensors* **2025**, *25*, 3348. https://doi.org/10.3390/s25113348
15. Noor, A.M.; Al Farisi, M.S.; Mazalan, M.; Ibrahim, N.F.A.; Wahab, A.A.; Zakaria, Z.; Rusli, N.I.; Sabani, N.; Manaf, A.A. Wearable Device for Continuous and Real-Time Monitoring of Human Sweat Sodium. *Sensors* **2025**, *25*, 3467. https://doi.org/10.3390/s25113467

## References

1. Penza, M. Low-cost sensors for outdoor air quality monitoring. In *Advanced Nanomaterials for Inexpensive Gas Microsensors*; Llobet, E., Ed.; Elsevier: Amsterdam, The Netherlands, 2020; pp. 235–288. [CrossRef]
2. Williams, R.; Duvall, R.; Kilaru, V.; Hagler, G.; Hassinger, L.; Benedict, K.; Rice, J.; Kaufman, A.; Judge, R.; Pierce, G.; et al. Deliberating performance targets workshop: Potential paths for emerging $PM_{2.5}$ and $O_3$ air sensor. *Atmos. Environ. X* **2019**, *2*, 100031. [CrossRef] [PubMed]

3. Borrego, C.; Ginja, J.; Coutinho, M.; Ribeiro, C.; Karatzas, K.; Sioumis, T.; Katsifarakis, N.; Konstantinidis, K.; De Vito, S.; Esposito, E.; et al. Assessment of air quality microsensors versus reference methods: The EuNetAir joint exercise-Part II. *Atmos. Environ.* **2018**, *193*, 127–142. [CrossRef]
4. World Meteorological Organization. *2021 WMO Report No. 1215; An Update on Low-Cost Sensors for the Measurement of Atmospheric Composition*; Peltier, R.E., Ed.; Lead Authors: Núria Castell, Andrea L Clements, Tim Dye, Christoph Hüglin, Jesse H Kroll, Shih-Chun Candice Lung, Zhi Ning, Matthew Parsons, Michele Penza, Fabienne Reisen, Erika von Schneidemesser; World Meteorological Organization: Geneva, Switzerland, 2021; ISBN 978-92-63-11215-6. Available online: https://library.wmo.int/index.php?id=21508&lvl=notice_display#.Yifnz3yZPIU (accessed on 1 August 2025).
5. Laera, A.M.; Penza, M. Chemiresistive materials for alcohol vapor sensing at room temperature. *Chemosensors* **2024**, *12*, 78. [CrossRef]
6. Laera, A.M.; Cassano, G.; Burresi, E.; Protopapa, M.L.; Penza, M. Flexible Humidity Sensor Based on Chemically Reduced Graphene Oxide. *Chemosensors* **2024**, *12*, 245. [CrossRef]
7. Dilonardo, E.; Palmisano, F.; Torsi, L.; Cioffi, N.; Penza, M.; Alvisi, M.; Di Franco, C. Evaluation of gas-sensing properties of ZnO nanostructures electrochemically doped with Au nanophases. *Beilstein J. Nanotechnol.* **2016**, *7*, 22–31. [CrossRef] [PubMed]
8. Penza, M.; Rossi, R.; Alvisi, M.; Serra, E.; Paolesse, R.; D'Amico, A.; Di Natale, C. Carbon nanotube films as a platform to transduce molecular recognition events in metalloporphyrins. *Nanotechnology* **2011**, *22*, 125502. [CrossRef] [PubMed]
9. Penza, M.; Rossi, R.; Alvisi, M.; Signore, M.A.; Cassano, G.; Dimaio, D.; Pentassuglia, R.; Piscopiello, E.; Serra, E.; Falconieri, M. Characterization of metal-modified and vertically-aligned carbon nanotube films for functionally enhanced gas sensor applications. *Thin Solid Films* **2009**, *517*, 6211–6216. [CrossRef]
10. Penza, M.; Pfister, V.; Suriano, D.; Dipinto, S.; Prato, M.; Cassano, G. Application of low-cost sensors in stationary and mobile nodes for urban air quality index monitoring. *Eng. Proc.* **2023**, *48*, 62. [CrossRef]
11. Brattoli, M.; de Gennaro, G.; de Pinto, V.; Demarinis-Loiotile, A.; Lovascio, S.; Penza, M. Odour detection methods: Olfactometry and chemical sensors. *Sensors* **2011**, *11*, 5290–5322. [CrossRef] [PubMed]
12. Suriano, D.; Abulude, F.O.; Penza, M. The Use of Low-Cost Gas Sensors for Air Quality Monitoring with Smartphone Technology: A Preliminary Study. *Chemosensors* **2025**, *13*, 189. [CrossRef]
13. Suriano, D.; Penza, M. Assessment of the Performance of a Low-Cost Air Quality Monitor in an Indoor Environment through Different Calibration Models. *Atmosphere* **2022**, *13*, 567. [CrossRef]
14. Suriano, D.; Cassano, G.; Penza, M. Design and Development of a Flexible, Plug-and-Play, Cost-Effective Tool for on-Field Evaluation of Gas Sensors. *J. Sens.* **2020**, *2020*, 8812025. [CrossRef]

*sensors*

MDPI

*Article*

# Comparison of Classical and Inverse Calibration Equations in Chemical Analysis

**Hsuan-Yu Chen [1] and Chiachung Chen [2,*]**

[1] Africa Industrial Research Center, National Chung Hsing University, Taichung 40227, Taiwan; wakaharu37@gmail.com

[2] Department of Bio-Industrial Mechatronics Engineering, National Chung Hsing University, Taichung 40227, Taiwan

* Correspondence: ccchen@dragon.nchu.edu.tw; Tel.: +886-4-22857562

**Abstract:** Chemical analysis adopts a calibration curve to establish the relationship between the measuring technique's response and the target analyte's standard concentration. The calibration equation is established using regression analysis to verify the response of a chemical instrument to the known properties of materials that served as standard values. An adequate calibration equation ensures the performance of these instruments. There are two kinds of calibration equations: classical equations and inverse equations. For the classical equation, the standard values are independent, and the instrument's response is dependent. The inverse equation is the opposite: the instrument's response is the independent value. For the new response value, the calculation of the new measurement by the classical equation must be transformed into a complex form to calculate the measurement values. However, the measurement values of the inverse equation could be computed directly. Different forms of calibration equations besides the linear equation could be used for the inverse calibration equation. This study used measurement data sets from two kinds of humidity sensors and nine data sets from the literature to evaluate the predictive performance of two calibration equations. Four criteria were proposed to evaluate the predictive ability of two calibration equations. The study found that the inverse calibration equation could be an effective tool for complex calibration equations in chemical analysis. The precision of the instrument's response is essential to ensure predictive performance. The inverse calibration equation could be embedded into the measurement device, and then intelligent instruments could be enhanced.

**Keywords:** calibration; classical equation; inverse equation; predictive performance

## 1. Introduction

Calibration is essential to ensuring the performance of sensors or instruments in chemical analysis. Many quantitative analytical techniques, such as high-performance liquid chromatography (HPLC), ultraviolet–visible spectroscopy (UV-Vis spectroscopy), gas chromatography/mass spectrometry (GC-MS), and electrophoresis, must establish a calibration equation to express the relationship between the measuring technique's response and the target analyzer's standard values [1,2].

In practical work, standard concentrations or environments are prepared. These values are called regressors or independent values, x. The responses from the measuring technique are called the dependent values, y. The data sets ($x_i$, $y_i$) model the relationship between x and y. The practical measuring case detects a sample with unknown conventions, $x_0$; the new response is $y_0$. The new concentration, $x_0$, will be calculated using the previously established calibration equation [3,4].

Two types of calibration equations were proposed. The first calibration equation is the classical calibration. The function of this equation is $y_i = f(x_i)$. If the dependent variable $y_i$ and the independent variance, $x_i$, exist in the linear relationship, the calibration model is [5]

$$y_i = b_0 + b_1 x_i + \varepsilon_i \tag{1}$$

where $b_0$ is the intercept, $b_1$ is the slope, and $\varepsilon_i$ represents random errors.

The assumptions of regression analysis are that the $\varepsilon_i$ represents random errors, independent, and normal distribution, and the $x_i$ values are non-measurement errors [6].

In a further application, the unknown $x_0$ was detected. The $y_0$ was detected by measuring instrument, and the predicted $\hat{x}_0$ value is calculated as

$$\hat{x}_0 = \frac{y_0 - b_0}{b_1} \tag{2}$$

The second calibration is called the inverse equation. The function of this equation is $x_i = g(y_i)$. In the inverse regression, $x_i$ is treated as the response, and $y_i$ is the regressor. If the linear relationship exists, the calibration model is

$$x_i = c_0 + c_1 y_i + \varepsilon_i \tag{3}$$

As the new response, $x_0$, is detected, the predicted $\hat{x}_0$ value is calculated directly using Equation (3).

$$\hat{x}_0 = c_0 + c_1 y_0 \tag{4}$$

One of the regression analysis assumptions is the negligible measurement error for the $x_i$ value. For the inverse equation, the assumption is invalid. However, different opinions are presented in different studies. Krutchkoff [7] compared the classical and inverse equations using the Monte Carlo methods and found the inverse equation had a lower mean square error value. Krutchkoff [8] found the inverse equation demonstrates better extrapolation performance. Centner [9] compared two calibration equations using Monte Carlo methods and two practical examples and found the inverse equation to be more reliable than the classical one. Tellinghuisen [10] evaluated two calibration equations for small data sets and claimed that the inverse equation is more efficient over an extensive range of the variable $x_i$. Shalabh [11] compared the measurement errors of two calibration equations with the balanced loss function. He suggested that the comparison should not be limited to linear equations and that more variables and nonlinear relationships should be considered.

Tellinghuisen [12] proposed sample algorithms for nonlinear calibration equations of the classical equation to calculate the prediction of new measurement values easily. However, this study did not mention the convenience of the inverse equation. Parker et al. [13] claimed that the inverse linear equation is simple and easy to use but violates some regression assumptions and found the inverse linear equation had more variability and bias in the prediction interval as the predicted value is away from the center of the data. In the study of Besalu [14], the inverse calibration has a better prediction ability than the classical calibration equation and gives lesser mean square error interpolations. Granovskii and Sirala [15] presented different conditions and included a known parameter to estimate experimental data and homo-variance when selecting classical or inverse calibration equations. Witkovsky and Wimmer [16] introduced a method to calculate the measurement uncertainty of polynomial equations for two calibration equations. Delgado [17] used the Beer–Lambert law to illustrate the misuse of the calibration equations for this calibration curve. To predict new measurements, the hypothesis of regressions does not necessarily have to be fulfilled for the nonlinear equation, and the inverse equation has a more extraordinary predictive ability than the classical equation.

Francois et al. [18] proposed two criteria to evaluate calibration equations: the maximum and the average prediction variance, and concluded that the predictive quality of both calibration equations was equal. Kannan et al. [19] compared the classical and inverse equations with the Pitman closeness criterion and showed that the inverse equation had better predictive ability than the classical equation as the calibration point is far from the average values of standards.

Most of the literature concerns only the linear relationship of the calibration equation. Delgado [17] mentions the misuse of linear equations in calibration equations. Chen and Chen [20] proposed six calibration equations for calibration curves in chemical analysis. The linear equation is one of their calibration equations. In this study, the calibration curves of two types of hygrometers were established. These data were divided into two sets: one for the model established and the other for evaluating the accuracy and precision of the adequate equation. Nine data sets were collected from previous studies to evaluate the predictive performance of two calibration equations. These data sets are listed in Table 1. Four criteria were proposed to evaluate the predictive ability of two calibration equations. The effect of the instrument's precision on the predictive ability was assessed.

**Table 1.** Published data in the literature for evaluating the predictive performance of two calibration equations.

| Study | Equipment | Target | Standard Range | Response Range | Calibration Equation | Statistic Criteria |
|---|---|---|---|---|---|---|
| Mulholland and Hibbert [21] | HPLC [1] | Daidzein | 0.162–10.96 mg/50 mL | 0.243–30.75 peak area | Linear $y = X^{1.1}$ | $R^2$, residual plot |
| Desimoni [22] | Flow injection analysis | Sulfides | 0.88–81.2 μm | 0.170–15.94 μA | Linear | $R^2$, residual plot |
| Lavagnini and Magno [23] | GC-MS [2] | Chloromethane | 0~4 μg/L | 0.111975~0.465813 peak area ratio | Linear polynomial | s, residual plot |
| Ortiz et al. [24] | Pulse polarography | Benzaldehyde | 0.0198~0.1740 mnol/L | 0.033~0.366 μA | Linear | Residual plots, S |
| Rawski et al. [25] | Spectrophotometry | Albumin | 0~20 μg/mL | 0~450 peak height × $10^{-3}$ | Linear | Lack of fit, $R^2$ |
| Desharnais et al. [26] | LC-MS [3] | Cocaine | 5~1000 ng/mL | 0.049~9.209 area ratio | Linear | Partial F-test |
| Martin et al. [27] | HPLC | Blood | 0~90 ng/mL | 0.002~0.272 area ratio | High-order polynomial | $R^2$ Residual plots |
| Martin et al. [28] | LC-QqQ-MS [4] | PrP | | 2150–3,054,469 | Linear | $R^2$ |
| Lavin et al. [29] | array BICELLS [5] | Anti-IgG | 1~100 μg/mL | 0.00~6.14 | Polynomial | AICs [5], $R^2$ |

Note: 1. HPLC: high-performance liquid chromatography; 2. GC-MS: gas chromatography/mass spectrometry; 3. LC-MS: liquid chromatography–tandem mass spectrometry; 4. LC-QqQ-MS: liquid chromatography–mass spectrometry; 5. BICELLS: biophotonic sensing cells.

## 2. Materials and Methods

### 2.1. Relative Humidity Sensors

This study used two types of humidity sensors: a capacitive Vaisala HMP-143A (Vaisala Oyj, Vantaa, Finland) and a resistive Shinyei THI-B141 (Shinyei Kaisha Technology, Kobe, Japan). The specifications of these sensors are listed in Table 2.

**Table 2.** The specifications of two humidity sensors.

| | Resistive Sensor | Capacitive Sensor |
|---|---|---|
| Name | THT-B121 | HMP 140A |
| Sensing element | Macro-molecule HPR-MQ | HUMICAP |
| Operating range | 0–60 °C | 0–50 °C |
| Measuring range | 10–99% RH | 0–100% |
| Nonlinearity and repeatability | ±0.25% RH | ±0.2% RH |

### 2.2. Saturated Salt Solutions

Eleven saturated salt solutions, made from LiCl, $CH_3COOK$, $MgCl_2$, $K_2CO_3$, $Mg(NO_3)_2$, NaBr, KI, NaCI, KCI, $KNO_3$, and $K_2SO_4$, were used to maintain the standard relative humidity values for calibration. The standard humidity values produced by these salt solutions were listed using Greenspan's equation [30].

### 2.3. Calibration of Humidity Sensors

Two types of humidity sensors were calibrated using saturated salt solutions. The procedures for producing saturated salt solutions were according to the OIMO R121 [31]. Chen's study [32] describes the detailed calibration procedures. Both data sets, the response data of sensors, and the standard values from saturated salt solutions for two types of humidity sensors were used to compare the classical and inverse calibration equations.

### 2.4. Establish the Calibration Equation

The standard humidity values of a saturated salt solution are called regressors or independent variables, $x_i$. The reading values of humidity sensors are called response or dependent variables, $y_i$.

#### 2.4.1. The Classical Equation

The form of this equation is a high-order polynomial equation.

$$y = b_0 + b_1x + b_2x^2 + \ldots + b_kx^k \quad (5)$$

where $b_0$, $b_1$, $b_2$, and $b_k$ are constants.

#### 2.4.2. The Inverse Equation

The form of the inverse equation is

$$x = c_0 + c_1y + c_2y^2 + \cdots ..c_ny^n \quad (6)$$

where $c_0$, $c_1$, $c_2$, and $c_n$ are constants.

### 2.5. The Evaluation Criteria for Calibration

The Criteria of Fitting Agreement

The quantitative criteria for assessing the fitting ability for different calibration equations of the same data sets are the coefficients of determination, $R^2$, and the standard error in the estimate errors, s. The qualitative criterion is the residual plot [6,33,34].

For the classical equation, the standard error in the estimate errors, $s_x$, is

$$s_x = \frac{\sqrt{\Sigma(y_i - \hat{y}_i)^2}}{n-p} \quad (7)$$

where $y_i$ is the dependent variable, $\hat{yi}$ is the predated value for this calibration equation, $n$ is the number of data points, and $p$ is the number of parameters.

For the inverse equation, the standard error in the estimate errors, $s_y$, is

$$s_y = \frac{\sqrt{\Sigma(x_i - \hat{x}_i)^2}}{n-p} \quad (8)$$

where $x_i$ is the dependent variable, $\hat{xi}$ is the predated value for this calibration equation, n is the number of data points, and p is the number of parameters.

The residual plots are used to assess the fitting ability of these equations. If the residual plots' distribution was uniform, this equation was adequate. If a fixed pattern was found for the residual plots, it showed that this equation was inadequate. More variables need to be considered.

### 2.6. Compare the Predictive Performance for Two Calibration Equations

#### 2.6.1. The Criteria for the Predictive Performance of Two Calibration Equations

Four criteria were proposed for the comparison of predictive performance [35].

The predictive errors were defined as the difference between predicted values from calibration equations and standard values:

$$e_i = x_{i0} - \hat{x}_{i0} \tag{9}$$

where $x_i$ is the standard value, and $\hat{x}_{i0}$ is the measurement value calculated by calibration equations.

1. The minimum $e_i$ value, $e_{i,min}$.
2. The maximum $e_i$ value, $e_{i,max}$.
3. Mean absolute error (MAE):

$$MAE = \frac{\Sigma|e_i|}{n} \tag{10}$$

where $|e_i|$ is the absolute $e_i$ value.

MAE is used to evaluate the accuracy of the equation. The smaller the MAE, the greater the accuracy of the predicted equation.

4. Root mean square error (RMSE):

$$RMSE = \left(\frac{\Sigma e_i^{\,2}}{n}\right)^{0.5} \tag{11}$$

RMSE is used to evaluate the precision of the equation. The smaller the RMSE, the greater the precision of the predicted equation.

#### 2.6.2. The Criteria for the Comparison of the Predictive Performance of Two Calibration Equations

Two criteria were proposed to compare the predictive performance of the two calibration equations for the same data sets:

$$RE_{MAE} = (MAE_{cla} - MAE_{inv})/MAE_{cla} \tag{12}$$

where $MAE_{cla}$ is the MAE value of the classical equation, and $MAE_{inv}$ is the MAE value of the inverse equation.

$$RE_{RMSE} = (RMSE_{cla} - RMSE_{inv})/RMSE_{cla} \tag{13}$$

where $RMSE_{cla}$ is the RMSE value of the classical equation, and the $RMSE_{inv}$ is the RMSE value of the inverse equation.

The $RE_{MAE}$ and $RE_{RMSE}$ compare the accuracy and precision of two calibration equations. If $MAE_{cla} > MAE_{inv}$, the inverse equation has better accuracy, and the $RE_{MAE}$ value is positive. The $RE_{MAE}$ values also showed the degree of the two calibration equations' different accuracy abilities.

If $RMSE_{cla} > RMSE_{inv}$, the inverse equation has better precision, and the $RE_{RMSE}$ value is positive. The $RE_{RMSE}$ values also showed the degree of the difference in precision between the two calibration equations.

### *2.7. Data Splitting*

All data measured from two humidity sensors or collected from the literature were divided into two data sets. The first data set was used to evaluate the adequate form of the calibration equation, and the second was used to evaluate the equation's predictive ability.

For example, sixty-six values ($x_i$, $y_i$) were collected from a humidity sensor. The thirty-three values from ($x_1$, $y_1$) to ($x_{33}$, $y_{33}$) were used to assess the adequate form of the classical equation $y_i = f(x_i)$ and the inverse equation $x_i = g(y_i)$. The response value, $y_i$, of the other 33 data points ($y_{34}$ to $y_{66}$) were substituted into $f(x_i)$ and $g(y_i)$ equations to calculate the predicted values ($\hat{x}_{34}$ to $\hat{x}_{66}$). The difference between $x_i$ (standard values) and $\hat{x}_i$ (calculated values) is the predicted error, $e_i$.

### 2.8. The Calculation of the New Measurement

After the calibration equations are established, the new observation from the sensors is $y_0$. The calculation of the new measurement values, $x_0$, is different for two calibration equations.

1. The inverse equation.

   The new measurement value $x_0$ is calculated directly from this equation.

$$\hat{x}_0 = c_0 + c_1 y_0 + c_2 {y_0}^2 + \ldots + c_n {y_0}^2 \tag{14}$$

2. The classical equation.

   For the linear equation, $y_0 = b_0 + b_1 x_0$, and $\hat{x}_0$ is calculated by

$$\hat{x}_0 = (y_0 - b_0)/b_1 \tag{15}$$

For the two other polynomial equations, $y_0 = b_0 + b_1 x_0 + b_2 {x_0}^2$, and $\hat{x}_0$ is calculated by

$$\hat{x}_0 = \frac{-b_1 \pm \left({b_1}^2 - 4(b_2(b_{0-}y_0))\right)^{0.5}}{2b_2} \tag{16}$$

For high-order polynomial equations, $y_0 = b_0 + b_1x + b_2{x_0}^2 + \ldots + b_kx^k$; the calculation of $x_0$ is very complex and solved by numeric software (The Jenkins-Traub UDF).

### 2.9. Data Source for Comparing Two Calibration Equations

Nine data sets from the literature were collected to compare the predictive ability between the classical and inverse equations. The published literature is shown in Table 2. All original data for the response of chemical analysis instrumentation was divided into two data sets: one for the model established and the other for evaluating the accuracy and precision of the adequate calibration equation.

Adequate classical and inverse equations were established with modeling data sets. Then, the evaluation data of the response were substituted into these equations to calculate the measurement value. The predicted errors were used to compare the performance.

The forms of the calibration equations for the literature data are as follows [20]:

1. Higher-order polynomial equation:

$$y = b_0 + b_1x + b_2x^2 + \ldots + b_kx^k \tag{17}$$

2. Exponential decay equation:

$$y = d_0{}^*\exp(-d_1{}^*x) \tag{18}$$

3. Power equation:

$$y = e_1x^{e2} \tag{19}$$

4. Exponential rise to maximum equations (ERTM equations):

$$y = f_1\,(1 - \mathrm{Exp}(-f_2x)) \tag{20}$$

## 3. Results

### 3.1. The Capacitive Humidity Sensor

#### 3.1.1. The Calibration Equation of Capacitive Humidity Sensors

Figure 1 presents the relationship between the reading values of capacitive humidity sensors and the standard values produced by the saturated salt solutions.

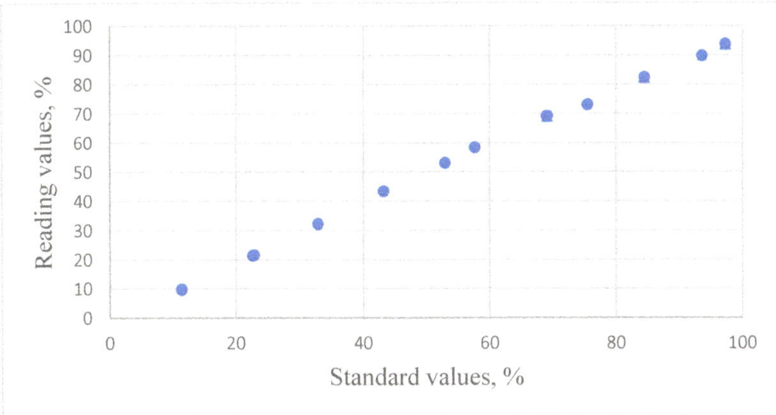


**Figure 1.** The distribution of the relative humidity data for reading values versus the standard humidity values for Vaisala HMP-143A capacitive sensors.

1. The classical equation

From the criteria, the adequate calibration equation established from the calibration data of the capacitive sensor is

$$y = -3.60919 + 1.1626x - 0.0017933\, x^2 \tag{21}$$

where $R^2 = 0.9996$, and $s_y = 0.5780$.

For the new response $y_0$, the corresponding values of the new measurement are calculated as follows:

$$y_0 = -3.60919 + 1.1626x_0 - 0.0017933{x_0}^2 \tag{22}$$

$$x_0 = 338.0968 \pm 290.811\left(1.326816 - 6.87731 \times 10^{-3}y_0\right)^{0.5} \tag{23}$$

2. The inverse equation of capacitive humidity sensors

From the criteria, the adequate calibration equation established from the calibration data of the capacitive sensor is

$$x = 3.5299 + 0.83231y + 0.001842y^2 \tag{24}$$

where $R^2 = 0.995$, and $s_x = 0.6344$.

For the new response of $y_0$ from sensors, the corresponding value of the new measurement is calculated directly.

$$x_0 = 3.5299 + 0.83231y_0 + 0.001842{y_0}^2 \tag{25}$$

3.1.2. The Evaluation of the Calibration Equation of Capacitive Humidity Sensors

The other data set, independent of the previous data sets used to establish the calibration equation, is used to evaluate the predictive performance of two types of calibration equations. The new data sets $(x_1', y_1')$, $(x_2', y_2')$, … $(x_n', y_n')$ were substituted into Equations (23) and (25).

The difference between the standard and calculated measurement values is $e_{iy}$ and $e_{ix}$. Table 3 lists four criteria: $e_{i,min}$, $e_{i,max}$, MAE, and RMSE.

**Table 3.** The predictive performance of two calibration equations for a capacitive humidity sensor.

| Criterion | Classical Equation | Inverse Equation |
|---|---|---|
| $e_{i,min}$ | −1.9395 | −1.0464 |
| $e_{i,max}$ | 0.2198 | 0.2414 |
| MAE | 0.9855 | 0.4944 |
| RMSE | 1.229 | 0.6064 |

In Table 3, the inverse equation has a smaller $e_{i,min}$ value and a larger $e_{i,max}$ value. The inverse equation's MAE and RMSE values are smaller than those of the classical equation. The MAE is the accuracy of the calibration equation, and the RMSE is its precision. The inverse calibration equation has better predictive performance than the classical equation for capacitive humidity sensors.

### *3.2. The Resistive Humidity Sensor*

The relationship between the reading values of resistive humidity sensors and the standard values is shown in Figure 2.

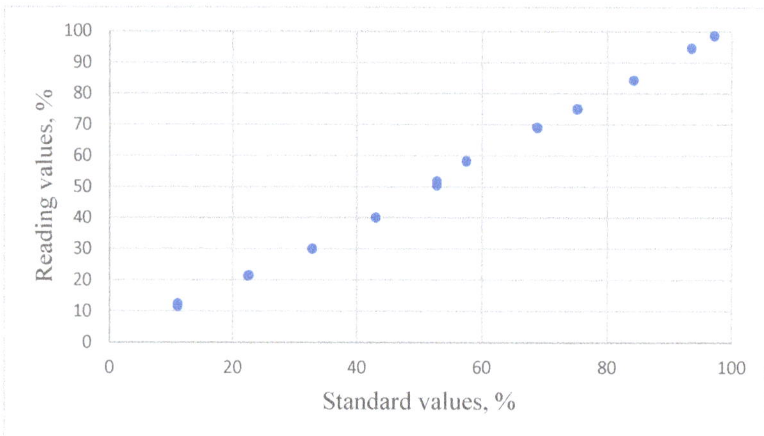


**Figure 2.** The distribution of the relative humidity data for reading values versus the standard humidity values for THT-B121 resistive sensors.

#### 3.2.1. The Calibration Equation of Resistive Humidity Sensors

The adequate calibrations established from the calibration data of the resistive sensor established by regression analysis are described below:

1. The classical equation.

$$y = 5.8524 + 0.4728x + 0.0099x^2 - 5.1025 \times 10^{-5}x^3 \quad (26)$$

where $R^2 = 0.996$, and $s_y = 0.5757$.

For the new response of this humidity sensor, $y_0$, the measurement value, $x_0$, is calculated by the following equation:

$$5.1025 \times 10^{-5}x_0{}^3 - 0.0099x_0{}^2 - 0.4729x_0 - 5.8524 + y_0 = 0 \quad (27)$$

Equation (27) is a three-order polynomial equation; the $x_0$ value needs to be solved by numeric software.

2. The inverse equation.

The adequate calibration equation established from the calibration data of the capacitive sensor is

$$x = -6.1370 + 1.5446y - 0.0103y^2 + 5.0 \times 10^{-5}\, y^3 \quad (28)$$

where $R^2 = 0.996$, and $s_x = 0.6344$.

For the new response, $y_0$, the measurement, $x_0$, can be calculated directly.

$$x_0 = -6.1370 + 1.5464y_0 - 0.0103{y_0}^2 + 5.3 \times 10^{-5}{y_0}^3 \quad (29)$$

3.2.2. The Evaluation of the Calibration Equation of Resistive Humidity Sensors

The other data set, independent of the data sets used to establish calibration equations, was used to evaluate the predictive ability.

The criteria for the evaluation of two equations are listed in Table 4.

**Table 4.** The predictive performance of two calibration equations for the resistive humidity sensor.

| Criterion | Classical Equation | Inverse Equation |
|---|---|---|
| $e_{i,min}$ | −1.9150 | −0.9950 |
| $e_{i,max}$ | 0.7311 | 0.8908 |
| MAE | 0.5431 | 0.5536 |
| RMSE | 0.4894 | 0.4807 |

The classical equation has smaller $e_{i,min}$, $e_{i,max}$, and MAE values. The inverse calibration equation has a smaller RMSE value. However, the differences in these criteria between the two calibration equations were limited. No significant difference could be found in the predictive performance of the two equations. However, the new measurement values of the inverse equation can be calculated directly. The calculation of new measurement values for the classical equation is very complex.

*3.3. The Evaluation of Two Calibration Equations from Previous Data in the Literature*

3.3.1. The Measurement of Chloromethane Concentration with GC-MS

Lavagnini and Magno [23] measured chloromethane concentration with GC-MC. There are nine concentration levels (μg/L) and ten replicates for each concentration. The first data set contained 45 data points for establishing the calibration equation. The other 45 data points were used to evaluate the predictive ability. The distribution between the response of the peak area and the standard concentration is shown in Figure 3. The regression analysis results of two calibration equations are listed in Table 5.

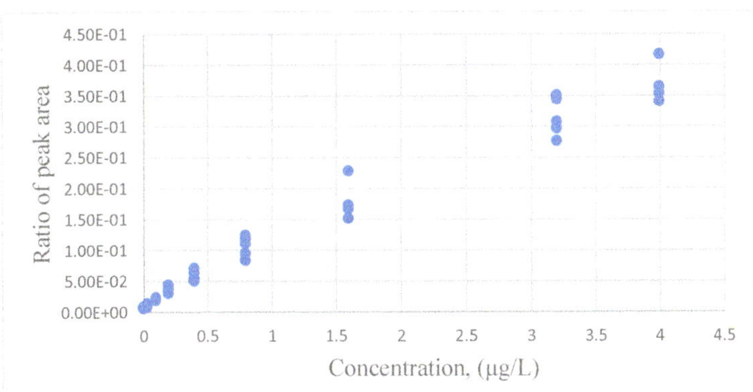


**Figure 3.** The distribution of the chloromethane data for the ratio of peak areas versus the standard concentrations for GC-MS.

**Table 5.** The criteria of the calibration regression equations for the chloromethane concentration ($x_i$) and the ratio peak area ($y_i$) for GC-MC.

| Classical Equation | $R^2$ | Residual Plots |
|---|---|---|
| 1. $y = 0.0187 + 0.089x$ | 0.9842 | F.P. |
| 2. $y = 0.0113 + 0.1158x - 0.0071x^2$ | 0.9893 | U.D. |
| 3. $y = 0.3033(1 - \exp(-0.2245x))$ | 0.9863 | U.D. |
| 4. $y = 0.0109 + 0.7477(1 - \exp(-0.1585x))$ | 0.9873 | U.D. |
| 5. $y = 0.1232x^{0.7874}$ | 0.9867 | F.P. |
| **Inverse Equation** | $R^2$ | **Residual Plots** |
| 1.$x = -0.1782 + 10.7494y$ | 0.9773 | F.D. |
| 2. $x = -0.1114 + 8.920y + 4.9491y^2$ | 0.9791 | U.D. |
| 3. $x = 1911.776(1 - \exp(-0.0053y))$ | 0.9691 | U.D. |
| 4. $x = -0.1899 + 925.325(1 - \exp(-0.012y))$ | 0.9841 | U.D. |
| 5. $x = 12.9265y^{1.2093}$ | 0.9793 | F.P. |

Note: F.P.: fixed pattern; U.D.: uniform distribution.

From the criteria for different equations, the adequate calibration equation for the classical equations is

$$y = 0.0109 + 0.7477(1 - \exp(-0.1585x)) \tag{30}$$

For the new response (the ratio of peak area), $y_0$, the measurements of two equations are calculated as follows:

$$x_0 = -6.30915Ln(1.01458 - 1.3374\, y_0) \tag{31}$$

The criteria for evaluating these calibration equations are listed in Table 6.

**Table 6.** The criteria for evaluating the predictive ability of these calibration equations for measuring chloromethane concentration with GC-MS.

| Criterion | Classical Equation | Inverse Equation |
|---|---|---|
| $e_{i,min}$ | −0.3859 | −0.3835 |
| $e_{i,max}$ | 1.4983 | 0.9350 |
| MAE | 0.4758 | 0.3328 |
| RMSE | 0.2695 | 0.2043 |

The results indicated that the inverse equation had smaller values for these criteria than the classical equation.

#### 3.3.2. Using Spectrophotometry to Measure Albumin

In the test conducted by Rawski et al. [25], there were 11 albumin standards (μg/mL) and three replicates for each concentration. The response of this instrument, $y_i$, was the peak height. The distribution between response and standard values of model development data is shown in Figure 4. The results of the regression analysis are listed in Table 7.

The adequate calibrations are listed as follows:

1. The classical equation is

$$y = 626.5993(1 - \exp(-0.0635x)) \tag{32}$$

For a new response, $y_0$, the new measurement of $x_0$ could be calculated by

$$x_0 = -15.748\ \mathrm{Ln}(1 - y_0/626.5933) \tag{33}$$

2. The inverse equation is

$$x = 0.2443 + 0.0164_y + 5.8629 \times 10^{-5}y^2 \tag{34}$$

The new measurement, $x_0$, can be calculated directly using an equation.

**Table 7.** The calibration regression equation and criteria for the albumin concentrations and the peak height with spectrophotometry.

| Classical Equation | $R^2$ | $s_y$ | Residual Plots |
|---|---|---|---|
| 1. $y - 43.5066 + 22.1861$ | 0.9651 | 29.081 | F.P. |
| 2. $y = 0.9519 + 36.726x - 0.7250x^2$ | 0.9980 | 5.5069 | F.P. |
| 3. $y = 626.5993(1 - \exp(-0.0635x))$ | 0.9985 | 6.0221 | U.D. |
| 4. $y = -1.9587 + 623.7546(1 - \exp(-0.0646x))$ | 0.9985 | 6.1770 | U.D. |
| 5. $y = 57.5278x^{0.6959}$ | 0.9939 | 12.184 | F.P. |
| **Inverse Equation** | $R^2$ | $s_x$ | **Residual Plots** |
| 1. $x = -1.5344 + 0.0435y$ | 0.9651 | 1.2876 | F.P. |
| 2. $x = 0.2443 + 0.0164y + 5.8629 \times 10^{-5}y^2$ | 0.994 | 0.5430 | U.D. |
| 3. $x = 9363.3Exp(1 - 4.18 \times 10^{-6}y)$ | 0.9522 | 1.5030 | F.P. |
| 4. $x = -1.5372 + 6695.5366(1 - \exp(-6.2466 \times 10^{-6}y))$ | 0.9646 | 1.3361 | F.P. |
| 5. $x = 0.0023y^{1.4809}$ | 0.992 | 0.6153 | U.D. |

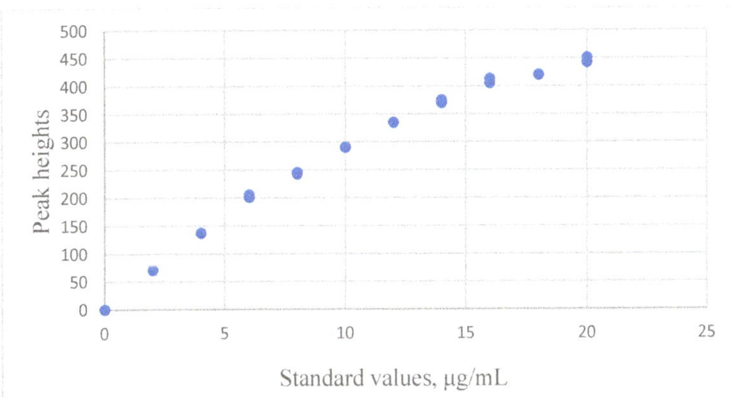


**Figure 4.** The distribution of the albumin concentration data for the peak heights versus the standard concentrations with spectrophotometry.

The criteria for evaluating predictive ability for two calibration equations are listed in Table 8.

**Table 8.** The criteria for evaluating the predictive ability of these calibration equations for measuring the albumin concentrations using spectrophotometry.

| Criterion | Classical Equation | Inverse Equation |
|---|---|---|
| $e_{i,min}$ | −0.4849 | −0.7791 |
| $e_{i,max}$ | 1.1608 | 1.3347 |
| MAE | 0.4770 | 0.5416 |
| RMSE | 0.4147 | 0.4214 |

The classical equation had better predictive ability. The MAE and RMSE values showed better accuracy and precision in the classical equation.

#### 3.3.3. The Measurement of Anti-IgG by Biophotonic Sensing Cells

The calibration of a biochip composed of a set of BICELLS (biophotonic sensing cells) for anti-IgG [29] was studied for adequate calibration equations. The response ($y_i$) is the transduction signal, and the standard concentration (µg/mL) is the regressor ($x_i$). The calculation regression results are as follows:

The classical equation is

$$y = -0.3135 + 2.109x - 0.002827x^2 + 1.335 \times 10^{-5}x^3 \tag{35}$$

where $R^2$ = 0.9863, and $s_y$ = 0.2331.

The inverse equation is

$$x = -0.2732 + 12.8733y - 5.259y^2 + 1.0162y^3 \quad (36)$$

where $R^2$ = 0.9833, and $s_x$ = 0.1957.

The comparison of the predictive performance of two calibration equations is listed in Table 9.

**Table 9.** The criteria for evaluating the predictive ability of these calibration equations for the measurement of anti-IgG by biophotonic sensing cells.

| Criterion | Classical Equation | Inverse Equation |
|---|---|---|
| $e_{i,min}$ | −8.0672 | −10.2422 |
| $e_{i\,max}$ | 4.3461 | 5.8671 |
| MAE | 2.2862 | 2.2715 |
| RMSE | 2.9587 | 3.3921 |

The classical equation performed better in terms of $e_{i,min}$, $e_{i,max}$, and RMSE (precision), and two calibration equations had similar predictive performance for the MAE values.

#### 3.3.4. The Measurement of Drug Concentration in Blood with an HPLC Assay

The drug concentration in blood ($x_i$) was measured with an HPLC assay, and the response ($y_i$) ranged from 0.002 to 0.272 [27].

The adequate calibration equations are as follows:

1. The classical equation is

$$y = 0.0199 + 1.3940(\ 1 - \exp(-0.5084x)) \quad (37)$$

where $R^2$ = 0.9921, and $s_y$ = 0.0348.

2. The inverse equation is

$$x = -2.7006 + 3.7037y^{0.5377} \quad (38)$$

where $R^2$ = 0.9912, and $s_x$ = 0.0896.

The predictive performance of the two calibration equations is listed in Table 10.

**Table 10.** The criteria for evaluating the predictive ability of these calibration equations for measuring drug concentration in blood with an HPLC assay.

| Criterion | Classical Equation | Inverse Equation |
|---|---|---|
| $e_{i,min}$ | −0.5011 | −0.475 |
| $e_{i,max}$ | 0.4713 | 0.482 |
| MAE | 0.1924 | 0.1847 |
| RMSE | 0.1171 | 0.1110 |

The class equation had smaller $e_{i,min}$ values, and the inverse equation had smaller $e_{i,max}$ values and better accuracy (smaller MAE) and precision (smaller RMSE). However, the predictive ability between two equations is not significantly different.

#### 3.3.5. Detection of EtP Compound by QqQ-MS

Martin et al. [28] reported determining EtP compound pollution by QqQ-MS. The instrument's response is area ($y_i$), and the standard concentration is $x_i$.

The adequate classical equation is

$$y = -18360.5421 + 1577.4319x - 0.1277x^2 \quad (39)$$

where $R^2$ = 0.9986, and $s_y$ = 28310.3.

The adequate inverse equation is

$$x = -0.5654 + 8.197 \times 10^{-4}y - 2.402 \times 10^{-10}y^2 + 9.719 \times 10^{-17}y^3 \quad (40)$$

where $R^2 = 0.9998$, and $s_x = 28310.3$.

The predictive performance of the two calibration equations is listed in Table 11.

**Table 11.** The criteria for evaluating the predictive ability of these calibration equations for measuring the EtP compound by QqQ-MS.

| Criterion | Classical Equation | Inverse Equation |
|---|---|---|
| $e_{i,min}$ | −33.0861 | −34.4825 |
| $e_{i,max}$ | 22.0372 | 23.6656 |
| MSE | 15.6931 | 15.5105 |
| RMSE | 12.6058 | 13.3721 |

The classical equation has lower values of $e_{i,min}$, $e_{i,max}$, and RMSE, which indicates better precision. The inverse equation has a lower MAE value.

### 3.3.6. The Measurement of Sulfides by Flow Injection Analysis

Desimoni [22] reported the calibration data set of sulfides in flow injection analysis at a palladium-vitreous carbon-modified electrode. The $x_i$ is the standard concentration, and the response $y_i$ is the current (μA).

The adequate classical equation is

$$y = 0.1658x^{1.0369} \quad (41)$$

where $R^2 = 0.9998$, and $s_y = 0.117$.

The adequate inverse equation is:

$$x = 596.3756(1 - Exp(-0.009238y)) \quad (42)$$

where $R^2 = 0.9999$, and $s_x = 0.295$.

The predictive performance of two calibration equations is listed in Table 12.

**Table 12.** The criteria for the evaluation of the predictive ability of these calibration equations for the measurement of sulfides by flow injection analysis.

| Criterion | Classical Equation | Inverse Equation |
|---|---|---|
| $e_{i,min}$ | −0.3948 | −0.1498 |
| $e_{i,max}$ | 0.4439 | 0.2781 |
| MSE | 0.2134 | 0.1355 |
| RMSE | 0.2449 | 0.1137 |

With smaller $e_{i,min}$, $e_{i,max}$, MAE, and RMSE values, the inverse equation performed better than the classical equation.

### 3.3.7. Measurement of Daidzein by HPLC Analysis

Mulholland and Mibbert [21] detected the daidzein concentration with a HPLC analyzer. The standard concentration is $x_i$ (mg/50 mL), and the HPLC response is the pear area ($y_i$).

The adequate classical equation is

$$y = -0.4955 + 3.222x^{0.9471} \quad (43)$$

where $R^2 = 0.9995$, and $s_y = 0.2459$.

The adequate inverse equation is

$$x = 0.1535 + 0.2992y^{1.055} \quad (44)$$

where $R^2$ = 0.9946, and $s_y$ = 0.0845.

The predictive performance of the two calibration equations is listed in Table 13.

**Table 13.** The criteria for evaluating the predictive ability of these calibration equations for the measurement of daidzein with HPLC analysis.

| Criterion | Classical Equation | Inverse Equation |
|---|---|---|
| $e_{i,min}$ | −0.0698 | −0.1562 |
| $e_{i,max}$ | 0.1701 | 0.0684 |
| MAE | 0.0837 | 0.0823 |
| RMSE | 0.0668 | 0.073 |

The classical equation has a lower $e_{i,min}$ and RMSE (precision), and the inverse equation has a lower $e_{i,max}$ and MAE (accuracy).

### 3.3.8. Measurement of Cocaine Concentration by LC-MS-MS

Desharnais et al. [26] measured cocaine concentrations with an LC-MS-MS instrument. The standard concentration is $x_i$, and the response, $y_i$, is the instrument's area ratio. The response data did not have a constant variance. These $x_i$ values were logarithmic transformations which require further analysis.

The adequate classical equation is

$$y = 0.1832 + 0.008991\ \ln(x)^{5.0619} \tag{45}$$

where $R^2$ = 0.9994, and $s_y$ = 0.1401.

The adequate inverse equation is

$$Lnx = -30.5041 + 34.2735\ y^{0.3119} \tag{46}$$

where $R^2$ = 0.9991, and $s_x$ = 0.0604.

The predictive performance of the two calibration equations is listed in Table 14.

**Table 14.** The criteria for evaluating the predictive ability of these calibration equations for measuring cocaine concentration by LC-MS-MS.

| Criterion | Classical Equation | Inverse Equation |
|---|---|---|
| $e_{i,min}$ | −71.2154 | −82.0002 |
| $e_{i,max}$ | 46.0926 | 36.4611 |
| MAE | 47.4945 | 27.3709 |
| RMSE | 35.4701 | 12.6038 |

The classical equation has lower $e_{i,min}$, and the inverse equation has lower $e_{i,max}$, MAE (accuracy), and RMSE (precision).

### 3.3.9. Measurement of Benzaldehyde Using Pulse Polarography

Ortiz et al. [24] reported the calibration data for determining benzaldehyde concentration by pulse polarography. The standard concentration (m/mol) is $x_i$, and the current response (μA) is $y_i$.

The adequate classical equation is

$$y = 2.0923 - 2.1131\ \mathrm{Exp}(-1.0328x) \tag{47}$$

where $R^2$ = 0.9762, and $s_y$ = 0.0204.

The adequate inverse equation is

$$x = 0.0308 - 0.01089y + 3.6389y^2 - 6.0929y^3 \tag{48}$$

where $R^2$ = 0.9821, and $s_x$ = 0.0037.

The predictive performance of the two calibration equations is listed in Table 15.

**Table 15.** The criteria for evaluating the predictive ability of these calibration equations for measuring benzaldehyde using pulse polarography.

| Criterion | Classical Equation | Inverse Equation |
| --- | --- | --- |
| $e_{i,min}$ | −0.02174 | −0.01608 |
| $e_{i,max}$ | 0.00858 | 0.007406 |
| MAE | 0.008197 | 0.006503 |
| RMSE | 0.009711 | 0.007014 |

The classical equation has lower $e_{i,max}$, and the inverse equation has lower $e_{i,min}$, MAE (accuracy), and RMSE (precision).

## 4. Discussion

The standard values ($x_i$) are prepared without measurement errors for the calibration procedure. One of the regression assumptions is the negligible error for the xi value. For the classical equation, the $x_i$ value of the standard values corresponded to this requirement. The instrument's response, $y_i$, is the regressor value. The measurement errors exist in the regressor variables. If the response variance is limited, the effect of measurement error on the regressor can be ignored with a minor effect [6].

The coefficient of variation (CV) can be used to evaluate the effect of measurement errors. The CV is defined as

$$CV = (s/y_{mean}) \times 100\% \quad (49)$$

where s is the standard deviation of the instrument's responses and $y_{mean}$ is the mean of the response values at the fixed standard value.

The results of comparing the predictive performance of two calibration equations for the measurement data of this study and the literature data are listed in Table 16. If the $RE_{MAE}$ and $RE_{RMSE}$ are positive, the inverse calibration equation has better accuracy and precision than the classical equation.

For the capacitive humidity sensor, the $RE_{MAE}$ is 49.83%, and the $RE_{RMSE}$ is 50.66%. This indicates that the inverse calibration equation has significantly better predictive performance than the classical calibration equation. The resistive humidity sensor's $RE_{MAE}$ is −1.93%, and the $RE_{RMSE}$ is 1.78%. This shows that the classical equation has better accuracy, and the inverse equation has better precision. However, these differences in the predictive performance of resistive humidity sensors are insignificant.

Figure 5 indicates the data distribution of the standard deviation of the responses of two humidity sensors in the same standard humidity environment. The capacitive sensor has excellent replicative ability, with standard deviations < 0.1%. However, the resistive sensor's standard deviations range from 0.14% to 0.44%. The larger standard deviations in the response in the same standard environment for resistive humidity sensors induced the problems. As the response ($y_i$) is recognized as the dependent variable (regressor) and the measurement errors are significant, the severity of the assumption of regression analysis influences the predictive performance of the inverse calibration equation. In other words, smaller standard deviations of the measurement values could ensure the better predictive performance of the inverse calibration equation.

**Table 16.** Comparison of the predictive performance of two calibration equations from this study and the literature data.

| | $RE_{MAE}$ [1] (Accuracy) | $RE_{RMSE}$ [2] (Precision) |
|---|---|---|
| I. Hygrometer | | |
| 1. Capacitive | 49.83% | 50.66% |
| 2. Resistance | −1.93% | 1.78% |
| II. Literature data | | |
| 1. GC-MS [23] | 48.56% | 46.53% |
| 2. Flow injection Analysis [22] | 36.5% | 53.57% |
| 3. LC-MS-MS [26] | 42.37% | 64.47% |
| 4. Pulse polarography [24] | 20.67% | 27.78% |
| 5. Spectrophotometry [25] | −13.54% | −1.62% |
| 6. BICELLS (biophotonic sensing cells) [29] | −0.64% | 14.65% |
| 7. HPLC [27] (drug in blood) | 4.0% | 5.2% |
| 8. LC-QqQ-MS [28] | 1.16% | −6.08% |
| 9. HPLC [21] (daidzein) | 1.67% | −9.28% |

Note: 1. $RE_{MAE} = (MAE_{cla} - MAE_{inv})/MAE_{cla}$, where MAEcla is the MAE value of the classical equation, and MAEinv is the MAE value of the inverse equation. The $RE_{MAE}$ is used to evaluate the accuracy. 2. $RE_{RMSE} = (RMSE_{cla} - RMSE_{inv})/RMSE_{cla}$, where $RMSE_{cla}$ is the RMSE value of the classical equation, and the $RMSE_{inv}$ is the RMSE value of the inverse equation. The $RE_{MAE}$ is used to evaluate the precision.

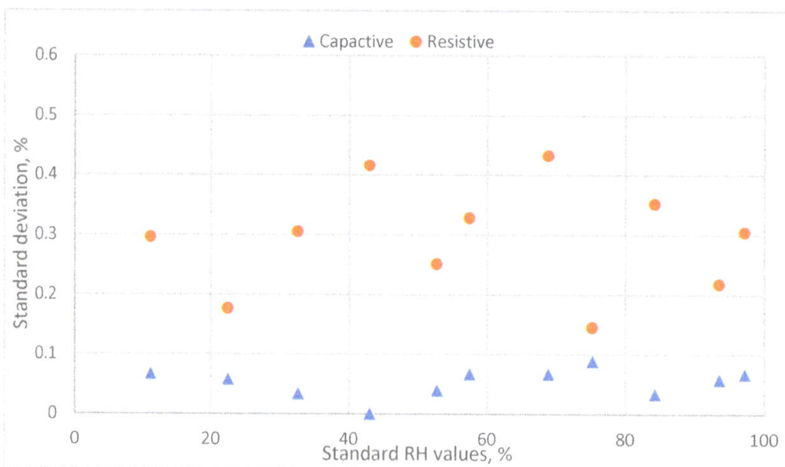


**Figure 5.** The distribution of the standard deviation values of the response and the standard relative humidity values for two humidity sensors.

The distribution of the CV values of the response and standard humidity values is shown in Figure 6. The CV values of the capacitive humidity sensor were <0.6%. However, the resistive humidity sensor's CV values range from 0.5 to 6.0%.

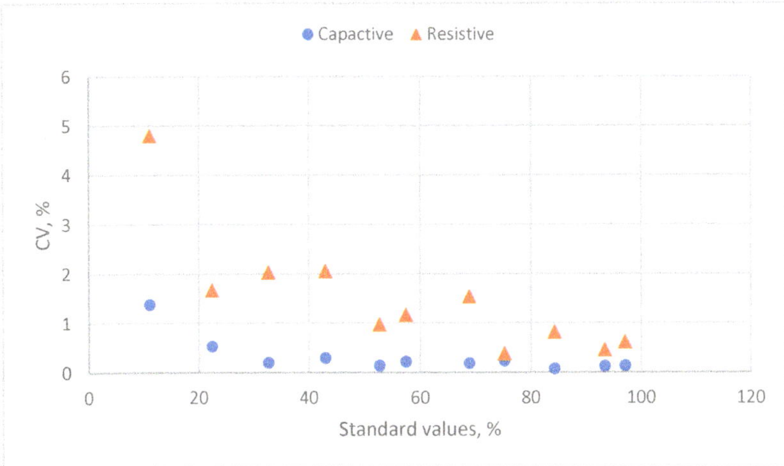


**Figure 6.** The distribution of the response's CV values and the standard relative humidity values for two humidity sensors.

The inverse equation was first proposed last century. In this equation, the response values of instruments serve as independent variables, violating the assumption of no errors for the regressor. For this reason, many researchers adopted the classical calibration equation. However, calculating new measurement values from the new response of the instrument values is needed to transform the original classical equation. So, only the linear equation was selected as the adequate calibration equation. However, much of the literature indicates that high-order polynomial and nonlinear equations are adequate equations [11,13,17,20]. The transform forms of these classical calibration equations are very complex and impractical with regard to calculating the measurement values.

Francois et al. [18] concluded that two calibration equations have equal predictive ability. In this study, we found some calibration data sets have similar results, especially for the response values with a higher coefficient of variance (CV).

Many studies have mentioned the better-predicted performance of the inverse calibration equation [9,10,13,14]. Their results were based on small data sets or theoretical discussions. This study used the measurement data sets from two kinds of humidity sensors and nine data sets collected from the literature to evaluate the predictive performance of two calibration equations and confirmed the same results. Shalabh [11] mentioned that measurement errors must be considered for prediction accuracy. This study found the same results with the evaluation of the MAE values. Center [9] found an improvement in the predictive ability of inverse calibration with the decrease in measurement variance. In this study's results, the inverse equation's predicted predictive ability is superior to the classical equation if the CV values of the instrument's response are smaller. Krutchkoff's [7,8] study showed the inverse equation has a more minor average error. This study proposed the criterion of the RMSE values to confirm this statement.

In Table 16, the inverse calibration equation had significant predictive ability for GC-MS, flow injection analysis, LC-MS-MS, and pulse polarography. The CV values of the GC-MS, flow injection analysis, LC-MS-MS, and pulse polarography ranged from 4.0 to 8.2%, 0.3 to 1.7%, 0.4 to 1.4%, and 2.5 to 3.5%, respectively. The classical calibration equation performed similarly to the inverse calibration equation for spectrophotometry, BICELLS, HPLC, and QqQ-MS calibration data. The CV values of the spectrophotometry, BICELLS, HPLC, and QqQ-MS ranged from 3 to 24%, 13.5 to 24.5%, and 30 to 34%, respectively.

The results indicated that the better predictive performance of the inverse calibration equation is more significant than that of the classical equation if the response of this instrument has excellent replicative ability. The CV values of the response in each standard environment could serve as an index to express its replicative ability. In the case when the CV < 5%, the inverse calibration equation has a better predictive performance. In the case when the CV > 5%, the classical calibration equation has a better or similar predictive performance.

Recently, measurements of intelligent instruments have been used to embed calibration equations into this device. The inverse calibration equation can work well and easily. That is, an adequate inverse calibration equation could be used for intelligent measurement.

In the sensor industry, most of the sensors are physical sensors, accounting for 90%, while chemical sensors account for 9%, and biological sensors account for 1% [36]. Research on chemical sensors is, therefore, declining. However, chemical sensor application objects include medicine, food, the chemical industry, semiconductor manufacturing processes, etc. The measurement performance of chemical sensors has a significant impact. Chemical sensing requires the establishment of a calibration equation. Therefore, this paper should be valued in academic research.

Recently, researchers have been concerned about the impact of calibration equations on measurement performance. Their measurement objects include solvent-independent molecular weight [37,38], tiny mass [39], soil moisture [40], and pesticide concentrations [41]. The results of this study for two calibration equations could be applied to different instruments to improve their measurement performance.

## 5. Conclusions

There are two kinds of calibration equations: classical equations and inverse equations. The classical equation is widely used, and the linear equation is the main one used for calculating new measurement values. The inverse equation can compute the new measurement directly and efficiently. However, some researchers suspect it violates the basic assumptions of regression analysis. In this study, the actual calibration data sets of two types of humidity sensors were collected. Nine calibration data sets of various instruments were collected from the literature. Four criteria were proposed to evaluate the predictive performance of the two calibration equations.

The results of this study show that the inverse equation has excellent predictive performance for the calibration equation of the capacitive humidity sensor. The classical equation has better accuracy, and the inverse equation has better precision for the predictive performance of resistive humidity sensors. If the instrument response has good repeatability, the inverse equation performs excellently for the nine data sets collected in the literature. If the repeatability of the instrument response is poor, two calibration equations have similar predictive performance. The CV value of measurement in standard conditions is used as a criterion. A CV value of less than 5% can be used as a threshold basis for the inverse calibration equation.

**Author Contributions:** Conceptualization, H.-Y.C. and C.C.; methodology, H.-Y.C. and C.C.; software, C.C.; formal analysis, H.-Y.C.; investigation, H.-Y.C. and C.C.; data curation, H.-Y.C.; writing—original draft preparation, H.-Y.C. and C.C.; writing—review and editing, H.-Y.C. and C.C.; visualization, C.C. supervision, C.C.; project administration, C.C. All authors have read and agreed to the published version of the manuscript.

**Funding:** This research received no external funding.

**Institutional Review Board Statement:** Not applicable.

**Informed Consent Statement:** Not applicable.

**Data Availability Statement:** Data is unavailable due to a statement is still required.

**Conflicts of Interest:** The authors declare no conflicts of interest.

## References


1. EURACHEM Working Group. The Fitness for Purpose of Analytical Methods. In *A Laboratory Guide to Method Validation and Related Topics*, 1st ed.; EURACHEM: London, UK, 1998.
2. IUPAC. Recommendation, guidelines for calibration in analytical chemistry. Part I. Fundamentals and single component calibration. *Pure Appl. Chem.* **1998**, *70*, 993–1014. [CrossRef]
3. Barwick, V. Preparation of Calibration Curves: A Guide to Best Practice; VAM, LGC/VAM/2003/032. 2003. Available online: https://www.researchgate.net/publication/334063221_Preparation_of_Calibration_Curves_A_Guide_to_Best_Practice (accessed on 10 June 2024).
4. Rozet, E.; Ceccato, A.; Hubert, C.; Ziemons, E.; Oprean, R.; Rudaz, S.; Boulanger, B.; Hubert, P. Analysis of recent pharmaceutical regulatory documents on analytical method validation. *J. Chromatogr. A* **2007**, *1158*, 111–125. [CrossRef] [PubMed]
5. Sanagi, M.M.; Nasir, Z.; Ling, S.L.; Hermawan, D.; Ibrahim, W.A.W.; Naim, A.A. A practical approach for linearity assessment of calibration curves under the International Union of Pure and Applied Chemistry (IUPAC) Guidelines for an in-house validation of method of analysis. *J. AOAC Intern.* **2010**, *93*, 1322–1330. [CrossRef]
6. Myers, R.H. *Classical and Modern Regression with Applications*, 2nd ed.; Duxbury Press: Monterey, CA, USA, 1990.
7. Krutchkoff, R.G. Classical and inverse regression methods of calibration. *Technometrics* **1967**, *9*, 25–439. [CrossRef]
8. Krutchkoff, R.G. Classical and inverse regression methods of calibration in extrapolation. *Technometrics* **1969**, *11*, 605–608. [CrossRef]
9. Centner, V.; Massart, D.L.; De Jong, S. Inverse calibration predicts better than classical calibration. *Fresenius' J. Anal. Chem.* **1998**, *361*, 2–9. [CrossRef]
10. Tellinghuisen, J. Inverse vs. classical calibration for small data sets. *Fresenius' J. Anal. Chem.* **2000**, *368*, 585–588. [CrossRef]
11. Shalabh. Least squares estimators in measurement error models under the balanced loss function. *Test* **2001**, *10*, 301–308. [CrossRef]
12. Tellinghuisen, J. Simple algorithms for nonlinear calibration by the classical and standard additions methods. *Analyst* **2005**, *130*, 370–378. [CrossRef]
13. Parker, P.A.; Vining, G.G.; Wilson, S.R.; Szarka, J.L., III; Johnson, N.G. The prediction properties of classical and inverse regression for the simple linear calibration problem. *J. Qual. Technol.* **2010**, *42*, 332–347. [CrossRef]
14. Besalú, E. The connection between inverse and classical calibration. *Talanta* **2013**, *116*, 45–49. [CrossRef] [PubMed]
15. Granovskii, V.; Siraia, T. Direct and inverse calibration curves of measuring instruments: Selection and fitting. In *16th International Congress of Metrology*; EDP Sciences: Les Ulis, France, 2013; p. 04006.
16. Witkovsky, V.; Wimmer, G. Inverse and direct prediction and its effect on measurement uncertainty in polynomial comparative calibration. In Proceedings of the 2019 12th International Conference on Measurement, Smolenice, Slovakia, 27–29 May 2019.
17. Delgado, R. Misuse of Beer-Lambert Law and other calibration curves. *R. Soc. Open Sci.* **2022**, *9*, 211103. [CrossRef] [PubMed]
18. François, N.; Govaerts, B.; Boulanger, B. Optimal designs for inverse prediction in univariate nonlinear calibration models. *Chemom. Intell. Lab. Syst.* **2004**, *74*, 283–292. [CrossRef]
19. Kannan, N.; Keating, J.P.; Mason, R.L. A comparison of classical and inverse estimators in the calibration problem. *Comm. Statist. Theory Methods* **2007**, *36*, 83–95. [CrossRef]
20. Chen, H.Y.; Chen, C. Evaluation of calibration equations by using regression analysis: An example of chemical analysis. *Sensors* **2022**, *22*, 447. [CrossRef]
21. Mulholland, M.; Hibbert, D.B. Linearity and the limitations of least squares calibration. *J. Chromatogr. A* **1997**, *762*, 73–82. [CrossRef]
22. Desimoni, E. A program for the weighted linear least-squares regression of unbalanced response arrays. *Analyst* **1999**, *124*, 1191–1196. [CrossRef]
23. Lavagnini, I.; Magno, F. A statistical overview on univariate calibration, inverse regression, and detection limits: Application to gas chromatography/mass spectrometry technique. *Mass Spectrom. Rev.* **2007**, *26*, 1–18. [CrossRef]
24. Ortiz, M.C.; Sánchez, M.S.; Sarabia, L.A. Quality of analytical measurements: Univariate regression. In *Comprehensive Chemometrics. Chemical and Biochemical Data Analysis*; Brown, S.D., Tauler, R., Walczak, B., Eds.; Elsevier: Amsterdam, The Netherlands, 2009; pp. 127–169.
25. Rawski, R.I.; Sanecki, P.T.; Kijowska, K.M.; Skital, P.M.; Saletnik, D.E. Regression analysis in analytical chemistry. Determination and validation of linear and quadratic regression dependencies. *S. Afr. J. Chem.* **2016**, *69*, 166–173. [CrossRef]
26. Desharnais, B.; Camirand-lemyre, F.; Mireault, P.; Skinner, C.D. Procedure for the selection and validation of a calibration model I—Description and application. *J. Anal. Toxicol.* **2017**, *41*, 261–268. [CrossRef]
27. Martin, J.; de Adana, D.D.R.; Asuero, A.G. Fitting models to data: Residual analysis, a primer. In *Uncertainty Quantification and Model Calibration*; Hessling, J.P., Ed.; IntechOpen Ltd.: London, UK, 2017; Chapter 7; p. 133.
28. Martin, J.; Gracia, A.R.; Asuero, A.G. Fitting nonlinear calibration curves: No models perfect. *J. Anal. Sci. Methods Instrum.* **2017**, *7*, 1–17. [CrossRef]
29. Lavín, Á.; Vicente, J.D.; Holgado, M.; Laguna, M.F.; Casquel, R.; Santamaría, B.; Maigler, M.V.; Hernández, A.L.; Ramírez, Y. On the determination of uncertainty and limit of detection in label-free biosensors. *Sensors* **2018**, *18*, 2038. [CrossRef]
30. Greenspan, L. Humidity fixed points of binary saturated aqueous solutions. *J. Res. Natl. Bur. Stand.* **1977**, *81A*, 89–96. [CrossRef]

31. OMIL. The Scale of Relative Humidity of Air Certified Against Saturated Salt Solutions. In *OMIL R 121*; Organization Internationale De Metrologie Legale: Paris, France, 1996.
32. Chen, H.Y.; Chen, C. Determination of optimal measurement points for calibration equations—Examples by RH sensors. *Sensors* **2019**, *19*, 1213. [CrossRef] [PubMed]
33. Weisberg, S. *Applied Linear Regression*, 4th ed.; Wiley: New York, NY, USA, 2013; p. 384.
34. Rawlings, J.O.; Pantula, S.G.; Dickey, D. Applied regression analysis. In *Springer Texts in Statistics*; Springer: New York, NY, USA, 1998.
35. Chen, C. Evaluation of resistance–temperature calibration equations for NTC thermistors. *Measurement* **2009**, *42*, 1103–1111. [CrossRef]
36. Javaid, M.; Haleem, A.; Singh, R.P.; Rab, S.; Suman, R. Significance of sensors for industry 4.0: Roles, capabilities, and applications. *Sens. Int.* **2021**, *2*, 100110. [CrossRef]
37. Voorter, P.J.; McKay, A.; Dai, J.; Paravagna, O.; Cameron, N.R.; Junkers, T. Solvent-independent molecular weight determination of polymers based on a truly universal calibration. *Angew. Chem. Int. Ed.* **2022**, *61*, e202114536. [CrossRef]
38. Visconti, G.; Boccard, J.; Feinberg, M.; Rudaz, S. From fundamentals in calibration to modern methodologies: A tutorial for small molecules quantification in liquid chromatography–mass spectrometry bioanalysis. *Anal. Chim. Acta* **2023**, *1240*, 340711. [CrossRef]
39. Olsson, C.O.A.; Igual-Muñoz, A.N.; Mischler, S. Methods for calibrating the electrochemical quartz crystal microbalance: Frequency to mass and compensation for viscous load. *Chemosensors* **2023**, *11*, 456. [CrossRef]
40. Gómez-Astorga, M.J.; Villagra-Mendoza, K.; Masís-Meléndez, F.; Ruíz-Barquero, A.; Rimolo-Donadio, R. Calibration of low-cost moisture sensors in a biochar-amended sandy loam soil with different salinity levels. *Sensors* **2024**, *24*, 5958. [CrossRef]
41. Veiga-del-Baño, J.M.; Oliva, J.; Cámara, M.Á.; Andreo-Martínez, P.; Motas, M. Matrix-matched calibration for the quantitative analysis of pesticides in pepper and wheat flour: Selection of the best calibration model. *Agriculture* **2024**, *14*, 1014. [CrossRef]

*sensors*

MDPI

*Article*

# Room-Temperature Ammonia Sensing Using Polyaniline-Coated Laser-Induced Graphene

**José Carlos Santos-Ceballos [1,2,3], Foad Salehnia [1,2,3,*], Frank Güell [1,4], Alfonso Romero [1,2,3], Xavier Vilanova [1,2,3] and Eduard Llobet [1,2,3,*]**

1 MINOS, School of Engineering, Universitat Rovira i Virgili, Avda. Països Catalans 26, 43007 Tarragona, Spain; josecarlos.santos@urv.cat (J.C.S.-C.); frank.guell@ub.edu (F.G.); alfonsojose.romero@urv.cat (A.R.); xavier.vilanova@urv.cat (X.V.)

2 IU-RESCAT, Research Institute in Sustainability, Climatic Change and Energy Transition, Universitat Rovira i Virgili, Joanot Martorell 15, 43480 Vila-seca, Spain

3 TecnATox—Centre for Environmental, Food and Toxicological Technology, Universitat Rovira i Virgili, Avda. Països Catalans 26, 43007 Tarragona, Spain

4 ENFOCAT, Facultat de Física, Universitat de Barcelona, C/Martí I Franquès 1, 08028 Barcelona, Spain

* Correspondence: foad.salehnia@urv.cat (F.S.); eduard.llobet@urv.cat (E.L.)

**Abstract:** The reliable detection of ammonia at room temperature is crucial for not only maintaining environmental safety but also for reducing the risks of hazardous pollutants. In this study, the electrochemical modification of laser-induced graphene (LIG) with polyaniline (PANI) led to the development of a chemo-resistive nanocomposite (PANI@LIG) for detecting ammonia levels at room temperature. The composite is characterized by field emission scanning electron microscopy, Fourier transforms infrared, and Raman and X-ray photoelectron spectroscopy. This work marks the first utilization of PANI@LIG for gas sensing and introduces a simple but effective approach for fabricating low-cost wearable gas sensors with high sensitivity and flexibility.

**Keywords:** laser-induced graphene; polyaniline; gas sensing; ammonia

## 1. Introduction

Graphene is a carbon nanomaterial with $sp^2$ bond structures forming a hexagonal honeycomb arrangement in a two-dimensional (2D) layer [1–3]. Due to its excellent electrical and thermal conductivity, high specific surface area, charge density, carrier mobility, strength, flexibility, and robustness, graphene is widely used in electronics applications like sensors, biosensors, batteries, supercapacitors and wearable devices [4–11]. The commonly used graphene synthesis processes, including chemical vapor deposition, micromechanical liquid phase exfoliation, electrochemical exfoliation and oxidation–reduction, involve high costs, complex processes with toxic chemicals, long processing times, difficulties for mass production and integration into flexible electronic devices, making graphene production challenging [12–17].

In 2014, Tour et al. introduced a viable alternative technology for graphene synthesis, a one-step simple, scalable, and low-cost method for producing three-dimensional (3D) porous graphene films from commercial polymer films prepared by direct laser writing with a $CO_2$ infrared laser; this nanomaterial is known as LIG [18–20]. LIG is valuable for its attributes, including high surface area, porosity, mechanical flexibility and excellent electrical conductivity [21]. These micro/nanostructures have hydrophilic/hydrophobic surfaces and outstanding electrochemical performance [19,22]. This material's simple fabrication process and properties enable its application in various fields, such as microfluidic systems, catalysis systems, water purification systems, electronic devices, sensors and biosensors [21,23–28].

In recent years, researchers have used LIG in chemo-resistive gas sensors, serving as electrodes [29–33], as a sensitive layer [34,35], or doped with other nanomaterials like metals, metal oxides (MOXs), transition metal dichalcogenides (TMDs) and conductive polymers [36–42]. These LIG composites provide the capability to detect multiple gas species like nitrogen oxides ($NO_x$) [29–32,35,37], volatile organic compounds (VOCs) [33], carbon dioxide ($CO_2$) [34] and ammonia ($NH_3$) [42]. In the case of $NH_3$, only few studies have been reported where LIG was used for its detection [43].

$NH_3$ measurement has broad applications such as air quality monitoring [44], agricultural and livestock practices regulation [45], human breath analysis for medical diagnostics [46], wastewater monitoring [47], and in the chemical industry [48]. Due to the hazards of $NH_3$, it is important to develop a real-time monitoring system to identify potential risks and ensure safety [49,50]. These sensors need to meet conditions such as being low-cost, having high precision, showing durability, and consuming low energy. The ability to detect ammonia at room temperature with no need for heating is important for low-power and portable devices. Traditional sensors often work at high temperatures, which limits their use in wearable and flexible applications.

Polyaniline's (PANI) ability to switch between its emeraldine and leucoemeraldine forms plays a critical role in its gas sensing performance. Upon exposure to ammonia, the protonation/deprotonation process causes a shift between these oxidation states, altering the electrical conductivity of the material. The emeraldine form (conductive state of PANI) undergoes deprotonation while exposed to $NH_3$, resulting in the formation of the leucoemeraldine form of PANI, which reduces the conductivity of the polymer [51,52]. The exploration of PANI as a functional material for gas sensors gained traction in the late 1980s, largely due to its unique doping/de-doping capability via protonation and deprotonation reactions [53]. This intrinsic property of PANI, allowing it to modulate its electrical conductivity in response to acidic or basic environments, makes it an ideal candidate for chemo-resistive sensing applications across various chemical analytes. Early breakthroughs in PANI-based ammonia sensors were reported by Hirata et al. (1994) and Kukla et al. (1996), demonstrating impressive gas response levels. These studies highlighted the material repeatability, room-temperature operation, and high environmental stability [53,54].

While there have been previous efforts to use PANI for ammonia detection, they usually involve complex chemical fabrication procedures or non-scalable fabrication methods. Graphene, as an inorganic material with conjugated $\pi$ electrons, can be combined with PANI to enhance sensitivity and selectivity at room temperature [55–59]. However, such combinations often involve multi-step syntheses or require additional materials for stability. In contrast to the previous study, our approach utilizes a direct electrochemical deposition of PANI onto LIG electrodes on a flexible substrate for the first time [60]. The choice of LIG as a substrate for PANI arises from its unique properties, including high conductivity, a large surface area for gas interactions, and mechanical flexibility. These characteristics complement PANI's intrinsic sensitivity to $NH_3$, creating a composite material that not only improves scalability and stability but also enables applications in flexible and wearable sensing platforms. This method is easy, cost-effective, and scalable compared to the chemical oxidative polymerization often employed in PANI–graphene composites. Moreover, the LIG itself is fabricated using a one-step laser patterning method, eliminating the need for high-cost, multi-step graphene synthesis methods such as chemical vapor deposition. This strategy provides fast and straightforward integration of the sensing material in flexible substrates suitable for wearable applications. The suggested method not only reduces the complexity of the fabrication process but also allows precise control over the thickness and morphology of the PANI layer via electrochemical polymerization. Based on this concept and our previous experience with LIG composites, this research marks the first time that an electrochemical deposition of PANI has been used to fabricate ammonia gas sensors on LIG electrodes, creating new possibilities for industrial-scale production of low-cost and flexible gas sensors [43].

## 2. Materials and Methods

### *2.1. Fabrication of LIG Electrode*

An LIG electrode was fabricated using a $CO_2$ pulsed infrared laser system (48-2, SYNRARD), with a wavelength of 10.6 µm and a max power of 25 W. The laser was focused on a 50 µm thick commercial polyimide film through a 74 mm focal length lens, and the laser beam was scanned at 200 mm/s, with a pulse frequency of 12 kHz and power of 12%. These are fundamental parameters in controlling the quality and characteristics of the LIG. They influence the LIG's conductivity, morphology, and suitability for electrochemical polymerization. These parameters were optimized using a digital twin tool that we previously reported [61], as they are key to controlling LIG quality, conductivity, and morphology [62]. The LIG electrode integrates with an 18 $mm^2$ sensing area of a 3D porous graphene layer, Ag-ink-coated contact pads, and LIG connection legs into a single structure.

### *2.2. Electrochemical Deposition of PANI on LIG Electrode*

A three-electrode system was used to carry out the electrochemical polymerization process (Figure 1a). The system consists of the LIG electrode as a working electrode, a platinum-wire auxiliary electrode, and an Ag/AgCl reference electrode with a salt bridge containing aqueous 3M NaCl. For the electrochemical polymerization, an aqueous electrolyte containing aniline (Sigma-Aldrich, St. Louis, MO, USA) with a concentration of 0.1 M, along with 1 M $H_2SO_4$ (Sigma-Aldrich), was prepared. The electrochemical polymerization of aniline was performed by using a potentiostat (pocketSTAT2, IVIUM Technologies, Eindhoven, The Netherlands) by cycling the potential between −0.5 and 1.3 V. Figure S1 (Supporting Materials) shows the cyclic voltammograms with a scan rate of 50 mV/s for 20 cycles. The electrodeposited LIG electrode was then rinsed with distilled water and dried in air.

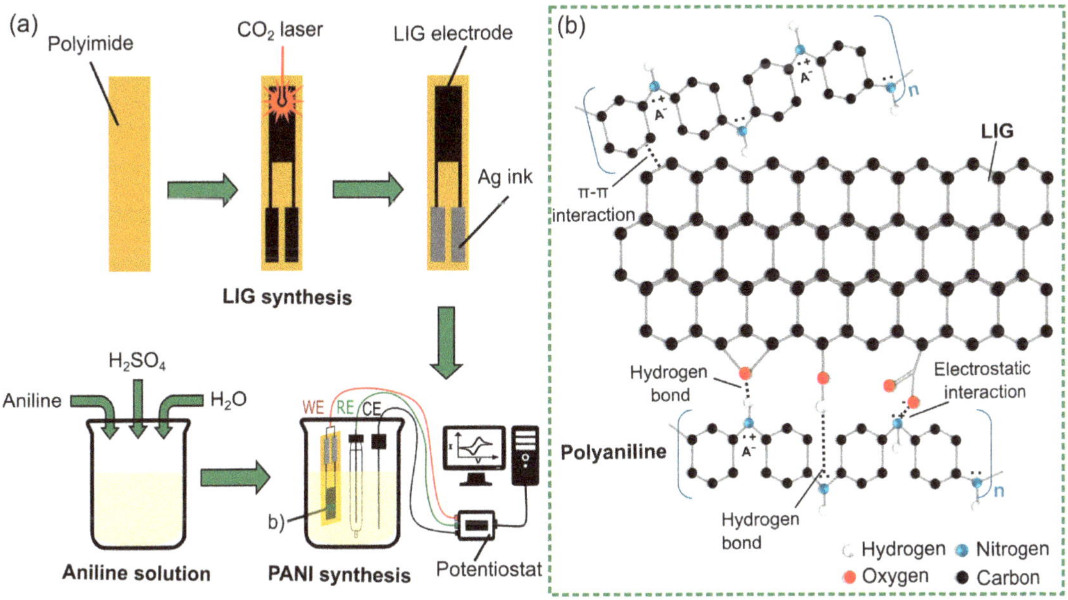


**Figure 1.** Schematic of (**a**) fabrication process of the PANI@LIG gas sensor and (**b**) interactions between PANI and LIG.

The possible interactions that may occur between aniline–LIG and PANI–LIG are shown in Figure 1b. The XPS analysis shows that the LIG has oxygenated functional groups (such as hydroxyl groups) on the surface and edges of the 3D porous graphene. By these functional groups, aniline monomers can be attached to LIG. Moreover, the LIG has good conductivity, which is required for electrochemical deposition. During the electrochemical deposition, PANI fibers grow on the surface and edges of LIG through electrostatic interaction, hydrogen bonding and π-π stacking between the two components [63]. It is also expected that hydrogen bonds are created between the hydroxyl groups on the LIG and PANI radicals; this causes the attraction of PANI chains on the LIG. The structure of the composite is further stabilized by π-π interaction between the PANI rings and π bonds of LIG [64,65].

### 2.3. Materials Characterization Techniques

Characterizations of the materials were performed in the Scientific and Technical Resources Service (SRCiT) of the University Rovira i Virgil (URV). The Raman analysis of the material was performed using a Renishaw InVia confocal Raman Spectrometer (Wotton-under-Edge, UK) with a coupled confocal microscope (Leica DM2500 Microsystems, Wetzlar, Germany). A 514 nm wavelength laser was employed, with the beam focused onto the LIG surface through a 50× objective lens. A Scios 2 DualBeam field emission scanning electron microscope (FESEM) was used to explore morphology. An optical microscope (Leica DMS300 Microsystems, Wetzlar, Germany). A 514 nm wavelength laser was employed, with the beam focused onto the LIG surface) was employed for visual inspection. The JASCO FT/IR 6700 (Asia portal) spectrophotometer (Tokyo, Japan) was employed for infrared spectroscopy analysis, and X-ray photoelectron spectroscopy (XPS) measurements were performed with ProvenX-NAP, SPECS (Berlin, Germany) using an AlKα—1486.7 eV X-ray monochromatic with μ-FOCUS 600, SPECS source.

### 2.4. Gas Sensing Tests

The fabricated PANI@LIG sensors were placed in a sealed Teflon chamber (35 $cm^3$), to evaluate their gas sensing performance at room temperature (Figure 2). The chamber was isolated from ambient humidity and had the capacity for four sensors. The sensor's electrical resistance was measured and recorded at a sampling frequency of 0.2 Hz, using a data acquisition system (34972A LXI, Keysight, Santa Rosa, CA, USA) controlled with a PC application (BenchLink Data Logger 3, Agilent Technologies, Santa Clara, CA, USA).

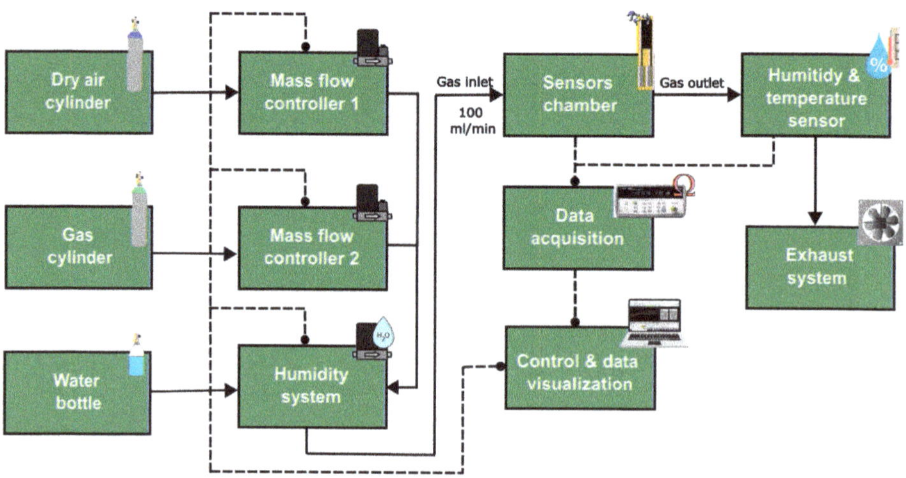


**Figure 2.** Schematic illustration of measurement system used for gas sensing tests.

Different gas concentrations were delivered into the chamber with the help of a mass-flow controller (MFC) system (EL-FLOW, Bronkhorst, Ruurlo, The Netherlands), controlled using PC applications (Flow View and Flow Plot, Bronkhorst). This system mixes gases from a zero-grade dry air cylinder as the carrier and a calibrated gas cylinder with 100 ppm of $NH_3$ (balanced in dry air) and delivers it to the chamber at a constant rate of 100 mL/min. The sensors were stabilized under dry air for 105 min and then exposed to cyclic exposure of 30 min of $NH_3$ (with different concentrations of 5, 10, 25, 50, and 100 ppm) and 75 min of dry air between each concentration. These concentration ranges were selected based on the $NH_3$ occupational exposure limits (OELs) according to the European Chemical Agency (ECHA) which specifies a long-term exposure limit (LTEL) of 20 ppm and a short-term exposure limit (STEL) of 50 ppm. The sensor's relative responses, expressed in percentage, were then calculated as a function of the resistance using the formula $\Delta R/R_0$, where $\Delta R$ represents $R$-$R_0$, $R$ corresponds to the value of the resistance after the target gas exposure and $R_0$ is defined as the sensor's baseline resistance in air [66–68]. The response time ($t_{resp}$) and recovery time ($t_{recov}$) were calculated as the time to reach 90% of total resistance change for $NH_3$ exposure and air re-exposure, respectively [67,68].

The gas sensing performance of sensors under a humid atmosphere was characterized using a controller evaporator mixer (W-202A, Bronkhorst) positioned in series between the gas mixer system and the chamber inlet. In addition, to monitor the environmental conditions during the measurements, a temperature and humidity sensor (SHT85, SENSIRION) was placed at the chamber outlet. The same gas measurement setup was employed to assess the sensor selectivity, replacing the $NH_3$ cylinder with a different gas cylinder. Other reducing species were used, including 100 ppm of carbon monoxide, 100 ppm of hydrogen, 20 ppm of ethanol, and aromatic volatile organic compounds such as 10 ppm of benzene and 10 ppm of toluene. Also, the sensor was exposed to 100 ppm of nitrogen dioxide as an oxidizing gas.

## 3. Results and Discussion

### 3.1. Structural and Morphological Characteristics

Figure 3 shows the FESEM images of bare LIG and PANI@LIG. The well-defined porous network seen at both 500 μm and 50 μm scales highlights the rough and interconnected structure of bare LIG demonstrated in Figure 3a,b. It depicts the surface morphology of LIG with a high surface area. The surface of LIG after the electrochemical polymerization is depicted in Figure 3c,d. The surface of the PANI@LIG composite appears to be less dense, with the PANI on the LIG surface covered. This can be seen particularly in the images (d) and (e) where the PANI has been extended in layers upon nanostructures, enhancing the material's porosity and surface area. Images (e) and (f) further zoom in on the PANI@LIG composite, showing the PANI fibrils forming a fine network at the nanoscale down to 500 nm. Also, Figure S2b,c (Supplementary Materials) comprises optical microscope images of the bare LIG and PANI@LIG, clearly showing the difference between them.

The Raman spectra of LIG and PANI@LIG are shown in Figure 4a. Several additional peaks appear in the PANI@LIG spectrum, showing PANI's successful polymerization on the LIG surface. Key peaks located around 1591, 1475, 1346, and 1162 $cm^{-1}$ were assigned to the C=C stretching of the quinoid and benzenoid rings [69], C–$N^+$ stretching modes, confirming the presence of conductive PANI. The peak at 1162 is for C–H bending and stretching of C–N, and some of the peaks in PANI@LIG at the lower frequency range are indicative of different stretching modes and ring deformations which are typical of PANI's molecular structure. For example, the peaks near 420 $cm^{-1}$ to 580 $cm^{-1}$ are associated with C–N stretching and ring deformation, while peaks near 810 $cm^{-1}$ to 1011 $cm^{-1}$ are linked to C–H out-of-plane bending and in-plane deformations. These low-frequency peaks provide additional evidence for the structural integrity of PANI and confirm its deposition on the LIG substrate. For the LIG spectrum, characteristic peaks are observed near 1573 $cm^{-1}$ and 1343 $cm^{-1}$, which correspond to the G-band and D-band, respectively. The G-band is associated with the in-plane vibrations of $sp^2$ carbon atoms in graphitic materials, while the D-band is linked to structural defects and disorder in the graphene structure. Additionally, a peak at approximately 2690 $cm^{-1}$ corresponds to the 2D band, which is another key feature of graphene and its multilayer formation. There is a slight shift in the positions of the G-band and D-band in the PANI@LIG spectrum compared to the pure LIG spectrum. This shift can indicate structural changes or interaction between PANI and LIG through π-π stacking and hydrogen bonding [70].

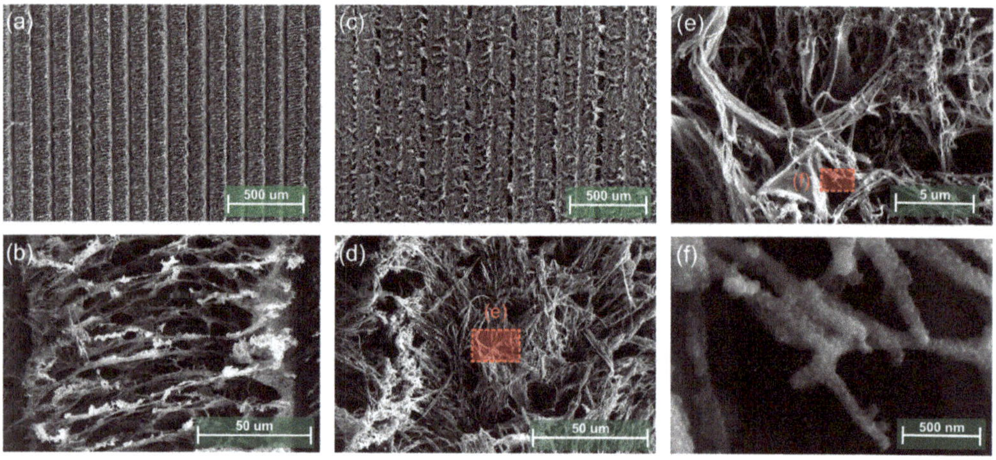


**Figure 3.** FESEM images of bare LIG (**a**,**b**), PANI@LIG (**c**–**f**).

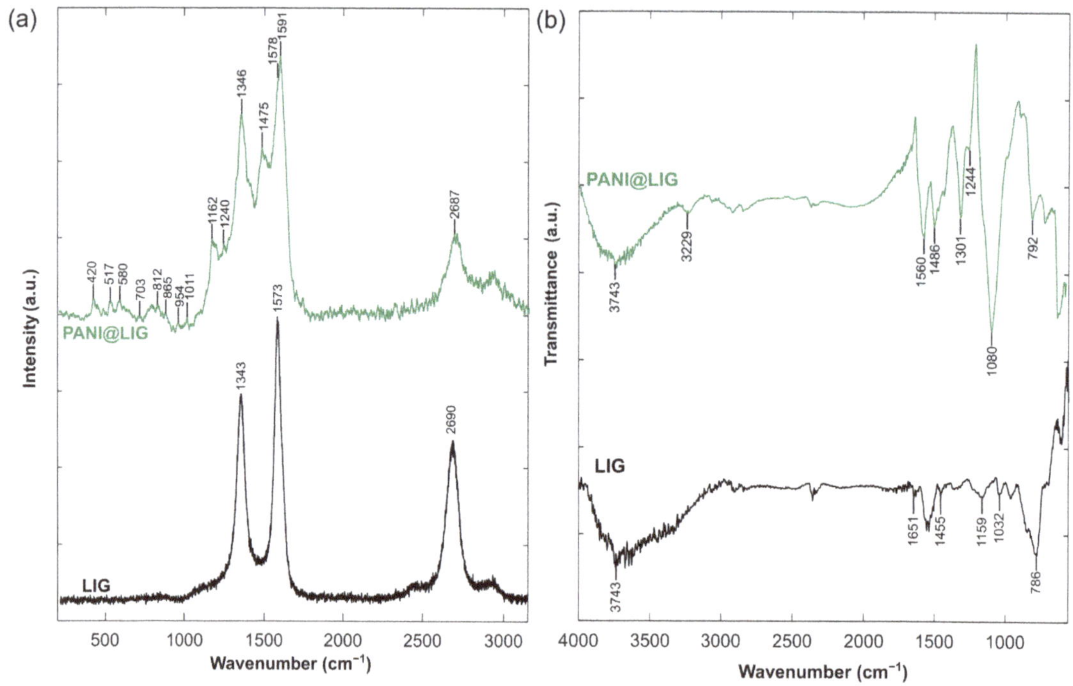


**Figure 4.** Raman spectra of bare LIG and PANI@LIG (**a**); ATR-FTIR spectra of bare LIG and PANI@LIG (**b**).

In order to investigate the infrared absorption properties, FTIR analysis was employed, and the ATR-FTIR spectra are shown in Figure 4b for LIG and PANI@LIG. The LIG sample (the black spectra) exhibits the large band in peak 3743 $cm^{-1}$, which can be attributed to O-H groups [71]. The intense absorption bands around 1651 $cm^{-1}$ are related to C=C stretching vibrations of the aromatic ring [72]. Peaks at 1593 $cm^{-1}$, 1455 $cm^{-1}$ and 1159 $cm^{-1}$ also correspond to different in-plane vibrations or deformation modes [73]. The peaks at around 1032 $cm^{-1}$ correspond to C-O stretching vibrations, indicating characteristics of O-containing functional groups on the LIG surface [60]. The peak at 786 $cm^{-1}$ is associated with C–H out-of-plane bending, which typically occurs in aromatic compounds, and it indicates some residual hydrogen or defects within the graphene structure [74]. The upper green spectra show the PANI@LIG spectrum, indicating distinctive absorption bands which appear to confirm the successful polymerization of PANI. The 3743 $cm^{-1}$ peak associated with the O-H group also appears in this spectrum, as well as the 3229 $cm^{-1}$ peak attributed to the N-H group [75]. Notably, peaks near 1560 $cm^{-1}$, 1486 $cm^{-1}$ and 1244 $cm^{-1}$ correspond to the C=C and C-N stretching of quinoid and benzenoid rings in PANI [76]. Peaks reaching the higher wave number 1080 $cm^{-1}$ with their shoulders are often assigned to N=Q=N stretching [77]. A peak located at 792 $cm^{-1}$ corresponds to the aromatic ring and arises from the out-of-plane bending vibration of C-H [78]. The differences between the spectra of LIG and PANI@LIG, especially the emergence of these specific peaks related to PANI, demonstrate the successful electrochemical polymerization of PANI onto the LIG substrate.

Figure 5a presents XPS analysis, giving us an idea of the surface chemistry of the PANI@LIG composite after being used as sensor. In the high-resolution C1s spectrum, shown in Figure 5b, the C-C/C=C bonds due to $sp^2$ hybridized carbon found in the graphene structure are represented by a peak at 284.5 eV [79]. The main peak at 285 eV was found to be associated with C-N bonds and represent PANI on the graphene surface [80]. The peak at 286.5 eV corresponds to C-O or C-N groups and can be attributed to either the oxygenated functional groups on the LIG or carbon–nitrogen interactions in PANI [79]. The peak at 287.3 eV is attributed to carbonyl groups, while the peak at 288.6 eV corresponds to carboxyl groups, more confirmation for the presence of oxygenated species on the LIG [81]. Moreover, the peak at 290.7 eV could be associated with $\pi$-$\pi^*$ infarction satellites, suggesting aromatic components in agreement with the conjugation of the graphene structure [79].

The N1s spectrum in Figure 5c further demonstrates nitrogen in the PANI@LIG composite. The peak centered at 398.8 eV is associated with imine groups originated from the structure of PANI, specifically when it adopts its oxidized emeraldine state [81]. The peak at 400.2 eV corresponds to the amine groups, which also originate from the PANI structure [80]. These two peaks confirm the coexistence of the two oxidation states of PANI. The peak at 402.2 eV is assigned to protonated nitrogen species, indicating PANI was in its conductive state known as emeraldine salt [80]. The higher binding energy peak at 405.5 eV is attributed to nitrogen oxide species, which may result from adsorbed $NO_2$ while checking the selectivity. The existence of oxygen-containing functional groups on the PANI@LIG composite is further proved by O1s spectrum displayed in Figure 5d. The peak at 530.4 eV is attributed to oxygen in carbonyl groups [82]. The peak at 531.6 eV is assigned to hydroxyl or ether groups, which are typically present in the edges of graphene and give them more water-wettability features [81]. The peak at 532.9 eV is related to carboxyl groups, indicating a further degree of oxidation on the surface [82].

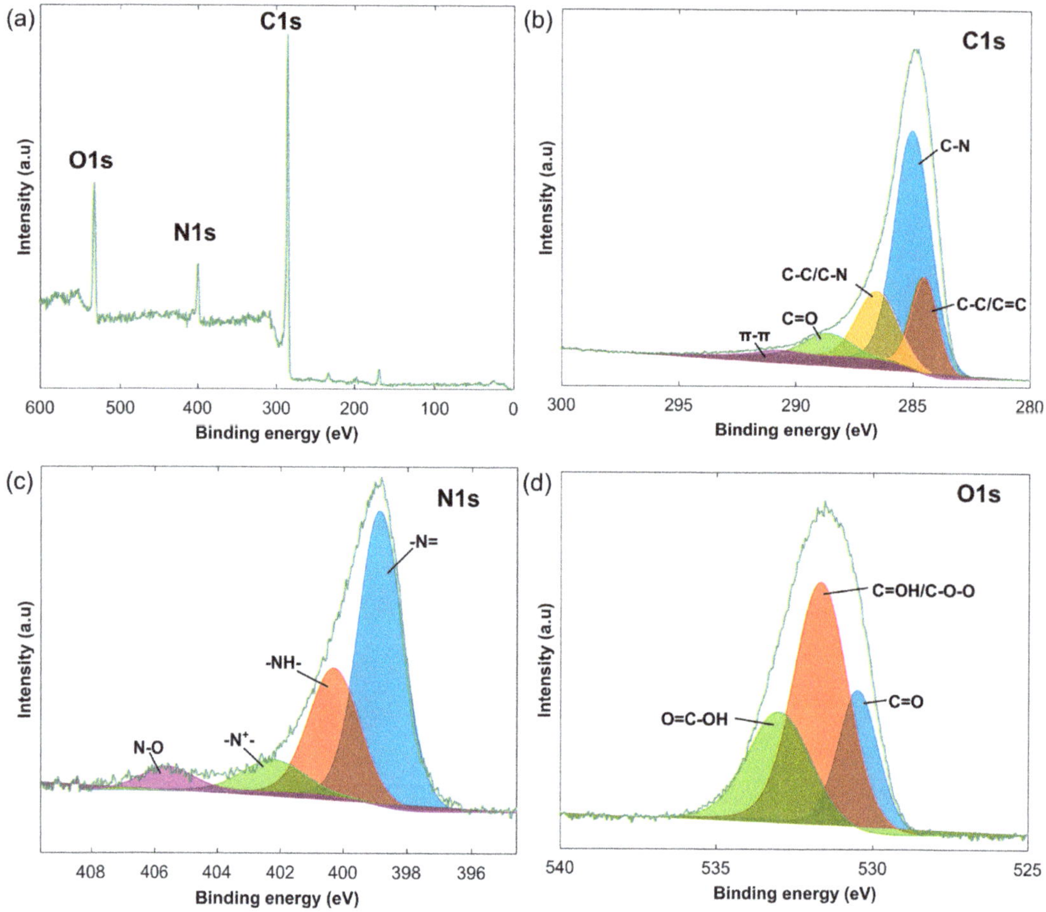


**Figure 5.** X-ray photoelectron spectroscopy (XPS) survey spectra of PANI@LIG (**a**), and high-resolution spectra fitting results of C1s (**b**), N1s (**c**) and O1s (**d**) of PANI@LIG.

### *3.2. Gas Sensing Performance Analysis*

The gas sensing performance was analyzed at varying concentrations of ammonia with a continuous electrical resistance measurement, in dry ambient conditions. Figure 6a shows the electrical resistance of the PANI@LIG gas sensor increased when exposed to varying concentrations of ammonia. This performance confirms that the PANI@LIG nanocomposite acts as a p-type material on exposure to a reducing gas, with ammonia donating electrons and neutralizing holes (positive charge carriers), thus reducing electrical conductivity. Moreover, the figure highlights the sensor's baseline stability and minimal noise levels. Figure 6b depicts the sensor regression model as a power function ($y(x) = 0.194{*}x^{0.532}$) of the calibration curve. This model was used to calculate the theoretical limit of detection (LOD), following a standard method consistent with the IUPAC definition. According to IUPAC, the LOD is the smallest concentration or absolute amount of analyte that has a signal significantly larger than the signal from a suitable blank. It is calculated based on Equation (1):

$$x_L = \overline{x}_B + ks_B \quad (1)$$

where $\overline{x}_B$ is the mean value of blank measurements (100 baseline points of the sensor's relative response were analyzed), $s_B$ is the standard deviation of the blank measures, and $k$ is a numerical factor ($k$= 3 for commonly used level of confidence = 99.7%). Then, LOD is estimated by Equation (2) as follows:

$$LOD = (ks_B/a)^{1/n} \quad (2)$$

where $a$ is the proportionality constant, and $n$ is the exponent from the power regression curve. The PANI@LIG sensor shows an LOD of 2.38 ppb for $NH_3$. This sensitivity demonstrates the potential efficacy of PANI@LIG for low-concentration detection, comparable to other high-performance materials used in gas sensing applications.

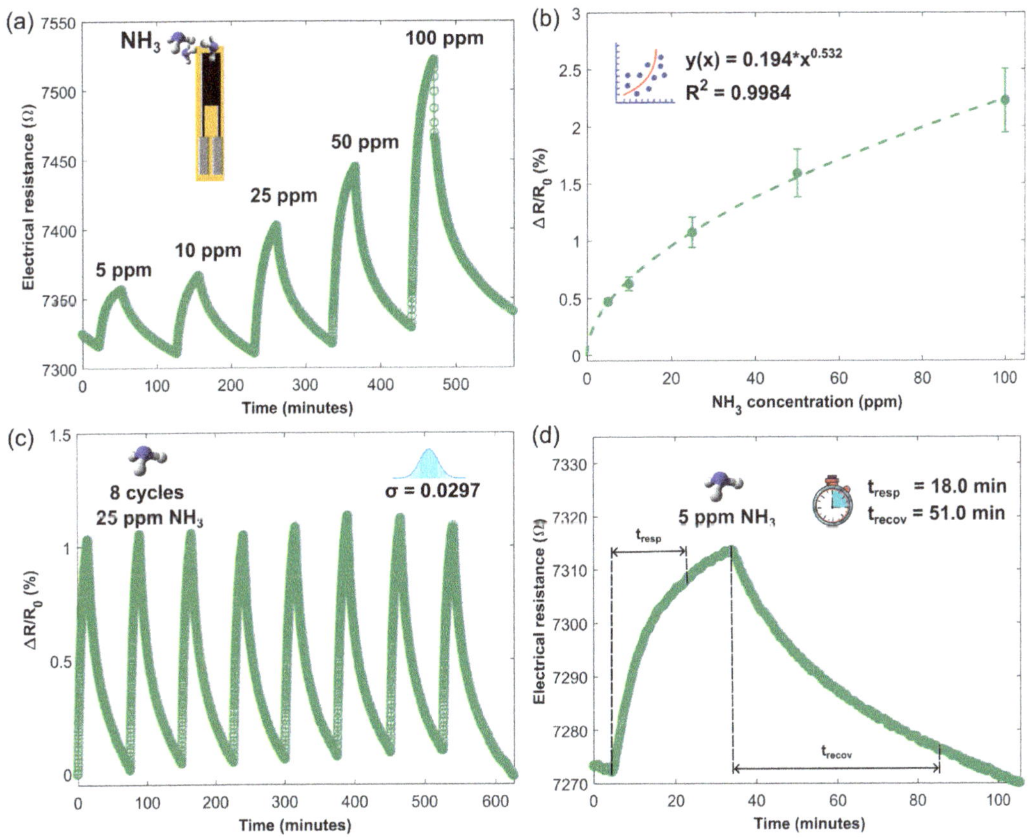


**Figure 6.** Gas sensing performance of PANI@LIG NCs gas sensors in dry ambient conditions. (**a**) Electrical resistance response to different concentrations (5, 10, 25, 50, and 100 ppm) of $NH_3$ at room temperature. (**b**) Regression curve. (**c**) Sensor repeatability testing at successive exposures of 25 ppm of $NH_3$. (**d**) Response to 5 ppm of $NH_3$ and analysis of response/recovery time.

Figure 6c shows the sensor repeatability; it was evaluated by applying eight successive cycles of 25 ppm of $NH_3$ for 30 min and recovery steps of 75 min between gas exposures, in which the PANI@LIG sensors present a standard deviation of about 0.029%. Figure 6d indicates the response time ($t_{resp}$ = 18.0 min) and recovery time ($t_{recov}$ = 51.0 min) of the sensor to 5 ppm of $NH_3$. Considering that this concentration is below the STEL defined by ECHA and no fast detectors are required, a response time in the order of minutes is sufficient. It can be concluded that these times are adequate for this sensor to be used in real-time monitoring of the environment.

Indeed, the influence of environmental moisture on gas sensor performance has emphasized humidity as a key parameter that essentially modulates sensor response. An experiment was carried out under 50% and 30% relative humidity (RH) levels, and the results were compared with those obtained under dry conditions. Figure S3b (Supporting Materials) indicates that variations in RH affect the sensor resistance. As RH increases, the sensor baseline decreases. This effect may be attributed to the further protonation of PANI through absorbed water or the generation of conductive $H_3O^+$. Figure 7a shows that the sensor's sensitivity improves with an increase in RH. Generally, the sensor response at 50% RH was greater than the responses at 30% RH and in dry conditions. The sensor response for 100 ppm at 50% RH was 2.9 times higher than in dry air. This performance hints that moisture promotes the $NH_3$ adsorption on the sensor's surface. This probably occurs through swelling of the active material by water molecules, causing conductive domains to move further apart, thus leading to higher overall resistance in the film [83]. Additionally, the humid environment may facilitate more efficient proton exchange between the water molecules and $NH_3$, further amplifying the sensor's response. Consequently, an increase in humidity will lead to a higher sensitivity for ammonium detection. Furthermore, sensor response/recovery times (Figure S3b) also improve (decrease) with rising RH.

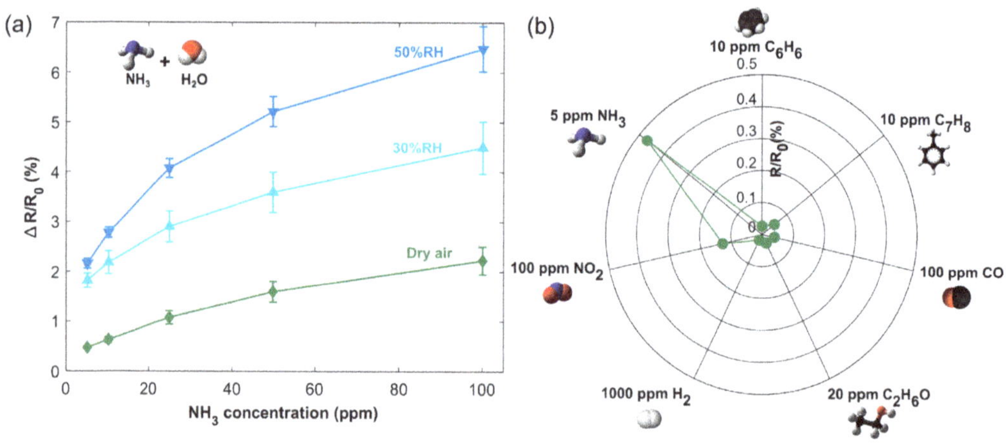


**Figure 7.** Calibration curves obtained for dry ambient conditions, 30%RH and 50%RH (**a**) and responses to different gas compounds (CO, $C_2H_6O$, $C_6H_6$, $C_7H_8$, $NH_3$, $H_2$, and $NO_2$) (**b**).

Selectivity is an important parameter to be considered in gas detection since it demonstrates the ability of the sensor to discriminate the target gas from interfering gases The selectivity was assessed (Figure 7b) by measuring high concentrations of other gases (benzene, toluene, carbon monoxide, ethanol, hydrogen, nitrogen dioxide). Therefore, the PANI@LIG gas sensor exhibited lower responses to these analytes compared with 5 ppm $NH_3$, and the fabricated PANI@LIG gas sensors showed acceptable performances for potential detection of environmental ammonia.

The sensor response was checked 45 days after the first test. During this period, the sensors were regularly exposed to ammonia and changing humidity conditions, and were exposed towards different gas species for selectivity tests. Figure S4a (Supporting Materials) shows the long-term stability of responses towards 50 ppm ammonia at room temperature. The sensor response decreased by 30% from day 1 to day 45. Also, the evolution of the baseline resistance can be found in Figure S4b, where the resistance value increases by 3%. We perceive that the decrease in the sensor response is due to the aging of the surface.

Table S1 (Supporting Materials) presents a comparison of various $NH_3$ gas sensors reported in the literature, all utilizing graphene–PANI nanocomposites capable of operating at room temperature. In general, the sensing performance was increased when the PANI was present in the sensitive material. However, many studies lack crucial details such as the applied flow rate, the type of carrier gas, and the effect of the influence of ambient humidity on performance. Although our sensor showed a lower response in comparison to those of other studies, its low LOD of the 2.38 ppb combined with the advantage of the electrochemical polymerization and a simple laser drawing method to produce graphene highlight its novelty. This is a simpler method in comparison to those used in most other studies and offers better control over both the thickness and morphology of PANI. Moreover, the sensor and ammonia detection operate at a lower flow rate and use synthetic air instead of nitrogen as the carrier gas, conditions closer to those needed in real-time monitoring for ammonia. Those experimental parameters are likely to influence the sensor response to some extent since higher flow rates or detection in a nitrogen atmosphere can boost the resistance changes.

### 3.3. Gas Sensing Mechanisms

Figure 8 shows the sensing mechanism of ammonia by the PANI@LIG composite. While the sensor is exposed to ammonia, the gas molecules are adsorbed on the surface. The porous structure of the PANI@LIG facilitates fast diffusion of ammonia on the sensor. During electrosynthesis in an acidic electrolyte, PANI molecules become protonated and exhibit p-type semiconductor properties as a result. When an ammonia molecule encounters PANI, it absorbs protons from the polymer so that ammonium ions form. This deprotonation changes the electronic structure of PANI, causing an increase in the sensor's resistance. The process is reversible; returning the sensor to air allows ammonium ions to convert back to ammonia and a proton, restoring the sensor to its original state [84]. The previous studies showed that the composite of graphene and PANI shows significantly enhanced sensing performance compared to pure PANI [85]. The high surface area of LIG increases the number of adsorption sites that are available for gas molecules and improves the sensor's sensitivity. LIG's three-dimensional porous structure with defects and high-energy binding sites helps efficient gas diffusion and adsorption. Additionally, the highly conductive nature of LIG sheets provides a rapid carrier transport network, facilitating swift electron transfer within the sensor. Also, $\pi$–$\pi$ interactions between PANI and LIG enhance electron mobility and create a synergistic effect that improves the ammonia sensing capability of the sensor. The I–V curve in Figure S5 (Supporting Materials) shows the linear relationship between the applied voltage and the flowing current through the sensor, which indicates that under the conditions of exposure, the sensor behaves as a resistive device in which an ohmic contact exists between the gas-sensitive film and the electrodes. The decrease in the slope of the curves when exposed to 50 ppm and 100 ppm $NH_3$ compared to synthetic air indicates an increase in the resistance of the sensor. It confirms that when $NH_3$ interacts with PANI@LIG, the number of charge carriers decreases due to deprotonation, leading to an increase in resistance.

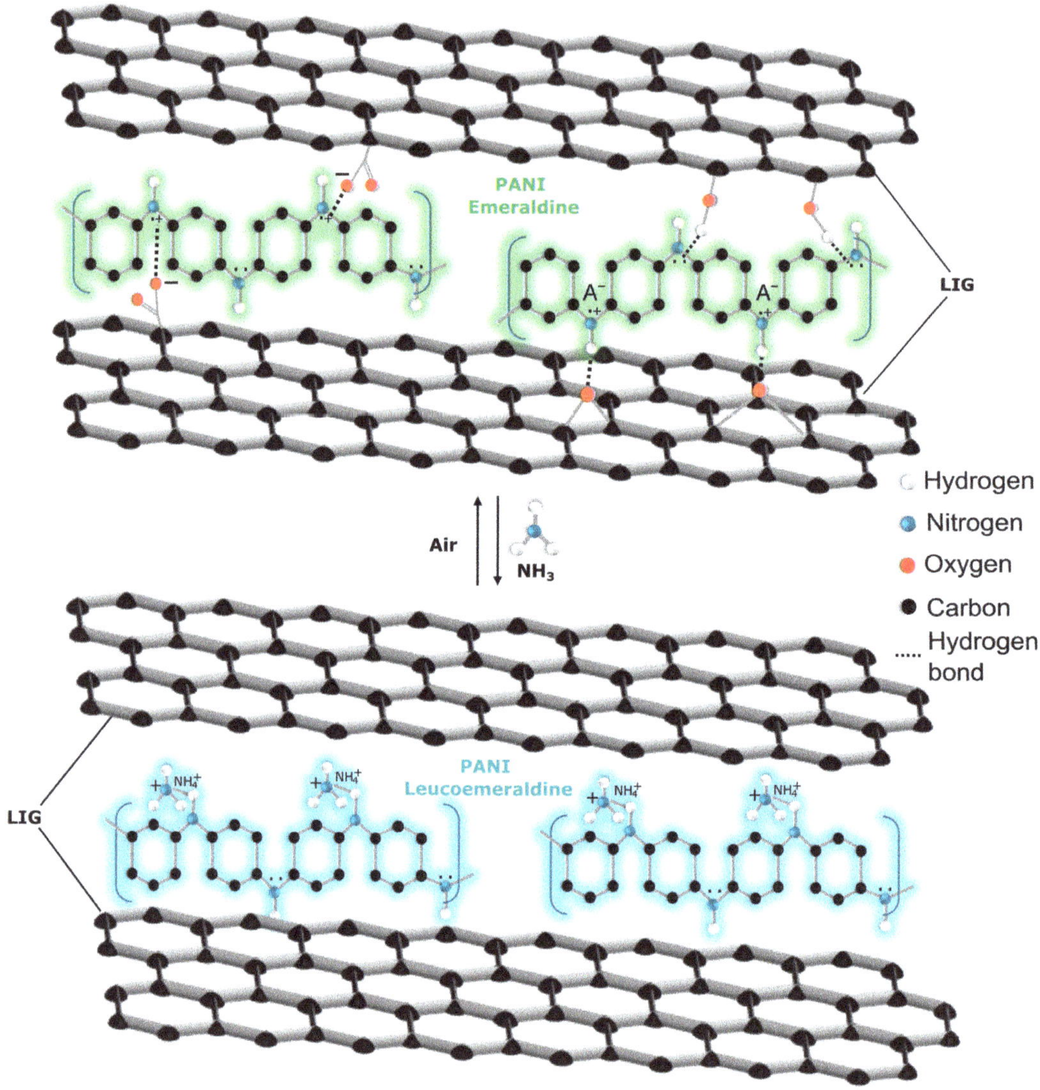


**Figure 8.** Schematic of the interaction between ammonia and PANI@LIG.

## 4. Conclusions

In summary, a flexible, low-cost and room-temperature-operating ammonia sensor with high sensitivity and selectivity was developed based on a PANI@LIG composite. The suggested method not only reduces the complexity of the fabrication process but also allows precise control over the thickness and morphology of the PANI layer via electrochemical polymerization. Wearable and portable low-cost devices can directly benefit from this simple fabrication process of a PANI/LIG sensor. While some sensors may offer higher response to ammonia, the convenience, low-cost scalability, low LOD of 2.38 ppb, as well as the sensor's performance in real-world conditions, make this sensor a candidate for applications such as environmental monitoring and industrial safety.

**Supplementary Materials:** The following supporting information can be downloaded at: https://www.mdpi.com/article/10.3390/s24237832/s1. Figure S1: Cyclic voltammograms; Figure S2: Picture of the PANI@LIG gas sensor and optical images of the bare LIG electrode and the PANI@LIG gas sensor; Figure S3: Relationship between sensor baseline resistance and relative humidity, and the correlation between sensor response/recovery times and relative humidity; Figure S4: long-term stability analysis; Figure S5: Current-voltage (I–V) characteristics of the gas sensor; Table S1: Comparison with previously reported results.

**Author Contributions:** Conceptualization, F.S. and J.C.S.-C.; methodology, F.S. and J.C.S.-C.; software, J.C.S.-C.; validation, F.S. and J.C.S.-C.; formal analysis, J.C.S.-C.; investigation, F.S. and J.C.S.-C.; data curation, F.S. and J.C.S.-C.; writing—original draft preparation, F.S. and J.C.S.-C.; writing—review and editing, E.L., F.G., A.R. and X.V.; visualization, F.S. and J.C.S.-C.; supervision, F.S., E.L., A.R. and X.V.; funding acquisition, E.L. and X.V. All authors have read and agreed to the published version of the manuscript.

**Funding:** This work was supported in part by the European Union (EU) in the Framework of H2020-MSCA-RISE-2018: Marie Sklodowska-Curie Actions under Grant 823895-PE, in part by the Ministerio de Ciencia, Innovación y Universidades (MICINN) under Grant PDC2022- 133967-I00 and Grant TED2021-131442B-C31, and in part by the Agència de Gestió d'Ajuts Universitaris i de Recerca (AGAUR) under Grant 2021 SGR 00147. The work of José Carlos Santos-Ceballos was supported by the AGAUR-Personal Investigador Predoctoral en Formació (FI) ajuts Predoctoral Program (Joan Oró of the Secretariat of Universities and Research of the Department of Research and Universities of the Generalitat of Catalonia and the European Social Plus Fund) under Grant 2023 FI-2 00180. The work of Foad Salehnia was supported in part by the Spanish Ministry of Universities, Recovery, Transformation, and Resilience Plan, and in part by the Maria Zambrano Grant funded by the European Union—NextGenerationEU under Grant 2021URV-MZ-13. The work of Eduard Llobet is supported by the Catalan Institution for Research and Advanced studies via the 2023 Editon of the ICREA Academia Award.

**Institutional Review Board Statement:** Not applicable.

**Informed Consent Statement:** Not applicable.

**Data Availability Statement:** Data are contained within the article.

**Acknowledgments:** The authors want to acknowledge the technicians of the Scientific and Technical Resources Service (SRCiT) of the URV, Rita Marimon Picó for obtaining the FESEM images, Ramon Guerrero Grueso for infrared spectroscopy analysis, and Daniel Ruano Sánchez for performing the XPS analysis and support its interpretation.

**Conflicts of Interest:** The authors declare no conflicts of interest.

## References

1. Novoselov, K.S.; Geim, A.K.; Morozov, S.V.; Jiang, D.; Zhang, Y.; Dubonos, S.V.; Grigorieva, I.V.; Firsov, A.A. Electric Field Effect in Atomically Thin Carbon Films. *Science (1979)* **2004**, *306*, 666–669. [CrossRef] [PubMed]
2. Urade, A.R.; Lahiri, I.; Suresh, K.S. Graphene Properties, Synthesis and Applications: A Review. *JOM* **2023**, *75*, 614–630. [CrossRef] [PubMed]
3. Mbayachi, V.B.; Ndayiragije, E.; Sammani, T.; Taj, S.; Mbuta, E.R.; Khan, A.U. Graphene Synthesis, Characterization and Its Applications: A Review. *Results Chem.* **2021**, *3*, 100163. [CrossRef]
4. Geim, A.K. Graphene: Status and Prospects. *Science (1979)* **2009**, *324*, 1530–1534. [CrossRef] [PubMed]
5. Singh, E.; Meyyappan, M.; Nalwa, H.S. Flexible Graphene-Based Wearable Gas and Chemical Sensors. *ACS Appl. Mater. Interfaces* **2017**, *9*, 34544–34586. [CrossRef]
6. Xiao, Y.; Pang, Y.X.; Yan, Y.; Qian, P.; Zhao, H.; Manickam, S.; Wu, T.; Pang, C.H. Synthesis and Functionalization of Graphene Materials for Biomedical Applications: Recent Advances, Challenges, and Perspectives. *Adv. Sci.* **2023**, *10*, 2205292. [CrossRef]
7. Weiss, N.O.; Zhou, H.; Liao, L.; Liu, Y.; Jiang, S.; Huang, Y.; Duan, X. Graphene: An Emerging Electronic Material. *Adv. Mater.* **2012**, *24*, 5782–5825. [CrossRef]
8. Han, T.H.; Kim, H.; Kwon, S.J.; Lee, T.W. Graphene-Based Flexible Electronic Devices. *Mater. Sci. Eng. R Rep.* **2017**, *118*, 1–43. [CrossRef]
9. Liu, J.; Bao, S.; Wang, X. Applications of Graphene-Based Materials in Sensors: A Review. *Micromachines* **2022**, *13*, 184. [CrossRef]
10. Syam Sundar, L.; Amin Mir, M.; Waqar Ashraf, M.; Djavanroodi, F. Synthesis and Characterization of Graphene and Its Composites for Lithium-Ion Battery Applications: A Comprehensive Review. *Alex. Eng. J.* **2023**, *78*, 224–245. [CrossRef]

11. Lakra, R.; Kumar, R.; Sahoo, P.K.; Thatoi, D.; Soam, A. A Mini-Review: Graphene Based Composites for Supercapacitor Application. *Inorg. Chem. Commun.* **2021**, *133*, 108929. [CrossRef]
12. Whitener, K.E.; Sheehan, P.E. Graphene Synthesis. *Diam. Relat. Mater.* **2014**, *46*, 25–34. [CrossRef]
13. Sun, B.; Pang, J.; Cheng, Q.; Zhang, S.; Li, Y.; Zhang, C.; Sun, D.; Ibarlucea, B.; Li, Y.; Chen, D.; et al. Synthesis of Wafer-Scale Graphene with Chemical Vapor Deposition for Electronic Device Applications. *Adv. Mater. Technol.* **2021**, *6*, 2000744. [CrossRef]
14. Kumar, N.; Salehiyan, R.; Chauke, V.; Joseph Botlhoko, O.; Setshedi, K.; Scriba, M.; Masukume, M.; Sinha Ray, S. Top-down Synthesis of Graphene: A Comprehensive Review. *FlatChem Chem. Flat Mater.* **2021**, *27*, 100224. [CrossRef]
15. Gutiérrez-Cruz, A.; Ruiz-Hernández, A.R.; Vega-Clemente, J.F.; Luna-Gazcón, D.G.; Campos-Delgado, J. A Review of Top-down and Bottom-up Synthesis Methods for the Production of Graphene, Graphene Oxide and Reduced Graphene Oxide. *J. Mater. Sci.* **2022**, *57*, 14543–14578. [CrossRef]
16. Madurani, K.A.; Suprapto, S.; Machrita, N.I.; Bahar, S.L.; Illiya, W.; Kurniawan, F. Progress in Graphene Synthesis and Its Application: History, Challenge and the Future Outlook for Research and Industry. *ECS J. Solid State Sci. Technol.* **2020**, *9*, 093013. [CrossRef]
17. Lin, L.; Peng, H.; Liu, Z. Synthesis Challenges for Graphene Industry. *Nat. Mater.* **2019**, *18*, 520–524. [CrossRef]
18. Lin, J.; Peng, Z.; Liu, Y.; Ruiz-Zepeda, F.; Ye, R.; Samuel, E.L.G.; Yacaman, M.J.; Yakobson, B.I.; Tour, J.M. Laser-Induced Porous Graphene Films from Commercial Polymers. *Nat. Commun.* **2014**, *5*, 5714. [CrossRef]
19. Ye, R.; James, D.K.; Tour, J.M. Laser-Induced Graphene. *Acc. Chem. Res.* **2018**, *51*, 1609–1620. [CrossRef]
20. Ye, R.; James, D.K.; Tour, J.M. Laser-Induced Graphene: From Discovery to Translation. *Adv. Mater.* **2019**, *31*, 1803621. [CrossRef]
21. Vivaldi, F.M.; Dallinger, A.; Bonini, A.; Poma, N.; Sembranti, L.; Biagini, D.; Salvo, P.; Greco, F.; Di Francesco, F. Three-Dimensional (3D) Laser-Induced Graphene: Structure, Properties, and Application to Chemical Sensing. *ACS Appl. Mater. Interfaces* **2021**, *13*, 30245–30260. [CrossRef] [PubMed]
22. Duy, L.X.; Peng, Z.; Li, Y.; Zhang, J.; Ji, Y.; Tour, J.M. Laser-Induced Graphene Fibers. *Carbon.* **2018**, *126*, 472–479. [CrossRef]
23. Cheng, L.; Guo, W.; Cao, X.; Dou, Y.; Huang, L.; Song, Y.; Su, J.; Zeng, Z.; Ye, R. Laser-Induced Graphene for Environmental Applications: Progress and Opportunities. *Mater. Chem. Front.* **2021**, *5*, 4874–4891. [CrossRef]
24. Le, T.D.; Phan, H.; Kwon, S.; Park, S.; Jung, Y.; Min, J.; Chun, B.J.; Yoon, H.; Ko, S.H.; Kim, S.; et al. Recent Advances in Laser-Induced Graphene: Mechanism, Fabrication, Properties, and Applications in Flexible Electronics. *Adv. Funct. Mater.* **2022**, *32*, 2205158. [CrossRef]
25. Movaghgharnezhad, S.; Kang, P. Laser-Induced Graphene: Synthesis Advances, Structural Tailoring, Enhanced Properties, and Sensing Applications. *J. Mater. Chem. C* **2024**, *12*, 6718–6742. [CrossRef]
26. Wan, Z.; Nguyen, N.-T.; Gao, Y.; Li, Q. Laser Induced Graphene for Biosensors. *Sustain. Mater. Technol.* **2020**, *25*, e00205. [CrossRef]
27. Zhu, J.; Huang, X.; Song, W. Physical and Chemical Sensors on the Basis of Laser-Induced Graphene: Mechanisms, Applications, and Perspectives. *ACS Nano* **2021**, *15*, 18708–18741. [CrossRef]
28. Wang, H.; Zhao, Z.; Liu, P.; Guo, X. Laser-Induced Graphene Based Flexible Electronic Devices. *Biosensors* **2022**, *12*, 55. [CrossRef]
29. Yan, W.; Yan, W.; Chen, T.; Xu, J.; Tian, Q.; Ho, D. Size-Tunable Flowerlike $MoS_2$ Nanospheres Combined with Laser-Induced Graphene Electrodes for $NO_2$ Sensing. *ACS Appl. Nano Mater.* **2020**, *3*, 2545–2553. [CrossRef]
30. Peng, Z.; Tao, L.-Q.; Zou, S.; Zhu, C.; Wang, G.; Sun, H.; Ren, T.-L. A Multi-Functional $NO_2$ Gas Monitor and Self-Alarm Based on Laser-Induced Graphene. *Chem. Eng. J.* **2022**, *428*, 131079. [CrossRef]
31. Tseng, S.-F.; Chen, P.-S.; Hsu, S.-H.; Hsiao, W.-T.; Peng, W.-J. Investigation of Fiber Laser-Induced Porous Graphene Electrodes in Controlled Atmospheres for ZnO Nanorod-Based $NO_2$ Gas Sensors. *Appl. Surf. Sci.* **2023**, *620*, 156847. [CrossRef]
32. Soydan, G.; Ergenc, A.F.; Alpas, A.T.; Solak, N. Development of an $NO_2$ Gas Sensor Based on Laser-Induced Graphene Operating at Room Temperature. *Sensors* **2024**, *24*, 3217. [CrossRef] [PubMed]
33. Li, D.; Shao, Y.; Zhang, Q.; Qu, M.; Ping, J.; Fu, Y.; Xie, J. A Flexible Virtual Sensor Array Based on Laser-Induced Graphene and MXene for Detecting Volatile Organic Compounds in Human Breath. *Analyst* **2021**, *146*, 5704–5713. [CrossRef]
34. Stanford, M.G.; Yang, K.; Chyan, Y.; Kittrell, C.; Tour, J.M. Laser-Induced Graphene for Flexible and Embeddable Gas Sensors. *ACS Nano* **2019**, *13*, 3474–3482. [CrossRef]
35. Yang, L.; Zheng, G.; Cao, Y.; Meng, C.; Li, Y.; Ji, H.; Chen, X.; Niu, G.; Yan, J.; Xue, Y.; et al. Moisture-Resistant, Stretchable $NO_x$ Gas Sensors Based on Laser-Induced Graphene for Environmental Monitoring and Breath Analysis. *Microsyst. Nanoeng.* **2022**, *8*, 78. [CrossRef]
36. Yang, L.; Yi, N.; Zhu, J.; Cheng, Z.; Yin, X.; Zhang, X.; Zhu, H.; Cheng, H. Novel Gas Sensing Platform Based on a Stretchable Laser-Induced Graphene Pattern with Self-Heating Capabilities. *J. Mater. Chem. A* **2020**, *8*, 6487–6500. [CrossRef]
37. Yi, N.; Cheng, Z.; Li, H.; Yang, L.; Zhu, J.; Zheng, X.; Chen, Y.; Liu, Z.; Zhu, H.; Cheng, H. Stretchable, Ultrasensitive, and Low-Temperature $NO_2$ Sensors Based on $MoS_2$@rGO Nanocomposites. *Mater. Today Phys.* **2020**, *15*, 100265. [CrossRef]
38. Yang, L.; Ji, H.; Meng, C.; Li, Y.; Zheng, G.; Chen, X.; Niu, G.; Yan, J.; Xue, Y.; Guo, S.; et al. Intrinsically Breathable and Flexible $NO_2$ Gas Sensors Produced by Laser Direct Writing of Self-Assembled Block Copolymers. *ACS Appl. Mater. Interfaces* **2022**, *14*, 17818–17825. [CrossRef] [PubMed]
39. Zhao, J.; Yi, N.; Ding, X.; Liu, S.; Zhu, J.; Castonguay, A.C.; Gao, Y.; Zarzar, L.D.; Cheng, H. In Situ Laser-Assisted Synthesis and Patterning of Graphene Foam Composites as a Flexible Gas Sensing Platform. *Chem. Eng. J.* **2023**, *456*, 140956. [CrossRef]
40. Zhang, Q.; Zhang, F.; Liu, X.; Yue, Z.; Chen, X.; Wan, Z. Doping of Laser-Induced Graphene and Its Applications. *Adv. Mater. Technol.* **2023**, *8*, 2300244. [CrossRef]

41. Kwak, D.; Kim, H.; Jang, S.; Kim, B.G.; Cho, D.; Chang, H.; Lee, J.O. Investigation of Laser-Induced Graphene (LIG) on a Flexible Substrate and Its Functionalization by Metal Doping for Gas-Sensing Applications. *Int. J. Mol. Sci.* **2024**, *25*, 1172. [CrossRef] [PubMed]
42. Santos-Ceballos, J.C.; Salehnia, F.; Romero, A.; Vilanova, X.; Llobet, E. Low Cost, Flexible, Room Temperature Gas Sensor: Polypyrrole-Modified Laser-Induced Graphene for Ammonia Detection. *IEEE Sens. J.* **2024**, *24*, 9366–9374. [CrossRef]
43. Santos-Ceballos, J.C.; Salehnia, F.; Romero, A.; Vilanova, X. Electrochemical Deposition of Polyaniline on Laser-Induced Graphene for Room Temperature Ammonia Sensing. In Proceedings of the EUROSENSORS XXXVI, Debrecen, Hungary, 1 September 2024; pp. 333–334.
44. Nair, A.A.; Yu, F. Quantification of Atmospheric Ammonia Concentrations: A Review of Its Measurement and Modeling. *Atmosphere* **2020**, *11*, 1092. [CrossRef]
45. Insausti, M.; Timmis, R.; Kinnersley, R.; Rufino, M.C. Advances in Sensing Ammonia from Agricultural Sources. *Sci. Total Environ.* **2020**, *706*, 135124. [CrossRef]
46. Lefferts, M.J.; Castell, M.R. Ammonia Breath Analysis. *Sens. Diagn.* **2022**, *1*, 955–967. [CrossRef]
47. Fang, X.; Guo, X.; Shi, H.; Cai, Q. Determination of Ammonia Nitrogen in Wastewater Using Electronic Nose. In Proceedings of the 2010 4th International Conference on Bioinformatics and Biomedical Engineering, Chengdu, China, 18–20 June 2010; IEEE: Piscataway, NJ, USA, 2010; pp. 1–4.
48. Timmer, B.; Olthuis, W.; Berg, A. van den Ammonia Sensors and Their Applications—A Review. *Sens. Actuators B Chem.* **2005**, *107*, 666–677. [CrossRef]
49. Yuliarti, R.; Khambali, K.; Rusmiati, R. Risk Analysis of Exposure to $NH_3$ And $H_2S$ Gas to Workers in The Small Industrial Environment of Magetan Regency in 2021. *Int. J. Adv. Health Sci. Technol.* **2022**, *2*, 169–174. [CrossRef]
50. Reis, T.; Moura, P.C.; Gonçalves, D.; Ribeiro, P.A.; Vassilenko, V.; Fino, M.H.; Raposo, M. Ammonia Detection by Electronic Noses for a Safer Work Environment. *Sensors* **2024**, *24*, 3152. [CrossRef]
51. Tanguy, N.R.; Thompson, M.; Yan, N. A Review on Advances in Application of Polyaniline for Ammonia Detection. *Sens. Actuators B Chem.* **2018**, *257*, 1044–1064. [CrossRef]
52. Wen, J.; Wang, S.; Feng, J.; Ma, J.; Zhang, H.; Wu, P.; Li, G.; Wu, Z.; Meng, F.; Li, L.; et al. Recent Progress in Polyaniline-Based Chemiresistive Flexible Gas Sensors: Design, Nanostructures, and Composite Materials. *J. Mater. Chem. A* **2024**, *12*, 6190–6210. [CrossRef]
53. Hirata, M.; Sun, L. Characteristics of an Organic Semiconductor Polyaniline Film as a Sensor for $NH_3$ Gas. *Sens. Actuators A Phys.* **1994**, *40*, 159–163. [CrossRef]
54. Kukla, A.L.; Shirshov, Y.M.; Piletsky, S.A. Ammonia Sensors Based on Sensitive Polyaniline Films. *Sens. Actuators B Chem.* **1996**, *37*, 135–140. [CrossRef]
55. Farooqi, B.A.; Ashraf, A.; Farooq, U.; Ayub, K. Comparative Study on Sensing Abilities of Polyaniline and Graphene Polyaniline Composite Sensors toward Methylamine and Ammonia. *Polym. Adv. Technol.* **2020**, *31*, 3351–3360. [CrossRef]
56. Wu, Z.; Chen, X.; Zhu, S.; Zhou, Z.; Yao, Y.; Quan, W.; Liu, B. Enhanced Sensitivity of Ammonia Sensor Using Graphene/Polyaniline Nanocomposite. *Sens. Actuators B Chem.* **2013**, *178*, 485–493. [CrossRef]
57. Guo, Y.; Wang, T.; Chen, F.; Sun, X.; Li, X.; Yu, Z.; Wan, P.; Chen, X. Hierarchical Graphene–Polyaniline Nanocomposite Films for High-Performance Flexible Electronic Gas Sensors. *Nanoscale* **2016**, *8*, 12073–12080. [CrossRef]
58. Gavgani, J.N.; Hasani, A.; Nouri, M.; Mahyari, M.; Salehi, A. Highly Sensitive and Flexible Ammonia Sensor Based on S and N Co-Doped Graphene Quantum Dots/Polyaniline Hybrid at Room Temperature. *Sens. Actuators B Chem.* **2016**, *229*, 239–248. [CrossRef]
59. Chang, J.; Zhang, X.; Wang, Z.; Li, C.; Hu, Q.; Gao, J.; Feng, L. Polyaniline-Reduced Graphene Oxide Nanosheets for Room Temperature $NH_3$ Detection. *ACS Appl. Nano Mater.* **2021**, *4*, 5263–5272. [CrossRef]
60. Tohidi, S.; Parhizkar, M.; Bidadi, H.; Mohamad-Rezaei, R. Electrodeposition of Polyaniline/Three-Dimensional Reduced Graphene Oxide Hybrid Films for Detection of Ammonia Gas at Room Temperature. *IEEE Sens. J.* **2020**, *20*, 9660–9667. [CrossRef]
61. Santos-Ceballos, J.C.; Salehnia, F.; Romero, A.; Vilanova, X. Application of Digital Twins for Simulation Based Tailoring of Laser Induced Graphene. *Sci. Rep.* **2024**, *14*, 10363. [CrossRef]
62. Santos-Ceballos, J.C.; Salehnia, F.; Romero, A.; Vilanova, X. Using Laser in the Fabrication of Graphene for Gas Sensing: A Digital Twin Approach. *J. Laser Micro/Nanoeng.* 2024; 19, 122–126. [CrossRef]
63. Xu, J.; Wang, K.; Zu, S.-Z.; Han, B.-H.; Wei, Z. Hierarchical Nanocomposites of Polyaniline Nanowire Arrays on Graphene Oxide Sheets with Synergistic Effect for Energy Storage. *ACS Nano* **2010**, *4*, 5019–5026. [CrossRef]
64. Mitra, M.; Kulsi, C.; Chatterjee, K.; Kargupta, K.; Ganguly, S.; Banerjee, D.; Goswami, S. Reduced Graphene Oxide-Polyaniline Composites—Synthesis, Characterization and Optimization for Thermoelectric Applications. *RSC Adv.* **2015**, *5*, 31039–31048. [CrossRef]
65. Wang, H.; Hao, Q.; Yang, X.; Lu, L.; Wang, X. Effect of Graphene Oxide on the Properties of Its Composite with Polyaniline. *ACS Appl. Mater. Interfaces* **2010**, *2*, 821–828. [CrossRef] [PubMed]
66. Casanova-Cháfer, J.; García-Aboal, R.; Atienzar, P.; Llobet, E. Gas Sensing Properties of Perovskite Decorated Graphene at Room Temperature. *Sensors* **2019**, *19*, 4563. [CrossRef] [PubMed]
67. Wang, Z.; Ni, L.; Zhang, X.; Feng, L. A Novel Flexible Substrate-Free $NH_3$ Sensing Membrane Based on PANI Covered RGO Functionalized Fiber. *Sens. Actuators B Chem.* **2023**, *380*, 133307. [CrossRef]

68. Wu, Q.; Shen, W.; Lv, D.; Chen, W.; Song, W.; Tan, R. An Enhanced Flexible Room Temperature Ammonia Gas Sensor Based on GP-PANI/PVDF Multi-Hierarchical Nanocomposite Film. *Sens. Actuators B Chem.* **2021**, *334*, 129630. [CrossRef]
69. Badi, N.; Khasim, S.; Roy, A.S. Micro-Raman Spectroscopy and Effective Conductivity Studies of Graphene Nanoplatelets/Polyaniline Composites. *J. Mater. Sci. Mater. Electron.* **2016**, *27*, 6249–6257. [CrossRef]
70. Ji, D.; Li, B.; Raj, B.T.; Li, X.; Zhang, D.; Rezeq, M.; Cantwell, W.; Zheng, L. In Situ Surface Polymerization of PANI/SWCNT Bilayer Film: Effective Composite for Improving Seebeck Coefficient and Power Factor. *Adv. Mater. Interfaces* **2024**, 2400566. [CrossRef]
71. Ajeel, K.I.; Kareem, Q.S. Synthesis and Characteristics of Polyaniline (PANI) Filled by Graphene (PANI/GR) Nano-Films. *J. Phys. Conf. Ser.* **2019**, *1234*, 012020. [CrossRef]
72. Yanilmaz, M.; Dirican, M.; Asiri, A.M.; Zhang, X. Flexible Polyaniline-Carbon Nanofiber Supercapacitor Electrodes. *J. Energy Storage* **2019**, *24*, 100766. [CrossRef]
73. Yang, S.; Zhu, S.; Hong, R. Graphene Oxide/Polyaniline Nanocomposites Used in Anticorrosive Coatings for Environmental Protection. *Coatings* **2020**, *10*, 1215. [CrossRef]
74. Visan, A.I.; Popescu-Pelin, G.; Gherasim, O.; Grumezescu, V.; Socol, M.; Zgura, I.; Florica, C.; Popescu, R.C.; Savu, D.; Holban, A.M.; et al. Laser Processed Antimicrobial Nanocomposite Based on Polyaniline Grafted Lignin Loaded with Gentamicin-Functionalized Magnetite. *Polymers* **2019**, *11*, 283. [CrossRef] [PubMed]
75. Butoi, B.; Groza, A.; Dinca, P.; Balan, A.; Barna, V. Morphological and Structural Analysis of Polyaniline and Poly(o-Anisidine) Layers Generated in a DC Glow Discharge Plasma by Using an Oblique Angle Electrode Deposition Configuration. *Polymers* **2017**, *9*, 732. [CrossRef] [PubMed]
76. Feng, X.M.; Li, R.M.; Ma, Y.W.; Chen, R.F.; Shi, N.E.; Fan, Q.L.; Huang, W. One-Step Electrochemical Synthesis of Graphene/Polyaniline Composite Film and Its Applications. *Adv. Funct. Mater.* **2011**, *21*, 2989–2996. [CrossRef]
77. Li, Z.L.; Yang, S.K.; Song, Y.; Xu, H.Y.; Wang, Z.Z.; Wang, W.K.; Dang, Z.; Zhao, Y.Q. In-Situ Modified Titanium Suboxides with Polyaniline/Graphene as Anode to Enhance Biovoltage Production of Microbial Fuel Cell. *Int. J. Hydrogen Energy* **2019**, *44*, 6862–6870. [CrossRef]
78. Trchová, M.; Stejskal, J. Polyaniline: The Infrared Spectroscopy of Conducting Polymer Nanotubes (IUPAC Technical Report). *Pure Appl. Chem.* **2011**, *83*, 1803–1817. [CrossRef]
79. Yang, X.; Qiu, Y.; Zhang, M.; Zhang, L.; Li, H. Facile Fabrication of Polyaniline/Graphene Composite Fibers as Electrodes for Fiber-Shaped Supercapacitors. *Appl. Sci.* **2021**, *11*, 8690. [CrossRef]
80. Oyetade, J.A.; Machunda, R.L.; Hilonga, A. Functional Impacts of Polyaniline in Composite Matrix of Photocatalysts: An Instrumental Overview. *RSC Adv.* **2023**, *13*, 15467–15489. [CrossRef] [PubMed]
81. Liu, Y.; Ma, Y.; Guang, S.; Ke, F.; Xu, H. Polyaniline-Graphene Composites with a Three-Dimensional Array-Based Nanostructure for High-Performance Supercapacitors. *Carbon* **2015**, *83*, 79–89. [CrossRef]
82. Goswami, S.; Maiti, U.N.; Maiti, S.; Nandy, S.; Mitra, M.K.; Chattopadhyay, K.K. Preparation of Graphene–Polyaniline Composites by Simple Chemical Procedure and Its Improved Field Emission Properties. *Carbon* **2011**, *49*, 2245–2252. [CrossRef]
83. Zhang, Y.; Zhang, J.; Jiang, Y.; Duan, Z.; Liu, B.; Zhao, Q.; Wang, S.; Yuan, Z.; Tai, H. Ultrasensitive Flexible $NH_3$ Gas Sensor Based on Polyaniline/$SrGe_4O_9$ Nanocomposite with Ppt-Level Detection Ability at Room Temperature. *Sens. Actuators B Chem.* **2020**, *319*, 128293. [CrossRef]
84. Kumar, L.; Rawal, I.; Kaur, A.; Annapoorni, S. Flexible Room Temperature Ammonia Sensor Based on Polyaniline. *Sens. Actuators B Chem.* **2017**, *240*, 408–416. [CrossRef]
85. Bai, S.; Zhao, Y.; Sun, J.; Tian, Y.; Luo, R.; Li, D.; Chen, A. Ultrasensitive Room Temperature $NH_3$ Sensor Based on a Graphene–Polyaniline Hybrid Loaded on PET Thin Film. *Chem. Commun.* **2015**, *51*, 7524–7527. [CrossRef] [PubMed]

sensors

MDPI

Article

# Non-Invasive Malaria Detection in Sub-Saharan Africa Using a DNA-Based Sensor System

Trine Juul-Kristensen [1,†], Celine Thiesen [1,2,†], Line Wulff Haurum [1], Josephine Geertsen Keller [1], Romeo Wenceslas Lendamba [3], Rella Zoleko Manego [3,4], Madeleine Eunice Betouke Ongwe [3,5], Birgitta Ruth Knudsen [1], Eduardo Pareja [6], Eduardo Pareja-Tobes [6], Rodrigo Labouriau [7], Ghyslain Mombo-Ngoma [3,4] and Cinzia Tesauro [1,*]

1 Department of Molecular Biology and Genetics, Universitetsbyen 81, Aarhus University, 8000 Aarhus, Denmark; tjk@mbg.au.dk (T.J.-K.); ceth@biomed.au.dk (C.T.); lwh@mbg.au.dk (L.W.H.); jgk@mbg.au.dk (J.G.K.); brk@mbg.au.dk (B.R.K.)
2 Department of Biomedicine, Høegh-Guldbergs Gade 10, Aarhus University, 8000 Aarhus, Denmark
3 Centre de Recherches Médicales de Lambaréné (CERMEL), Lambaréné BP 242, Gabon; wenceslas.lendamba@cermel.org (R.W.L.); rella.zoleko@cermel.org (R.Z.M.); m.e.betouke_ongwe@lumc.nl (M.E.B.O.); ghyslain.mombongoma@cermel.org (G.M.-N.)
4 Department of Implementation Research, Bernhard Nocht Institute for Tropical Medicine, & I. Dep. of Medicine, University Medical Centre Hamburg-Eppendorf, Bernhard-Nocht-Strasse 74, 20359 Hamburg, Germany
5 Institut de Recherche en Ecologie Tropicale, CENAREST, Libreville BP 13354, Gabon
6 Nisonide SL, 18008 Granada, Spain; eduardo.pareja@nisolab.com (E.P.); edu@nisolab.com (E.P.-T.)
7 Department of Mathematics, Aarhus University, Ny Munkegade 118, 8000 Aarhus, Denmark; rodrigo.labouriau@math.au.dk
* Correspondence: ctesauro@mbg.au.dk
† These authors contributed equally to this work.

**Abstract:** Malaria poses a serious global health problem, with half the world population being at risk. Regular screening is crucial for breaking the transmission cycle and combatting the disease spreading. However, current diagnostic tools relying on blood samples face challenges in many malaria-epidemic areas. In the present study, we demonstrate the detection of the malaria-causing *Plasmodium* parasite in non-invasive saliva samples (N = 61) from infected individuals by combining a DNA-based Rolling-circle-Enhanced-Enzyme-Activity-Detection (REEAD) sensor system with a chemiluminescence readout that could be detected with an in house developed affordable and battery-powered portable reader. We successfully transferred the technology to sub-Saharan Africa, where the malaria burden is high, and demonstrated a proof of concept in a small study (N = 40) showing significant differences ($p < 0.00001$) between malaria-positive individuals (N = 33) and presumed asymptomatic negative individuals (N = 7) all collected in Gabon. This is the first successful application of the REEAD sensor system for the detection of malaria in saliva in a high-epidemic area and holds promise for the potential future use of REEAD for malaria diagnosis or surveillance based on non-invasive specimens in sub-Saharan Africa.

**Keywords:** malaria; diagnosis; rolling circle amplification; saliva; topoisomerase 1

## 1. Introduction

Malaria is among the most serious global health issues. It is a major cause of death and illness in many low- and middle-income countries with an estimated 249 million cases and 608,000 deaths in 2022, particularly among young children and pregnant women [1]. Today, nearly half of the world population is at risk of malaria. These numbers may even increase with climate change that alters the geographic distribution of malaria by altering the habitats suitable for malaria-parasite-carrying mosquitos [2–4]. This development underlines the importance of regular screening in populations at risk of infection and in areas where malaria is currently under control. The World Health Organization (WHO) has set ambitious goals for malaria elimination in 26 countries by 2025 [5], bringing the number of certified malaria-free countries to 69 [6]. Screening and treating asymptomatic carriers allow us to break the transmission cycle and are vital steps toward these targets. Moreover, regular screening and surveillance of malaria can provide valuable data on how climate change impacts the pattern of malaria transmission.

Malaria is caused by infections with *Plasmodium* parasites, which typically reside in red blood cells [7,8]. Consequently, malaria is currently diagnosed in blood. The gold standards are light microscopy of thick or thin blood smears or polymerase chain reaction (PCR) [9,10]. These methods are often either unavailable or limited in low-resource settings where malaria is prevalent. Here, antigen rapid diagnostic tests (RDTs) are commonly used. However, the sensitivity of RDTs is low for patients with low-level parasitemia, and they fail to detect all malaria-causing *Plasmodium* species [11]. Due to these limitations and cultural reluctance to give blood in many malaria-endemic areas, new diagnostic tools allowing for sensitive and specific detection of all malaria-causing *Plasmodium* species in non-invasive samples at low-resource settings are highly needed [12].

DNA sensors or sensor systems composed of DNA may present attractive solutions to this challenge. DNA sensors for the sensitive and specific detection of biomarkers benefit markedly from advances of this century in the chemical synthesis of modified DNA oligonucleotides. During the past 15+ years, a large number of DNA sensors or sensor systems for the detection of different biomarkers, including small molecules [13–16], proteins [17,18], or enzyme activities [18–26], have been described. Besides the ease by which the secondary structure and functionality of DNA molecules can be manipulated for the detection of specific biomarkers [25,27,28], DNA can be amplified by polymerases, making it an excellent material for detection systems with enhanced sensitivity [29–31]. DNA sensor systems that rely on isothermal amplification systems such as rolling circle amplification (RCA) catalyzed by highly processive polymerases such as phi29 polymerase offer the additional advantage of being directly quantitative. When employing an enzyme activity as a biomarker for detection, the sensitivity is further increased as each enzyme generates many products that can each be amplified before detection without being consumed by the process (see schematic outline in Figure 1A).

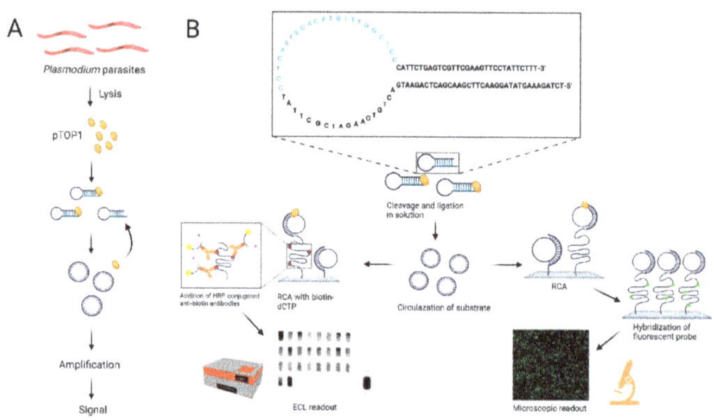


**Figure 1.** The REEAD sensor system. (**A**) The advantage of using pTOP1 as a biomarker for the detection of *Plasmodium* infections. Each parasite contains a high number of pTOP1 enzymes that each generate multiple DNA products without being consumed in the process. (**B**) The top panel shows the sequence and structure of the pTOP1-specific DNA substrate with the primer annealing site shown in blue. Cleavage–ligation by pTOP1 converts the substrate to a closed circle that is hybridized to a glass slide and amplified by RCA in the presence of (i) dNTPs with biotin-conjugated dCTPs for chemiluminescence readout (**left lower panel**) or (ii) without modified dNTPs followed by hybridization to fluorescently labeled probes for readout in a fluorescence microscope (**right lower panel**).

By taking advantage of linear DNA substrates that could be converted to a closed DNA circle and act as a template for phi29-mediated amplification only by their specific target enzyme, we previously reported Rolling-circle-Enzyme-Enhanced-Amplification-Detection (REEAD) DNA sensor systems for detecting disease-causing human pathogens such as *Plasmodium*, *Mycobacteria*, and HIV [26,32–35]. For the detection of *Plasmodium*, we utilized a single-stranded DNA substrate that folds into a hairpin structure with a primer annealing sequence in the single-stranded loop region and a recognition site that can be cleaved and ligated specifically by the life-essential *Plasmodium*-expressed enzyme topoisomerase1 (pTOP1) [36,37]. As illustrated in Figure 1B, pTOP1 converts the substrate to a single-stranded DNA circle that can function as a template for RCA in a solid support format or solution. Detection can be achieved by fluorescence labeling or coupling with Horse Radish Peroxidase (HRP), which enables different readouts [38]. Using this setup, we have demonstrated the specific detection of all human malaria-causing *Plasmodium* species in blood and saliva with estimated detection limits of 0.06 and 2 parasites/µL, respectively [39]. However, all previous testing of the malaria-specific REEAD was performed in high-resource laboratory facilities using samples from confirmed positives collected in sub-Saharan Africa or Asia that were tested against negative samples collected in Denmark [33,39]. Moreover, all previous tests relied on a rather complicated microfluidics-based extraction system to release active enzymes from the *Plasmodium* parasites, while a simple colorimetric readout was applied to a few tests with promising results [33,39].

In the present study, we demonstrate the detection of *Plasmodium* parasites in saliva from confirmed infected individuals in a malaria clinic in Gabon. This was achieved using an extraction method combined with a chemiluminescent-based readout for enhanced detection. For readout, we present an in-house built battery-driven affordable 3D printed portable reader optimized for the detection of chemiluminescence REEAD signals. Using 3D printing enables the rapid, cost-effective manufacturing of lightweight devices, making it ideal for low-resource settings by addressing affordability and logistical challenges, including ease of transport and deployment in remote areas. Using this reader for testing saliva for the presence of *Plasmodium* parasites in Lambaréné, Gabon, we demonstrate significant ($p < 0.00001$) differences between a population of confirmed malaria-positive (N = 33) and a population of presumed asymptomatic negative individuals (N = 7) all locally collected in Gabon. These results hold promise for the applicability of the malaria-specific REEAD in sub-Saharan Africa where the reported detection of *Plasmodium* parasites in saliva may also be relevant for surveillance programs.

## 2. Materials and Methods

### 2.1. Reagents

All chemicals were purchased from Sigma Aldrich, Søborg, Denmark. CodeLink Activated HD slides (#DHD1-0023) were from SurModics (Saint Paul, MN, USA), Pertex glue (#00801) was from Histolab (Askim, Sweden), a PAP Pen was from Super HT (Japan), a silicone Wellmaker grid was custom-made by Grace bio-labs (Bend, OR, USA), Vectashield without DAPI (#H-1000) was from Vector Laboratories (Burlington, ON, Canada), exonuclease I and III were from Thermo Scientific (Roskilde, Denmark), and glass beads 150–212 µm (#70-100 U.S. sieve) were from Sigma Aldrich (Steinheim, Germany).

### 2.2. DNA Oligonucleotides

DNA oligonucleotides were synthesized by Sigma Aldrich, Søborg, Denmark. The sequences were as follows:

- 5′-Amine REEAD primer:
  5′-[AmC6] CCAACCAACCAACCAAGGAGCCAAACATGTGCATTGAGG-3′;
- pTOP1substrate:
  5′-TCTAGAAAGTATAGGAACTTCGAACGACTCAGAATGACTGTGAAGA TCGCTTATCCTCAATGCACATGTTTGGCTCCCATTCTGAG TCGTTCGAAGTTCCTATTCTTT-3′;

- hTOP1substrate:
  5′-AGAAAAATTTTAAAAAAACTGTGAAGATCGCTTATTTTTTTAAAAAT TTTTCTAAGTCTTTTAGATCCCTCAATGCACATGTTTGGCTCCGATCTAAAA GACTTAGA-3′;
- Fluorescent probe:
  5′-[FAM] CCT CAA TGC ACA TGT TTG GCT CC-3′.

### 2.3. *Saliva and Blood Samples from Malaria Patients and Uninfected Individuals*

Saliva samples from patients diagnosed with malaria were obtained at the CERMEL, Albert Schweitzer Hospital, Lambaréné, Gabon [40]. The patients were diagnosed first using a rapid diagnostic test (RDT) in matching blood, followed by thick smear microscopy. The study was conducted in accordance with the Declaration of Helsinki, and samples were collected and analyzed following local regulations and guidelines. Patients' samples were obtained from the malaria screening activities conducted at CERMEL according to ethical clearance granted for ongoing studies and informed consent provided by each individual or their representatives. Thick smear microscopy was performed by two independent readers using the Lambaréné method [41]. A third reading was performed if required (e.g., if the first and second readings differed more than 50% of parasite count or if the same slide was read positive/negative by different readers). The number of counted parasites per μL blood was recorded. The final parasitemia was determined as the average of the first and second readings (or, if a third reading was performed, the average of the two closest results). In the case of testing by RDTs, the "Paracheck Pf" test was used.

### 2.4. *REEAD*

#### 2.4.1. Preparation of Slides

A CodeLink Activated HD slide was cut into an appropriate size. The slide was either attached to an object glass by Pertex glue followed by drawing squared-shaped fields with a Mini PAP Pen (for the fluorescence microscope readout) or attached to a custom-made silicone grid to create delimited rectangular-shaped wells of 1.2 mm × 2.8 mm dimensions termed the Wellmaker. Then, 10 μM of the 5′-Amine REEAD primer in 300 mM $Na_3PO_4$ pH 8 was coupled to the slide. The slide was incubated overnight in a humidity chamber with saturated NaCl. The slide was subsequently blocked in blocking buffer (50 mM Tris, 50 mM Tris-HCl, and 50 mM ethanolamine, pH 9) for 30 min at 50 °C and washed twice in $ddH_2O$ before it was washed in wash buffer 1 (4× SSC and 0.1% SDS) for 30 min at 50 °C and finally washed twice in $ddH_2O$.

#### 2.4.2. Extraction of Saliva Samples

Frozen saliva samples were thawed on ice. Subsequently, 200 μL of the saliva was homogenized by passing through a 0.5 mL tube with a hole in the bottom by centrifugation at 1000× *g* for 3 min. The homogenized saliva was collected in a 1.5 mL tube containing 25 mM HEPES pH 7.9, 30 mM NaCl, 0.02% Triton x-100, and 1 mM DTT. The samples were incubated for 15 min on ice before the addition of glass beads (1:1 vol/vol). The samples were vortexed using an automated vortex (Scientific Industries #SI-D258) for 30 s with a 1 min break on ice. Vortexing was repeated 2–5 times as described.

#### 2.4.3. Circularization with Saliva Samples

Circularization of the pTOP1 substrate was carried out by incubating 4 μL of the extracted saliva with 0.5 μM of the substrate in the presence of a buffer containing 250 mM NaCl, 10 mM Tris-HCl pH 7.5, and 1 mM EDTA for 1 h at 37 °C in a total volume of 20 μL. The reaction was stopped by heat inactivation at 95 °C for 5 min. The circles were hybridized to the 5′-Amine REEAD primer coupled slides overnight in a humidity chamber at room temperature and subsequently washed in 100 mM Tris-HCl pH 7.5, 150 mM NaCl, 0.3% SDS for 1 min, followed by washing in 100 mM Tris-HCl pH 7.5, 150 mM NaCl, 0.05% Tween20 for 1 min, and finally dehydrated for 1 min in 70% EtOH.

#### 2.4.4. Circularization with *P. falciparum* TOP1 Spiked in Saliva

Saliva was extracted as described above and mixed with purified pfTOP1 (different concentrations as indicated) before being added into a buffer containing 250 mM NaCl, 10 mM Tris-HCl pH 7.5, and 1 mM EDTA. The circularization reaction with the pTOP1 substrate incubated was carried out for 1 h at 37 °C and subsequently stopped by heat inactivation at 95 °C. The circles were hybridized to the 5′-Amine REEAD primer coupled slides as described above.

#### 2.4.5. Circularization with Purified *P. falciparum* TOP1 and Human TOP1

Circularization of the substrate with either pfTOP1 or hTOP1 substrate was carried out by incubating 1 ng/µL pfTOP1 or 120 ng/µL hTOP1 with 0.5 µM of the specific substrate in the presence of a buffer containing 10 mM Tris-HCl pH 7.5 and 5 mM EDTA and supplemented with 250 mM NaCl (pfTOP1) or 50 mM NaCl (hTOP1) for 1 h at 37 °C. The reaction was stopped either by heat inactivation at 95 °C for 5 min (pfTOP1) or by adding NaCl to a final concentration of 250 mM (hTOP1). The circles were hybridized to the 5′-Amine REEAD primer coupled slides as described above.

Before hybridizing some of the circles, the circularization product was digested with 40 units of exonuclease III and 4 units of exonuclease I for 1 h at 37 °C. Exonuclease digestion was heat inactivated at 95 °C for 5 min.

#### 2.4.6. Rolling Circle Amplification and Detection Using Fluorescence Microscope

The rolling circle amplification reaction was performed in 1× Phi29 buffer (50 mM Tris-HCl pH 7.5, 10 mM $MgCl_2$, 10 mM $(NH_4)_2SO_4$, 4 mM DTT) supplemented with 0.2 µg BSA, 1 mM dNTP, and 1 unit of Phi29 polymerase. The reaction was carried out for 1 h at 37 °C in a humidity chamber, followed by washing as previously described. The rolling circle products were detected by hybridizing 2 µM of the fluorescent probe in a buffer containing 2× SSC, 20% formamide, and 5% glycerol for 30 min in a humidity chamber at 37 °C. The slides were washed as previously described, mounted with Vectashield, and covered with cover glass. The slides were analyzed in an Olympus IX73 fluorescence microscope with a 60× objective. Twelve images of the fluorescent signals were acquired for each sample and quantified using Image J.

#### 2.4.7. Rolling Circle Amplification and Detection Using CCD Camera or VPCIReader

Rolling circle amplification was performed in 1× Phi29 buffer (50 mM Tris-HCl pH 7.5, 10 mM $MgCl_2$, 10 mM $(NH_4)_2SO_4$, 4 mM DTT) supplemented with 0.2 µg BSA, 100 µM dATP, 100 µM dTTP, 100 µM dGTP, 90 µM dCTP, 10 µM biotin-dCTP, and 1 unit Phi29 polymerase. The reaction was carried out for 2 h in a humidity chamber at 37 °C. The Wellmaker was then washed as previously described for the microscope slide. Subsequently, HRP-conjugated anti-biotin antibody was diluted 1:300 in a buffer containing 1x TBST (20 mM Tris-HCl pH 9, 150 mM NaCl, 0.05% Tween20 pH 9) supplemented with 5% skimmed dry milk, and 5% BSA and added to the wells of the Wellmaker for 1 h at room temperature. The Wellmaker was washed 3 × 5 min in 1x TBST, and finally, 2 µL of 1:1 ECL mixture was added to allow chemiluminescence readout using a CCD camera or the VPCIReader.

#### 2.4.8. VPCIReader Usage

The VPCIReader is operated through a web application. Upon connecting to the VPCIReader, the device undergoes calibration to a tray that does not contain a slide, utilizing a dark frame image for this process. Following calibration, the slide is inserted into the plate assemblers located within the tray. With the slide placed in the VPCIReader, the light intensity emitted by each sample can be measured using the web application. The VPCIReader delivers an image of the slide along with a corresponding table that presents quantitative values of light intensity. For a comprehensive overview of the usage, see Supplementary Materials S2.

### 2.5. Protein Purifications

#### 2.5.1. *P. falciparum* TOP1

Purified as previously described [39].

#### 2.5.2. Human TOP1

Purified as previously described [42].

#### 2.5.3. Phi29 Polymerase

Purified as previously described [42].

### 2.6. Statistical Analysis

For statistical description, see Supplementary Materials (S1A and S1B).

## 3. Results and Discussion

### 3.1. Semiquantitative Detection of the Plasmodium-SPECIFIC Biomarker pTOP1 by the Use of Chemiluminescence Readout

In the original pTOP1-specific REEAD setup, the hybridization of fluorescently labeled probes to the generated RCA products allowed for the visualization of each product as a fluorescent spot in a fluorescence microscope [33]. As each RCA product corresponded to one circle generated by one pTOP1 cleavage–ligation event, the assay was directly quantitative, allowing for measurement of pTOP1 activity at the single-catalytic-event level. This readout is highly sensitive but tedious to perform and requires a rather sophisticated fluorescence microscope. Hence, it is unsuited for diagnostic purposes in low-resource settings. More suitable for user-friendly detection in rural areas, we here demonstrate semiquantitative detection of pTOP1 using a chemiluminescent readout and describe the design and construction of a portable reader for quantitative detection of signals.

To enable a chemiluminescent readout, biotin-conjugated nucleotides were incorporated in the RCA products. This allowed for the binding of HRP-conjugated anti-biotin antibodies and subsequent chemiluminescent visualization of the products upon HRP-catalyzed conversion of luminol to 3-aminophthalate that emits light at 425 nm (see Figure 1B). This protocol was previously used to demonstrate semiquantitative detection of human Topoisomerase 1 (hTOP1) activities with a detection limit approximately 10x higher than the detection limit observed in a fluorescence microscope [43]. In the present study, the detection of purified recombinant pTOP1 using chemiluminescence versus fluorescence REEAD was investigated. To mimic clinical specimens, pTOP1 was diluted in saliva from uninfected individuals as described in the materials and methods section and incubated with the pTOP1-specific REEAD substrate in concentrations ranging from 1 ng/μL to 0.015625 ng/μL. The results of the chemiluminescence or fluorescence readout of the same samples are shown in Figure 2A,B. As evident from Figure 2A,B, the detection limit of the ECL readout is approximately 10x higher than the detection limit of the microscopic readout. This is similar to what we observed for purified hTOP1 [43].

Moreover, comparing the chemiluminescence or microscope reading of signals obtained with the same concentration of pTOP1 in a reaction mixture with or without added saliva demonstrated that the addition of saliva reduced the intensity of the obtained REEAD signals by approximately 50%. As the reaction steps succeeding the circularization reaction only contained trace amounts of saliva due to repeated dilutions, the reduced REEAD signals were most probably the result of inhibition of the pTOP1 cleavage–ligation reaction. Although the REEAD signals will increase with increasing pTOP1 concentrations, the inhibitory effect of the sample specimen, saliva, poses a limitation to the sample volume that can be added to the circle reaction mixture. We restricted the addition of the saliva to reaction mixtures to 20% vol/vol based on the effect observed in the current experiment where 20% vol/vol saliva was added to the circle reactions where indicated (Figure 2A,B).

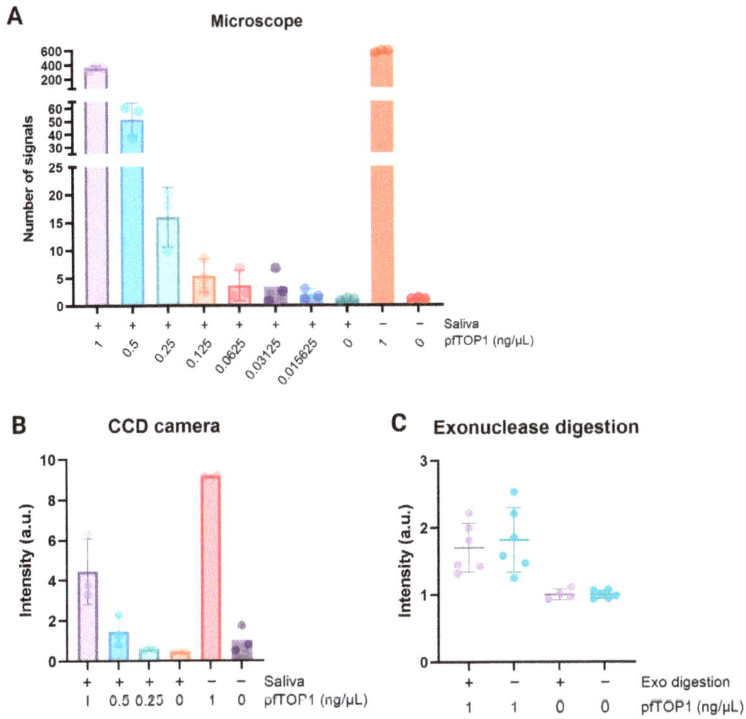


**Figure 2.** Detection of pfTOP1 with chemiluminescence REEAD. (**A**,**B**) Bar charts showing the results of analyzing 1, 0.5, 0.25, 0.125, 0.0625, 0.03125 ng/µL pfTOP1 spiked in saliva by REEAD using the fluorescence readout detected using a microscope (**A**) or the chemiluminescence readout detected using a commercial CCD camera (**B**). As controls, samples without pfTOP1 (one with and one without saliva) were included. As a positive control, a sample with 1 ng/µL pfTOP1 without saliva was used. The identity of the samples is indicated below the bar charts. The experiments were performed in triplicates (indicated by each dot). To compensate for slide-to-slide variations, the signals obtained by either the microscope or chemiluminescence readout were normalized to the average of the samples with saliva and 0 ng/µL pfTOP1 and plotted as mean +/− standard deviation (SD). (**C**) Graphic depiction of the results obtained when testing the effect of removing unreacted DNA substrate before chemiluminescence readout. The identity of the samples is shown below the graph. Each experiment was repeated four to six times (indicated by dots). To compensate for slide-to-slide variations, the chemiluminescence REEAD signals were normalized to the average intensity of samples without pfTOP1 without exonuclease digestion (Exo) and plotted as mean +/− SD.

Detection of pTOP1 by REEAD is based on the principle that a signal is only generated when the pTOP1-specific linear DNA substrate is converted to a circle and amplified by RCA as illustrated in Figure 1. The chemiluminescence readout is based on the labeling of RCA products by biotins via the incorporation of biotin-conjugated dCTPs each time the polymerase encounters a Guanine in the template. As evident from Figure 1, Phi29 elongation from the primer annealing site of the pTOP1 substrate to the 5′-end of the unreacted substrate will result in the incorporation of 12 biotin-conjugated dCTPs in a truncated amplification product. In principle, this can allow the binding of an HRP-conjugated anti-biotin antibody and result in chemiluminescence background signals. Such potential background signals could be avoided by removing unreacted pTOP1 substrate by exonuclease digestion before RCA. To investigate if the incorporation of biotin-conjugated dCTPs in truncated amplification products templated by unreacted substrate resulted in unspecific background signals high enough to create a problem for malaria detection, we investigated the effect of exonuclease digestion of the circles before adding Phi29 polymerase. This treatment removes un-circularized substrate and is expected to eliminate potential background signals. Figure 2C shows the results of subjecting a DNA circle mixture generated by 1 ng/μL of pTOP1 (Figure 2B, light purple bar) to RCA and chemiluminescence readout with or without exonuclease digestion. As evident from the figure, we observed no difference between the two samples. Likewise, a negative control sample without pTOP1 resulted in a minimal background signal, which was not affected by exonuclease digestion. For simplicity, we therefore decided to continue our studies using a protocol that does not involve exonuclease digestion before RCA.

### *3.2. Detection of Plasmodium in Saliva from Malaria-Positive Individuals in High-Resource Laboratory Settings*

Detection of *Plasmodium* parasites in clinical samples using the pTOP1-specific REEAD depends on the release of active enzymes from the parasites in a sample preparation step. We previously demonstrated effective lysis of parasites and circle generation using a droplet microfluidic setup [39]. However, even though this procedure could be performed with a simplified handheld device [33], it was difficult to handle, required extensive training, and was not suitable for routine testing of many samples. In the present study, we investigated the possibilities of releasing active enzymes from *Plasmodium* using a simple vortex protocol that is easy to perform and suitable for testing several samples simultaneously in modestly equipped test laboratories. Saliva from symptomatic malaria-positive individuals (confirmed by RDT in matching blood collected in Gabon) and from asymptomatic malaria-negative individuals collected in Denmark was vortexed repeatedly with glass beads as described in the materials and methods. The resulting lysates were mixed with the pTOP1 substrate to generate DNA circles that were subsequently amplified by RCA and detected using chemiluminescent readout. As evident from Figure 3A, two vortex repetitions of the samples did not result in any detectable difference in signals obtained from positive versus negative samples, probably due to insufficient enzyme extraction; some difference could be observed after three to four vortex repetitions, but only after five vortex repetitions was a clear difference between the positive and negative samples evident. Increasing the number of vortex repetitions did not add to signals obtained in positive samples, and seven vortex repetitions even decreased the obtained REEAD signal. We therefore continued the studies using five vortex repetitions for 30 s. Note that due to difficulties in obtaining sufficient sample volumes from the same individual, the experiment was only repeated twice. Also, due to the lack of sufficient sample volume, experiments investigating more vortex treatments were performed on samples from other patients and are, therefore, not included in the graph shown in Figure 3A. It has been a matter of debate if saliva from malaria-positive individuals contains intact *Plasmodium* parasites. Traces of the parasite, including DNA and proteins, have been observed by us and others [33,39,44], but it has been unclear if these components were remnants of already lysed parasites or if they resided inside intact parasites present in the saliva. The necessity of five vortex repetitions with glass beads before pTOP1 activity could be detected, as demonstrated in the present study, strongly argues for the presence of intact *Plasmodium* parasites in saliva from malaria-positive individuals. Our studies do, however, not provide any information on the viability of such parasites.

To address if the procedure including extraction of pTOP1 by repeated vortex combined with chemiluminescence REEAD detection could distinguish between malaria-positive and negative individuals in saliva specimens, we tested 30 saliva samples collected in Gabon from confirmed malaria-positive individuals (based on RDT analysis in matching blood) in comparison to 31 saliva samples from asymptomatic presumed malaria negative individuals collected in Denmark. The results were detected using a commercial CCD camera and the signal intensity was plotted as a function of malaria status, as shown in Figure 3B. As evident from the plot, most of the samples from malaria-positive individuals resulted in signals higher than signals obtained from samples from malaria-negative individuals, and the two populations of samples were significantly different ($p < 0.0001$) as demonstrated by Kruskal–Wallis test (see Supplementary Materials S1A). These results support the applicability of the protocol for the detection of malaria in saliva. Approximately three false positives and three false negatives could be observed. The false negatives could be the result of long-time storage and transportation of the samples collected in Gabon and/or insufficient extraction of pTOP1 from these samples. The molecular background for the false positive samples is less clear and requires further investigation.

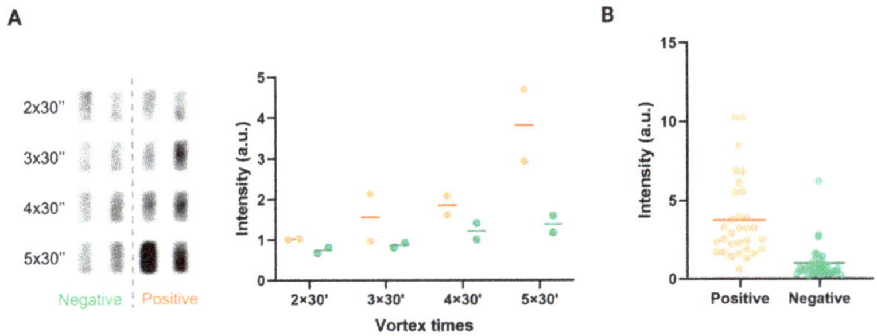


**Figure 3.** Detection of *Plasmodium* in clinical saliva samples. (**A**) (**Left panel**) Graphical depiction of chemiluminescence REEAD results obtained when measuring extracts from two saliva samples from confirmed malaria positives and two saliva samples from presumed negative individuals prepared by 2–5 vortex repetitions with glass beads. The average of the results from two individual experiments is shown by horizontal lines. (**Right panel**) Raw data obtained with a CCD camera. (**B**) The results were obtained by analyzing 30 saliva samples from confirmed malaria-positive individuals and 31 saliva samples from presumed malaria negatives using chemiluminescence REEAD. The average of the results is shown by horizontal lines. Statistics are shown in Supplementary Materials S1A. To compensate for slide-to-slide variations, the chemiluminescence REEAD signals were normalized to the average of the signals obtained by analyzing negative samples vortexed 2 times (**A**) or to the average of the signals obtained by analyzing negative samples (**B**).

### 3.3. Development and Proof-of-Concept Testing of an Affordable Portable Reader for Measuring Chemiluminescence REEAD Signals

The detection of chemiluminescence REEAD signals shown above was performed using a commercial CCD camera. This option is expensive and not suited for most test laboratories in malaria-endemic areas. Therefore, to enable REEAD detection of malaria in low-resource areas, we designed and constructed a miniaturized, cost-efficient, and portable chemiluminescence assay reader (termed VPCIReader) that provides high-precision image acquisition and analysis capabilities, specifically designed for chemiluminescence REEAD (see Figure 4 and Supplementary Materials S2 for specifications and construction details). The reader was constructed with an integrated battery with an autonomy of approximately 6–8 h and designed to be easily operated through a web application.

The software architecture was developed with four modules: i. the Image Analysis module that interacts with the sensor for image acquisition to finally quantify the REEAD signal, ii. the Data Storage module, iii. the Web Application module that serves as the user interface and links image acquisition and analysis with data storage, iv. the System Configuration.

The performance of the portable chemiluminescence VPCIReader was compared to a commercial CCD camera (Amersham Imager 600) concerning quantitative and sensitive detection of REEAD signals. For this purpose, we used a serial dilution of test circles split in two aliquots each subjected to RCA on two different test slides using the standard chemiluminescence readout protocol as outlined in Figure 1. ECL was applied to each of the slides before they were visualized using the commercial CCD camera (Figure 4B) and the portable VPCIReader (Figure 4C), respectively.

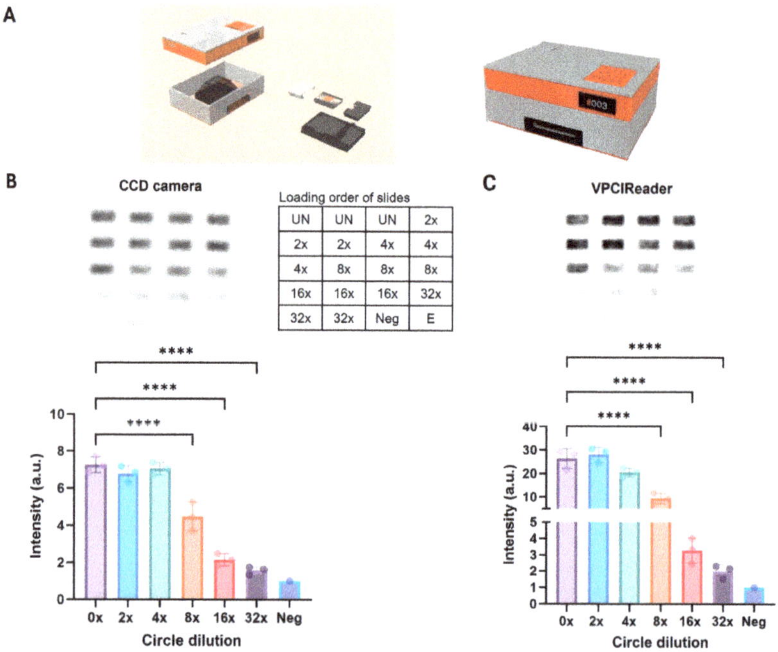


**Figure 4.** Comparison of CCD camera and portable chemiluminescence reader (VPCIReader). (**A**) Left panel, schematic showing the construction of the chemiluminescence VPCIReader. Right panel, photo of the VPCIReader. (**B**,**C**) Bar charts showing the results of analyzing titrations of test DNA circles (diluted 0 to 32 times and indicated) by capturing the results of chemiluminescence REEAD by a commercially available CCD camera (**B**) or by the developed chemiluminescence VPCIReader. The readings of each of the three individual experiments are shown by dots. The sample marked "Neg" contains non-circularized DNA with a sequence matching the test DNA circles. To compensate for slide-to-slide variations, the chemiluminescence REEAD signals were normalized to the "Neg" sample and plotted as mean +/− SD. **** = $p < 0.0001$, ordinary one-way ANOVA. "E" refers to an empty well only containing 5′-Amine REEAD primer.

As demonstrated by the graphical depiction of the results of three repetitive experiments (Figure 4B,C, lower panel), the commercial CCD camera and the portable reader gave comparable results when comparing quantifiability and detection limit with the utilized dilutions (note that only the relative readings and not the absolute numbers provided by the two systems can be compared). Concerning the signal-to-noise ratio, a tendency of the portable reader to perform slightly better than the CCD camera was indeed observed. The same tendency was observed when directly comparing the images generated by the two reader systems, where the portable reader produced considerably sharper pictures than the CCD camera (see representative images Figure 4B,C, top panel). Additionally, the VPCIReader web application provides direct access to the intensity values of each rectangular area, while the CCD camera only captures images that necessitate further processing through software-based densitometric analysis. Based on these results, we proceeded to investigate the performance of the portable VPCIReader for the detection of malaria by chemiluminescence REEAD in sub-Saharan Africa.

### *3.4. Detection of Malaria in Saliva from Infected Individuals in Sub-Saharan Africa*

To investigate the feasibility of chemiluminescence REEAD in combination with a readout using the above-described custom-made portable reader in sub-Saharan Africa, we tested saliva from 33 malaria-positive individuals (confirmed positive with RDT in matching blood) and seven asymptomatic presumed negative individuals at the CERMEL test site, Lambaréné in Gabon [40]. The results shown in Figure 5 demonstrate that most of the saliva samples collected from malaria-positive individuals resulted in signals above signals obtained from presumed negative individuals, and the two groups were found to be significantly different ($p < 0.0001$) using a Kruskal–Wallis test (Supplementary Materials S1B). As evident from the plot in Figure 5 and the table showing the exact readings (Supplementary Materials S1B), four of the samples from presumed negative individuals and approximately seven of the samples of confirmed individuals resulted in readings within the same interval. The relatively low readings in some of the samples from confirmed positive individuals may correlate with lower parasitemia in these individuals. However, as the malaria status of most of the symptomatic individuals was confirmed only by RDT in matching blood, the parasite number is not known. Incomplete extraction of active pTOP1 or suboptimal conditions during transport of some of the samples from rural areas may also explain the relatively low readings. Moreover, the relatively high readings in some of the asymptomatic presumed negative individuals may be the result of these individuals having low *Plasmodium* infection numbers, undetected by the RTD. This is often the case for asymptomatic individuals in populations living in malaria-endemic areas [45].

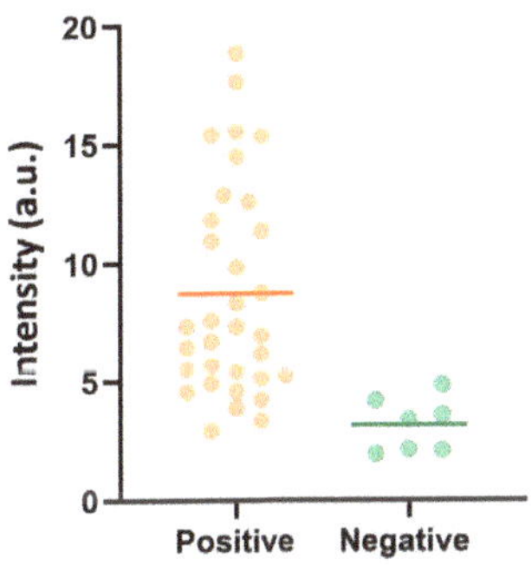


**Figure 5.** Detection of *Plasmodium* in clinical saliva samples by using VPCIReader. The results obtained by analyzing 33 saliva samples from confirmed malaria-positive individuals and 7 saliva samples from presumed malaria negatives using chemiluminescence REEAD. The average of the results is shown by horizontal lines. Statistics are shown in Supplementary Materials S1B. To compensate for slide-to-slide variations, the chemiluminescence REEAD signals were normalized to a well only containing 5′-Amine REEAD primer.

## 4. Conclusions

In the present study, we present a modified protocol for pTOP1-specific REEAD based on a chemiluminescence readout that allows for the testing of malaria in saliva from suspected individuals in moderately equipped test laboratories. Moreover, we describe the construction and validation of a cost-effective portable reader custom-made for measuring *Plasmodium*-specific chemiluminescence signals obtained by the modified REEAD protocol in malaria-endemic areas. A direct comparison of readings obtained from the same samples using a commercially available CCD camera and the custom-made reader demonstrated that the performance of the developed reader was comparable to the CCD camera for image acquisition and quantification of the REEAD results. Consistently, the portable reader, in combination with the simplified REEAD protocol, was successfully employed to detect the presence of *Plasmodium* parasites in saliva from confirmed malaria-positive individuals in sub-Saharan Africa. Hence, the chemiluminescence REEAD in combination with the availability of an affordable, simple, and portable reader presents a potential new attractive method for diagnosis or screening of malaria using non-invasive samples.

Readings from some of the samples from presumed negative individuals were in the same range as some of the lower readings obtained from samples from confirmed positives. The reason for this overlap is not clear. It could be due to low-level infections in the persons presumed negative in the current study; it could be that some of the samples from positive individuals were damaged during transportation, or that the active enzyme was not extracted efficiently. To address these potential issues, future studies would benefit from improved logistics, such as the use of portable refrigeration or temperature-controlled transport, to ensure the stability of the samples. Additionally, refining enzyme extraction protocols with increased sample size will help reduce potential variability in results.

This study was conducted as a pilot investigation; more comprehensive field trials will be necessary to validate the diagnostic accuracy of the method, assessing its sensitivity and specificity. Future studies should aim to determine the robustness of this system across diverse malaria-endemic settings, investigate its performance in larger, more heterogeneous populations, and optimize the protocol for broader application. With these advancements, the chemiluminescence REEAD method, coupled with the portable reader, could become a valuable tool for malaria screening using non-invasive sample specimens, in resource-limited environments.

**Supplementary Materials:** The following supporting information can be downloaded at https://www.mdpi.com/article/10.3390/s24247947/s1: Material S1A: Statistical analysis of data shown in Figure 3B; Material S1B: Statistical analysis of data shown in Figure 5; Material S2: Description and web application User Manual of the portable reader for chemiluminescence REEAD.

**Author Contributions:** Conceptualization, C.T. (Cinzia Tesauro), B.R.K., J.G.K., and G.M.-N., methodology, T.J.-K., C.T. (Celine Thiesen), L.W.H., J.G.K., and C.T. (Cinzia Tesauro); software, E.P.-T. and E.P., validation, T.J.-K. and C.T. (Celine Thiesen); formal analysis, R.L., C.T. (Cinzia Tesauro), and B.R.K., investigation, R.W.L., R.Z.M., M.E.B.O., and G.M.-N., writing—original draft preparation, C.T. (Cinzia Tesauro) and G.M.-N., writing—review and editing, C.T. (Cinzia Tesauro), B.R.K., J.G.K., G.M.-N., T.J.-K., C.T. (Celine Thiesen), L.W.H., E.P.-T., E.P., R.L., R.W.L., R.Z.M., and M.E.B.O., supervision, C.T. (Cinzia Tesauro), B.R.K., and G.M.-N. All authors have read and agreed to the published version of the manuscript.

**Funding:** This research was funded by Aase og Ejnar Danielsen Fond Grant to Birgitta Knudsen. The VPCIReader conceptualization and production were funded by VPCIR biosciences.

**Institutional Review Board Statement:** The study was conducted in accordance with the Declaration of Helsinki, and samples were collected and analyzed following local regulations and guidelines at CERMEL, Albert Schweitzer Hospital, Lambaréné, Gabon.

**Informed Consent Statement:** Informed consent was obtained from all subjects involved in the study.

**Data Availability Statement:** Dataset included in the Supplementary Materials.

**Acknowledgments:** We are thankful to laboratory technician Noriko Hansen for skillful technical assistance including enzyme purification. The experimental work was funded by Rajagopal Foundation and Aase and Ejnar Danielsen Foundation and partly sponsored by VPCIR biosciences, who financed the development and construction of the VPCIReader.

**Conflicts of Interest:** The authors B.R.K. and C.T. (Cinzia Tesauro) are shareholders of VPCIR biosciences. E.P.-T. is the owner of NISONIDE SL; E.P. works as a consultant for NISONIDE SL. The other authors declare that they have no competing interests.

## References

1. WHO. *World Malaria Report 2023*; WHO: Geneva, Switzerland, 2023.
2. Wanjala, C.L.; Kweka, E.J. Impact of Highland Topography Changes on Exposure to Malaria Vectors and Immunity in Western Kenya. *Front. Public Health* **2016**, *4*, 227. [CrossRef]
3. Rodó, X.; Martinez, P.P.; Siraj, A.; Pascua, M. Malaria trends in Ethiopian highlands track the 2000 'slowdown' in global warming. *Nat. Commun.* **2021**, *12*, 1555. [CrossRef] [PubMed]
4. Carlson, C.J.; Bannon, E.; Mendenhall, E.; Newfield, T.; Bansal, S. Rapid range shifts in African Anopheles mosquitoes over the last century. *Biol. Lett.* **2023**, *19*, 20220365. [CrossRef] [PubMed]
5. Gasimov, E. *Update on Malaria Elimination, Including Zoonotic Malaria in Malaria Policy Advisory Group Meeting*; World Health Organization: Yaoundé, Cameroon, 2024.
6. WHO. Countries and Territories Certified Malaria-Free by WHO. 2024. Available online: https://www.who.int/teams/global-malaria-programme/elimination/countries-and-territories-certified-malaria-free-by-who (accessed on 10 July 2024).

7. Cooke, B.M.; Mohandas, N.; Coppel, R.L. The malaria-infected red blood cell: Structural and functional changes. *Adv. Parasitol.* **2001**, *50*, 1–86. [PubMed]
8. Gilson, P.R.; Crabb, B.S. Morphology and kinetics of the three distinct phases of red blood cell invasion by *Plasmodium falciparum* merozoites. *Int. J. Parasitol.* **2009**, *39*, 91–96. [CrossRef]
9. Warhurst, D.C.; Williams, J.E. ACP Broadsheet no 148. July 1996. Laboratory diagnosis of malaria. *J. Clin. Pathol.* **1996**, *49*, 533–538. [CrossRef]
10. Tangpukdee, N.; Duangdee, C.; Wilairatana, P.; Krudsood, S. Malaria diagnosis: A brief review. *Korean J. Parasitol.* **2009**, *47*, 93–102. [CrossRef]
11. Murray, C.K.; Gasser, R.A., Jr.; Magill, A.J.; Miller, R.S. Update on rapid diagnostic testing for malaria. *Clin. Microbiol. Rev.* **2008**, *21*, 97–110. [CrossRef]
12. Gitta, B.; Kilian, N. Diagnosis of Malaria Parasites *Plasmodium* spp. in Endemic Areas: Current Strategies for an Ancient Disease. *Bioessays* **2020**, *42*, e1900138. [CrossRef]
13. Baker, B.R.; Lai, R.Y.; Wood, M.S.; Doctor, E.H.; Heeger, A.J.; Plaxco, K.W. An electronic, aptamer-based small-molecule sensor for the rapid, label-free detection of cocaine in adulterated samples and biological fluids. *J. Am. Chem. Soc.* **2006**, *128*, 3138–3139. [CrossRef]
14. Halder, S.; Krishnan, Y. Design of ultrasensitive DNA-based fluorescent pH sensitive nanodevices. *Nanoscale* **2015**, *7*, 10008–10012. [CrossRef] [PubMed]
15. Liu, J.; Wang, C.; Jiang, Y.; Hu, Y.; Li, J.; Yang, S.; Li, Y.; Yang, R.; Tan, W.; Huang, C.Z. Graphene signal amplification for sensitive and real-time fluorescence anisotropy detection of small molecules. *Anal. Chem.* **2013**, *85*, 1424–1430. [CrossRef] [PubMed]
16. Zayats, M.; Huang, Y.; Gill, R.; Ma, C.A.; Willner, I. Label-free and reagentless aptamer-based sensors for small molecules. *J. Am. Chem. Soc.* **2006**, *128*, 13666–13667. [CrossRef] [PubMed]
17. Bi, S.; Li, L.; Zhang, S. Triggered polycatenated DNA scaffolds for DNA sensors and aptasensors by a combination of rolling circle amplification and DNAzyme amplification. *Anal. Chem.* **2010**, *82*, 9447–9454. [CrossRef] [PubMed]
18. Zuccaro, L.; Tesauro, C.; Kurkina, T.; Fiorani, P.; Yu, H.K.; Knudsen, B.R.; Kern, K.; Desideri, A.; Balasubramanian, K. Real-Time Label-Free Direct Electronic Monitoring of Topoisomerase Enzyme Binding Kinetics on Graphene. *ACS Nano* **2015**, *9*, 11166–11176. [CrossRef] [PubMed]
19. Andersen, F.F.; Stougaard, M.; Jørgensen, H.L.; Bendsen, S.; Juul, S.; Hald, K.; Andersen, A.H.; Koch, J.; Knudsen, B.R. Multiplexed detection of site specific recombinase and DNA topoisomerase activities at the single molecule level. *ACS Nano* **2009**, *3*, 4043–4054. [CrossRef]
20. Hardin, A.H.; Sarkar, S.K.; Seol, Y.; Liou, G.F.; Osheroff, N.; Neuman, K.C. Direct measurement of DNA bending by type IIA topoisomerases: Implications for non-equilibrium topology simplification. *Nucleic AcidsRes.* **2011**, *39*, 5729–5743. [CrossRef]
21. Jakobsen, A.-K.; Stougaard, M. Combining a Nanosensor and ELISA for Measurement of Tyrosyl-DNA Phosphodiesterase 1 (TDP1) Activity and Protein Amount in Cell and Tissue Extract. *Nano Life* **2015**, *5*, 1541001. [CrossRef]
22. Jensen, P.W.; Falconi, M.; Kristoffersen, E.L.; Simonsen, A.T.; Cifuentes, J.B.; Marcussen, L.B.; Frøhlich, R.; Vagner, J.; Harmsen, C.; Juul, S.; et al. Real-time detection of TDP1 activity using a fluorophore-quencher coupled DNA-biosensor. *Biosens. Bioelectron.* **2013**, *48*, 230–237. [CrossRef]
23. Jepsen, M.L.; Harmsen, C.; Godbole, A.A.; Nagaraja, V.; Knudsen, B.R.; Ho, Y.P. Specific detection of the cleavage activity of mycobacterial enzymes using a quantum dot based DNA nanosensor. *Nanoscale* **2016**, *8*, 358–364. [CrossRef]
24. Kristoffersen, E.L.; Jørgensen, L.A.; Franch, O.; Etzerodt, M.; Frøhlich, R.; Bjergbæk, L.; Stougaard, M.; Ho, Y.P.; Knudsen, B.R. Real-time investigation of human topoisomerase I reaction kinetics using an optical sensor: A fast method for drug screening and determination of active enzyme concentrations. *Nanoscale* **2015**, *7*, 9825–9834. [CrossRef] [PubMed]
25. Stougaard, M.; Juul, S.; Andersen, F.F.; Knudsen, B.R. Strategies for highly sensitive biomarker detection by Rolling Circle Amplification of signals from nucleic acid composed sensors. *Integr. Biol.* **2011**, *3*, 982–992. [CrossRef] [PubMed]
26. Tesauro, C.; Juul, S.; Arnò, B.; Nielsen, C.J.; Fiorani, P.; Frøhlich, R.F.; Andersen, F.F.; Desideri, A.; Stougaard, M.; Petersen, E.; et al. Specific detection of topoisomerase I from the malaria causing *P. falciparum* parasite using isothermal rolling circle amplification. In Proceedings of the 2012 Annual International Conference of the IEEE Engineering in Medicine and Biology Society, San Diego, CA, USA, 28 August–1 September 2012; pp. 2416–2419.
27. Jäger, S.; Rasched, G.; Kornreich-Leshem, H.; Engeser, M.; Thum, O.; Famulok, M. A versatile toolbox for variable DNA functionalization at high density. *J. Am. Chem. Soc.* **2005**, *127*, 15071–15082. [CrossRef] [PubMed]
28. Liu, Y.; Tuleouva, N.; Ramanculov, E.; Revzin, A. Aptamer-based electrochemical biosensor for interferon gamma detection. *Anal. Chem.* **2010**, *82*, 8131–8136. [CrossRef]
29. Blanco, L.; Bernad, A.; Lázaro, J.M.; Martín, G.; Garmendia, C.; Salas, M. Highly efficient DNA synthesis by the phage phi 29 DNA polymerase. Symmetrical mode of DNA replication. *J. Biol. Chem.* **1989**, *264*, 8935–8940. [CrossRef] [PubMed]
30. Fujita, H.; Kataoka, Y.; Tobita, S.; Kuwahara, M.; Sugimoto, N. Novel One-Tube-One-Step Real-Time Methodology for Rapid Transcriptomic Biomarker Detection: Signal Amplification by Ternary Initiation Complexes. *Anal. Chem.* **2016**, *88*, 7137–7144. [CrossRef]
31. Wei, H.; Tang, S.; Hu, T.; Zhao, G.; Guan, Y. Production of dumbbell probe through hairpin cleavage-ligation and increasing RCA sensitivity and specificity by circle to circle amplification. *Sci. Rep.* **2016**, *6*, 29229. [CrossRef]

32. Franch, O.; Han, X.; Marcussen, L.B.; Givskov, A.; Andersen, M.B.; Godbole, A.A.; Harmsen, C.; Nørskov-Lauritsen, N.; Thomsen, J.; Pedersen, F.S.; et al. A new DNA sensor system for specific and quantitative detection of mycobacteria. *Nanoscale* **2019**, *11*, 587–597. [CrossRef]
33. Hede, M.S.; Fjelstrup, S.; Lötsch, F.; Zoleko, R.M.; Klicpera, A.; Groger, M.; Mischlinger, J.; Endame, L.; Veletzky, L.; Neher, R.; et al. Detection of the Malaria causing *Plasmodium* Parasite in Saliva from Infected Patients using Topoisomerase I Activity as a Biomarker. *Sci. Rep.* **2018**, *8*, 4122. [CrossRef]
34. Keller, J.G.; Petersen, K.V.; Mizielinski, K.; Thiesen, C.; Bjergbæk, L.; Reguera, R.M.; Pérez-Pertejo, Y.; Balaña-Fouce, R.; Trejo, A.; Masdeu, C.; et al. Gel-Free Tools for Quick and Simple Screening of Anti-Topoisomerase 1 Compounds. *Pharmaceuticals* **2023**, *16*, 657. [CrossRef]
35. Wang, J.; Liu, J.; Thomsen, J.; Selnihhin, D.; Hede, M.S.; Kirsebom, F.C.; Franch, O.; Fjelstrup, S.; Stougaard, M.; Ho, Y.P.; et al. Novel DNA sensor system for highly sensitive and quantitative retrovirus detection using virus encoded integrase as a biomarker. *Nanoscale* **2017**, *9*, 440–448. [CrossRef] [PubMed]
36. Singh, P.; Rani, K.; Gotmare, A.; Bhattacharyya, S. A tale of topoisomerases and the knotty genetic material in the backdrop of *Plasmodium* biology. *Biosci. Rep.* **2022**, *42*, BSR20212847. [CrossRef] [PubMed]
37. Tosh, K.; Cheesman, S.; Horrocks, P.; Kilbey, B. *Plasmodium falciparum*: Stage-related expression of topoisomerase I. *Exp. Parasitol.* **1999**, *91*, 126–132. [CrossRef] [PubMed]
38. Keller, J.G.; Mizielinski, K.; Petersen, K.V.; Stougaard, M.; Knudsen, B.R.; Tesauro, C. Simple and Fast Rolling Circle Amplification-Based Detection of Topoisomerase 1 Activity in Crude Biological Samples. *J. Vis. Exp.* **2022**, e64484. [CrossRef]
39. Juul, S.; Nielsen, C.J.; Labouriau, R.; Roy, A.; Tesauro, C.; Jensen, P.W.; Harmsen, C.; Kristoffersen, E.L.; Chiu, Y.L.; Frøhlich, R.; et al. Droplet microfluidics platform for highly sensitive and quantitative detection of malaria-causing *Plasmodium* parasites based on enzyme activity measurement. *ACS Nano* **2012**, *6*, 10676–10683. [CrossRef]
40. Ramharter, M.; Agnandji, S.T.; Adegnika, A.A.; Lell, B.; Mombo-Ngoma, G.; Grobusch, M.P.; McCall, M.; Muranaka, R.; Kreidenweiss, A.; Velavan, T.P.; et al. Development of sustainable research excellence with a global perspective on infectious diseases: Centre de Recherches Médicales de Lambaréné (CERMEL), Gabon. *Wien. Klin. Wochenschr.* **2021**, *133*, 500–508. [CrossRef]
41. Planche, T.; Krishna, S.; Kombila, M.; Engel, K.; Faucher, J.F.; Ngou-Milama, E.; Kremsner, P.G. Comparison of methods for the rapid laboratory assessment of children with malaria. *Am. J. Trop. Med. Hyg.* **2001**, *65*, 599–602. [CrossRef]
42. Juul-Kristensen, T.; Keller, J.G.; Borg, K.N.; Hansen, N.Y.; Foldager, A.; Ladegaard, R.; Ho, Y.P.; Loeschcke, V.; Knudsen, B.R. Topoisomerase 1 Activity Is Reduced in Response to Thermal Stress in Fruit Flies and in Human HeLa Cells. *Biosensors* **2023**, *13*, 950. [CrossRef]
43. Josephine Geertsen, K.; Vandsø, P.K.; Birgitta, R.K.; Cinzia, T. Simple and Fast DNA-Based Tool to Investigate Topoisomerase 1 Activity, a Biomarker for Drug Susceptibility in Colorectal Cancer. In *Recent Understanding of Colorectal Cancer Treatment*; Keun-Yeong, J., Ed.; IntechOpen: Rijeka, Croatia, 2022; Chapter 3.
44. Putaporntip, C.; Buppan, P.; Jongwutiwes, S. Improved performance with saliva and urine as alternative DNA sources for malaria diagnosis by mitochondrial DNA-based PCR assays. *Clin. Microbiol. Infect.* **2011**, *17*, 1484–1491. [CrossRef]
45. McMorrow, M.L.; Aidoo, M.; Kachur, S.P. Malaria rapid diagnostic tests in elimination settings—Can they find the last parasite? *Clin. Microbiol. Infect.* **2011**, *17*, 1624–1631. [CrossRef]

sensors

MDPI

Article

# Graphene/$TiO_2$ Heterostructure Integrated with a Micro-Lightplate for Low-Power $NO_2$ Gas Detection

Paniz Vafaei [1], Margus Kodu [1], Harry Alles [1], Valter Kiisk [1], Olga Casals [2], Joan Daniel Prades [3] and Raivo Jaaniso [1,*]

1 Institute of Physics, University of Tartu, EE-50411 Tartu, Estonia; paniz.vafaei@ut.ee (P.V.); margus.kodu@ut.ee (M.K.); harry.alles@ut.ee (H.A.); valter.kiisk@ut.ee (V.K.)
2 MIND-IN2 UB, Department of Electronics and Biomedical Engineering, Universitat de Barcelona, E-08028 Barcelona, Spain; olga.casals@ub.edu
3 Laboratory for Emerging Nanometrology (LENA), Institute of Semiconductor Technology (IHT), Technische Universität Braunschweig, Hans-Sommer Str. 66, 38106 Braunschweig, Germany; daniel.prades@tu-braunschweig.de
* Correspondence: raivo.jaaniso@ut.ee

**Abstract:** Low-power gas sensors that can be used in IoT (Internet of Things) systems, consumer devices, and point-of-care devices will enable new applications in environmental monitoring and health protection. We fabricated a monolithic chemiresistive gas sensor by integrating a micro-lightplate with a 2D sensing material composed of single-layer graphene and monolayer-thick $TiO_2$. Applying ultraviolet (380 nm) light with quantum energy above the $TiO_2$ bandgap effectively enhanced the sensor responses. Low (<1 μW optical) power operation of the device was demonstrated by measuring $NO_2$ gas at low concentrations, which is typical in air quality monitoring, with an estimated limit of detection < 0.1 ppb. The gas response amplitudes remained nearly constant over the studied light intensity range (1–150 mW/cm$^2$) owing to the balance between the photoinduced adsorption and desorption processes of the gas molecules. The rates of both processes followed an approximately square-root dependence on light intensity, plausibly because the electron–hole recombination of photoinduced charge carriers is the primary rate-limiting factor. These results pave the way for integrating 2D materials with micro-LED arrays as a feasible path to advanced electronic noses.

**Keywords:** gas sensor; $NO_2$; micro-lightplate; graphene; $TiO_2$

## 1. Introduction

Air quality assurance and environmental protection require monitoring of toxic and polluting gases with different spatial resolutions, from satellite surveillance to ground IoT networks and personal gas detection, which provide the most localized sensing [1,2]. Microsensors embedded in consumer devices, including mobile phones and wearables, would enable personal compliance with the cleanliness of the environment and, if exhaled air is detected, monitoring of the health situation [3]. A small footprint and low power consumption are of paramount importance for wearables and self-powered IoT devices. In this regard, chemiresistor-type gas sensors stand out for their potential for miniaturization and mass production in the semiconductor industry. Although the lowest power consumption can be achieved with sensors capable of reversible operation at room temperature [4], such devices frequently have very slow signal recovery and may not function stably enough in outdoor conditions without a miniature heater or light source.

Metal oxide-based sensors on MEMS platforms with micro-hotplates have reached a footprint of a few $mm^2$ and power consumption of 10 mW [5,6]. The energy consumption is due to the heating required to accelerate the desorption of gases and bring the response and recovery times of the sensor within acceptable limits. An alternative to providing external energy for the amplification and acceleration of sensor responses is to use light instead of heat [7,8]. Light, especially if its quantum energy surpasses the bandgap energy of the sensor material, effectively produces electronic excitations that facilitate surface reactions.

Recently, it was demonstrated that instead of using a separate light source, an effective way to couple the light energy into the sensing material is to integrate it with a micro-LED (μLED) or, by analogy with a micro-hotplate, with a micro-lightplate (μLP) [9]. With ZnO nanoparticles coated on the isolating layer on top of a blue (455 nm) μLED's active area (190 μm × 250 μm), $NO_2$ gas was detected at 25 ppb with power consumption as low as 30 μW [10]. With a different design of a monolithic GaN-based μLED and sensing material composed of $In_2O_3$ nanowires, the power level was further reduced below the microwatt level, albeit at the expense of higher (1 ppm) $NO_2$ concentration [11]. In Ref. [12], ultraviolet (UV) μLP (390 nm) was used, and a significant increase in power efficiency was achieved by reducing the size of the μLED from 200 μm to 30 μm. A $NO_2$ limit of detection (LoD) of 15 ppb was demonstrated at a power consumption of ~200 μW.

The major driving forces for μLED development are the lighting and display markets, with the trends being a reduction in power consumption and pixel size of the arrays for micro-displays [13]. Beyond display technology, the integration of LED nitride and CMOS technologies may open avenues for groundbreaking applications such as highly efficient nanosensors and miniaturized neuromorphic networks [14]. Chemical sensors and sensor arrays [15] may benefit from the miniaturization trend if the technology for forming sensor materials on top of the LED array keeps track. The thickness of the metal oxide (MOX) layers in the μLP devices described above was 100s nm, and it is typically even more in commercial hotplate-based solutions. For light-assisted sensors, the power can be relatively easily reduced by shrinking the area of the sensor. This, in turn, means that thinner sensing layers are needed to match the small μLED size ($\leq$3 μm [16]) as well as to ensure efficient light absorption in the sensing layer.

An excellent opportunity for this could be the use of 2D materials [17,18], which are not only atomically thin and have a large surface-to-volume ratio, an essential prerequisite for the material's high gas sensitivity, but can also be technologically handled for mass production [19]. Graphene is an excellent transducer of environmental perturbations because of its low density of charge carriers, and it has the high electrical conductivity needed for microscopic sensing areas [20]. Pristine graphene, however, has low gas sensitivity because of its inertness (e.g., the binding energy of $NO_2$ is only approximately 0.06 eV [21]) and has to be functionalized by introducing defects, functional groups, or nanoparticles [22–25]. It was demonstrated that the sensitivity of graphene could be increased by two orders of magnitude by adding a nanolayer of $ZrO_2$ via pulsed laser deposition [26]. In such a layered heterostructure, graphene acts as a transducer, owing to its low electron density but high conductivity, and the metal oxide serves as a receptor for gas molecules. Recently, an even more efficient heterostructure of single-layer graphene with $TiO_2$, a well-known photocatalytic material [27], was shown to be perfectly suitable for air quality applications, owing to its optimal sensitivity and high stability [28]. An efficient charge transfer from graphene to $NO_2$ molecules adsorbed on titania was demonstrated.

In this study, we developed a monolithic gas sensor by integrating a graphene-$TiO_2$ heterostructure (Gr/$TiO_2$) with UV μLP and demonstrated the detection of $NO_2$ gas at typical concentrations for air quality monitoring (20–150 ppb) at low power levels (electrical power down to 100 μW, optical power < 1 μW). We analyzed the performance characteristics related to gas response power dependence and discussed the underlying processes.

## 2. Experimental Section

### 2.1. Device and Sensor Material Fabrication

The InGaN-based μLPs had a layout similar to that in Ref. [8] with a μLED and interdigitated electrode (IDE) area of 190 μm × 250 μm. A schematic cross-section and a photograph of the device are shown in Figure 1a,b, respectively. The μLPs grown on sapphire wafers were diced in pairs, and the dimensions of a die with two devices on it are 4.05 × 5.6 mm$^2$. Single-layer CVD graphene grown on a polycrystalline copper foil (Graphenea, San Sebastian, Spain) was transferred onto the μLP die using a wet transfer procedure [Figure 1c]. After cutting the graphene/Cu/graphene sheet with an appropriate size and covering it with a layer of PMMA, argon plasma treatment (Diener Tetra 30/50, Plasma Surface Technology, Ebhausen, Germany) was applied to remove the graphene from the uncovered side of the sheet. The next step was to remove copper from the PMMA/graphene/Cu sheet by keeping it in a 1 M ammonium persulfate (Sigma-Aldrich, Steinheim, Germany) solution for two hours. The PMMA/graphene film was then rinsed several times with deionized water, transferred onto the μLP in water, and left to dry overnight. Finally, the μLP with the transferred PMMA/graphene layer was baked at 120 °C for an hour and then placed in pure acetone (≥99.8%, Thermo Fisher Scientific, Seelze, Germany) for 2 h to dissolve the polymer layer. After drying the device, a part of the graphene was selectively removed from its surface to galvanically separate the LED and sensor circuits (see Figure 1b). The electrical isolation between the sensor and LED circuits was accomplished in two stages: by a short 1 min Ar plasma etching while masking the middle area around the interdigitated electrodes (IDE) and then by laser cutting of graphene near the IDE. The black rectangle in Figure 1b shows the masked area during plasma etching for graphene removal, and the dotted lines show the laser-cutting trajectory. A femtosecond laser (ORIGAMI O-05LP, Istanbul, Turkiye) with a pulse energy density of 0.7 mJ/cm$^2$ was used to cut a few μm wide path through graphene without damaging the underlying μLP. Thereafter, pulsed laser deposition (PLD) was applied to produce a functionalizing layer on top of the graphene. Before deposition, the μLPs with graphene were heated in a PLD chamber at 150 °C in a vacuum ($10^{-6}$ mbar) for 1.5 h. PLD was performed at 45 °C in 0.05 mbar $N_2$ gas using a KrF excimer laser (COMPexPro 205, Coherent Lambda Physic GmbH, Göttingen, Germany) with a pulse frequency of 5 Hz and an energy density of 5 J/cm$^2$ for target ablation. The number of laser pulses used for target ablation was 100, which resulted in a $TiO_2$ layer with a thickness of about 0.5 nm [28].

### 2.2. Gas Response Measurement Set-Up

The experimental setup for the gas sensitivity measurements is shown in Figure 2. The test gas was prepared from cylinder gases ($N_2$, $O_2$, $NO_2/N_2$ of 99.999% purity, AS Linde Gas, Tallinn, Estonia), which passed through mass flow controllers (model SLA5820, Brooks Instrument, Hatfield, PA, USA) into a 180 cm$^3$ micro-probe chamber (Nextron, Busan, Korea). The total gas flow rate through the chamber was maintained at 200 sccm. The $O_2$ content in the gas mixture was kept constant at 21% to simulate a typical atmospheric composition. The relative humidity of the testing gas was held either at 0, 20, or 40% by bypassing part of the $N_2$ through the water bubbler. In the cross-sensitivity tests, the gases were supplied from cylinders, except for ozone, which was produced with a UV lamp-based generator (SOG-1, UPV/Analytic Jena, Jena, Germany) and monitored using an analyzer (model 430, Teledyne API, San Diego, CA, USA). The electrical conductance of the sensors was measured with a Keithley 2400 source measure unit, using a typical 50 mV bias voltage.

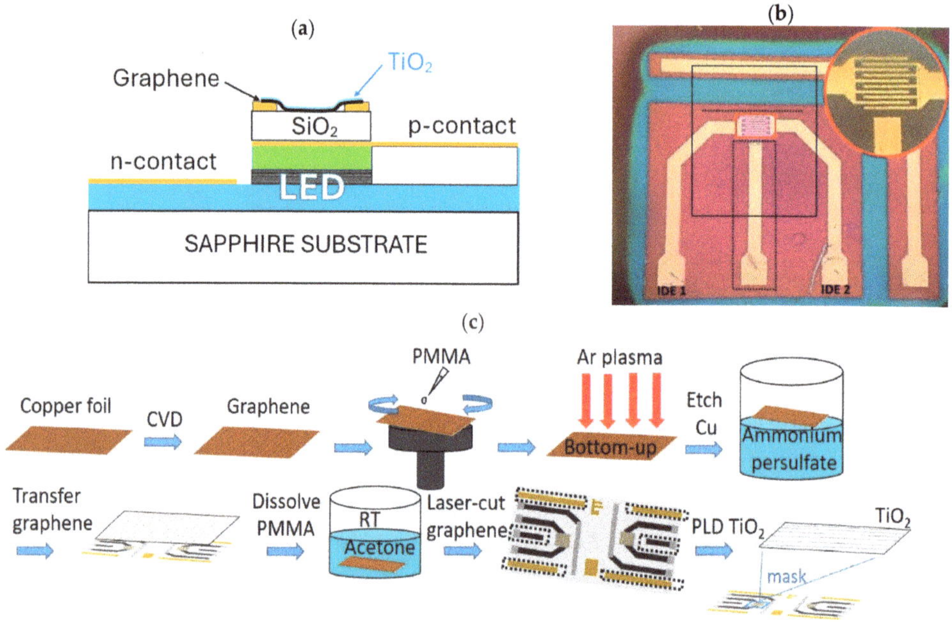


**Figure 1.** (**a**) Schematic cross-section of the device, (**b**) photograph of the μLP with a magnified area of interdigitated electrodes, and (**c**) sequence of sensor layer fabrication on the μLP.

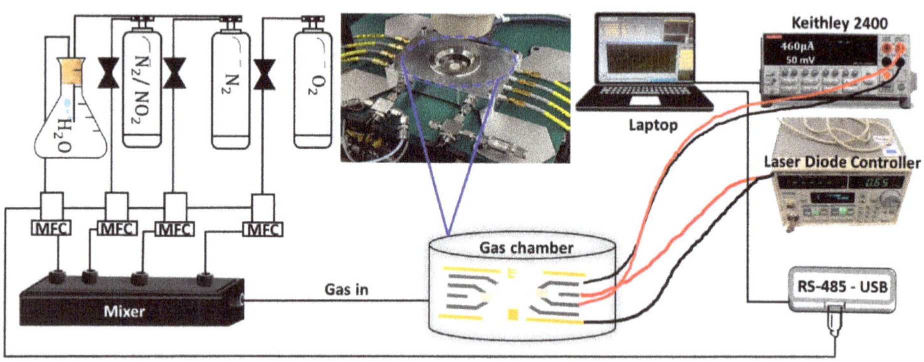


**Figure 2.** Schematic illustration of the gas sensing setup.

*2.3. Characterization Methods and Instruments*

The sensor materials were characterized by scanning electron microscopy (Nova NanoSEM 450, FEI, Hillsboro, OR, USA) and Raman spectroscopy (inVia, Renishaw, Gloucestershire, UK; 514.5 nm excitation). The electrical current of the μLP was regulated with a laser diode controller LDC500 (SRS, Sunnyvale, CA, USA, and the optical power was measured using a model 1918-C power meter (Newport, Irvine, CA, USA). The μLP optical power was evaluated from the measured data by assuming that the source was Lambertian. The electroluminescence spectrum was recorded using FLAME-T-XR1-ES spectrometer (Ocean Optics, Largo, FL, USA; spectral resolution < 2 nm).

# 3. Results and Discussion

## *3.1. Characterization*

The characteristics of the μLP are shown in Figure 3, where (a) shows the working device with microprobe contacts and panel (b) plots its volt–ampere characteristic curve. The peak emission wavelength of the μLP was 380 nm [see the spectrum in Figure 3c]. Figure 3d shows the optical power and intensity on the sensor area versus the consumed electrical power. The external quantum and wall-plug efficiencies had incidentally very similar numerical values, being 0.4 ± 0.04% near the threshold current of 40 μA and increasing to 7% at higher currents.

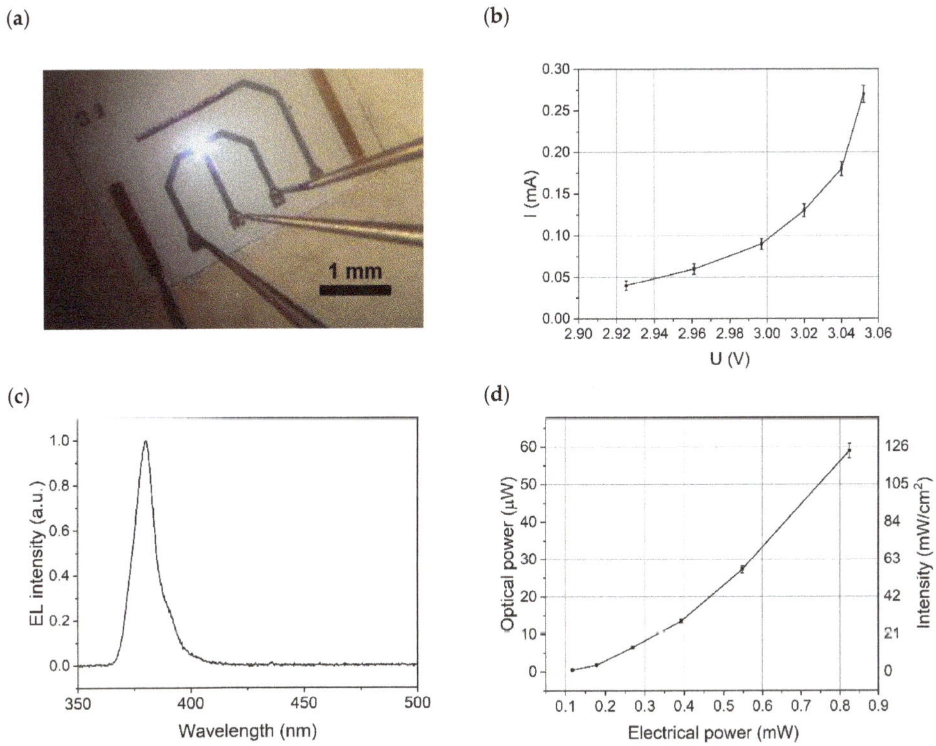


**Figure 3.** (**a**) Image of the μLP with a working LED, its above-threshold (**b**) volt–ampere characteristic, (**c**) electroluminescence (EL) spectrum, and (**d**) dependence of the μLP optical power and surface intensity on the applied electrical power.

SEM images of graphene in the electrode gap of the μLP are shown in Figure 4a,b. The spots that are darker in color in Figure 4a are commonly observed in CVD graphene and are related to the multilayer graphene. The dark linear features, which are also commonly observed, are due to the topography of the polycrystalline Cu-foil used in the synthesis process and the grain boundaries of graphene. The Raman spectrum [Figure 4c] contains mainly two bands, a G-band at 1577 cm$^{-1}$ and a 2D-band at 2664 cm$^{-1}$, with an intensity ratio of 1:3, which is characteristic of single-layer graphene [29]. The XPS spectra of the graphene used in the present study were recently analyzed in detail in Ref. [30]. Deconvolution of the C1s peak resolved sp2 carbon (87%), sp3 carbon (7.5%), and oxidized (C–O, C=O, O=C–O) species (5–6%). After the PLD of $TiO_2$ on top of graphene, a defect-related band D (1344 cm$^{-1}$) emerges in the Raman spectrum. The SEM image [Figure 4d] shows the granular morphology of the deposited material, uniformly coating the graphene. The linear features observed in the pristine graphene image "shine" through a thin $TiO_2$ coating and are also visible in Figure 4d. For comparison, Figure 4b shows an image of pristine graphene at exactly the same scale, which does not have a granular oxide layer on top but instead shows characteristic streaks formed by the copper substrate during growth.

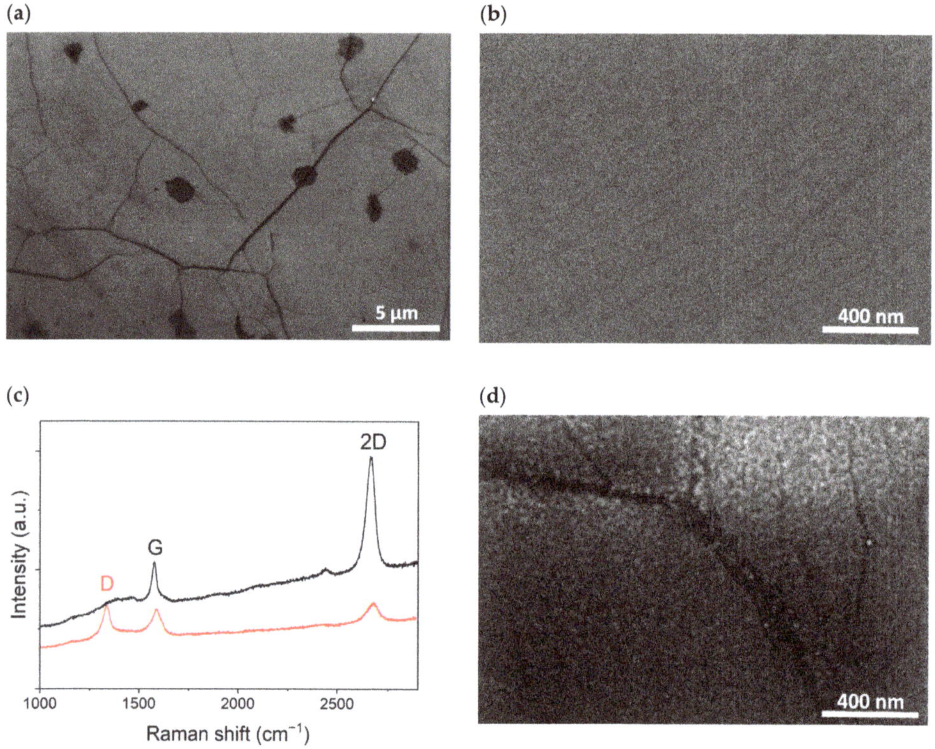


**Figure 4.** (**a**,**b**) SEM images of CVD graphene on µLP. (**c**) Raman spectra of graphene before and after the PLD of $TiO_2$. (**d**) SEM image of the sensor material after coating the graphene with a $TiO_2$ nanolayer.

### 3.2. Gas Sensing Performance

The sequences of conductance measurements when the sensor was exposed to a series of $NO_2$ concentrations in the dark and under µLED illumination are presented in Figure 5a,b for pristine graphene and Gr/$TiO_2$, respectively. The measured conductance is solely due to graphene in both cases, and the approximately monolayer-thick $TiO_2$ (bandgap 3.2 eV [25]) plays a negligible role. $NO_2$ gas was injected at three different concentrations (20, 50, and 150 ppb), each for 5 min, followed by 5 min intervals in clean air. The responses of the device with pristine graphene [Figure 5a] were relatively small (~1%) even under high intensity (~1 W/cm$^2$) of UV light. The insensitivity of pristine (defect-free) graphene to toxic gases has been observed in several previous publications [22,28,31].

The sequence in Figure 5b starts with $NO_2$ gas exposures in the dark; three relatively small raising steps in conductance can be seen. CVD graphene is p-doped by adsorbed oxygen and water molecules in air [32,33], and MOX-coated graphene retains this conductivity type [26]. Because $NO_2$ acts as an electron acceptor during adsorption, hole doping is further promoted in the presence of $NO_2$, increasing the conductivity. However, the gas response in the dark was slow and showed prolonged recovery. After switching on the μLED at its lowest power (see Figure 4b; 70 min after the start of the measurement), the conductance of the samples held in synthetic air starts to decrease and stabilizes at a lower level. Such a persistent photoresistance effect has also been observed in pristine graphene and is explained by the photoinduced removal of oxygen and the associated decrease in the density of electron holes in graphene [34]. It can be seen from the figure that under the influence of UV radiation, the gas responses become faster, and the recovery is improved with increasing optical power. The values of the applied optical power are indicated on top in Figure 5b as being directly relevant to the photoinduced phenomena; the values of the electrical power applied were 0.12, 0.18, and 0.24 mW, respectively.

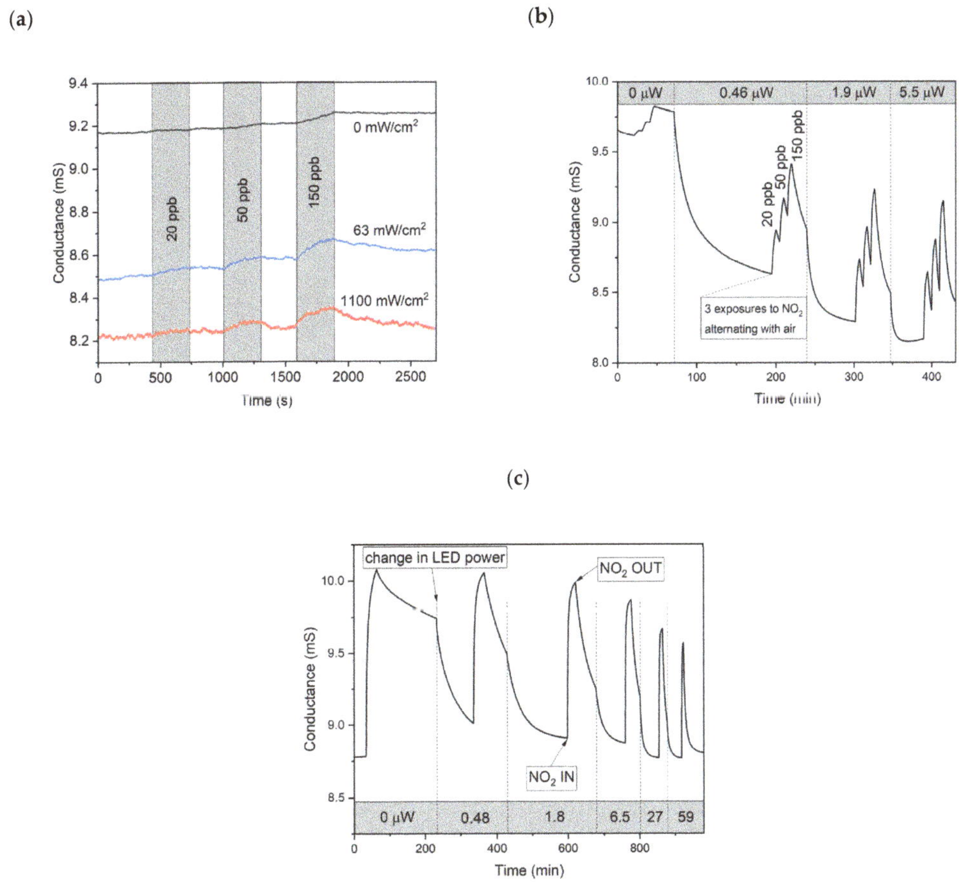


**Figure 5.** (**a**) Dynamic responses of pristine graphene to $NO_2$ gas at concentrations of 20, 50, and 150 ppb at different irradiation intensities on μLP. (**b**) The same for the Gr/$TiO_2$ μLP sensor, recorded in the dark and under incremental UV illumination with the μLED optical power of 0.46, 1.9, and 5.5 μW (corresponding to 0.8, 3.3, and 9.7 mW/cm$^2$). Synthetic air was used as the background gas. (**c**) Sensor conductance during the exposures to 150 ppb of $NO_2$ gas at different levels of μLP optical power. The power levels in μW units are shown in the gray area at the bottom. Synthetic air with a relative humidity (RH) of 20% was used as the background gas.

To investigate the dependence of photoinduced processes on light intensity, measurements were made with the same $NO_2$ concentration but at different irradiation powers

[Figure 5c]. Longer clean air intervals between the gas injections were used for more complete signal recovery. This figure shows a relatively strong gas response under dark conditions, unlike that shown in Figure 5b. The difference is that the sensor was exposed to UV shortly (an hour) before the measurements, whereas in the case of Figure 5b, the sensor was held in the dark for several days. Persistent photoresistance can, indeed, last many hours [33] and seemingly also affects the surface properties of $Gr/TiO_2$ with respect to $NO_2$. The processes occurring in the sensor material can be summarized in simple terms as follows:

$$\mathrm{Gr/TiO_2 + h\nu \rightarrow e^- + h^+} \quad (1)$$

$$\mathrm{O_2^-(ads) + h^+ \rightarrow O_2(gas)} \quad (2)$$

After the photoelectrons and holes are formed, the previously chemisorbed oxygen is partly released, and the hole conductivity decreases. As a result, more sites were becoming available for $NO_2$ adsorption, written as follows:

$$\mathrm{NO_2(gas) + e^- \rightarrow NO_2^-(ads)} \quad (3)$$

This process is more dominant than oxygen adsorption because of the significantly larger electron affinity of $NO_2$ (2.273 eV) than that of $O_2$ (0.450 eV) [35]. The recovery of conductance was due to the $NO_2$ desorption process, written as follows:

$$\mathrm{NO_2^-(ads) + h^+ \rightarrow NO_2(gas)} \quad (4)$$

Clearly, both the response and recovery processes became faster with increasing light power, as seen in Figure 5.

For quantitative characterization, the response and recovery curves were approximated using biexponential kinetics, written as follows:

$$F(t) = A_0 + A_1(1 - \exp(-k_1 t)) + A_2(1 - \exp(-k_2 t)) \quad (5)$$

The average response and recovery rate $k_{av}$ was defined as follows:

$$k_{av} = \frac{A_1 k_1 + A_2 k_2}{A_1 + A_2} \quad (6)$$

We used average parameters, allowing us to compare the data in cases where the fitting procedure converged to only a single exponent in Equation (5). Moreover, there is a distribution of rates on an amorphous metal oxide surface [36], rather than just one or two discrete values. The two kinetic components are still a reasonable approximation, although they are somewhat dependent on the time span of the approximated curve.

Initially, the response rates to the step-like $NO_2$ gas injection increase rapidly with the increasing light intensity (Figure 6). The rate dependence can be fitted with the power dependence, but it clearly reaches a plateau above 60 $mW/cm^2$. Saturation can be considered an artifact because it is actually determined by the instrumental gas exchange rate, which can be estimated from the chamber volume and the flow rate to be slightly above 1 $min^{-1}$. The recovery process was slower and could be fitted with a power law for the entire range of light intensities. Both approximations using the power law led to exponents close to 0.5.

This dependence can be explained as follows. Let us assume that the surface processes involving $NO_2$ are described by the Langmuir kinetic model, written as follows:

$$\frac{d\theta}{dt} = k_a(1 - \theta) - k_d\theta \quad (7)$$

where $k_a$ and $k_d$ are adsorption and desorption rates, respectively, and $\theta$ is the coverage of occupied adsorption sites. The processes (3) and (4) imply the following:

$$k_a = a \cdot n_e \cdot p \tag{8}$$

$$k_d = d \cdot n_h \tag{9}$$

where $n_e$ and $n_h$ are the density of electrons and holes, respectively, and $p$ is the gas pressure. The parameters $a$ and $d$ have constant values at a given temperature.

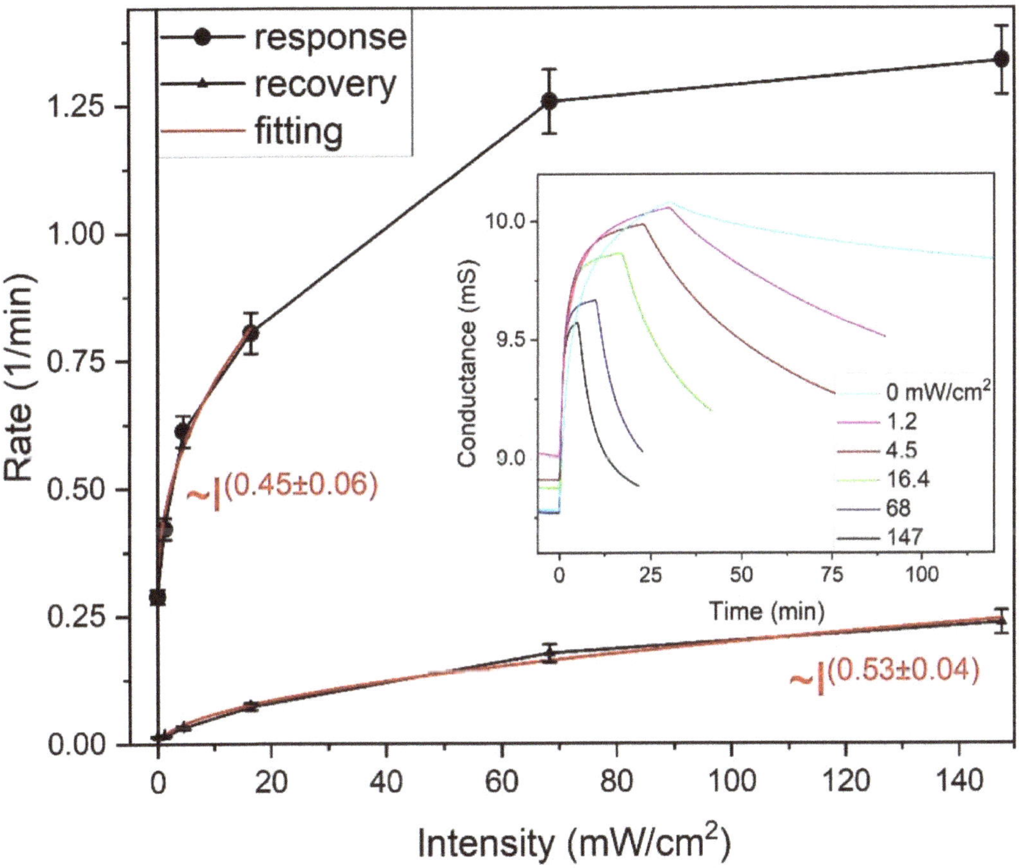


**Figure 6.** The dependence of average response and recovery rates on light intensity. Approximations with power functions and power exponents of the intensity (I) dependence are shown in red. The inset shows the response curves approximated with Equation (5).

The general solution of Equation (7) is written as follows:

$$\theta(t) = \theta(0) \cdot e^{-kt} + \frac{k_a}{k} \cdot \left(1 - e^{-kt}\right) \tag{10}$$

where

$$k = k_a + k_d = a \cdot n_e \cdot p + d \cdot n_h \tag{11}$$

and $\theta$ (0) is the initial coverage at $t$ = 0.

If, in addition, we assume that only photogenerated charge carriers participate in adsorption–desorption processes and the carriers are annihilated by the bimolecular process [37] much faster than the adsorption–desorption processes, then we obtain the following:

$$\frac{dn_e}{dt} = \frac{dn_h}{dt} = I\kappa - bn_e n_h \tag{12}$$

where $I$ is photon flux, $\kappa$ is the absorption constant, and $b$ is the annihilation constant. As the equilibrium in Equation (12) is established much faster than in Equation (7), we can assume the following:

$$0 = \frac{dn_e}{dt} = \frac{dn_h}{dt} = I\kappa - bn_e n_h \tag{13}$$

and evaluate the quasi-static values of charge carrier densities for Equations (7)–(11):

$$n_e = n_h = \sqrt{\frac{I\kappa}{b}} \tag{14}$$

Consequently, the rates $k_a$, $k_d$, and $k$ in Equations (8), (9) and (11) are proportional to the square root of the light intensity. This result persists for a more general case of inhomogeneous Langmuir adsorption, for example for two (or more) adsorption sites, and consequently, the average rates (Equation (6)) behave similarly in accordance with the experimental results shown in Figure 6.

The total response amplitude $A_1 + A_2$ obtained by fitting decreased by approximately 25% in the full light intensity range spanning from 1 to 150 mW/cm$^2$, mainly at the expense of the slower component. This differs from the behavior of MOX-based sensors coated directly on the μLP, which showed a bell-shaped (log-normal) dependence of the response amplitude on light intensity [10,12]. Obviously, an optimal intensity exists in the latter case corresponding to the bell-curve maximum. For the sensor studied in this work, there seems to be no optimal intensity owing to the physics of the device; the optimum intensity is determined by practical considerations (a trade-off between the response and recovery rates vs. power consumption).

Another difference between graphene-based materials and semiconducting metal oxides is the significantly higher conductivity of the former. As a result, the electrodes can be accommodated in a much smaller area, and measurements can be made with a higher signal-to-noise ratio. In our measurements, the sensitivity was S = 25 μS/ppb at the lowest studied concentration, whereas the rms noise amplitude (N) was typically only 0.2 μS within a 1 Hz bandwidth. If estimating the level of detection (LoD) of $NO_2$ with a common rule LoD = 3N/S ($\geq$99% confidence level), the result is LoD = 0.024 ppb.

Finally, to characterize the selectivity, we checked the cross-sensitivity to several other toxic gases and humidity (Figure 7). Panel (a) compares the dynamic relative responses (relative change in conductance) to $NO_2$ gas at three different RH levels at the irradiation intensity of 60 mW/cm$^2$. The responses appeared to be somewhat stronger and faster in the humid air. The impact of humidity is a common and complex phenomenon in chemiresistive sensors [38]. In a recent study [39], the dependence of the light-assisted response to $NO_2$ on humidity was ascribed to the decrease in active sites and increase in carrier concentration owing to adsorbed water molecules and OH groups. The latter factor also explains our results through the deactivation of the recombination centers by adsorbed water molecules.

The relative conductance changes during 15 min of gas exposure to different toxic gases in dry air under 5 mW/cm$^2$ irradiations are shown in Figure 7b. As expected, the sensitivity to $O_3$ was similar to that of $NO_2$, as both molecules are strong oxidizers, with electron affinities of 2.103 and 2.273 eV, respectively [34]. In the case of the reducing toxic gases CO, $NH_3$, and $H_2S$, the concentrations had to be several orders of magnitude higher in order to observe sizeable effects.

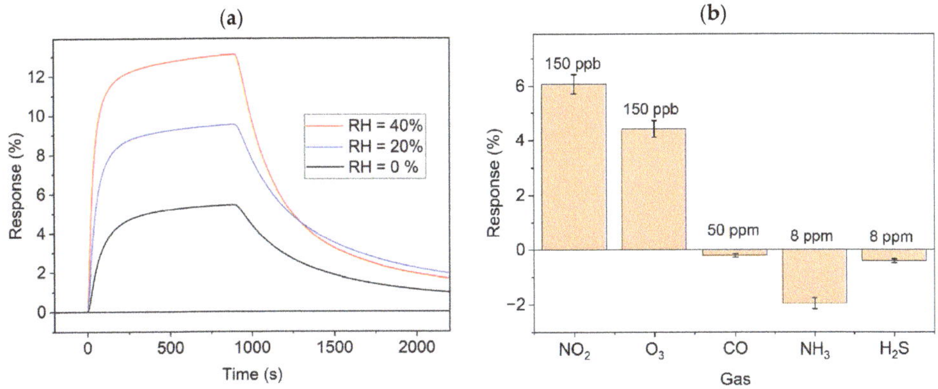


**Figure 7.** Relative responses to (**a**) 150 ppb of $NO_2$ at different levels of relative humidity and (**b**) different toxic gases at concentrations as indicated.

## 4. Conclusions

In summary, we assembled monolithic gas microsensors by integrating a UV micro-lightplate with a 2D sensing material made of CVD graphene and a less than a nanometer thick layer of $TiO_2$. The low-power (0.5 μW optical, 100 μW electrical) operation of the device was demonstrated while detecting $NO_2$ concentrations typical of air quality monitoring with an extrapolated limit of detection of 0.02 ppb. The gas response amplitude was nearly constant over the studied light intensity range (1–150 mW/cm$^2$) because of the balance between the photoinduced adsorption and desorption processes. The rates of both the response and the recovery processes followed an approximately square-root dependence on the light intensity, implying that bimolecular electron–hole recombination is the primary mechanism of the photoinduced charge carrier relaxation and the rate-limiting factor of the sensor. In practical terms, such a dependence on light intensity is useful because there is relatively little loss in reaction speed when reducing power. The integration of 2D materials with high-density μLED arrays can provide a feasible path for advanced electronic noses with large sensor arrays.

**Author Contributions:** P.V.: investigation, visualization, writing—original draft; M.K.: investigation, and writing—original draft; H.A.: validation, and writing—review and editing; V.K.: methodology, and writing—review and editing; O.C.: resources, and writing—review and editing; J.D.P.: conceptualization, funding acquisition, and writing—review and editing; R.J.: conceptualization; formal analysis, funding acquisition, supervision, and writing—review and editing. All authors have read and agreed to the published version of the manuscript.

**Funding:** This work was supported by the Estonian Research Council grants PRG1580, PRG753, and TEM-TA110. The work of J. Daniel Prades was supported in part by the Alexander von Humboldt Professorship of the Humboldt Foundation and in part by the Federal Ministry for Education and Research, Germany.

**Data Availability Statement:** Data will be made available upon request.

**Acknowledgments:** The authors would like to thank Tauno Kahro for the help with graphene transfer, Elyad Damerchi with SEM measurements, and Indrek Renge for valuable comments.

**Conflicts of Interest:** The authors declare no conflicts of interest.

## References


1. Schneider, P.; Vogt, M.; Haugen, R.; Hassani, A.; Castell, N.; Dauge, F.R.; Bartonova, A. Deployment and evaluation of a network of open low-cost air quality sensor systems. *Atmosphere* **2023**, *14*, 540. [CrossRef]
2. Wang, A.; Mora, S.; Machida, Y.; de Souza, P.; Paul, S.; Oyinlola, O.; Duarte, F.; Ratti, C. Hyperlocal Environmental data with a mobile platform in urban environments. *Sci. Data* **2023**, *10*, 524. [CrossRef]
3. Shan, B.; Broza, Y.Y.; Li, W.; Wang, Y.; Wu, S.; Liu, Z.; Wang, J.; Gui, S.; Wang, L.; Zhang, Z.; et al. Multiplexed nanomaterial-based sensor array for detection of COVID-19 in exhaled breath. *ACS Nano* **2020**, *14*, 12125–12132. [CrossRef]
4. Tang, Y.; Zhao, Y.; Liu, H. Room-Temperature Semiconductor Gas Sensors: Challenges and Opportunities. *ACS Sens.* **2022**, *7*, 3582–3597. [CrossRef]
5. Briand, D.; Courbat, J. Micromachined semiconductor gas sensors. In *Semiconductor Gas Sensors*; Elsevier: Amsterdam, The Netherlands, 2020; pp. 413–464. ISBN 978-0-08-102559-8.
6. Guha, P.K.; Santra, S.; Gardner, J.W. Integrated CMOS-based sensors for gas and odor detection. In *Semiconductor Gas Sensors*; Elsevier: Amsterdam, The Netherlands, 2020; pp. 465–487. ISBN 978-0-08-102559-8.
7. Espid, E.; Taghipour, F. UV-LED photo-activated chemical gas sensors: A review. *Crit. Rev. Solid State Mater. Sci.* **2017**, *42*, 416–432. [CrossRef]
8. Zhang, Q.; Chen, C.; Liu, Y.; Xu, M.; Xie, G.; Tai, H.; Du, X.; Jiang, Y.; Su, Y. Room-temperature light-activated chemical sensors for gas monitoring and applications: A review. *J. Phys. D Appl. Phys.* **2022**, *55*, 213001. [CrossRef]
9. Markiewicz, N.; Casals, O.; Fabrega, C.; Gràcia, I.; Cané, C.; Wasisto, H.S.; Waag, A.; Prades, J.D. Micro light plates for low-power photoactivated (gas) sensors. *Appl. Phys. Lett.* **2019**, *114*, 053508. [CrossRef]
10. Casals, O.; Markiewicz, N.; Fabrega, C.; Gràcia, I.; Cané, C.; Wasisto, H.S.; Waag, A.; Prades, J.D. A parts per billion (ppb) sensor for $NO_2$ with microwatt ($\mu$W) power requirements based on micro light plates. *ACS Sens.* **2019**, *4*, 822–826. [CrossRef]
11. Cho, I.; Sim, Y.C.; Lee, K.; Cho, M.; Park, J.; Kang, M.; Chang, K.S.; Jeong, C.B.; Cho, Y.; Park, I. Nanowatt-level photoactivated gas sensor based on fully-integrated visible microLED and plasmonic nanomaterials. *Small* **2023**, *19*, 2207165. [CrossRef]
12. Cho, I.; Sim, Y.C.; Cho, M.; Cho, Y.-H.; Park, I. Monolithic micro light-emitting diode/metal oxide nanowire gas sensor with microwatt-level power consumption. *ACS Sens.* **2020**, *5*, 563–570. [CrossRef] [PubMed]
13. Behrman, K.; Kymissis, I. Micro light-emitting diodes. *Nat. Electron.* **2022**, *5*, 564–573. [CrossRef]
14. Prades, J.D.; Meierhofer, F.; Diéguez, A.; Waag, A. MicroLED arrays—A perspective beyond displays. *Appl. Phys. Lett.* **2024**, *125*, 150504. [CrossRef]
15. Chen, Z.; Chen, Z.; Song, Z.; Ye, W.; Fan, Z. Smart gas sensor arrays powered by artificial intelligence. *J. Semicond.* **2019**, *40*, 111601. [CrossRef]
16. Liu, Y.; Feng, F.; Zhang, K.; Jiang, F.; Chan, K.-W.; Kwok, H.-S.; Liu, Z. Analysis of size dependence and the behavior under ultrahigh current density injection condition of GaN-based micro-LEDs with pixel size down to 3 Mm. *J. Phys. D Appl. Phys.* **2022**, *55*, 315107. [CrossRef]
17. Meng, Z.; Stolz, R.M.; Mendecki, L.; Mirica, K.A. Electrically-transduced chemical sensors based on two-dimensional nanomaterials. *Chem. Rev.* **2019**, *119*, 478–598. [CrossRef] [PubMed]
18. Buckley, D.J.; Black, N.C.G.; Castanon, E.G.; Melios, C.; Hardman, M.; Kazakova, O. Frontiers of graphene and 2D material-based gas sensors for environmental monitoring. *2D Mater.* **2020**, *7*, 032002. [CrossRef]
19. Soikkeli, M.; Murros, A.; Rantala, A.; Txoperena, O.; Kilpi, O.-P.; Kainlauri, M.; Sovanto, K.; Maestre, A.; Centeno, A.; Tukkiniemi, K.; et al. Wafer-scale graphene field-effect transistor biosensor arrays with monolithic CMOS readout. *ACS Appl. Electron. Mater.* **2023**, *5*, 4925–4932. [CrossRef] [PubMed]
20. Recum, P.; Hirsch, T. Graphene-based chemiresistive gas sensors. *Nanoscale Adv.* **2024**, *6*, 11–31. [CrossRef] [PubMed]
21. Leenaerts, O.; Partoens, B.; Peeters, F.M. Adsorption of $H_2O$, $NH_3$, CO, $NO_2$, and NO on graphene: A first-principles study. *Phys. Rev. B* **2008**, *77*, 125416. [CrossRef]
22. Kumar, B.; Min, K.; Bashirzadeh, M.; Farimani, A.B.; Bae, M.-H.; Estrada, D.; Kim, Y.D.; Yasaei, P.; Park, Y.D.; Pop, E.; et al. The role of external defects in chemical sensing of graphene field-effect transistors. *Nano Lett.* **2013**, *13*, 1962–1968. [CrossRef]
23. Eriksson, J.; Puglisi, D.; Kang, Y.H.; Yakimova, R.; Lloyd Spetz, A. Adjusting the electronic properties and gas reactivity of epitaxial graphene by thin surface metallization. *Phys. B Condens. Matter* **2014**, *439*, 105–108. [CrossRef]
24. Ricciardella, F.; Vollebregt, S.; Polichetti, T.; Miscuglio, M.; Alfano, B.; Miglietta, M.L.; Massera, E.; Di Francia, G.; Sarro, P.M. Effects of graphene defects on gas sensing properties towards $NO_2$ detection. *Nanoscale* **2017**, *9*, 6085–6093. [CrossRef] [PubMed]
25. Chen, Z.; Wang, J.; Wang, Y. Strategies for the performance enhancement of graphene-based gas sensors: A review. *Talanta* **2021**, *235*, 122745. [CrossRef]
26. Kodu, M.; Berholts, A.; Kahro, T.; Avarmaa, T.; Kasikov, A.; Niilisk, A.; Alles, H.; Jaaniso, R. Highly sensitive $NO_2$ sensors by pulsed laser deposition on graphene. *Appl. Phys. Lett.* **2016**, *109*, 113108. [CrossRef]
27. Batzill, M.; Diebold, U. The surface and materials science of tin oxide. *Prog. Surf. Sci.* **2005**, *79*, 47–154. [CrossRef]

28. Berholts, A.; Kodu, M.; Rubin, P.; Kahro, T.; Alles, H.; Jaaniso, R. Layered heterostructure of graphene and $TiO_2$ as a highly sensitive and stable photoassisted $NO_2$ Sensor. *ACS Appl. Mater. Interfaces* **2024**, *16*, 43827–43837. [CrossRef] [PubMed]
29. Ferrari, A.C.; Meyer, J.C.; Scardaci, V.; Casiraghi, C.; Lazzeri, M.; Mauri, F.; Piscanec, S.; Jiang, D.; Novoselov, K.S.; Roth, S.; et al. Raman spectrum of graphene and graphene Layers. *Phys. Rev. Lett.* **2006**, *97*, 187401. [CrossRef] [PubMed]
30. Kahro, T.; Tarre, A.; Käämbre, T.; Piirsoo, H.-M.; Kozlova, J.; Ritslaid, P.; Kasikov, A.; Jõgiaas, T.; Vinuesa, G.; Dueñas, S.; et al. Hafnium oxide/graphene/hafnium oxide-stacked nanostructures as resistive switching media. *ACS Appl. Nano Mater.* **2021**, *4*, 5152–5163. [CrossRef]
31. Dan, Y.; Lu, Y.; Kybert, N.J.; Luo, Z.; Johnson, A.T.C. Intrinsic response of graphene vapor sensors. *Nano Lett.* **2009**, *9*, 1472–1475. [CrossRef] [PubMed]
32. Yang, Y.; Brenner, K.; Murali, R. The influence of atmosphere on electrical transport in graphene. *Carbon* **2012**, *50*, 1727–1733. [CrossRef]
33. Berholts, A.; Kahro, T.; Floren, A.; Alles, H.; Jaaniso, R. Photo-activated oxygen sensitivity of graphene at room temperature. *Appl. Phys. Lett.* **2014**, *105*, 163111. [CrossRef]
34. Biswas, C.; Güneş, F.; Loc, D.D.; Lim, S.C.; Jeong, M.S.; Pribat, D.; Lee, Y.H. Negative and positive persistent photoconductance in graphene. *Nano Lett.* **2011**, *11*, 4682–4687. [CrossRef] [PubMed]
35. Lide, D.R. *CRC Handbook of Chemistry and Physics*, 89th ed.; Taylor & Francis: Oxfordshire, UK, 2008; ISBN 1-4200-6679-X.
36. Rudzinski, W.; Panczyk, T.; Plazinski, W. Kinetics of isothermal gas adsorption on heterogeneous solid surfaces: Equations based on generalization of the statistical rate theory of interfacial transport. *J. Phys. Chem. B* **2005**, *109*, 21868–21878. [CrossRef]
37. Zhang, Z.; Yates, J.T. Direct observation of surface-mediated electron−hole pair recombination in $TiO_2$ (110). *J. Phys. Chem. C* **2010**, *114*, 3098–3101. [CrossRef]
38. Bârsan, N.; Weimar, U. Understanding the fundamental principles of metal oxide based gas sensors; the example of CO sensing with $SnO_2$ sensors in the presence of humidity. *J. Phys. Condens. Matter* **2003**, *15*, R813–R839. [CrossRef]
39. Nasriddinov, A.; Rumyantseva, M.; Konstantinova, E.; Marikutsa, A.; Tokarev, S.; Yaltseva, P.; Fedorova, O.; Gaskov, A. Effect of humidity on light-activated NO and $NO_2$ gas sensing by hybrid materials. *Nanomaterials* **2020**, *10*, 915. [CrossRef] [PubMed]

sensors

MDPI

*Article*

# Holey Carbon Nanohorns-Based Nanohybrid as Sensing Layer for Resistive Ethanol Sensor

**Bogdan-Catalin Serban [1], Niculae Dumbravescu [1], Octavian Buiu [1], Marius Bumbac [2,3,*], Mihai Brezeanu [4], Cristina Pachiu [1], Cristina-Mihaela Nicolescu [3], Oana Brancoveanu [1] and Cornel Cobianu [5,6]**

1 National Institute for Research and Development in Microtechnologies, IMT-Bucharest, 126 A Str. Erou Iancu Nicolae, 077190 Voluntari, Romania; bogdan.serban@imt.ro (B.-C.S.); niculae.dumbravescu@imt.ro (N.D.); octavian.buiu@imt.ro (O.B.); cristina.pachiu@imt.ro (C.P.); oana.brincoveanu@imt.ro (O.B.)
2 Sciences and Advanced Technologies Department, Faculty of Sciences and Arts, Valahia University of Târgoviște, Aleea Sinaia, nr 13, 130004 Târgoviște, Romania
3 Institute of Multidisciplinary Research for Science Technology, Valahia University of Târgoviște, Aleea Sinaia, nr 13, 130004 Târgoviște, Romania; cristina.nicolescu@valahia.ro
4 Faculty of Electronics, Telecommunications and IT, National University of Science and Technology Politehnica Bucharest, Bd Iuliu Maniu 1-3, 061071 Bucharest, Romania; mbrezeanu@upb.ro
5 Academy of Romanian Scientists (ARS), Str. Ilfov Nr. 3, Sector 5, 050044 Bucharest, Romania; cornel.cobianu@gmail.com
6 eBio-Hub Center of Excellence in Bioengineering, National University of Science and Technology Politehnica Bucharest, Blvd. Iuliu Maniu Nr. 6, Sector 6, 061344 Bucharest, Romania
* Correspondence: marius.bumbac@valahia.ro; Tel.: +40-72-1219270

**Abstract:** The study presents the ethanol vapor sensing performance of a resistive sensor that utilizes a quaternary nanohybrid sensing layer composed of holey carbon nanohorns (CNHox), graphene oxide (GO), $SnO_2$, and polyvinylpyrrolidone (PVP) in an equal mass ratio of 1:1:1:1 ($w/w/w/w$). The sensing device includes a flexible polyimide substrate and interdigital transducer (IDT)-like electrodes. The sensing film is deposited by drop-casting on the sensing structure. The morphology and composition of the sensitive film are analyzed using scanning electron microscopy (SEM), Energy Dispersive X-ray (EDX) Spectroscopy, and Raman spectroscopy. The manufactured resistive device presents good sensitivity to concentrations of alcohol vapors varying in the range of 0.008–0.16 mg/cm$^3$. The resistance of the proposed sensing structure increases over the entire range of measured ethanol concentration. Different types of sensing mechanisms are recognized. The decrease in the hole concentration in CNHox, GO, and CNHox due to the interaction with ethanol vapors, which act as electron donors, and the swelling of the PVP are plausible and seem to be the prevalent sensing pathway. The hard–soft acid-base (HSAB) principle strengthens our analysis.

**Keywords:** ethanol sensor; holey carbon nanohorns; graphene oxide; swelling; HSAB principle

## 1. Introduction

Ethyl alcohol is a volatile, flammable, colorless liquid with a typical wine-like aroma and pungent taste [1]. It is a common industrial raw material encountered in many chemical industrial processes and consumer products [2]. Thus, ethanol is extensively used in medicine (manufacturing processes of pharmaceutical preparations such as lotions, tonics, rubbing compounds, as well as antidote for both methanol and ethylene glycol poisoning) [3,4], cosmetics (mouthwash products, soaps, perfumes) [5], chemical industry (excellent solvent for oils, fats, resins, dyes, inks, waxes, platform molecule for the synthesis

 https://doi.org/10.3390/s25051299

of key chemical compounds such as acetaldehyde, acetic acid, ethene, butadiene, elastomers, biofuel synthesis processes, etc.) [6–8], as well as food and drink industry (used as a natural product to extract and concentrate flavors and aromas, antimicrobial agent in pizza crust) [9]. A large amount of ethanol vapors present in the air may cause adverse central nervous system effects such as headache, mental excitement or depression, unconsciousness, and coma. Monitoring the level of ethanol is an essential process in various fields, such as environmental monitoring (air pollution caused by ethanol emissions) [10], food quality assessment (ethanol, along with $CO_2$, are the primary spoilage metabolites of the freshly cut fruit) [11], traffic management (breath alcohol concentration is the routine test for Measurement of drunkenness level for drivers) [12,13], agriculture (measuring ethanol levels in crops) [14], medical field (breathalyzer) [15], alcoholic beverage industry [16], and ethanol production in biofuel plants [17]. Thereby, in recent decades, a lot of sensing principles and technologies such as conductometric [18], resistive [19,20], field effect transistor [21], FTIR, RAMAN, UV-VIS [22,23], surface acoustic wave [24], optical fiber [25], electrochemical [26], capacitive [27] have been developed or improved to manufacture ethanol sensor with superior performances.

Apart from the working principle of sensors and their design, the materials selected as the sensing layer appear as a key element in developing ethanol gas sensors with improved characteristics such as sensitivity, selectivity, response time, recovery time, hysteresis, and repeatability. Consequently, several materials have been explored as sensing layers within the design of ethanol sensors: metal–organic frameworks [28], metal oxide semiconductors [29], porous silicon [30], conducting polymers [31], metal sulfides [32], dielectric polymers [33], porphyrins [34], or SiC [35]. However, most of these sensors operate at high temperatures, leading to high energy consumption. Only a few studies have demonstrated metal oxide-based ethanol sensors functioning at room temperature.

Furthermore, many carbon-based materials are extensively used as sensing layers within the design of the ethanol gas sensors. Outstanding properties of these materials, such as large specific surface area, ability to interact with target molecule at room temperature, high mechanical strength, fast charge transfer, high stability, versatile covalent and noncovalent functionalization, and environmentally friendly have triggered substantial research to investigate their potential as ethanol gas sensing materials [36–41]. Carbon nanotubes [36], graphene oxide [37], reduced graphene oxide [38], carbon nanofiber [39], carbon nanodots [40], and graphene [41] are some of the carbonaceous nanomaterials studied for ethanol sensing applications. In recent years, carbon nanohorns (nanostructures composed of $sp^2$-hybridized carbon atoms forming a conical shape with diameters of 2–5 nm and lengths of 30–50 nm [42]) have garnered growing attention for gas sensing applications. This interest is driven by their exceptional properties, including clean synthesis methods, the availability of high-purity samples, excellent chemical and thermal stability, large specific surface area, and low toxicity. Thus, both pristine and functionalized carbon nanohorns and their nanocomposites/nanohybrids were used as sensing layers within the design of resistive sensors for the detection and monitoring of relative humidity [43–51], ammonia [52], and ozone [53]. Moreover, holey carbon nanohorns (CNHox) were used as a sensing layer to detect and monitor ethanol gas [54,55]. In recent years, scientists have increasingly focused on designing room-temperature ethanol sensors due to their potential for real-time monitoring and low power consumption. This study aims to explore the room-temperature ethanol sensing properties of a novel nanohybrid, to discover new ways to reduce electric power consumption in next-generation, environmentally friendly sensors for Internet of Things (IoT) applications.

This paper presents the ethanol detection response of a resistive sensor based on a novel sensitive layer, a quaternary nanohybrid comprising CNHox, GO, $SnO_2$, and PVP at 1/1/1/1 *w*/*w*/*w*/*w* mass ratio. The proposed nanohybrid sensing layer for resistive detection of ethanol vapors includes four components whose chemical, physical, and electrical properties recommend them as key elements for the sensitive material. CNHox exhibits outstanding properties, such as high conductivity, high dispersibility, uniform size, excellent porosity, thermal and chemical stability, high adsorption capacity, superior permeability, exceptional catalytic properties, large specific surface area, low toxicity, and clean synthesis process (no metal catalyst is involved in their synthesis; thus, the produced CNHox are free of metal impurities) [55].

## 2. Materials and Methods

### 2.1. Materials

All the chemicals used in the sensing experiments were bought from Sigma Aldrich (Burlington, MA, USA). The chemicals used were of the highest available purity and were utilized without further modification. The metal-free CNHox powder (0% metallic compound) is characterized by a specific surface area of around 1300–1400 $m^2/g$ (according to the Brunauer–Emmett–Teller evaluation method) with lengths between 40 nm and 50 nm and diameters between 2 and 5 nm (Figure 1a).

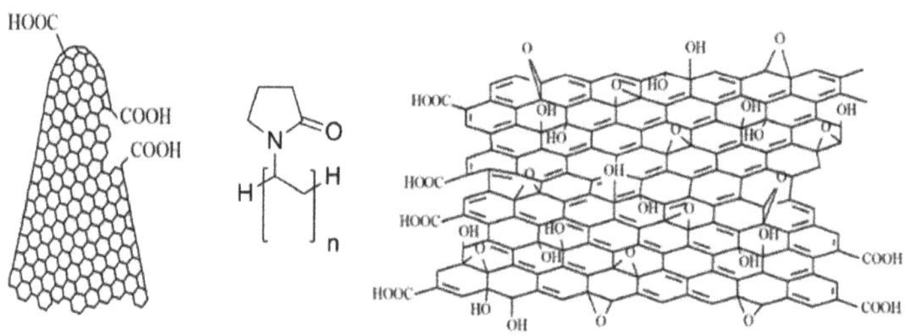


(a) (b) (c)

**Figure 1.** Structure of: (**a**) holey carbon nanohorns, (**b**) polyvinylpyrrolidone, and (**c**) graphene oxide.

According to the supplier, the amorphous graphite of CNHox cannot exceed 10%. PVP has an average molar weight of 29,000 Da (Figure 1b). CNHox powder is used as received. GO (4–10% edge-oxidized, average number of layers 15–20) is used as a dispersion in water, 2 mg/mL (Figure 1c). Tin (IV) oxide ($SnO_2$) purchased as a powder (99.9% purity) has an average nanoparticle size lower than 100 nm and a specific surface area of around 10–25 $m^2/g$ 2-Propanol used in the synthesis is anhydrous (99.5%).

### 2.2. Sensing Layer Characterization

Raman spectra were recorded at room temperature using a Witec Raman spectrometer (Alpha-SNOM 300 S, WiTec GmbH, Ulm, Germany) with 532 nm excitation. A 532 nm diode-pumped solid-state laser, delivering 145 mW of power, was focused onto the sample with a 6 mm working distance objective on a Thorlabs MY100X-806 (Newton, NJ, USA) microscope, producing a laser spot size of ~1.0 µm. The spectra were collected in back-scattering geometry with 600 grooves/mm grating, using an exposure time of 20 s per accumulation. Calibration was performed with the 520 $cm^{-1}$ Raman line of a silicon wafer; data acquisition and processing were conducted using WiTec Project Five software (WITec Project 5.1).

Scanning electron microscopy (SEM) was used to examine the surface topography of the sensing films. Surface visualization was performed using a field emission gun scanning electron microscope (FEG-SEM), Nova NanoSEM 630 (Thermo Scientific, Waltham, MA, USA), offering high-resolution imaging at low voltages and excellent surface sensitivity.

The samples were analyzed without any preparation, and a measurement current of 1 nA was applied.

*2.3. Preparation of the Sensing Layer*

The synthesis of the sensing layer based on nanohybrid CNHox/GO/PVP/$SnO_2$ = 1/1/1/1 (*w*/*w*/*w*/*w*) was conducted as described in the following section [56]. PVP solution was prepared by dissolving 6 mg of polymer in 10 mL of 2-propanol under stirring in the ultrasonic bath. Then, 6 mg of CNHox is added slowly to the PVP solution, under stirring. The suspension homogenization was achieved by employing a mild sonication bath at 42 kHz, with an output power of 70 W. Then, 3 mL of GO water dispersion was added to the prepared alcoholic PVP and CNHox suspension solution and stirred in the ultrasonic bath for 6 h at room temperature. In the last step of the dispersion preparation, 6 mg of $SnO_2$ was added and stirred for 3 h in an ultrasonic bath at room temperature. An annealing process was performed for solvent evaporation by heating the prepared dispersion for 12 h at 100 °C in a vacuum oven under low pressure (1 mbar). This procedure yields a uniform distribution of the CNHox, GO carbon nanoparticles, and the metal oxide semiconductor ($SnO_2$) nanoparticles in the hydrophilic PVP polymer network.

Microelectronic lift-off processes prepared the flexible polyimide substrate containing the metallic interdigitated structure; firstly, a photoresist layer was deposited, thermally treated, and patterned on the flexible substrate. Then, the Au/Cr layer was deposited on the entire substrate by electron gun evaporation, with chromium deposited first as an adherence layer for the Au layer. The Au/Cr layer adhesion test using tape confirmed that the Au/Cr layer adheres strongly to the polyimide substrate. Then, the photoresist and the metal deposited over it were removed by the well-known photoresist development process. Thus, the Au/Cr metal layer remained only in the desired layout regions, as shown in Figure 2, where the entire geometry of the IDT was present. The metal stripes of the Interdigitated Transducer (IDT) consisted of chromium with a thickness of 10 nm and gold with a thickness of 100 nm. The width and spacing of the digits were both 10 microns, and there is a separation of 0.6 mm between the digits and the bus bar. Finally, after masking the contact areas on the IDT structure from Figure 2, dispersing and sonicating the above final quaternary nanocomposite in isopropyl alcohol (IPA), the drop-casting method followed by 60 °C drying was used to generate the sensing film to be used in the ethanol detection experiments.

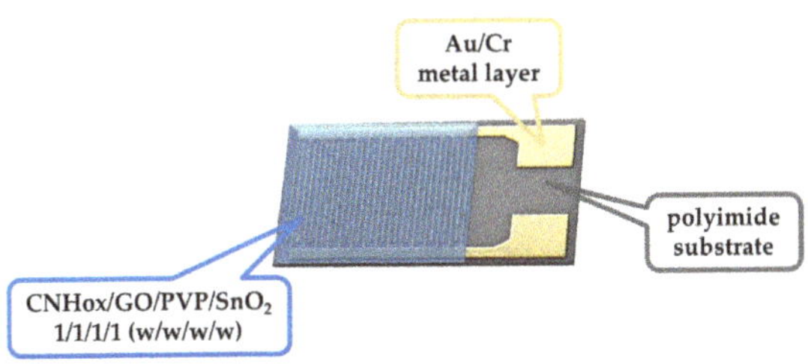


**Figure 2.** Layout of IDT sensing structure (chip area: 5 × 7 $mm^2$).

The ethanol chemiresistive sensor's performance was evaluated by exposing it to varying ethanol concentrations in a 0.27 L testing chamber. To minimize moisture interference, after sealing the testing box, it was purged with nitrogen until the relative humidity (RH) was below 2%, as indicated by the commercial sensor SHT31 Sensirion (±2% RH accuracy). Ethanol was introduced in a controlled manner: a micropipette (0.1–2 µL range) dispensed volumes of 0.5–2.5 µL for low concentrations. For higher concentrations, a micro-syringe delivered 2.4 mg ethanol drops [56]. A magnetic stirrer (500 rpm for 8 min) ensured complete ethanol evaporation and homogeneous vapor distribution before measurements began. Ethanol drops were repeatedly weighed to minimize pipetting and syringe dispensing errors, with the mean drop weight used for calculations. Minimal measurement errors in three repeated cycles confirmed a strong correlation between the calculated ethanol concentration and the sensor's resistance variations.

Before each test, nitrogen purging continued until the sensor's resistance returned to its baseline value ($R_i$ = 308 ohms), ensuring a zero relative variation in the measured resistance. The 8 min ethanol evaporation time was determined through preliminary tests to mitigate vaporization and diffusion errors, guaranteeing a stable resistance reading over time. Magnetic stirring was employed to prevent non-uniform ethanol distribution and adsorption onto chamber walls, as the latter contributed to sensor response variability at ethanol concentrations above 0.15 mg/cm$^3$. All sensing experiments were conducted at room temperature in a temperature-controlled white chamber to prevent fluctuations. The high CNHox content (25% *w*/*w*) in the quaternary nanocomposite, well above the percolation threshold, ensured low and easily measurable electrical resistance, allowing for ultralow-power operation.

## 3. Results

The Raman spectra of the composite material CNHox/GO/$SnO_2$/PVP reflect the individual contributions of these materials and provide insights into their interactions, structure, as well as modification in intensities due to the composite formation (Figure 3). PVP introduces new peaks, particularly in the C-H and C=O regions, as its Raman bands overlap with the bands of GO and CNHox.

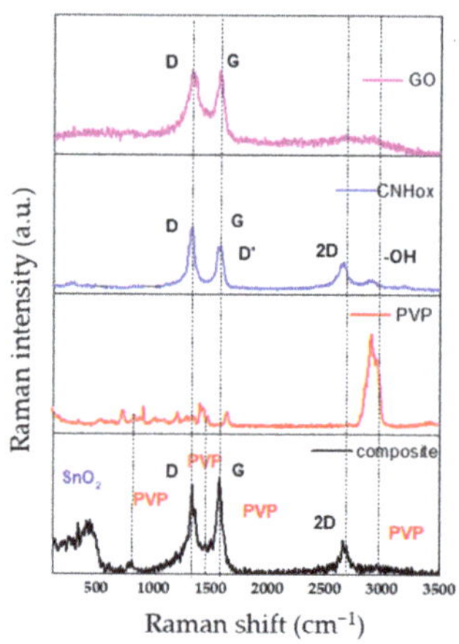


**Figure 3.** Raman spectra of the CNHox/GO/$SnO_2$/PVP sensing layer deposited on a silicon substrate.

The positions and analysis of the vibration modes of each material in the composite were summarized in Table 1 [57–61].

The discontinuous character of particles' distribution on the sensitive layer's surface can be observed in all recorded SEM (Figure 4). The nanoscale images show a particle size distribution in the range of 10–100 nm, similar to the size of synthesis precursors, proving an excellent dispersion process in the selected solvent. In addition, a very high porosity of the quaternary nanocomposite is revealed, supporting the high sensitivity of the sensing layer at room temperature.

**Table 1.** Analysis of the vibration modes of each material in the composite.

| Material | Vibration Mode | Position ($cm^{-1}$) | Analysis |
|---|---|---|---|
| CNHox [45,57] | D | ~1350 | Defects in the carbon network, oxygenated groups (epoxide, carboxyl) |
| | G | ~1580 | C-C vibrations in the carbon network ($sp^2$ bond) |
| | 2D | ~2700 | C-C stretching vibration in the graphite network |
| | D′ | ~1620 | Defect vibrations, interactions with oxygenated groups |
| | OH (hydroxyl) | ~3000–3200 | Signal associated with -OH groups on nanohorns |
| GO [58,59] | D | ~1350 | Defects and discontinuities due to oxygenated groups |
| | G | ~1580 | C-C vibration in the carbon network |
| | 2D | ~2700 | C-C stretching vibration |
| | D′ | ~1620 | Structural defects and interactions with oxygenated groups |
| | OH (hydroxyl) | ~3400 | -OH groups on graphene oxide |
| | COOH (carboxyl) | ~1700 | C=O vibration of carboxyl groups |
| | Epoxid (C-O-C) | ~1050 | C-O vibration of epoxide groups |
| $SnO_2$ | $A_1g$ | ~630 | Symmetric Sn-O stretching vibration in the crystalline network |
| | $B_2g$ | ~450 | Asymmetric Sn-O bond vibration |
| | $B_1g$ | ~750 | Vibration around the tin atom in the tetragonal network |
| | $E_1g$ | ~200 | In-plane vibration of the network |
| | Lattice vibration | ~120–180 | Related to collective movements of oxygen and tin atoms throughout the entire crystalline network |
| PVP [60,61] | C-H | ~1300 | Stretching and deformation of C-H bonds in methylene groups |
| | C-C | ~750–900~1350 | Bending of C-C bonds in the polymer chain |
| | C-N | ~1200–1250 | Interactions between carbon and nitrogen atoms in the amide group |
| | C=O | ~1650 | C=O stretching vibration in the amide group |
| | C-H | ~2800–3000 | Stretching vibrations of C-H bonds in methylene ($-CH_2$) and methyl ($-CH_3$) groups |

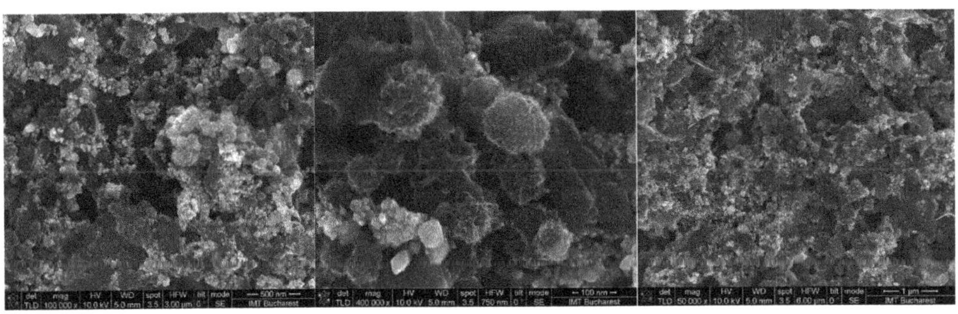

(a) (b) (c)

**Figure 4.** SEM images of the CNHox/GO/$SnO_2$/PVP = 1/1/1/1 at 1:1:1:1 *w*/*w*/*w*/*w* ratio: (**a**) magnification 100,000; (**b**) magnification 400,000; (**c**) magnification 50,000.

SEM-EDX analysis results are shown in Figure 5, which reveals the elemental composition of the sensitive film (symbolized in both atomic and weight percentages). The method confirmed the presence of the elements C, Sn, O, N, and Si in agreement with the chemical composition of the synthesized nanohybrid. However, these quantitative evaluations may undergo errors specific to the EDX analysis due to the unavoidable inhomogeneous distribution of CNHox, GO, and $SnO_2$ in the PVP network.

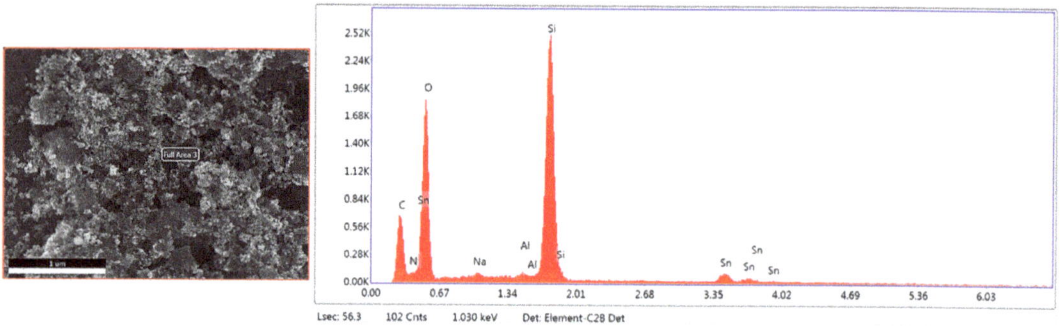


| Element | Weight % | Atomic % | Net Int. | Error % | Kratio | Z | R | A | F |
|---|---|---|---|---|---|---|---|---|---|
| C K | 17.25 | 29.90 | 64.32 | 11.90 | 0.0467 | 1.1525 | 0.9366 | 0.2348 | 1.0000 |
| N K | 0.00 | 0.01 | 0.02 | 99.99 | 0.0000 | 1.1198 | 0.9485 | 0.3015 | 1.0000 |
| O K | 27.17 | 35.35 | 223.00 | 8.05 | 0.1449 | 1.0914 | 0.9589 | 0.4886 | 1.0000 |
| NaK | 1.01 | 0.92 | 8.01 | 18.99 | 0.0070 | 0.9799 | 0.9849 | 0.6993 | 1.0065 |
| AlK | 0.93 | 0.72 | 9.34 | 14.67 | 0.0080 | 0.9539 | 0.9991 | 0.8836 | 1.0230 |
| SiK | 41.87 | 31.04 | 398.98 | 3.60 | 0.3798 | 0.9722 | 1.0053 | 0.9291 | 1.0040 |
| SnL | 11.76 | 2.06 | 17.47 | 18.00 | 0.0790 | 0.6409 | 1.1703 | 1.0253 | 1.0219 |

**Figure 5.** Surface composition of a quaternary nanohybrid comprising CNHox, GO, $SnO_2$, and PVP (1:1:1:1 $w/w/w/w$) obtained by EDX spectroscopy coupled with SEM.

The sensing results are presented in Figures 6 and 7, showing that the sensor's response is reproducible for the three measurement cycles in the 0–0.16 mg/cm$^3$ concentration range.

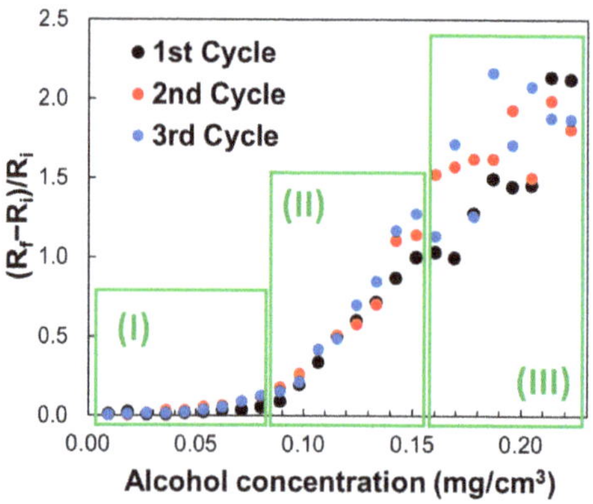


**Figure 6.** Graphical representation of the tested sensing device on three different cycles.

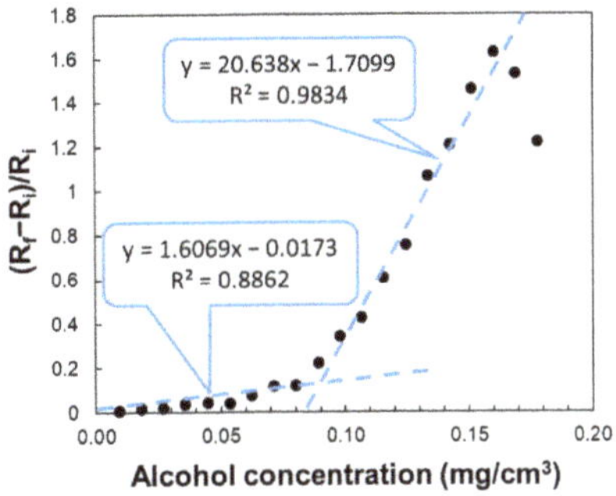


**Figure 7.** The measured response of the chemiresistive sensor to ethanol concentrations below 0.16 mg/cm$^3$.

The transfer function of the sensor was presented by calculating the electrical resistance change relative to the initial resistance ($(R_f - R_i)/R_i$) as a function of ethanol concentration in the testing chamber (Figures 6 and 7).

This makes the results dimensionless and more manageable to compare across different experiments, regardless of the absolute resistance values. Another reason for using relative change in the resistance is directly related to the sensitivity of the chemiresistive sensor. Using a relative change emphasizes how significant the response was compared to the sensor's baseline performance. Uncontrollable factors, such as variations in fabrication, temperature, or environmental conditions, can influence absolute resistance values. The relative change minimizes the effect of variations induced during the fabrication process, with a focus on the dynamic response caused by the stimulus. At the same time, proper monitoring and control of the environment (temperature and relative humidity) was implemented throughout the experimental work.

In the representation of the relative variation in resistance as a function of ethanol concentration, three response zones of the sensor are highlighted (Figure 6). In the 0–0.088 mg/cm$^3$ concentration range, the relative resistance variation in the sensor shows a quasi-linear increase, the slope of the trend line being lower than that associated with the linear response for the 0.088–0.16 mg/cm$^3$ ethanol range (Figure 7).

Figure 6 presents the variation of $(R_f - R_i)/R_i$ with ethanol vapors concentration, where $R_i$ is the resistance of the sensing film before exposure to ethanol, and $R_f$ is the resistance after ethanol exposure. The chemiresistive sensing structure exhibits a good response to alcohol vapor concentrations varying from 0.008 to 0.16 mg/cm$^3$, with resistance increasing over almost the entire range of ethanol concentrations, showing a maximum for ethanol concentration of 0.16 mg/cm$^3$. From the transfer function as shown in Figure 7, it is evident that the sensor response is not linear, but it can be linearized for ethanol concentrations. Thus, one can observe that for low ethanol concentration (0.008 to 0.08 mg/cm$^3$), the linearity is good ($R^2$ = 0.88), with a slope of 1.606 (Figure 7), while for larger ethanol concentration (0.088 to 0.160 mg/cm$^3$), the linearity is very good ($R^2$ = 0.98), with a slope of 20.638 (Figure 7). The analysis of the experimental results also showed that the sensitivity for the range of high ethanol concentration (0.088 to 0.160 mg/cm$^3$) was approximately 12 times higher than for low ethanol concentration (0.008 to 0.08 mg/cm$^3$). The resistance of the tested sensor was measured after its readings were stabilized. In the concentration range of 0 to 0.089 mg/cm$^3$, the stabilization time ranged from 80 to 120 s. However, when the ethanol concentration in the test chamber exceeded 0.089 g/cm$^3$, the sensor's equilibration time increased by 10 to 20 s. The performance of already tested ethanol sensors varies significantly. Some studies report response times ranging from a few seconds to tens of seconds, while others indicate that response times can decrease from tens of seconds to just a few seconds. Similarly, recovery times can drop from several hours to just a few minutes, depending on the sensor's configuration, detection method, sensing layer, and operating temperature [62].

At concentrations higher than 0.16 $mg/cm^3$ (inset III in Figure 6), the sensor's response shows an increase in the relative variation in resistance value compared to the ethanol concentration. Still, the calculated values no longer follow a linear trend, and the response is not reproducible. Saturation effects are likely the primary reason for this result. Additionally, capillary condensation may influence the performance of the manufactured sensor at elevated ethanol concentrations.

## 4. Discussion

Considering the possible interactions between ethanol and the materials within the quaternary nanohybrid-based sensing layer employed, three distinct ethanol detection mechanisms can be identified and analyzed:

### *4.1. Ethyl Alcohol Acts as an Electron Donor for CNHox and GO*

The first mechanism explaining the ethanol detection considers that CNHox and GO exhibit typical electrical behavior for p-type semiconductor materials (the movement of holes inside them causes the main current flow in such nanocarbon materials). At an interaction with CNHox and GO, ethanol molecules donate their electron pairs to these p-type semiconductor materials, which recombine with the holes from their valence band, and thus, the number of holes (majority carriers) is decreasing in both carbon structures. Accordingly, the sensitive layer becomes less conductive, which agrees with the experimental result from Figure 6 in the ethanol concentration range from 0 to 0.16 $mg/cm^3$. Moreover, the interaction of ethanol molecules with CNHox and GO can be analyzed from the perspective of the hard–soft acid-base (HSAB) theory. Examples of hard, soft, and borderline acids and bases are given in Table 2. According to this theory, the chemical species react preferentially with similar hardness or softness species. As a result, hard bases tend to interact with hard acids, soft bases favor interactions with soft acids, and borderline acids generally bond with borderline bases. Like other chemical species containing oxygen atoms with lone electron pairs, ethanol molecules are classified as hard bases.

**Table 2.** Examples of hard, soft, and borderline acids and bases according to the HSAB theory.

| | Hard | Borderline | Soft |
|---|---|---|---|
| Bases | $C_2H_5OH$, $RO^-$, $HO^-$, $(CH_3)_2O$, $H_2O$, $N_2H_4$, $R\text{-}NH_2$, $CO_3{}^{2-}$, $F^-$, $Cl^-$ | $C_6H_5NH_2$, $C_5H_5N$(pyridine), $N_2$, $Br^-$, $N_3{}^-$, $Cl^-$ | RSH, $R_2S$, $H_2S$, $C_2H_4$, CO, $CN^-$, RCN, $H_2{}^-$, $R_3P$, $C_6H_6$, $RS^-$, $I^-$ |
| Acids | $BCl_3$, $SO_3$, $H_3O^+$, $Li^+$, $Mg^{2+}$, $BF_3$, $Al^{3+}$, $Co^{3+}$, $Sn^{4+}$, $Ti^{4+}$, $La^{3+}$, $CR_3{}^+$, $In^{3+}$, $Zr^{4+}$, $CO_2$ | $Bi^{3+}$, $Ni^{2+}$, $Zn^{2+}$, $Fe^{2+}$, $Pb^{2+}$, $Cu^{2+}$, $Pb^{2+}$, $BMe_3$ | $Cd^{2+}$, $Cu^+$, $Ag^+$, carbenes, $I_2$, $Hg^{2+}$, $NO_2$, bulk metals |

Table 2 shows carbocations (positively charged carbon ions) categorized as hard acids. Consequently, the holes within the structures of CNHox and GO can be regarded as hard acids. These holes neutralize each other through multiple recombination processes [63,64]. Finally, according to HSAB theory, $Sn^{4+}$ ions are classified as hard acids, making a "hard acid–hard base" interaction between ethanol and $SnO_2$ highly probable.

Thus, HSAB theory supports the feasibility of interactions between all components of the sensitive layer and ethanol, as well as the recombination of electrons and holes in the p-type semiconductor materials, CNHox and GO, aligning with the reasoning above based on the charge carrier recombination principle in semiconductors.

### 4.2. Electron Trapping and Generation from the $SnO_2$-Oxygen-Ethyl Alcohol Interaction

$SnO_2$ is a semiconducting metal oxide with n-type conductivity, meaning the electron concentration is much higher than the hole concentration for this material. Therefore, the electrical conductivity is mainly performed by electrons. It is well known that, at the $SnO_2$ grain boundaries between different $SnO_2$ nanoparticles, there are depletion regions due to interface defects trapping the electrons, and these depletion regions act as energy barriers in the electrical conduction process.

During the detection process, the following chemical processes occur as follows: (i) firstly, gaseous residual oxygen molecules are adsorbed on the surface of the exposed $SnO_2$ nanoparticles; (ii) secondly, the electrons from the conduction band of the $SnO_2$ nanoparticles are attracted by these adsorbed oxygen molecules; and then (iii) converted to oxygen anions ($O_2^-$) according to the set of reactions shown below:

$$O_2\ (gas) \rightarrow O_2\ (ads) \quad (1)$$

$$O_2\ (ads) + e^- \rightarrow O_2^-\ (ads) \quad (2)$$

$$O_2^-\ (ads) + e^- \rightarrow 2O^- \quad (3)$$

$$O^- + e^- \rightarrow O^{2-} \quad (4)$$

Therefore, according to Equations (2)–(4), the concentration of free electrons in $SnO_2$ will decrease, expanding the depletion regions. Subsequently, (ii) the adsorbed ethanol as a reducing gas will remove the oxygen anions from the $SnO_2$ surface by generating $CO_2$ and $H_2O$ (Equations (5) and (6) from below), and finally, the electrons from these oxidation reactions return to the body of the $SnO_2$ nanoparticles, and the Fermi level of $SnO_2$ also returns to a near-initial state. Simultaneously, the potential energy barrier is lowered, and the depletion layer becomes thinner. As a result, electrons can move more easily between nanoparticles, decreasing the sensing structure's resistance [65].

$$CH_3CH_2OH\ (ads) + 6O^-(ads) \rightarrow 2CO_2\ (g) + 3H_2O\ (g) + 6e^- \quad (5)$$

$$CH_3CH_2OH\ (ads) + 6O_2^-\ (ads) \rightarrow 2CO_2\ (g) + 3H_2O\ (g) + 12e^- \quad (6)$$

Based on this plausible explanation, the sensitive layer is expected to become more conductive due to the interaction between $SnO_2$ nanoparticles and ethanol molecules. However, according to Figure 6, the experimental chemiresistive sensor response to the ethanol concentration increase shows a continuous rise in the electrical resistance with the concentration of ethanol in the range of 0–0.16 $mg/cm^3$, which indicates that this mechanism is not prevailing, at least in this concentration range.

### 4.3. Swelling of PVP

According to Zereshki et al. [66], ethanol molecules have a strong interaction with PVP chains, and the dielectric polymer swells; the higher the ethanol concentration is, the higher the swelling is. The generation of the local water molecules, as described by Equations (5) and (6) above, may also contribute to the PVP swelling [67]. However, when the swelling process is developing, the distances between the conductive CNHox and low energy bandgap GO nanoparticles increase, and the number of electrical percolating pathways [68] decreases, as shown in Figure 8, and the overall electrical resistance increases.

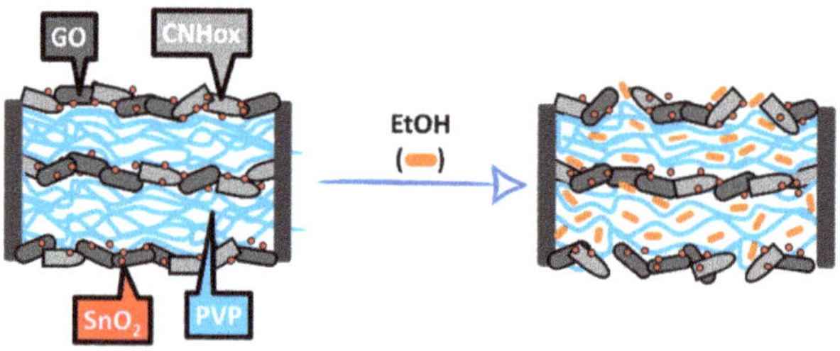


**Figure 8.** The swelling of PVP upon contact with ethanol molecules disrupts the percolating pathways of the CNHox and GO.

According to this perspective, the sensitive layer is expected to become more resistive as the ethanol concentration steadily increases. The high sensitivity value in the ethanol concentration range of 0.08–0.16 mg/cm$^3$ can be best explained by the increased hoping distance between these conductive segments, which may provide a much-increased value of the slope of the resistance dependence in Figures 6 and 7. Therefore, the swelling phenomenon explains the experimental results in the ethanol concentration range of 0.08–0.16 mg/cm$^3$. As we can see, when the ethanol concentration is higher than 0.16 mg/cm$^3$, some regions where the sensitive layer becomes more conductive are identified.

Two alternative hypotheses can be taken into account. First, electron trapping and generation from the $SnO_2$-oxygen-ethyl alcohol interaction (mechanism II, discussed above) may become prevalent. Secondly, the ionization of some water generated from reactions (5) and (6) may decrease the resistance of the sensing layer. Last but not least, capillary condensation can affect the behavior of the manufactured sensor at higher ethanol concentrations.

Besides these described mechanisms, the mutual interactions between nanohybrid constituents can be considered. Thus, $\pi$-$\pi$ stacking interaction between GO and the nanohorns, as well as the hydrogen bonds between both nanocarbon materials and PVP, yield a possible supramolecular organization as presented in Figure 9.

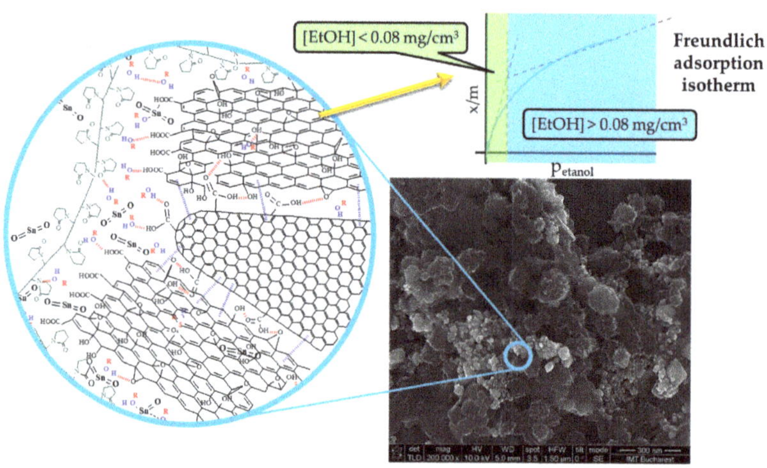


**Figure 9.** Possible architecture for supramolecular structure generated by CNHox, GO, $SnO_2$, and PVP.

It is also reasonable to assume the formation of islands of p-n semiconductor heterojunctions between graphene oxide, holey carbon nanohorns, and $SnO_2$, which diminishes mutual interaction between organic constituents of nanohybrid. As a consequence, the specific surface area will be increased. Last but not least, mutual interaction between metal oxide semiconducting, yields changes in the pore distribution, which increases the specific surface area of nanohybrid exposed to ethanol vapors, affecting the number of active sensing sites. Considering that the resistance of the sensing layer increases across nearly the entire range of measured ethanol concentrations, it can be inferred that the p-type semiconductor behavior of both nanocarbon materials, along with the swelling of PVP, are the primary mechanisms responsible for the overall resistance of the sensitive layer when exposed to ethanol.

Each component of the quaternary nanohybrid used for resistive ethanol monitoring played a specific role. Unlike conventional ethanol sensors, which primarily rely on metal oxides and rare elements [69–72], CNHox and GO are derived from carbon-based materials recognized for their excellent electrical properties [73]. CNHox demonstrated exceptional properties, including increased conductivity (as a p-type semiconductor), high uniformity, a large surface area, easy synthesis (without metallic compounds), and sensitivity to alcohol molecules. These characteristics make CNHox a strong candidate for monitoring ethanol vapors at room temperature. GO offers several advantages, such as being a good charge carrier, enabling scalable fabrication, and serving as an effective dispersant for functionalized carbon nanohorns. Through intermolecular hydrogen bonding and $\pi$–$\pi$ stacking interactions, GO can act as a dispersant for CNHox, helping to redisperse bundles of oxidized carbon nanohorns. Additionally, GO is a p-type material that shows reduced electrical conduction when exposed to ethanol. Polyvinylpyrrolidone (PVP) is an electrically insulating polymer with excellent binding properties and strong interactions with ethanol molecules.

A potential quantitative model to describe ethanol adsorption onto the sensing layer is the Freundlich adsorption isotherm. This model defines the relationship between the amount of gas adsorbed on a solid surface and the gas pressure. The Freundlich isotherm describes multilayer adsorption on a heterogeneous surface with varying binding energies, leading to the formation of multiple layers. Since adsorption sites have different affinities, the adsorption energy decreases as surface coverage increases. A key feature of the Freundlich model is the absence of a maximum adsorption capacity, meaning the adsorbent can continue adsorbing indefinitely, though at a diminishing rate [74].

Figure 9 illustrates the adsorption model. At ethanol concentrations below 0.08 mg/cm$^3$, the ratio of $[EtOH]_{adsorbed}/m_{adsorbent}$ (x/m) increases rapidly due to the abundance of free adsorption sites. This results in a slow change in resistance, as only a small number of ethanol molecules are adsorbed at the interface, as shown in Figures 6 and 7. When the ethanol concentration exceeds 0.08 mg/cm$^3$, an inflection point appears in the graph showing the sensor's measured response. This inflection point aligns with the one observed in the Freundlich isotherm. Beyond 0.08 mg/cm$^3$, adsorption slows at the interface because the sensing layer surface becomes saturated with ethanol molecules. At this stage, the swelling of PVP accelerates, leading to a rapid increase in resistance as presented in Figures 6 and 7.

This behavior demonstrates a clear correlation between the ethanol concentration and the Freundlich adsorption isotherm, where the rapid adsorption at lower concentrations and the subsequent saturation at higher concentrations align with the principles of the Freundlich model, explaining the dynamics of the sensor response dynamics observed in Figures 6 and 7.

## 5. Conclusions

This study introduces a quaternary nanohybrid sensing layer (CNHox/GO/$SnO_2$/PVP in a 1/1/1/1 mass ratio) designed explicitly for resistive ethanol vapor detection. Unlike conventional ethanol sensors, which primarily rely on metal oxides and rare elements, this innovative approach combines the synergistic properties of its components to enhance performance. CNHox provides high conductivity and porosity, improving electron transport and gas diffusion, while GO increases surface area and introduces functional groups that enhance ethanol interaction. $SnO_2$ further strengthens ethanol adsorption and sensing response, and PVP ensures structural integrity and dispersion stability. Beyond performance, this sensor offers a cost-effective and environmentally friendly alternative to traditional designs. GO is a more affordable material than CNHox, helping to reduce production costs without compromising sensor quality, while $SnO_2$ remains a low-cost yet effective complement to both GO and CNHox. By integrating these materials, the proposed sensor benefits from the high sensitivity of carbon-based nanomaterials while maintaining affordability and sustainability, making it a practical and scalable solution for ethanol-sensing applications. Last but not least, low power consumption (below 2 mW) makes these sensors promising alternatives in wireless sensor networks for Internet of Things applications, where energy constraint is one of the biggest challenges.

The sensing structure includes a flexible polyimide substrate and IDT-like electrodes. The sensing film used is a quaternary nanohybrid comprising two types of nanocarbon materials, CNHox and GO, a metal oxide semiconducting ($SnO_2$) and a hydrophilic polymer (PVP), in a 1/1/1/1 mass ratio. The experimental setup was manufactured to investigate the response of the sensing layer deposited onto the substrate to various concentrations of ethanol vapors in a dry atmosphere. The developed resistive sensing structure showed good sensitivity to ethanol vapors over a broad range of concentrations (0.008–0.16 mg/cm$^3$). In addition, the linearity of the sensor's response was investigated at different ethanol concentration intervals, and the best sensing results were found in the range of ethanol concentrations between 0.088 and 0.160 mg/cm$^3$. The analysis of the experimental sensing results emphasized that the sensitivity (i.e., the relative variation in the sensor resistance) for the range of high vapors ethanol concentration (0.088–0.16) mg/cm$^3$ was approximately 12 times higher than for low analyte concentration (0.008 to 0.088 mg/cm$^3$). Different mechanisms for explaining the ethanol vapors detection behavior of the developed sensing structure were discussed and assessed. Based on the obtained experimental data, it was argued that the p-type semiconductor behavior of CNOs and GO in conjunction with swelling of PVP are the dominant sensing mechanisms and yield an overall increase in the sensitive layer electrical resistance with ethanol vapor concentration. The HSAB principle is an additional solid argument that supports our interpretations regarding the sensing mechanism

**Author Contributions:** Conceptualization, B.-C.S., O.B. (Octavian Buiu) and M.B. (Marius Bumbac); methodology, B.-C.S., N.D., O.B. (Octavian Buiu) and C.C.; validation, B.-C.S., O.B. (Octavian Buiu) and M.B. (Marius Bumbac); formal analysis, B.-C.S. and M.B. (Marius Bumbac); investigation, N.D., C.P., O.B. (Oana Brancoveanu), C.-M.N. and M.B. (Mihai Brezeanu); resources, B.-C.S. and O.B. (Octavian Buiu); data curation, B.-C.S. and M.B. (Marius Bumbac); writing—original draft preparation all authors; writing—review and editing, B.-C.S. and M.B. (Marius Bumbac); visualization, B.-C.S., O.B. (Octavian Buiu) and M.B. (Marius Bumbac); supervision, B.-C.S. and M.B. (Marius Bumbac); project administration, O.B. (Octavian Buiu); funding acquisition, O.B. (Octavian Buiu). All authors have read and agreed to the published version of the manuscript.

**Funding:** IMT authors would like to acknowledge the financial support provided by UEFISCDI contract no 364PED—23 October 2020 and the Romanian Ministry of Education and Research via the "MICRO-NANO-SIS PLUS" Nucleu program, grant number PN 19 16.

**Institutional Review Board Statement:** Not applicable.

**Informed Consent Statement:** Not applicable.

**Data Availability Statement:** The data that support the findings of this study are available from the corresponding author, M.B., upon reasonable request.

**Conflicts of Interest:** The authors declare no conflicts of interest.

## References

1. Wilkinson, P.K. Pharmacokinetics of ethanol: A review. *Alcohol. Clin. Exp. Res.* **1980**, *4*, 6–21. [CrossRef] [PubMed]
2. Baras, J.K.; Gaćeša, S.B.; Pejin, D.J. Ethanol is a strategic raw material. *Hem. Ind.* **2002**, *56*, 89–104. [CrossRef]
3. Lachenmeier, D.W. Safety evaluation of topical applications of ethanol on the skin and inside the oral cavity. *J. Occup. Med. Toxicol.* **2008**, *3*, 26. [CrossRef] [PubMed]
4. Le Dare, B.; Gicquel, T. Therapeutic applications of ethanol: A review. *J. Pharm. Pharm. Sci.* **2019**, *22*, 525–535. [CrossRef]
5. Ferreira, L.M.; Binz, T.; Yegles, M. The influence of ethanol containing cosmetics on ethyl glucuronide concentration in hair. *Forensic Sci. Int.* **2012**, *218*, 123–125. [CrossRef]
6. Viikari, L.; Vehmaanperä, J.; Koivula, A. Lignocellulosic ethanol: From science to industry. *Biomass Bioenergy* **2012**, *46*, 13–24. [CrossRef]
7. Hill, J.; Nelson, E.; Tilman, D.; Polasky, S.; Tiffany, D. Environmental, economic, and energetic costs and benefits of biodiesel and ethanol biofuels. *Proc. Natl. Acad. Sci. USA* **2006**, *103*, 11206–11210. [CrossRef]
8. Takei, T.; Iguchi, N.; Haruta, M. Synthesis of acetaldehyde, acetic acid, and others by the dehydrogenation and oxidation of ethanol. *Catal. Surv. Asia* **2011**, *15*, 80–88. [CrossRef]
9. Alzeer, J.; Abou Hadeed, K. Ethanol and its Halal status in food industries. *Trends Food Sci. Technol.* **2016**, *58*, 14–20. [CrossRef]
10. Tavares, J.R.; Sthel, M.S.; Campos, L.S.; Rocha, M.V.; Lima, G.R.; da Silva, M.G.; Vargas, H. Evaluation of pollutant gases emitted by ethanol and gasoline-powered vehicles. *Procedia Environ. Sci.* **2011**, *4*, 51–60. [CrossRef]
11. Reimann, H.; Zimmermann, B.; Eckert, E.; Lassek, E. Risk assessment of low-dose ethanol in food. *Food Chem. Toxicol.* **2023**, *173*, 113633. [CrossRef] [PubMed]
12. Celaya-Padilla, J.M.; Romero-González, J.S.; Galvan-Tejada, C.E.; Galvan-Tejada, J.I.; Luna-García, H.; Arceo-Olague, J.G.; Gamboa-Rosales, H. In-vehicle alcohol detection using low-cost sensors and genetic algorithms to aid in the drinking and driving detection. *Sensors* **2021**, *21*, 7752. [CrossRef] [PubMed]
13. Wakana, H.; Yamada, M. Portable alcohol detection system for driver monitoring. In *2019 IEEE Sensors*; IEEE: Piscataway, NJ, USA, 2019; pp. 1–4.
14. Li, S.; Simonian, A.; Chin, B.A. Sensors for agriculture and the food industry. *Electrochem. Soc. Interface* **2010**, *19*, 41. [CrossRef]
15. Mojumder, S.; Das, T.; Das, S.; Chakraborty, N.; Saha, D.; Pal, M. Y and Al co-doped ZnO-nanopowder-based ultrasensitive trace ethanol sensor: A potential breath analyzer for fatty liver disease and drunken driving detection. *Sens. Actuators B Chem.* **2022**, *372*, 132611. [CrossRef]
16. Angeloni, R.; Tomassetti, M.; Castrucci, M.; Campanella, L. Ethanol determination in alcoholic beverages using two different amperometric enzyme sensors. *Curr. Anal. Chem.* **2015**, *11*, 56–67. [CrossRef]
17. Dharmalingam, G.; Sivasubramaniam, R.; Parthiban, S. Quantification of ethanol by metal-oxide-based resistive sensors: A review. *J. Electron. Mater.* **2020**, *49*, 3009–3024. [CrossRef]
18. Su, C.; Zhang, L.; Han, Y.; Ren, C.; Chen, X.; Hu, J.; Zeng, M.; Hu, N.; Su, Y.; Zhou, Z.; et al. Controllable synthesis of crescent-shaped porous NiO nanoplates for conductometric ethanol gas sensors. *Sens. Actuators B Chem.* **2019**, *296*, 126642. [CrossRef]
19. Wang, Q.; Bai, J.; Hu, Q.; Hao, J.; Cheng, X.; Li, J.; Xie, E.; Wang, Y.; Pan, X. W-doped NiO as a material for selective resistive ethanol sensors. *Sens. Actuators B Chem.* **2020**, *308*, 127668. [CrossRef]
20. Gai, L.Y.; Lai, R.P.; Dong, X.H.; Wu, X.; Luan, Q.T.; Wang, J.; Lin, H.F.; Ding, W.H.; Wu, G.L.; Xie, W.F. Recent advances in ethanol gas sensors based on metal oxide semiconductor heterojunctions. *Rare Met.* **2022**, *41*, 1818–1842. [CrossRef]
21. Wongrat, E.; Ponhan, W.; Choopun, S. Room temperature ethanol sensing properties of FET sensors based on ZnO nanostructures. *Ceram. Int.* **2017**, *43*, S520–S524. [CrossRef]
22. Sänze, S.; Hess, C. Ethanol gas sensing by indium oxide: An operando spectroscopic Raman-FTIR study. *J. Phys. Chem. C* **2014**, *118*, 25603–25613. [CrossRef]
23. Elger, A.K.; Hess, C. Elucidating the mechanism of working SnO2 gas sensors using combined operando UV/Vis, Raman, and IR spectroscopy. *Angew. Chem. Int. Ed.* **2019**, *58*, 15057–15061. [CrossRef] [PubMed]
24. Giffney, T.J.; Ng, Y.H.; Aw, K.C. A surface acoustic wave ethanol sensor with zinc oxide nanorods. *Smart Mater. Res.* **2011**, *2012*, 210748. [CrossRef]
25. Sharifpour-Boushehri, S.; Hosseini-Golgoo, S.M.; Sheikhi, M.H. A low-cost and reliable fiber optic ethanol sensor based on nano-sized $SnO_2$. *Opt. Fiber Technol.* **2015**, *24*, 93–99. [CrossRef]

26. Pargoleti, E.; Tricoli, A.; Pifferi, V.; Orsini, S.; Longhi, M.; Guglielmi, V.; Cerrato, G.; Falciola, L.; Cappelletti, G. An electrochemical outlook upon the gaseous ethanol sensing by graphene oxide-$SnO_2$ hybrid materials. *Appl. Surf. Sci.* **2019**, *483*, 1081–1089. [CrossRef]
27. Li, X.J.; Chen, S.J.; Feng, C.Y. Characterization of silicon nanoporous pillar array as room-temperature capacitive ethanol gas sensor. *Sens. Actuators B Chem.* **2007**, *123*, 461–465. [CrossRef]
28. Hosseini, M.S.; Zeinali, S.; Sheikhi, M.H. Fabrication of capacitive sensor based on Cu-BTC (MOF-199) nanoporous film for detection of ethanol and methanol vapors. *Sens. Actuators B Chem.* **2016**, *230*, 9–16. [CrossRef]
29. Fu, Q.; Lu, K.; Li, N.; Dong, Z. Advances in the development of MOS-based sensors for detection of ethanol: A review. *Mater. Res. Bull.* **2023**, *168*, 112457. [CrossRef]
30. Kim, H.J.; Kim, Y.Y.; Lee, K.W. Multiparametric sensor based on DBR porous silicon for detection of ethanol gas. *Curr. Appl. Phys.* **2010**, *10*, 181–183. [CrossRef]
31. Hwang, B.J.; Yang, Y.J.; Lin, C.W. A microscopic gas-sensing model for ethanol sensors based on conductive polymer composites from polypyrrole and poly (ethylene oxide). *J. Electrochem. Soc.* **1999**, *146*, 1231. [CrossRef]
32. Roshan, H.; Kuchi, P.S.; Sheikhi, M.H.; Mirzaei, A. Enhancement of room temperature ethanol sensing behavior of PbS–SnS2 nanocomposite by Au decoration. *Mater. Sci. Semicond. Process.* **2010**, *127*, 105742. [CrossRef]
33. Aria, M.M.; Irajizad, A.; Astaraei, F.R.; Shariatpanahi, S.P.; Sarvari, R. Ethanol sensing properties of PVP electrospun membranes studied by quartz crystal microbalance. *Measurement* **2016**, *78*, 283–288. [CrossRef]
34. Timm, R.A.; Falla, M.P.; Huila, M.F.; Peres, H.E.; Ramirez-Fernandez, F.J.; Araki, K.; Toma, H.E. Vanadium oxide-porphyrin nanocomposites as gas sensor interfaces for probing low water content in ethanol. *Sens. Actuators B Chem.* **2010**, *146*, 61–68. [CrossRef]
35. Milovanov, Y.S.; Skryshevsky, V.A.; Gavrilchenko, I.V.; Kostiukevych, O.M.; Gryn, S.V.; Alekseev, S.A. Ethanol gas sensing performance of electrochemically anodized freestanding porous SiC. *Diam. Relat. Mater.* **2019**, *91*, 84–89. [CrossRef]
36. Young, S.J.; Lin, Z.D. Ethanol gas sensors based on multi-wall carbon nanotubes on oxidized Si substrate. *Microsyst. Technol.* **2018**, *24*, 55–58. [CrossRef]
37. Zhu, X.; Zhang, J.; Xie, Q.; Hou, Z.L. High-sensitivity and ultrafast-response ethanol sensors based on graphene oxide. *ACS Appl. Mater. Interfaces* **2020**, *12*, 38708–38713. [CrossRef]
38. Tiwary, P.; Chatterjee, S.G.; Singha, S.S.; Mahapatra, R.; Chakraborty, A.K. Room temperature ethanol sensing by chemically reduced graphene oxide film. *FlatChem* **2021**, *30*, 100317. [CrossRef]
39. Shooshtari, M.; Sacco, L.N.; van Ginkel, J.; Vollebregt, S.; Salehi, A. Enhancement of room temperature Ethanol sensing by optimizing the density of Vertically Aligned Carbon Nanofibers decorated with Gold nanoparticles. *Materials* **2022**, *15*, 1383. [CrossRef]
40. Tian, F.; Ma, G.; Zhao, X.; Gao, J.; Zhang, J.; Suo, H.; Zhao, C. The Influence of Carboxyl-Functionalized Carbon Dots on Ethanol Selectivity in Gas Sensing. *Chemosensors* **2023**, *11*, 370. [CrossRef]
41. Pienutsa, N.; Roongruangsree, P.; Seedokbuab, V.; Yannawibut, K.; Phatoomvijitwong, C.; Srinives, S. $SnO_2$-graphene composite gas sensor for a room temperature detection of ethanol. *Nanotechnology* **2020**, *32*, 115502. [CrossRef]
42. Iijima, S.; Yudasaka, M.; Yamada, R.; Bandow, S.; Suenaga, K.; Kokai, F.; Takahashi, K. Nano-aggregates of single-walled graphitic carbon nano-horns. *Chem. Phys. Lett.* **1999**, *309*, 165–170. [CrossRef]
43. Selvam, K.P.; Nakagawa, T.; Marui, T.; Inoue, H.; Nishikawa, T.; Hayashi, Y. Synthesis of solvent-free conductive and flexible cellulose–carbon nanohorn sheets and their application as a water vapor sensor. *Mater. Res. Express* **2020**, *7*, 056402. [CrossRef]
44. Serban, B.C.; Buiu, O.; Dumbravescu, N.; Cobianu, C.; Avramescu, V.; Brezeanu, M.; Bumbac, M.; Nicolescu, C.M. Oxidized Carbon Nanohorns as Novel Sensing Layer for Resistive Humidity Sensor. *Acta Chim. Slov.* **2020**, *67*, 469–475. [CrossRef] [PubMed]
45. Serban, B.C.; Buiu, O.; Dumbravescu, N.; Cobianu, C.; Avramescu, V.; Brezeanu, M.; Bumbac, M.; Pachiu, C.; Nicolescu, C.M. Oxidized carbon nanohorn-hydrophilic polymer nanocomposite as the resistive sensing layer for relative humidity. *Anal. Lett.* **2021**, *54*, 527–540. [CrossRef]
46. Serban, B.C.; Cobianu, C.; Buiu, O.; Bumbac, M.; Dumbravescu, N.; Avramescu, V.; Nicolescu, C.M.; Brezeanu, M.; Pachiu, C.; Craciun, G.; et al. Ternary nanocomposites based on oxidized carbon nanohorns as sensing layers for room temperature resistive humidity sensing. *Materials* **2021**, *14*, 2705. [CrossRef]
47. Serban, B.C.; Cobianu, C.; Buiu, O.; Bumbac, M.; Dumbravescu, N.; Avramescu, V.; Nicolescu, C.M.; Brezeanu, M.; Radulescu, C.; Craciun, G.; et al. Quaternary Holey Carbon Nanohorns/$SnO_2$/ZnO/PVP Nano-Hybrid as Sensing Element for Resistive-Type Humidity Sensor. *Coatings* **2021**, *11*, 1307. [CrossRef]
48. Serban, B.C.; Cobianu, C.; Buiu, O.; Bumbac, M.; Dumbravescu, N.; Avramescu, V.; Radulescu, C.; Craciun, G.; Nicolescu, C.M.; Romanitan, C.; et al. Ternary Holey Carbon Nanohorns/TiO2/PVP Nanohybrids as Sensing Films for Resistive Humidity Sensors. *Coatings* **2021**, *11*, 1065. [CrossRef]

49. Serban, B.C.; Cobianu, C.; Buiu, O.; Bumbac, M.; Dumbravescu, N.; Avramescu, V.; Radulescu, C.; Craciun, G.; Nicolescu, C.M.; Romanitan, C.; et al. Quaternary Oxidized Carbon Nanohorns Based Nanohybrid as Sensing Coating for Room Temperature Resistive Humidity Monitoring. *Coatings* **2021**, *11*, 530. [CrossRef]
50. Serban, B.C.; Buiu, O.; Dumbravescu, N.; Cobianu, C.; Avramescu, V.; Brezeanu, M.; Bumbac, M.; Nicolescu, C.M. Optimization of Sensing Layers Selection Process for Relative Humidity Sensors. *Sci. Technol.* **2020**, *23*, 93–104.
51. Serban, B.C.; Cobianu, C.; Dumbravescu, N.; Buiu, O.; Avramescu, V.; Bumbac, M. Electrical percolation threshold in oxidized single-wall carbon nanohorn-polyvinylpyrrolidone nanocomposite: A possible application for high sensitivity resistive humidity sensor. In Proceedings of the 2020 International Semiconductor Conference (CAS), Sinaia, Romania, 7–9 October 2020; pp. 239–242.
52. Sano, N.; Kinugasa, M.; Otsuki, F.; Suehiro, J. Gas sensor using single-wall carbon nanohorns. *Adv. Powder Technol.* **2007**, *18*, 455–466. [CrossRef]
53. Sano, N.; Ohtsuki, N.F. Carbon nanohorn sensor to detect ozone in water. *J. Electrost.* **2007**, *65*, 263–268. [CrossRef]
54. Cobianu, C.; Serban, B.C.; Dumbravescu, N.; Buiu, O.; Avramescu, V.; Nicolescu, C.M. Room temperature chemiresistive ethanol detection by ternary nanocomposites of oxidized single-wall carbon nanohorn (ox-SWCNH). In Proceedings of the 2020 International Semiconductor Conference (CAS), Sinaia, Romania, 7–9 October 2020; pp. 13–16.
55. Cobianu, C.; Serban, B.C.; Dumbravescu, N.; Buiu, O.; Avramescu, V.; Pachiu, C. Organic–inorganic ternary nanohybrids of single-walled carbon nanohorns for room temperature chemiresistive ethanol detection. *Nanomaterials* **2020**, *10*, 2552. [CrossRef] [PubMed]
56. Serban, B.C.; Dumbravescu, N.; Buiu, O.; Bumbac, M.; Brezeanu, M.; Cobianu, C.; Marinescu, R.; Pachiu, C.; Nicolescu, C.M.; Comanescu, F. Quaternary Oxidized Carbon Nanohorns-Based Nanohybrid as Sensing Layer for Room Temperature Resistive Ethanol Sensing. In Proceedings of the 2023 International Semiconductor Conference (CAS), Sinaia, Romania, 11–13 October 2023; pp. 67–70.
57. Jorio, A.; Saito, R. Raman spectroscopy for carbon nanotube applications. *J. Appl. Phys.* **2021**, *129*, 021102. [CrossRef]
58. Ferrari, A.C.; Meyer, J.C.; Scardaci, V.; Casiraghi, C.; Lazzeri, M.; Mauri, F.; Piscanec, S.; Jiang, D.; Geim, A.K. Raman spectrum of graphene and graphene layers. *Phys. Rev. Lett.* **2006**, *97*, 187401. [CrossRef]
59. Bokobza, L.; Bruneel, J.L.; Couzi, M. Raman spectra of carbon-based materials (from graphite to carbon black) and of some silicone composites. *C* **2015**, *1*, 77–94. [CrossRef]
60. Taylor, L.S.; Langkilde, F.W.; Zografi, G. Fourier transform Raman spectroscopic study of the interaction of water vapor with amorphous polymers. *J. Pharm. Sci.* **2001**, *90*, 888–901. [CrossRef]
61. Dumitraşcu, M.; Albu, M.G.; Vîrgolici, M.; Vancea, C.; Meltzer, V. Characterization of electron beam irradiated polyvinylpyrrolidone-dextran (PVP/DEX) blends. *Solid State Phenom.* **2012**, *188*, 102–108. [CrossRef]
62. Ponzoni, A. A statistical analysis of response and recovery times: The case of ethanol chemiresistors based on pure $SnO_2$. *Sensors* **2022**, *2*, 6346. [CrossRef]
63. Pearson, R.G. The HSAB principle—More quantitative aspects. *Inorganica Chim. Acta* **1995**, *240*, 93–98. [CrossRef]
64. Serban, B.C.; Brezeanu, M.; Cobianu, C.; Costea, S.; Buiu, O.; Stratulat, A.; Varachiu, N. Materials selection for gas sensing. An HSAB perspective. In Proceedings of the 2014 International Semiconductor Conference (CAS), Sinaia, Romania, 13–15 October 2014; IEEE: Piscataway, NJ, USA, 2014; pp. 21–30.
65. Choi, K.S.; Park, S.; Chang, S.P. Enhanced ethanol sensing properties based on $SnO_2$ nanowires coated with $Fe_2O_3$ nanoparticles. *Sens. Actuators B Chem.* **2017**, *238*, 871–879. [CrossRef]
66. Zereshki, S.; Figoli, A.; Madaeni, S.S.; Simone, S.; Esmailinezhad, M.; Drioli, E. Effect of polymer composition in PEEKWC/PVP blends on pervaporation separation of ethanol/cyclohexane mixture. *Sep. Purif. Technol.* **2010**, *75*, 257–265. [CrossRef]
67. Shooshtari, M.; Salehi, A.; Vollebregt, S. Effect of humidity on gas sensing performance of carbon nanotube gas sensors operated at room temperature. *IEEE Sens. J.* **2020**, *21*, 5763–5770. [CrossRef]
68. Serban, B.C.; Cobianu, C.; Dumbravescu, N.; Buiu, O.; Bumbac, M.; Nicolescu, C.M.; Pachiu, C.; Brezeanu, M. Electrical Percolation Threshold and Size Effects in Polyvinylpyrrolidone-Oxidized Single-Wall Carbon Nanohorn Nanocomposite: The Impact for Relative Humidity Resistive Sensors Design. *Sensors* **2021**, *21*, 1435. [CrossRef] [PubMed]
69. Li, Z.; Liu, J.; Yi, X.; Wu, W.; Li, F.; Zhu, Z.; Li, H.; Shi, J.; Xu, Y.; Zhou, F.; et al. Metal–Organic Frameworks-Based Fabry– Pérot Cavity Encapsulated TiO2 Nanoparticles for Selective Chemical Sensing. *Adv. Funct. Mater.* **2022**, *32*, 2109541. [CrossRef]
70. Li, Z.; Liu, J.; Feng, L.; Pan, Y.; Tang, J.; Li, H.; Cheng, G.; Li, Z.; Shi, J.; Xu, Y.; et al. Monolithic MOF-based metal–insulator–metal resonator for filtering and sensing. *Nano Lett.* **2023**, *23*, 637–644. [CrossRef]
71. Liu, J.; Feng, L.; Li, Z.; Wu, Y.; Zhou, F.; Xu, Y. Plasma-etching on monolithic MOFs-based MIM filter boosted chemical sensing. *Nano Res.* **2024**, *17*, 2800–2807. [CrossRef]
72. Li, Z.; Tian, L.; Wu, W.; Feng, L.; Khaniyev, B.; Mukhametkarimov, Y.; Ibraimov, M.; Zhou, F.; Liu, W.; Liu, J. Colorimetric Fabry-Pérot Sensor with Hetero-Structured Dielectric for Humidity Monitoring. *Small Methods* **2024**, 2401485. [CrossRef]

73. Shooshtari, M. Ammonia gas sensors based on multi-wall carbon nanofiber field effect transistors by using gate modulation. *Colloids Surf. A Physicochem. Eng. Asp.* **2025**, *704*, 135563. [CrossRef]
74. Vigdorowitsch, M.; Pchelintsev, A.; Tsygankova, L.; Tanygina, E. Freundlich isotherm: An adsorption model complete framework. *Appl. Sci.* **2021**, *11*, 8078. [CrossRef]

sensors

MDPI

*Article*

# Synergistic Enhancement of Chemiresistive $NO_2$ Gas Sensors Using Nitrogen-Doped Reduced Graphene Oxide (N-rGO) Decorated with Nickel Oxide (NiO) Nanoparticles: Achieving sub-ppb Detection Limit

**Chiheb Walleni [1,2,3,4,5], Mounir Ben Ali [5,6], Mohamed Faouzi Ncib [4,5] and Eduard Llobet [1,2,3,*]**

1 MINOS, School of Engineering, Universitat Rovira i Virgili, Avda. Països Catalans 26, 43007 Tarragona, Spain; chiheb.walleni@estudiants.urv.cat

2 IU-RESCAT, Research Institute in Sustainability, Climatic Change and Energy Transition, Universitat Rovira i Virgili, Joanot Martorell 15, 43480 Vila-seca, Spain

3 TecnATox—Centre for Environmental, Food and Toxicological Technology, Universitat Rovira i Virgili, Avda. Països Catalans 26, 43007 Tarragona, Spain

4 Higher School of Sciences and Technologies of Hammam Sousse, University of Sousse, Hammam Sousse 4011, Tunisia; mohamed.faouzi.ncib@gmail.com

5 NANOMISENE Laboratory, LR16CRMN01, Center of Research on Microelectronics and Nanotechnology (CRMN), Technopole of Sousse, B.P334, Sahloul 4054, Tunisia; mounirbenali@crmn.mesrs.tn

6 Higher Institute of Applied Science and Technology of Sousse, University of Sousse, Sousse 4003, Tunisia

* Correspondence: eduard.llobet@urv.cat; Tel.: +34-977-558-502

**Abstract:** Detecting low nitrogen dioxide concentrations ($NO_2$) is crucial for environmental monitoring. In this paper, we report the synergistic effect of decorating nitrogen-doped reduced graphene oxide (N-rGO) with nickel oxide (NiO) nanoparticles for developing highly selective and sensitive chemiresistive $NO_2$ gas sensors. The N-rGO/NiO sensor was synthesized straightforwardly, ensuring uniform decoration of NiO nanoparticles on the N-rGO surface. Comprehensive characterization using SEM, TEM, XRD, and Raman spectroscopy confirmed the successful integration of NiO nanoparticles with N-rGO and revealed key structural and morphological features contributing to its enhanced sensing performance. As a result, the NiO/N-rGO nanohybrids demonstrate a significantly enhanced response five orders of magnitude higher than that of N-rGO toward low $NO_2$ concentrations (<1 ppm) at 100 °C. Moreover, the present device has an outstanding performance, high sensitivity, and low limit of detection (<1 ppb). The findings pave the way for integrating these sensors into advanced applications, including environmental monitoring and IoT-enabled air quality management systems.

**Keywords:** N-rGO; NiO; $NO_2$; gas sensing

## 1. Introduction

In addition to promoting green energy, safeguarding the environment and human health from harmful gases such as $NO_x$, $NH_3$, $CO_x$, $H_2S$, and VOCs is a critical priority in advancing a sustainable lifestyle. Among air pollutants, $NO_2$ has garnered significant attention due to its substantial impact on air quality and its association with the formation of ground-level ozone, acid rain, and climate change through disruptions in atmospheric chemical balance. Furthermore, $NO_2$ poses serious health risks, including respiratory disorders, chronic bronchitis, asthma, and various cardiovascular and pulmonary diseases, depending on exposure duration [1,2].

To address these challenges, the development of highly sensitive $NO_2$ gas sensors has become indispensable. Researchers and industries alike share the responsibility of advancing this technology, which has found increasing relevance in cutting-edge applications such as the Internet of Things (IoT) and Micro-Electro-Mechanical Systems (MEMS) [3]. Effective $NO_2$ sensors require several critical attributes, including high selectivity and stability, low power consumption, optimal device architecture, and, most importantly, a carefully chosen sensing material.

Chemiresistive gas sensors have emerged as a preferred choice due to their cost-effectiveness, miniaturization potential, and compatibility with $NO_2$ detection [4,5]. Among sensitive materials, graphene-based substances have gained prominence, thanks to their remarkable properties, including a large surface area, high carrier mobility, and excellent electrical and thermal conductivity [6–8]. Nitrogen-doped reduced graphene oxide (N-rGO) stands out among graphene derivatives for applications such as energy storage, catalysis, and gas sensing [9–12]. Its superior performance in gas sensing compared to pristine graphene stems from features like enhanced bandgap engineering, increased active sites for catalytic activity, a high surface area, a defect-rich structure, and low operating temperature. However, a few studies have reported its application in chemiresistive $NO_2$ sensors [13–15].

In parallel, transition metal oxides (TMOs), known for their semiconducting properties, have demonstrated considerable potential as sensitive materials for chemiresistive gas sensors. TMOs offer advantages such as ease of production, high sensitivity, excellent stability, and rapid response/recovery times [16–18]. Despite these benefits, challenges such as high operating temperatures and limited selectivity persist [19,20]. TMOs can be categorized into n-type oxides (e.g., $SnO_2$, ZnO, $In_2O_3$, $WO_3$, and $TiO_2$) and p-type oxides (e.g., NiO and CuO). In gas sensing, n-type oxides typically exhibit resistance changes upon exposure to reducing or oxidizing gases, while p-type oxides demonstrate the opposite behavior [17].

Nickel oxide (NiO), a p-type metal oxide, boasts unique physical properties, including a wide bandgap (3.6–4 eV), excellent thermal conductivity, and stability. These characteristics make it a versatile material for energy storage, optoelectronics, and gas sensing applications [21–23]. However, its role as a sensitive material for chemiresistive gas sensors remains relatively rare, even in combination with other materials like graphene derivatives. Studies have suggested synergistic effects when combining NiO with graphene-based materials for detecting gases such as $H_2$, $H_2S$, $NH_3$, CO, and VOCs [24–28]. Nonetheless, its potential for $NO_2$ detection under optimized conditions has not been widely discussed.

This study aims to bridge this gap by leveraging the complementary properties of N-rGO and NiO to develop enhanced chemiresistive $NO_2$ gas sensors. The sensor materials are synthesized through straightforward methods and evaluated for their performance at sub-ppm $NO_2$ concentrations under varying conditions. Their structural and functional attributes are characterized using techniques such as SEM, TEM, XRD, and Raman spectroscopy. Finally, the gas sensing results and mechanisms are thoroughly discussed, highlighting their significance for future applications.

## 2. Materials and Methods

### 2.1. Synthesis of Materials

The synthesis of nitrogen-doped reduced graphene oxide (N-rGO) began with the preparation of graphene oxide (GO) as a precursor, which was synthesized using a modified Hummer's method [29]. In brief, 50 mg of GO powder was dispersed in 50 mL of distilled water and subjected to ultrasonication for 2 h to ensure complete dispersion. Subsequently, an appropriate quantity of urea, serving as the nitrogen dopant source, was added to the GO solution and stirred for 30 min. The resulting mixture was transferred to a Teflon-lined autoclave and subjected to hydrothermal treatment at 180 °C for 12 h. This process not only reduced the GO to reduced graphene oxide (rGO) but also incorporated nitrogen dopants into the graphene structure, yielding N-rGO (Figure 1a).

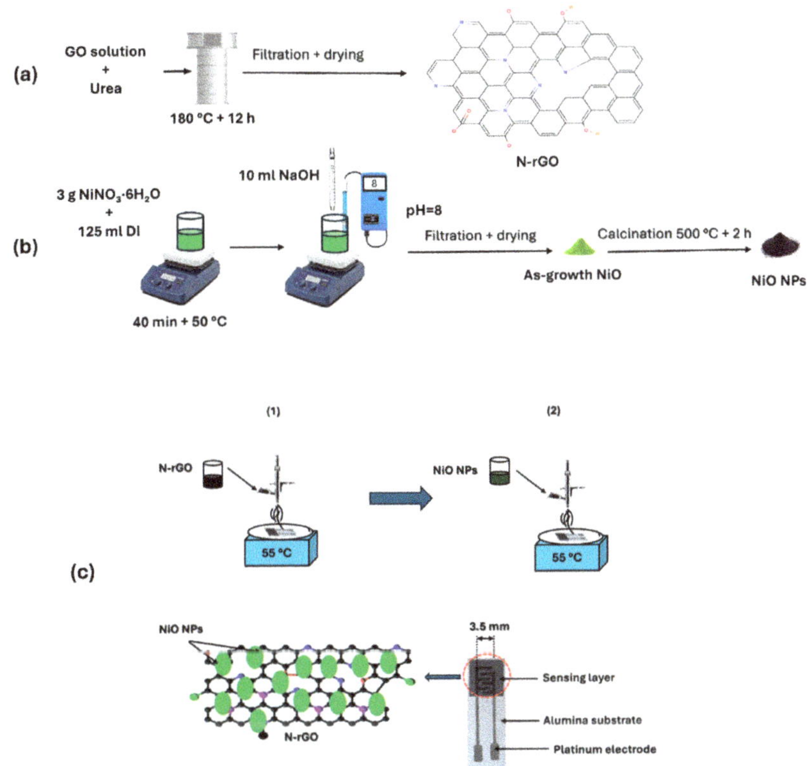


**Figure 1.** (**a**) Preparation of N-rGO and (**b**) NiO nanoparticles. (1) and (2) illustrate the airbrushing process used to coat the electrode area of the transducer substrate. (**c**) fabrication of the NiO/N-rGO sensor device.

The synthesis of nickel oxide (NiO) nanoparticles was carried out using the co-precipitation method (Figure 1b). Initially, 3 g of hydrated nickel nitrate ($NiNO_3{\cdot}6H_2O$) was dissolved in 125 mL of distilled water and stirred at 50 °C for 40 min. Subsequently, 10 mL of 0.1 M NaOH solution was added dropwise to the mixture until the pH reached 8. The resulting precipitate was thoroughly washed and dried at 80 °C. Finally, the dried product was calcined at 400 °C for 3 h to obtain nickel oxide nanoparticles (NiO NPs).

### 2.2. Preparation of Sensors

In this stage, the airbrushing technique was employed to decorate nitrogen-doped reduced graphene oxide (N-rGO) with nickel oxide nanoparticles (NiO NPs). The sensor fabrication process is illustrated in Figure 1c. To begin, 5 mg of N-rGO powder was dispersed in 10 mL of ethanol and subjected to ultrasonication for 1 h to create a stable suspension. Separately, 5 mg of NiO NPs was dispersed in 10 mL of ethanol. The deposition process was carried out using airbrushing at a temperature of approximately 55 °C, with nitrogen ($N_2$) serving as the carrier gas. Initially, the N-rGO suspension was airbrushed onto platinum screen-printed electrodes (alumina substrates from CeramTech GmbH, Plochingen, Germany) as a chemiresistive gas sensor. Subsequently, NiO NPs were deposited onto the N-rGO layer to complete the decoration process. The film resistance was monitored continuously using a multimeter, ensuring improved reproducibility across devices. The final thickness of the deposited sensing layers was measured to be 50 ± 5 μm.

*2.3. Gas Sensing Measurements*

In the final stage of the experiment, the as-fabricated NiO/N-rGO sensors were positioned within an airtight Teflon test chamber with a volume of 35 $cm^3$, designed with two openings to accommodate UV lamps. The chamber was equipped with an inlet for gas delivery and an outlet connected to an exhaust system. The entire setup was integrated into a fully automated gas flow measurement system capable of supplying diluted gas mixtures via mass flow controllers (Bronkhorst High-Tech B.V., Ruurlo, The Netherlands). For gas sensing experiments, calibrated gas cylinders balanced in dry synthetic air (Air Premier, purity: 99.999%) were utilized. The sensors' operating temperatures were regulated by connecting their meander heaters to an external power supply (Agilent U8002A, Santa Clara, CA, USA).

The sensor responses were recorded using an Agilent 34972A data acquisition system, which continuously monitored the resistance of the sensing materials during exposure to varying concentrations of target gases, including $NO_2$, ethanol, $NH_3$, and $CO_2$. Additionally, the effect of humidity on sensor performance was assessed using a controller evaporator mixer (CEM) to generate a controlled humidity level of 70% RH, simulating real environmental conditions at the optimal operating temperature. The overall gas sensing measurement setup is illustrated in Figure 2.

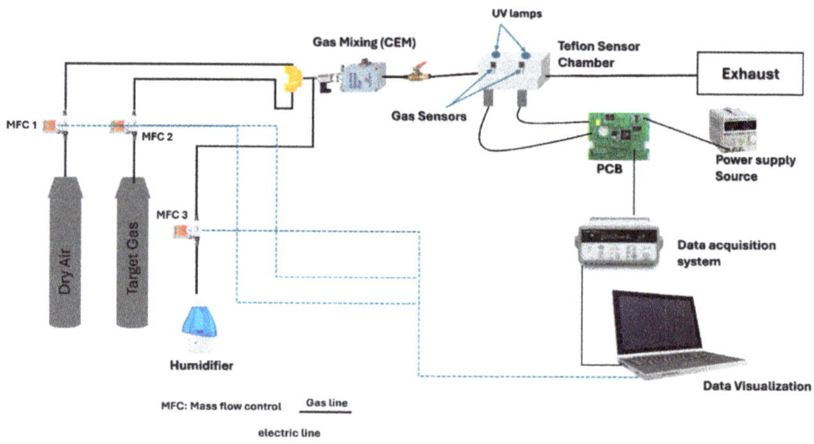


**Figure 2.** Experimental setup of gas sensing measurements.

To optimize power consumption and simulate realistic testing conditions, the total gas flow rate was maintained at a low rate of 100 mL/min. Before measurements, the sensors were stabilized under synthetic dry air for 1 h at both room temperature and 100 °C. Following stabilization, the sensors were exposed to the target gases diluted in synthetic dry air for 15 min, followed by a 1 h recovery period in pure dry air. Throughout the measurements, the temperature inside the test chamber was maintained at 25 °C, with a residual ambient humidity of 4% relative humidity (R.H.), equivalent to approximately 1250 ppm of water vapor.

Sensor response, R (%), was defined as follows:

$$R\ (\%) = [\,|R - R_0|\,/R_0] \times 100$$

where $R_0$ is the resistance under dry air, and R is the resistance measured during exposure to the target gas.

### 2.4. Material Characterization

The morphological characteristics of the samples were examined using scanning electron microscopy (SEM) with a FEI Quanta 450 instrument from FELMI-ZFE (Graz, Austria). Furthermore, transmission electron microscopy (TEM) imaging was performed using an ultra-high-resolution transmission electron microscope (UHR-TEM), specifically the Libra® 200MC model (Zeiss, Jena, Gemrnay). Structural analysis was carried out through X-ray diffraction (XRD) measurements using the Shimadzu Corporation LabX XRD-600 instrument (Midland, ON, Canada), equipped with CuK$\alpha$ radiation ($\lambda$ = 1.54056 Å). The XRD patterns were recorded at room temperature over a 2$\theta$ range of 10° to 80°. Raman spectroscopy was employed to analyze the structural features of the samples within a wavelength range of 100–3000 $cm^{-1}$, using a Renishaw inVia Raman Microscope (Changchun New Industries Optoelectronics Technology Co., Ltd., Changchun, China).

## 3. Results

### 3.1. Characterization

SEM micrographs provide detailed insights into the morphological characteristics of N-rGO and NiO NPs and NiO/N-rGO deposited film. As shown in Figure 3a, N-rGO exhibits a distinctive folded 2D graphene flake structure, reflecting the effects of the reduction and nitrogen doping processes [14]. The observed wrinkles and corrugations are likely due to the intercalation of nitrogen atoms within the graphene layers [9]. In contrast, the NiO surface, presented in Figure 3b, reveals a distribution of spherical nanoparticles with comparable sizes. Additional details on the size distribution of NiO nanoparticles are presented in the histogram shown in Figure S2 (Supplementary Information). Figure 3c confirms the successful attachment of the deposited NiO NPs onto N-rGO sensing film. This is further supported by TEM analysis results.

Figure 4 showcases the detailed morphological analysis of N-rGO and NiO/N-rGO using high-resolution transmission electron microscopy (HRTEM), supported by Energy Dispersive Spectroscopy (EDS) data (additional EDS results are provided in Figure S1, Supporting Information). To prepare the samples, a copper grid was immersed in a NiO/N-rGO suspension dispersed in ethanol and treated with ultrasonication to ensure even distribution.

The HRTEM image in Figure 4a reveals the characteristic two-dimensional nanostructures of graphene, displaying thin, flake-like layers folded around darker regions. In contrast, Figure 4b,c illustrates the successful and uniform decoration of nickel oxide nanoparticles across the N-rGO surface. The NiO nanoparticles exhibit a nanocrystalline structure with an interplanar spacing of 2.28 Å, corresponding to the (200) plane. This spacing, slightly offset by approximately 0.2 nm compared to the standard value (ICDD card number: 73-1519), is attributed to lattice distortions induced by interactions with the N-rGO substrate, as confirmed by XRD analysis [30].

The oxygen functional groups and nitrogen dopants present on the N-rGO surface serve as active sites for heterogeneous nucleation, enabling the formation of densely packed and monodisperse nickel oxide nanoparticles [31]. This distinctive morphological configuration of the NiO/N-rGO hybrid not only highlights the structural integrity of the material but also emphasizes its enhanced reactivity, making it a promising candidate for $NO_2$ gas sensing applications.

Figure 5a shows the XRD diffractograms for the NiO, N-rGO, and GO samples. The X-ray diffraction (XRD) pattern of the nanocomposite displays distinctive broad peaks, which are localized at 37.08° (111), 43.32° (200), 62.75° (220) and 75.15° (311) [32–35]. These diffraction peaks correspond to the ones observed in the case of NiO nanoparticles. They match with a cubic phase of NiO (ICDD card number: 73-1519), with a lattice constant a = 4.168 Å belonging to the Fm-3m space [34]. Conversely, the diffractogram of GO displays a broad peak at about 2θ = 11° (001). This peak indicates the presence of oxygen functional groups in the GO structure after the oxidation of graphite [35]. Nevertheless, this distinctive peak of GO vanishes completely in the XRD pattern of the N-rGO nanomaterial, and a new characteristic peak appears at 25° (002), This means that most of the oxygen-containing groups in GO were efficiently eliminated after the reduction and doping with nitrogen [36]. This also indicates that the π-conjugated structure of graphene has been restored considerably at the produced rGO. For the NiO/N-rGO deposited film, XRD analysis was performed to examine its crystalline structure. Given the thin nature of this film on top of the alumina substrate, $Al_2O_3$ peaks highly interfere with the ones that are characteristic for NiO and N-rGO in the XRD diffractogram pattern (see Figure S3).

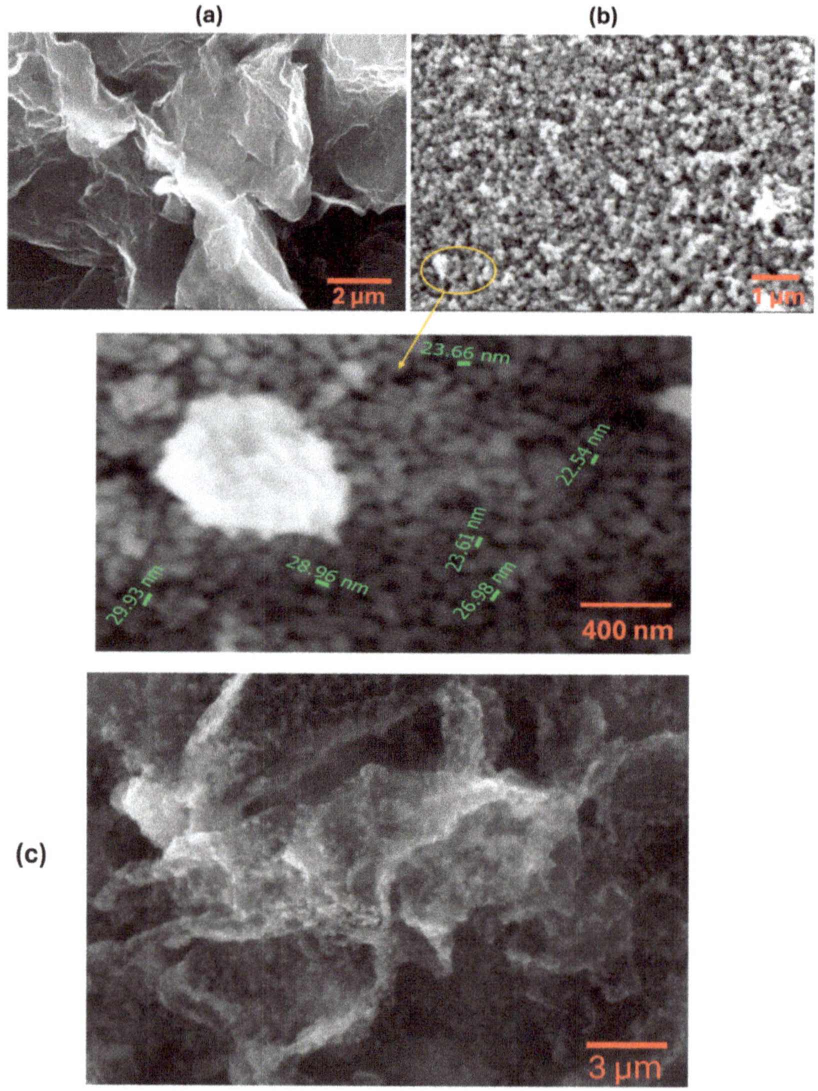


**Figure 3.** SEM images of (**a**) N-rGO and (**b**) NiO nanoparticles. (**c**) NiO/N-rGO deposited film.

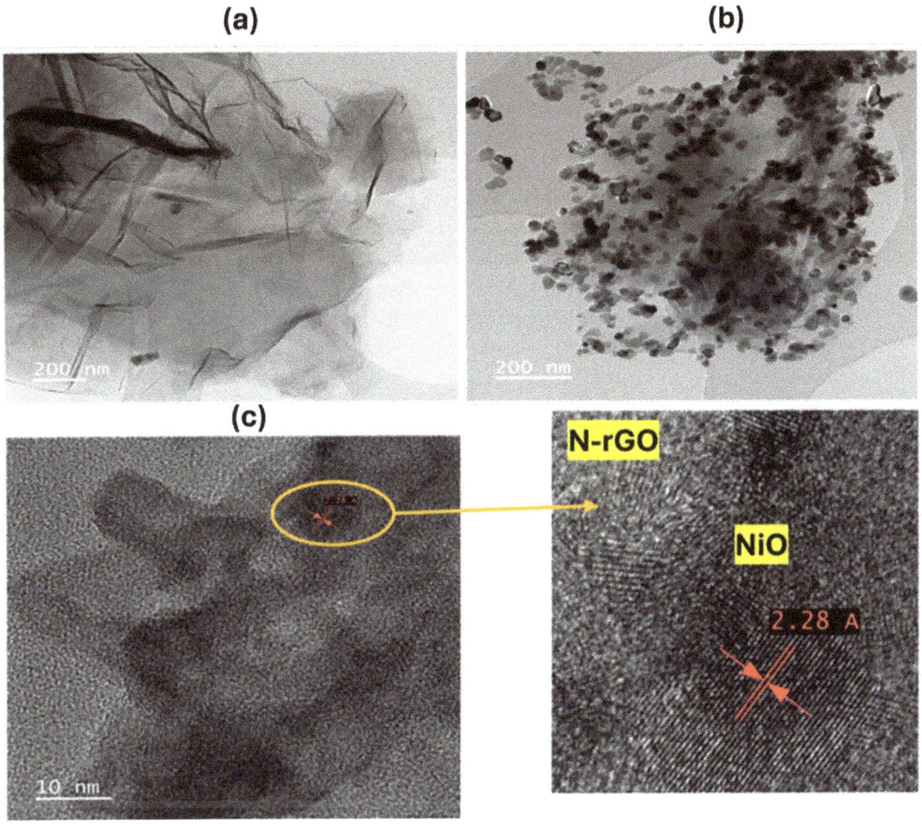


**Figure 4.** TEM images of (**a**) N-rGO and (**b**,**c**) NiO/N-rGO. The lower panel on the right shows a magnification of the circled area as indicated in panel c.

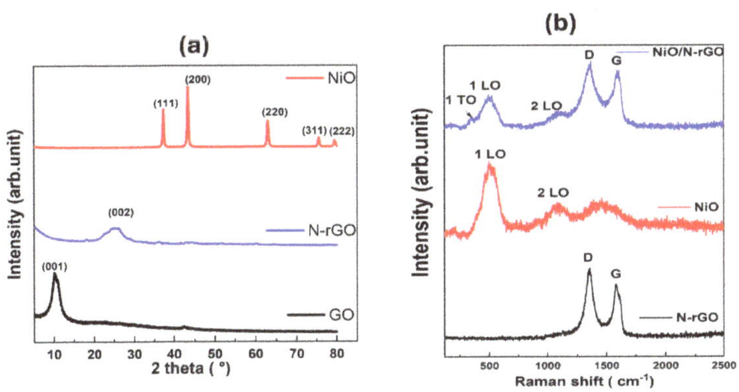


**Figure 5.** (**a**) XRD pattern of GO, N-rGO, and NiO. (**b**) Raman Spectra of N-rGO, NiO, and NiO/N-rGO.

Raman spectroscopy, a vital technique for identifying structural fingerprints, was utilized to gain detailed insights into the structural characteristics of samples. Figure 5b depicts the Raman spectra of N-rGO, NiO, and NiO/N-rGO nanomaterials within the range of 250 to 2500 $cm^{-1}$. The N-rGO spectrum exhibits a D band at 1339 $cm^{-1}$ and a G band at 1570 $cm^{-1}$ [37]. The intensity ratio of the D band to the G band ($I_D/I_G$) indicates a higher intensity of the D band, confirming the presence of structural defects associated with oxygen functional groups and nitrogen doping [9]. In the case of NiO, two prominent peaks are observed at 518 $cm^{-1}$ and 1058 $cm^{-1}$, corresponding to the Ni-O stretching mode (1LO) and the two-phonon vibration mode (2LO), respectively [38]. The 1LO mode represents the longitudinal optical phonon vibrations of atoms within the crystal lattice, while the 2LO mode involves the simultaneous vibration of two phonons [39]. The Raman spectrum of the NiO/N-rGO nanocomposite integrates all characteristic peaks from both NiO and N-rGO, highlighting the successful formation of the composite material. Additionally, the peak at 518 $cm^{-1}$ is asymmetric, featuring a shoulder on the left side around 400 $cm^{-1}$, corresponding to the 1TO peak of the first-order phonon. It is typically absent in an ideal cubic NiO structure. Its presence, however, is attributed to lattice distortions and defect states, resulting in non-stoichiometry within the Ni-O framework, such as nickel and oxygen vacancies [40].

### 3.2. Gas Sensing Characterization

The as-fabricated N-rGO and NiO/N-rGO sensors were evaluated for nitrogen dioxide ($NO_2$) detection by monitoring their relative resistance changes when exposed to varying $NO_2$ concentrations. As seen in Figure 6a, initial gas sensing measurements were conducted at $NO_2$ concentrations of 800 ppb under room temperature (RT), 100 °C, and 150 °C conditions over multiple cycles. The results underscore the significant role of operating temperature in enhancing the sensitivity and reliability of the NiO/N-rGO sensor for $NO_2$ detection. Upon exposure to $NO_2$, an oxidizing gas, both sensors exhibited behavior consistent with p-type semiconductors, characterized by a decrease in resistance (Figure 6b,c). This response aligns with previous studies on $NO_2$ sensing [9], where the adsorption of $NO_2$ molecules onto the N-rGO surface triggers charge transfer. Specifically, $NO_2$ molecules accept electrons from the N-rGO, leading to a reduction in resistance. Both sensors demonstrated stable and reproducible responses to the target gas, with effective baseline recovery after each exposure cycle.

At room temperature, while both sensors exhibited resistance changes (Figure S4), these changes were not distinguishable from baseline drift. The response magnitudes of the as-fabricated N-rGO and NiO/N-rGO sensors were calculated as 5.71% and 7.28%, respectively. This indicates that the incorporation of NiO nanoparticles into N-rGO offers limited improvement in response performance at room temperature. This modest enhancement is attributed to the high activation energy required for the NiO/N-rGO sensor [41]. In contrast, when operated at 100 °C, both sensors demonstrated significantly improved responses and a more stable resistance baseline (Figure 6b,c). The calculated responses were 7.28% for N-rGO and 28.25% for NiO/N-rGO. Notably, the NiO/N-rGO sensor exhibited a response magnitude approximately three times higher than that observed at room temperature, indicating a substantial enhancement in sensing performance. This improved response is ascribed to the low activation energy of the NiO/N-rGO sensor influenced by the rise in temperature [41]. However, at 150 °C (Figure S4), both sensors exhibited a decreased response to $NO_2$ gas, further confirming that the optimal operating temperature for achieving excellent sensing performance is 100 °C. Meanwhile, the decrease observed at 150 °C may be attributed to the increased desorption rate of $NO_2$ molecules from the sensor surface, which reduces the overall sensor response.

The response and recovery times were determined to be ($t_{resp}$ = 11 min, $t_{rec}$ = 45 min), respectively, for the NiO/N-rGO sensor and ($t_{resp}$ = 10 min, $t_{rec}$ = 44 min) for the N-rGO sensor, as shown in Figure S5. The response time is defined as the duration required for the sensor to reach 90% of its final stable resistance upon exposure to the target gas. Conversely, the recovery time refers to the time needed for the sensor to return to 10% of its baseline resistance after the removal of the target gas.

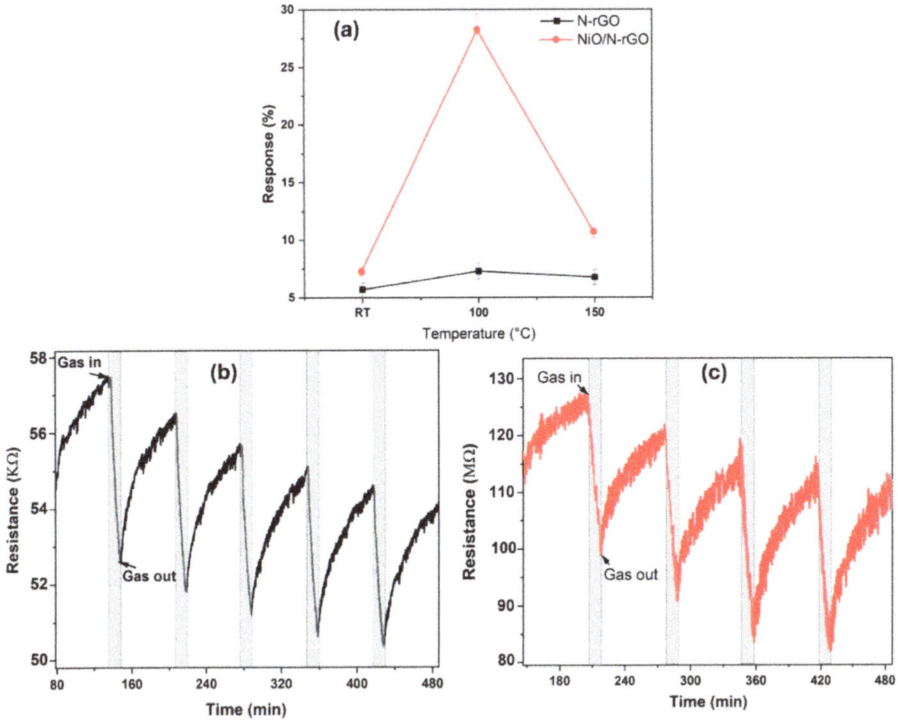


**Figure 6.** (**a**) Sensor responses as a function of their operating temperature toward 800 ppb of $NO_2$. Dynamic response and recovery curves for repeated exposure cycles to 800 ppb $NO_2$ for (**b**) N-rGO and (**c**) NiO/N-rGO. The sensor operating temperature was 100 °C.

The dynamic resistance changes in the as fabricated sensors were further recorded across a wide range of $NO_2$ gas concentrations—50, 100, 250, 500, 800, and 1000 ppb—at an operating temperature of 100 °C, as illustrated in Figure 7a,b. The calculated responses for the N-rGO sensor were 1.18%, 1.75%, 3%, 5.9%, 7.28%, and 7.56%, respectively, while the NiO/N-rGO sensor demonstrated responses of 10%, 12.42%, 15.8%, 22.1%, 28.25%, and 33.63%, respectively. Both sensors exhibited stable and reproducible sensing responses, as evidenced by their small standard deviation errors on the order of $10^{-4}$ [42].

Notably, as depicted in Figure 7c, both sensors demonstrated the capability to detect $NO_2$ concentrations below 50 ppb, a value significantly lower than the threshold limit of 0.2 ppm (200 ppb) for an 8 h time-weighted average (TWA) recommended by the American Conference of Industrial Hygienists (ACGIH). The NiO/N-rGO sensor showed a noticeably enhanced response for each exposure to $NO_2$, outperforming the N-rGO sensor by approximately five orders of magnitude. This significant improvement highlights the remarkable influence of NiO nanoparticles on enhancing the responsiveness of the N-rGO sensing film to $NO_2$ gas.

The enhanced sensing performance of the NiO/N-rGO sensor can be attributed to the large surface area of the NiO/N-rGO nanohybrids, which promotes efficient charge carrier transfer between $NO_2$ molecules and the hybrid material. Additionally, XRD and Raman's analysis confirmed the excellent crystallinity of NiO, along with the presence of oxygen vacancies, which may enhance electron mobility and strengthen NiO–$NO_2$ interactions [43].

This synergistic effect highlights the significant role of NiO nanoparticles in improving gas sensing capabilities, particularly for the detection of trace levels of $NO_2$.

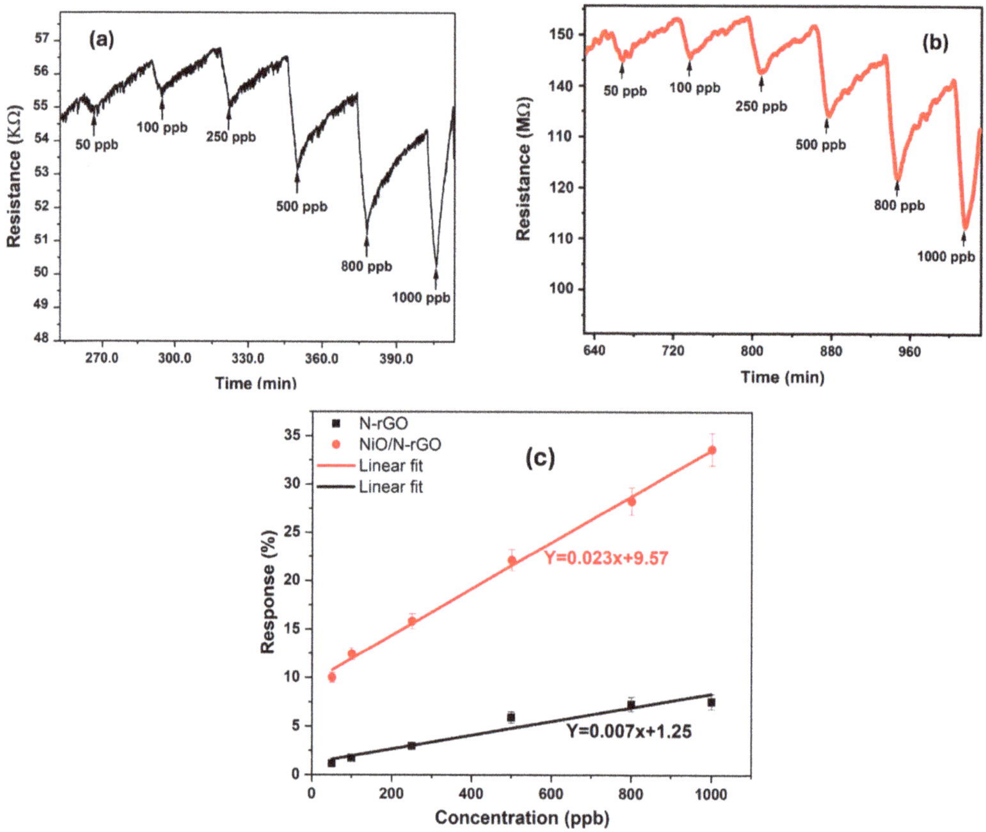


**Figure 7.** (**a**) N-rGO and (**b**) NiO/N-rGO gas sensing films' resistance changes as a function of time toward different $NO_2$ concentrations at 100 °C. (**c**) N-rGO and NiO/N-rGO sensors' response as a function of $NO_2$ concentration at 100 °C.

The sensitivity and limit of detection (*LOD*) are calculated following the expression shown below:

$$LOD = 3 \times \frac{RMS_{noise}}{b}$$

where $b$ is the slope of the calibration curve (sensitivity) and $RMS_{noise}$ is the root-mean-square deviation at the baseline [44]. $RMS_{noise}$ is the standard deviation of the noise level. Under dry air, the RMS noise values were determined from 100 baseline data points before $NO_2$ exposure, yielding values of $9.22 \times 10^{-2}$ for N-rGO and $0.77 \times 10^{-2}$ for NiO/N-rGO, respectively. Consequently, the calculated limit of detection (LOD) was approximately 39 ppb for the N-rGO sensor and 0.96 ppb for the NiO/N-rGO sensor.

As demonstrated in Table 1, the as-fabricated NiO/N-rGO sensor exhibits significantly improved sensitivity and a lower limit of detection (LOD) (<1 ppb) compared to its pristine N-rGO counterpart. This enhancement underscores the pivotal role of NiO nanoparticles in augmenting the gas sensing properties of N-rGO, particularly for $NO_2$ detection.

The gas sensing performance of N-rGO and NiO-decorated N-rGO was also evaluated for other gases, including $CO_2$, $NH_3$, and ethanol. The typical resistance response dynamics for 100 ppm $CO_2$, 20 ppm ethanol, and 10 ppm $NH_3$ are presented in Figure S6, respectively (Supplementary Information). A summary of the sensing results for each gas is provided in the histogram in Figure 8a. The results clearly demonstrate that the incorporation of NiO into N-rGO significantly enhances the response to $NO_2$ while effectively reducing cross-sensitivity to $CO_2$, $NH_3$, and ethanol. NiO was reported as a highly sensitive material for VOCs such as ethanol at high operating temperatures (<300 °C) [22]. Therefore, this temperature range is unsuitable for the current application. Operating at such high temperatures could compromise the stability of the N-rGO film and contradict the goal of achieving low power consumption for the device [45].

**Table 1.** Sensitivity and LOD values.

| Sensors | N-rGO | NiO/N-rGO |
| --- | --- | --- |
| Sensitivity ($10^{-2}$ $ppm^{-1}$) | 709 | 2398 |
| LoD (ppb) | 39 | <1 |

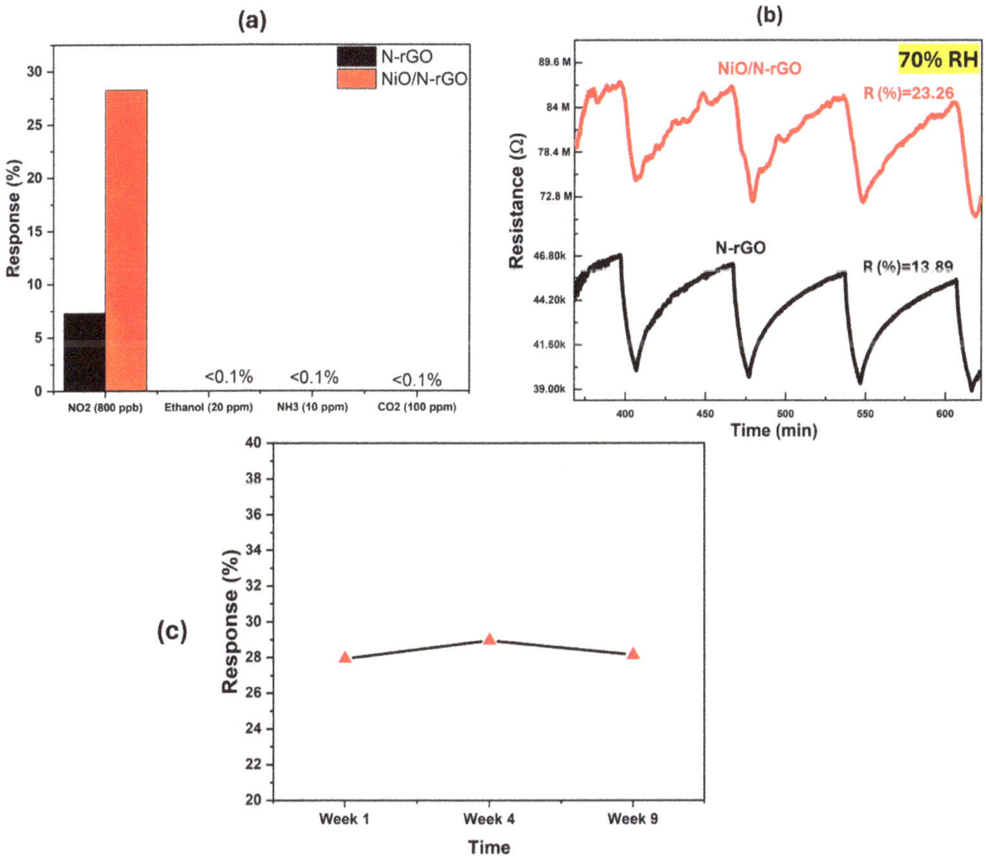


**Figure 8.** (**a**) Response histogram of N-rGO and NiO/N-rGO gas sensors to $NO_2$ (800 ppb), Ethanol (20 ppm), $NH_3$ (10 ppm), and $CO_2$ (100 ppm). (**b**) N-rGO and NiO/N-rGO sensors' resistance variations to 800 ppb of $NO_2$ under 70% RH. Sensors operated at 100 °C. (**c**) NiO/N-rGO sensor response to 800 ppb of $NO_2$ gas at 100 °C over a 9-week period.

Ambient moisture interference is a critical parameter in evaluating the sensitivity of gas sensors under working conditions. To investigate this, the prepared sensors were tested in a highly humid environment (70% RH) at 100 °C, alongside an 800 ppb $NO_2$ concentration. As shown in Figure 8b, the N-rGO sensor exhibited an increased response to ambient moisture, rising from 7.28% under dry conditions to 13.89% at 70% RH. In contrast, the NiO/N-rGO sensor showed a decreased response in the humid environment compared to its dry condition response of 23.26%.

This observation suggests that the p-type sensitivity of N-rGO improves in the presence of $NO_2$, likely due to water molecules promoting the adsorption of $NO_2$ through redox reactions. During this process, nitrogen dioxide dissociates into $NO_2^-$ and $H^+$ ions, facilitating the movement of $H^+$ protons. This proton mobility decreases the resistance, resulting in an enhanced electrical response [46]. On the other hand, the notable decrease in sensitivity for the NiO/N-rGO sensor under high humidity can be attributed to two factors. Firstly, the presence of water molecules inhibits the chemisorption of oxygen molecules by covering the surface of NiO/N-rGO with adsorbed moisture. Secondly, the interaction of water molecules with oxygen species leads to the formation of hydroxyl groups on the sensor surface, which partially hinders the adsorption of $NO_2$ molecules. These factors significantly alter the sensor's resistance and diminish its response to $NO_2$ gas in a highly humid environment [47,48].

All measurements were conducted over a period of 9 weeks in which, even though a drift appeared in the value of the baseline resistance (see Figure S7 in Supporting Information), no significant changes were observed in sensor responses (see Figure 8). This confirms the stability of both N-rGO and NiO/N-rGO sensors in $NO_2$ detection.

As presented in Table 2, the as-fabricated NiO/N-rGO sensor exhibits a markedly superior response to trace levels of $NO_2$ gas compared to previously reported MO/graphene-based sensors. This improvement emphasizes the effectiveness of the synergistic interaction between nitrogen-doped reduced graphene oxide and nickel oxide nanoparticles in enhancing $NO_2$ gas sensing performance.

**Table 2.** Comparison of the performance in the detection of $NO_2$ between this work and previously reported results.

| Material-Based Sensor | T (°C) | $NO_2$ (ppm) | Response (%) | Tresp/Trec (s) | LOD (ppb) | Ref. |
|---|---|---|---|---|---|---|
| NiO NPs/N-rGO | 100 | 0.8 | 28.25 | 660/2700 | <1 | This work |
| N-rGO | 100 | 0.8 | 7.28 | 600/2640 | 39 | This work |
| $SnO_2$/rGO | 150 | 100 | 97.24 | 14/509 | 1000 | [49] |
| NiO NS/rGO | 200 | 1 | ~670 | - | - | [50] |
| ZnO/rGO | 110 | 2.5 | 33.11 | 182/234 | 5 | [51] |
| NiO Honeycomb | 200 | 20 | 57.3 | - | 20 | [52] |
| CuO/rGO | RT | 20 | 58.1 | 30/- | 1000 | [53] |
| ZnO/$SnO_2$/rGO | RT | 5 | 141 | 33/92 | - | [54] |

## 4. Discussion

The interaction mechanism between the NiO/N-rGO sensing film and $NO_2$ gas molecules is further elucidated in Figure 9.

To begin, it is essential to describe the systematic interaction between pristine N-rGO and $NO_2$ molecules. Upon exposure to an oxidizing gas like $NO_2$, the resistance curves exhibit a p-type response, consistent with the intrinsic behavior of reduced graphene oxide (rGO) as a p-type semiconductor [55]. Nitrogen doping, as an element from Group V, enhances n-type conduction in rGO by incorporating pyrrolic, pyridinic, and graphitic bonds within the graphene basal plane and donates one p-electron to the aromatic $\pi$ system [10].

This nitrogen doping introduces excess electrons, promoting stronger interactions with $NO_2$. Furthermore, nitrogen atoms can bind with the oxygen atoms of $NO_2$ molecules [56]. In contrast, this study demonstrates that N-rGO exhibits a p-type response toward $NO_2$, confirming its p-type semiconducting behavior. This phenomenon can be attributed to the interaction between $NO_2$ molecules and the aromatic $\pi$-system of N-rGO, where nitrogen dopants (as verified by EDS analysis) play a crucial role. It is hypothesized that $NO_2$ molecules withdraw electrons from the $\pi$-system at the nitrogen-doped sites, thereby reinforcing the p-type semiconducting behavior of N-rGO. As a result, when N-rGO reacts with $NO_2$, its Fermi level shifts closer to the valence band, increasing the number of holes in the valence band and thus causing a significant decrease in sensor resistance upon exposure to nitrogen dioxide.

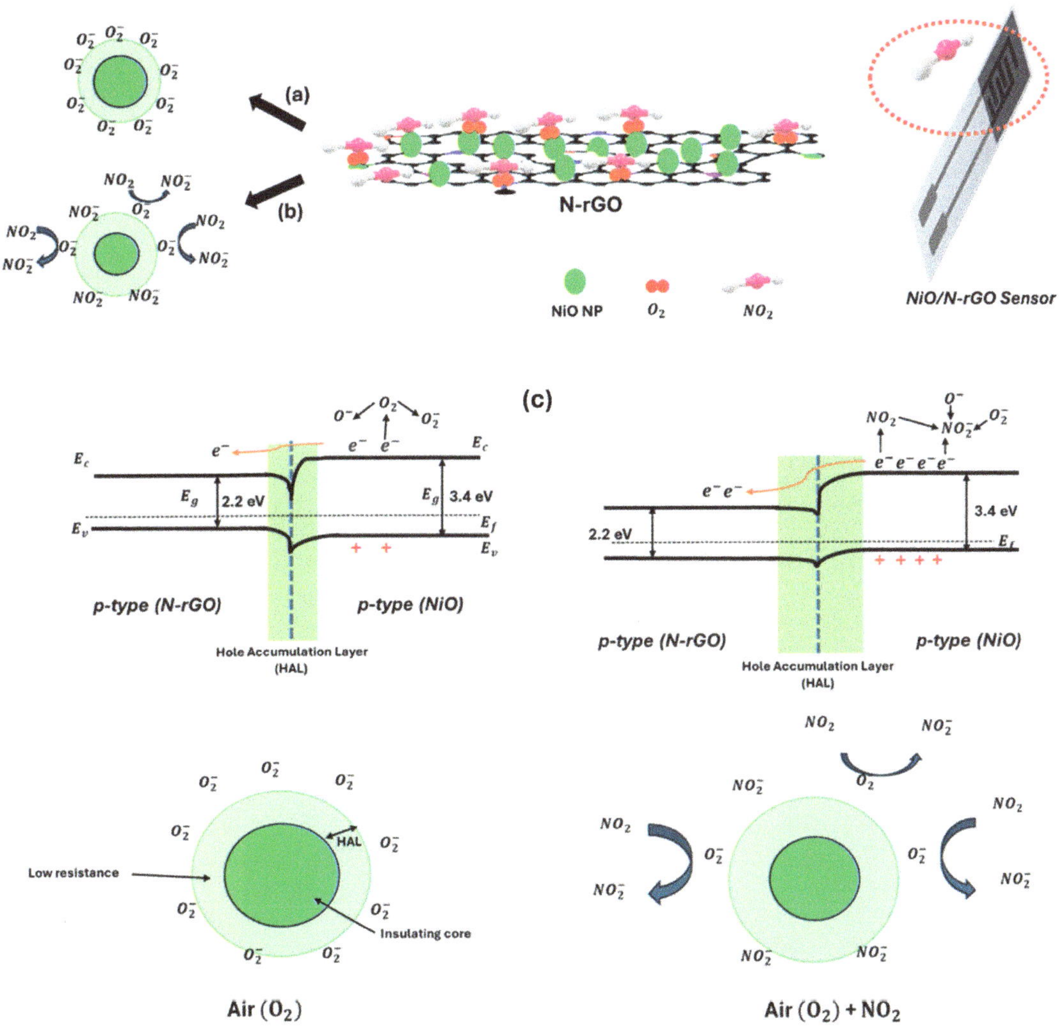


**Figure 9.** Formation of core–shell structures of charge carriers in NiO to (**a**) air ($O_2$), (**b**) $NO_2$, and (**c**) the sensing mechanism of NiO/N-rGO sensitivity to $NO_2$ at 100 °C. At the interface between p-type N-rGO and p-type NiO, a Hole Accumulation Layer (HAL) forms in both cases under air or in the presence of $NO_2$.

Turning to the NiO/N-rGO sensing film mechanism, Figure 9 illustrates the sequential interactions. Initially, as shown in Figure 9a, when the sensing film is exposed to air, oxygen molecules are adsorbed onto the surface of NiO nanoparticles (NPs). At an operating temperature of 100 °C, these oxygen molecules are ionized into $O_2^-$ and $O^-$ species. On the p-type NiO surface, these ionized oxygen species attract the majority carriers (holes), forming a well-defined Hole Accumulation Layer (HAL).

In the presence of $NO_2$ gas, as depicted in Figure 9b, the adsorbed $NO_2$ molecules act as strong electron acceptors and capture electrons from the p-type NiO, forming $NO_2^-$. This process significantly increases the hole concentration in the NiO NPs, leading to an expansion of the HAL. Consequently, the potential barrier height decreases due to the high electron affinity of $NO_2$, which surpasses that of oxygen.

Furthermore, as illustrated in Figure 9c, the hybrid configuration of N-rGO sheets and NiO nanoparticles facilitates the formation of a p-p junction. N-rGO may donate electrons to the NiO NPs, enhancing the binding with $NO_2$ and $O_2$ and further widening the depletion layer [57]. Additionally, nitrogen-doped rGO contains active sites, such as nitrogen dopants not bound to NiO NPs, which provide further interaction opportunities with $NO_2$ gas. N-rGO also serves as an efficient charge transport channel due to its high carrier mobility, enabling faster electron transfer to the electrodes for collection.

These attributes collectively endow the NiO/N-rGO nanohybrids with exceptional $NO_2$ sensing performance and enhanced sensitivity, stability, and carrier transport efficiency.

## 5. Conclusions

In summary, we have successfully prepared a NiO/N-rGO nanohybrid-based $NO_2$ gas sensor based on an alumina transducer substrate using facile and inexpensive routes. The sensing materials were widely investigated for their main structures and analyzed using SEM, TEM, XRD, and Raman techniques. Interestingly, NiO NPs were well attached to the surface of N-rGO, giving insights into the sensing mechanism, as confirmed by characterization results. Accordingly, the NiO/N-rGO sensor exhibited excellent sensitivity toward trace amounts of nitrogen dioxide ($NO_2$) at a moderate operating temperature of 100 °C. This represents the first demonstration of a high-yield synthesis technique that produces NiO/N-rGO nanohybrids, which can be readily deposited on a wide range of substrate materials. In the first test, the sensor showed a very high selectivity toward $NO_2$ (the other gaseous species tested were $CO_2$, ethanol, and $NH_3$). Consequently, NiO NPs proved their potential for boosting the sensitivity of N-rGO toward $NO_2$ gas, thanks to the p-p junctions created that facilitate carrier conduction, as explained by the underlying sensing mechanism. The nanomaterial presented robust performances such as a high sensitivity and very low limit of detection, showing high prospects for being integrated in the next generation of advanced chemoresistive sensors.

**Supplementary Materials:** The following supporting information can be downloaded at: https://www.mdpi.com/article/10.3390/s25051631/s1. Figure S1: EDS analysis of (a,b) N-rGO and (c,d) NiO/N-rGO; Figure S2: size distribution histogram of NiO nanoparticles; Figure S3: XRD pattern of N-rGO, NiO, and NiO/N-rGO; Figure S4: Resistance curves of N-rGO and NiO/N-rGO sensors toward 800 ppb of $NO_2$ at room temperature and 150 °C; Figure S5: Response and recovery times of N-rGO and NiO/N-rGO sensors toward 800 ppb of $NO_2$ at 100 °C; Figure S6: Resistance curves of N-rGO and NiO/N-rGO sensors toward 20 ppm ethanol, 100 ppm of $CO_2$, and 10 ppm $NH_3$ at 100 °C; Figure S7: Dynamic response and recovery curves of the NiO/N-rGO sensor over 9 weeks of measurements. Repeated exposures to 800 ppb $NO_2$ at an operating temperature of 100 °C.

**Author Contributions:** Conceptualization, M.B.A., M.F.N. and E.L.; methodology, C.W.; validation, E.L.; formal analysis, C.W.; investigation, C.W.; resources, M.B.A., M.F.N. and E.L.; data curation, C.W.; writing—original draft preparation, C.W.; writing—review and editing, M.B.A., M.F.N. and

E.L.; visualization, C.W.; supervision, M.B.A., M.F.N. and E.L.; funding acquisition, M.B.A., M.F.N. and E.L. All authors have read and agreed to the published version of the manuscript.

**Funding:** C. Walleni would like to thank the University of Sousse for its financial support. E. Llobet is supported by the Catalan Institution for Research and Advanced Studies (ICREA) via the 2023 Edition of the ICREA Academia Award. Supported in part via Grant *PID2022-142451OB-C21* funded by MICIU/AEI/ https://doi.org/10.13039/501100011033 and by "ERDF/EU".

**Institutional Review Board Statement:** Not applicable.

**Informed Consent Statement:** Not applicable.

**Data Availability Statement:** Data can be obtained from the authors upon request.

**Conflicts of Interest:** The authors declare no conflicts of interest.

## References

1. Salonen, H.; Salthammer, T.; Morawska, L. Human Exposure to $NO_2$ in School and Office Indoor Environments. *Environ. Int.* **2019**, *130*, 104887. [CrossRef] [PubMed]
2. Manisalidis, I.; Stavropoulou, E.; Stavropoulos, A.; Bezirtzoglou, E. Environmental and Health Impacts of Air Pollution: A Review. *Front. Public Health* **2020**, *8*, 14. [CrossRef] [PubMed]
3. Mei, H.; Zhang, F.; Zhou, T.; Zhang, T. Pulse-Driven MEMS $NO_2$ Sensors Based on Hierarchical $In_2O_3$ Nanostructures for Sensitive and Ultra-Low Power Detection. *Sensors* **2024**, *24*, 7188. [CrossRef] [PubMed]
4. Recum, P.; Hirsch, T. Graphene-Based Chemiresistive Gas Sensors. *Nanoscale Adv.* **2024**, *6*, 11–31. [CrossRef]
5. Sharma, A.; Eadi, S.B.; Noothalapati, H.; Otyepka, M.; Lee, H.-D.; Jayaramulu, K. Porous Materials as Effective Chemiresistive Gas Sensors. *Chem. Soc. Rev.* **2024**, *53*, 2530–2577. [CrossRef]
6. Ma, J.; Zhang, M.; Dong, L.; Sun, Y.; Su, Y.; Xue, Z.; Di, Z. Gas Sensor Based on Defective Graphene/Pristine Graphene Hybrid towards High Sensitivity Detection of $NO_2$. *AIP Adv.* **2019**, *9*, 075207. [CrossRef]
7. Yuan, W.; Shi, G. Graphene-Based Gas Sensors. *J. Mater. Chem. A Mater.* **2013**, *1*, 10078. [CrossRef]
8. Ghosh, R.; Aslam, M.; Kalita, H. Graphene Derivatives for Chemiresistive Gas Sensors: A Review. *Mater. Today Commun.* **2022**, *30*, 103182. [CrossRef]
9. Walleni, C.; Malik, S.B.; Missaoui, G.; Alouani, M.A.; Nsib, M.F.; Llobet, E. Selective $NO_2$ Gas Sensors Employing Nitrogen- and Boron-Doped and Codoped Reduced Graphene Oxide. *ACS Omega* **2024**, *9*, 13028–13040. [CrossRef]
10. Wang, H.; Maiyalagan, T.; Wang, X. Review on Recent Progress in Nitrogen-Doped Graphene: Synthesis, Characterization, and Its Potential Applications. *ACS Catal.* **2012**, *2*, 781–794. [CrossRef]
11. Ouyang, W.; Zeng, D.; Yu, X.; Xie, F.; Zhang, W.; Chen, J.; Yan, J.; Xie, F.; Wang, L.; Meng, H.; et al. Exploring the Active Sites of Nitrogen-Doped Graphene as Catalysts for the Oxygen Reduction Reaction. *Int. J. Hydrogen Energy* **2014**, *39*, 15996–16005. [CrossRef]
12. Ariharan, A.; Viswanathan, B.; Nandhakumar, V. Nitrogen Doped Graphene as Potential Material for Hydrogen Storage. *Graphene* **2017**, *6*, 41–60. [CrossRef]
13. Shaik, M.; Rao, V.K.; Gupta, M.; Murthy, K.S.R.C.; Jain, R. Chemiresistive Gas Sensor for the Sensitive Detection of Nitrogen Dioxide Based on Nitrogen Doped Graphene Nanosheets. *RSC Adv.* **2016**, *6*, 1527–1534. [CrossRef]
14. Chang, Y.-S.; Chen, F.-K.; Tsai, D.-C.; Kuo, B.-H.; Shieu, F.-S. N-Doped Reduced Graphene Oxide for Room-Temperature NO Gas Sensors. *Sci. Rep.* **2021**, *11*, 20719. [CrossRef]
15. Mirzaei, A.; Bharath, S.P.; Kim, J.-Y.; Pawar, K.K.; Kim, H.W.; Kim, S.S. N-Doped Graphene and Its Derivatives as Resistive Gas Sensors: An Overview. *Chemosensors* **2023**, *11*, 334. [CrossRef]
16. Wang, C.; Yin, L.; Zhang, L.; Xiang, D.; Gao, R. Metal Oxide Gas Sensors: Sensitivity and Influencing Factors. *Sensors* **2010**, *10*, 2088–2106. [CrossRef]
17. Yang, S.; Lei, G.; Xu, H.; Lan, Z.; Wang, Z.; Gu, H. Metal Oxide Based Heterojunctions for Gas Sensors: A Review. *Nanomaterials* **2021**, *11*, 1026. [CrossRef]
18. Masroor, S. Basics of Metal Oxides: Properties and Applications. In *Inorganic Anticorrosive Materials*; Elsevier: Amsterdam, The Netherlands, 2022; pp. 85–94.
19. Khomarloo, N.; Mohsenzadeh, E.; Gidik, H.; Bagherzadeh, R.; Latifi, M. Overall Perspective of Electrospun Semiconductor Metal Oxides as High-Performance Gas Sensor Materials for NO*x* Detection. *RSC Adv.* **2024**, *14*, 7806–7824. [CrossRef]
20. Fine, G.F.; Cavanagh, L.M.; Afonja, A.; Binions, R. Metal Oxide Semi-Conductor Gas Sensors in Environmental Monitoring. *Sensors* **2010**, *10*, 5469–5502. [CrossRef]

21. Ayyala, S.K.; Covington, J.A. Nickel-Oxide Based Thick-Film Gas Sensors for Volatile Organic Compound Detection. *Chemosensors* **2021**, *9*, 247. [CrossRef]
22. Deshpande, M.P.; Patel, K.N.; Gujarati, V.P.; Patel, K.; Chaki, S.H. Structural, Thermal and Optical Properties of Nickel Oxide (NiO) Nanoparticles Synthesized by Chemical Precipitation Method. *Adv. Mater. Res.* **2016**, *1141*, 65–71. [CrossRef]
23. Walleni, C.; Hamdaoui, N.; Malik, S.B.; Nsib, M.F.; Llobet, E. High-Performance Uv Photodetector Based on Nickel Oxide Loaded with Low Amount of Nitrogen and Boron Co-Doped Reduced Graphene Oxide for Bias-Switchable Photoconductance. *J. Alloys Compd.* **2024**, *976*, 173248. [CrossRef]
24. Khaleed, A.A.; Bello, A.; Dangbegnon, J.K.; Madito, M.J.; Ugbo, F.U.; Akande, A.A.; Dhonge, B.P.; Barzegar, F.; Momodu, D.Y.; Mwakikunga, B.W.; et al. Gas Sensing Study of Hydrothermal Reflux Synthesized NiO/Graphene Foam Electrode for CO Sensing. *J. Mater. Sci.* **2017**, *52*, 2035–2044. [CrossRef]
25. Kamal, T. High Performance NiO Decorated Graphene as a Potential $H_2$ Gas Sensor. *J. Alloys Compd.* **2017**, *729*, 1058–1063. [CrossRef]
26. Srirattanapibul, S.; Nakarungsee, P.; Issro, C.; Tang, I.-M.; Thongmee, S. Performance of NiO Intercalated RGO Nanocomposites for NH3 Sensing at Room Temperature. *Mater. Sci. Semicond. Process* **2022**, *137*, 106221. [CrossRef]
27. Shanavas, S.; Ahamad, T.; Alshehri, S.M.; Acevedo, R.; Anbarasan, P.M. A Facile Microwave Route for Fabrication of NiO/RGO Hybrid Sensor with Efficient $CO_2$ and Acetone Gas Sensing Performance Using Clad Modified Fiber Optic Method. *Optik* **2021**, *226*, 165970. [CrossRef]
28. Shanmugasundaram, A.; Chinh, N.D.; Jeong, Y.-J.; Hou, T.F.; Kim, D.-S.; Kim, D.; Kim, Y.-B.; Lee, D.-W. Hierarchical Nanohybrids of B- and N-Codoped Graphene/Mesoporous NiO Nanodisks: An Exciting New Material for Selective Sensing of H 2 S at near Ambient Temperature. *J. Mater. Chem. A Mater.* **2019**, *7*, 9263–9278. [CrossRef]
29. Muzyka, R.; Kwoka, M.; Smędowski, Ł.; Díez, N.; Gryglewicz, G. Oxidation of Graphite by Different Modified Hummers Methods. *New Carbon. Mater.* **2017**, *32*, 15–20. [CrossRef]
30. Xu, C.; Li, Y.; Adams, R.A.; Pol, V.G.; Xiao, Y.; Varma, A.; Chen, P. One-Step Combustion Synthesis of Carbon-Coated NiO/Ni Composites for Lithium and Sodium Storage. *J. Alloys Compd.* **2021**, *884*, 160927. [CrossRef]
31. Li, X.; Zhong, A.; Wei, S.; Luo, X.; Liang, Y.; Zhu, Q. Polyelectrolyte Functionalized Gold Nanoparticles-Reduced Graphene Oxide Nanohybrid for Electrochemical Determination of Aminophenol Isomers. *Electrochim. Acta* **2015**, *164*, 203–210. [CrossRef]
32. Salavati-Niasari, M.; Davar, F.; Fereshteh, Z. Synthesis of Nickel and Nickel Oxide Nanoparticles via Heat-Treatment of Simple Octanoate Precursor. *J. Alloys Compd.* **2010**, *494*, 410–414. [CrossRef]
33. Saikia, J.P.; Paul, S.; Konwar, B.K.; Samdarshi, S.K. Nickel Oxide Nanoparticles: A Novel Antioxidant. *Colloids Surf. B Biointerfaces* **2010**, *78*, 146–148. [CrossRef] [PubMed]
34. Deep Yadav, D.; Jha, R.; Singh, S.; Kumar, A. Synthesis and Characterisation of Nickel Oxide Nanoparticles Using CTAB as Capping Agent. *Mater. Today Proc.* **2023**, *73*, 333–336. [CrossRef]
35. Rochman, R.A.; Wahyuningsih, S.; Ramelan, A.H. Preparation of Nitrogen and Sulphur Co-Doped Reduced Graphene Oxide (RGO-NS) Using N and S Heteroatom of Thiourea. *IOP Conf. Ser. Mater. Sci. Eng.* **2019**, *509*, 012119. [CrossRef]
36. Duan, X.; Indrawirawan, S.; Sun, H.; Wang, S. Effects of Nitrogen-, Boron-, and Phosphorus-Doping or Codoping on Metal-Free Graphene Catalysis. *Catal Today* **2015**, *249*, 184–191. [CrossRef]
37. Beams, R.; Gustavo Cançado, L.; Novotny, L. Raman Characterization of Defects and Dopants in Graphene. *J. Phys. Condens. Matter* **2015**, *27*, 083002. [CrossRef]
38. Dharmaraj, N.; Prabu, P.; Nagarajan, S.; Kim, C.H.; Park, J.H.; Kim, H.Y. Synthesis of Nickel Oxide Nanoparticles Using Nickel Acetate and Poly(Vinyl Acetate) Precursor. *Mater. Sci. Eng. B* **2006**, *128*, 111–114. [CrossRef]
39. Kaschner, A.; Hoffmann, A.; Thomsen, C. Resonant Raman Scattering on Free and Bound Excitons in GaN. *Phys. Rev. B* **2001**, *64*, 165314. [CrossRef]
40. Usharani, N.J.; Bhattacharya, S.S. Effect of Defect States in the Optical and Magnetic Properties of Nanocrystalline NiO Synthesised in a Single Step by an Aerosol Process. *Ceram. Int.* **2020**, *46*, 5671–5680. [CrossRef]
41. Chu, S.-Y.; Wu, M.-J.; Yeh, T.-H.; Lee, C.-T.; Lee, H.-Y. Investigation of High-Sensitivity $NO_2$ Gas Sensors with $Ga_2O_3$ Nanorod Sensing Membrane Grown by Hydrothermal Synthesis Method. *Nanomaterials* **2023**, *13*, 1064. [CrossRef]
42. Malik, S.B.; Annanouch, F.E.; D'Souza, R.; Bittencourt, C.; Todorović, M.; Llobet, E. High-Yield $WS_2$ Synthesis through Sulfurization in Custom-Modified Atmospheric Pressure Chemical Vapor Deposition Reactor, Paving the Way for Selective $NH_3$ Vapor Detection. *ACS Appl. Mater. Interfaces* **2024**, *16*, 48585–48597. [CrossRef] [PubMed]
43. Tan, G.; Tang, D.; Wang, X.; He, L.; Mu, T.; Li, G. Preparation of NiO Thin Films and Their Application for $NO_2$ Gas Detection. *Int. J. Electrochem. Sci.* **2022**, *17*, 220551. [CrossRef]
44. Li, J.; Lu, Y.; Ye, Q.; Cinke, M.; Han, J.; Meyyappan, M. Carbon Nanotube Sensors for Gas and Organic Vapor Detection. *Nano Lett.* **2003**, *3*, 929–933. [CrossRef]
45. Verma, R.; Chauhan, M.S.; Pandey, S.; Dandia, A. Reduced Graphene Oxide/NiO Based Nano-Composites for the Efficient Removal of Alizarin Dye, Indigo Dye and Reduction of Nitro Aromatic Compounds. *Heliyon* **2023**, *9*, e17162. [CrossRef]

46. Yan, W.; Worsley, M.A.; Pham, T.; Zettl, A.; Carraro, C.; Maboudian, R. Effects of Ambient Humidity and Temperature on the $NO_2$ Sensing Characteristics of $WS_2$/Graphene Aerogel. *Appl. Surf. Sci.* **2018**, *450*, 372–379. [CrossRef]
47. Song, X.; Li, L.; Chen, X.; Xu, Q.; Song, B.; Pan, Z.; Liu, Y.; Juan, F.; Xu, F.; Cao, B. Enhanced Triethylamine Sensing Performance of $\alpha$-$Fe_2O_3$ Nanoparticle/ZnO Nanorod Heterostructures. *Sens. Actuators B Chem.* **2019**, *298*, 126917. [CrossRef]
48. Bai, H.; Guo, H.; Wang, J.; Dong, Y.; Liu, B.; Xie, Z.; Guo, F.; Chen, D.; Zhang, R.; Zheng, Y. A Room-Temperature $NO_2$ Gas Sensor Based on CuO Nanoflakes Modified with RGO Nanosheets. *Sens. Actuators B Chem.* **2021**, *337*, 129783. [CrossRef]
49. Choudhari, A.; Bhanvase, B.A.; Saharan, V.K.; Salame, P.H.; Hunge, Y. Sonochemical Preparation and Characterization of RGO/$SnO_2$ Nanocomposite: Electrochemical and Gas Sensing Performance. *Ceram. Int.* **2020**, *46*, 11290–11296. [CrossRef]
50. Hoa, L.T.; Tien, H.N.; Luan, V.H.; Chung, J.S.; Hur, S.H. Fabrication of a Novel 2D-Graphene/2D-NiO Nanosheet-Based Hybrid Nanostructure and Its Use in Highly Sensitive $NO_2$ Sensors. *Sens. Actuators B Chem.* **2013**, *185*, 701–705. [CrossRef]
51. Cao, P.; Cai, Y.; Pawar, D.; Navale, S.T.; Rao, C.N.; Han, S.; Xu, W.; Fang, M.; Liu, X.; Zeng, Y.; et al. Down to Ppb Level $NO_2$ Detection by ZnO/RGO Heterojunction Based Chemiresistive Sensors. *Chem. Eng. J.* **2020**, *401*, 125491. [CrossRef]
52. Gomaa, M.M.; Sayed, M.H.; Patil, V.L.; Boshta, M.; Patil, P.S. Gas Sensing Performance of Sprayed NiO Thin Films toward $NO_2$ Gas. *J. Alloys Compd.* **2021**, *885*, 160908. [CrossRef]
53. Jyoti; Varma, G.D. Morphology-Dependent Room Temperature $NO_2$ Detection of CuO Nanostructure/RGO Composites. *Appl. Phys. A* **2020**, *126*, 143. [CrossRef]
54. Wang, Z.; Gao, S.; Fei, T.; Liu, S.; Zhang, T. Construction of ZnO/$SnO_2$ Heterostructure on Reduced Graphene Oxide for Enhanced Nitrogen Dioxide Sensitive Performances at Room Temperature. *ACS Sens.* **2019**, *4*, 2048–2057. [CrossRef] [PubMed]
55. Alouani, M.A.; Casanova-Chafer, J.; de Bernardi-Martín, S.; García-Gómez, A.; Salehnia, F.; Santos-Ceballos, J.C.; Santos-Betancourt, A.; Vilanova, X.; Llobet, E. The Effect of Doping RGO with Nanosized $MnO_2$ on Its Gas Sensing Properties. *Chemosensors* **2024**, *12*, 256. [CrossRef]
56. Adjizian, J.-J.; Leghrib, R.; Koos, A.A.; Suarez-Martinez, I.; Crossley, A.; Wagner, P.; Grobert, N.; Llobet, E.; Ewels, C.P. Boron- and Nitrogen-Doped Multi-Wall Carbon Nanotubes for Gas Detection. *Carbon* **2014**, *66*, 662–673. [CrossRef]
57. Bai, S.; Han, J.; Meng, J.C.; Sun, L.; Sun, J.; Zhao, Y.; Tang, P.; Luo, R.; Li, D.; Chen, A. NiO/ZnO Composite Decorated on RGO for Detection of $NO_2$. *Sens. Actuators B Chem.* **2021**, *339*, 129720. [CrossRef]

*sensors*

MDPI

*Article*

# Membrane Permeability Monitoring to Antipsychotic Olanzapine Using Platinum Black-Modified Electrodes

**Murugaiya Sridar Ilango [1,†], Dayananda Desagani [1,†], Srikanth Jagadeesan [2], Alexander Snezhko [1], Gad Vatine [2] and Hadar Ben-Yoav [1,*]**

[1] Nanobioelectronics Laboratory, Department of Biomedical Engineering, Ben-Gurion University of the Negev, Beer Sheva 8410501, Israel; niranjansridar@gmail.com (M.S.I.); dayanandabits@gmail.com (D.D.)
[2] Department of Physiology and Cell Biology, Faculty of Health Sciences, Ben-Gurion University of the Negev, Beer Sheva 8410501, Israel; vatineg@bgu.ac.il
* Correspondence: benyoav@bgu.ac.il
† These authors contributed equally to this work.

**Abstract:** The blood–brain barrier (BBB) is key to the regular functioning of the central nervous system. The dysfunction of the BBB has been described in various neurological disorders, including schizophrenia. Schizophrenia (SCZ) is a chronic psychiatric disorder described by hallucinations, delusions, and negative symptoms. The Olanzapine (OLZ) drug is an electroactive species, and its levels can be monitored using electrochemical sensors. The detection of OLZ was demonstrated previously by using electrochemical sensors, and this technique can be used to monitor the levels of OLZ in real time. The challenge is to identify the permeability of OLZ through the BBB, so a replica model was designed with the BBB based on a Transwell membrane seeded with endothelial cells. A microfabricated electrode consisting of a 3 mm Au disk was modified with platinum black; this enables higher selectivity of electrochemical signals from OLZ. The dose–response of OLZ was characterized in phosphate buffer saline solution (10 mM, pH 7.4) by adding 20–200 nM (in steps 20) of OLZ stock solution. The observed chronoamperometric electrochemical signals showed an increasing current at 0.45 V vs. Ag/AgCl with an increasing OLZ concentration. The controls for the experiments were performed in phosphate-buffered saline solution (10 mM, pH 7.4). The detection limit was calculated as $9.96 \pm 7.35 \times 10^{-6}$ nM from the calibration curve. The membrane permeability of the OLZ drug tested with five SCZ patients was monitored by studying the TEER measurements and permeability rate constant data.

**Keywords:** olanzapine; platinum black; blood–brain barrier; electrochemical sensors; schizophrenia

## 1. Introduction

Olanzapine (2-Methyl-4-(4-methyl-1-piperazinyl)-10H-thieno[2,3-b][1,5]benzodiazepine) is an antipsychotic drug used for schizophrenia (SCZ) diseases such as alogia, anhedonia, avolition, and other psychotic diseases [1–3]. Olanzapine (OLZ) is a widely used second-generation antipsychotic drug for the treatment of schizophrenia and bipolar disorder due to its broad-spectrum efficacy and fewer side effects compared to first-generation antipsychotics [2,4–6]. Schizophrenia is a multifactorial disease that can be treated most effectively using drugs that interact with multiple neurotransmitter systems [3,6,7]. Olanzapine is one of these drugs and shows a high affinity for neurotransmitters such as dopamine D1, D2, D4, serotonin 5-HT2A, 5-HT2C, 5-HT3, α1-adrenergic histamine H1, and muscarinic receptors [8–11].

The pharmacological action of OLZ largely depends on its ability to cross the blood–brain barrier (BBB) efficiently and interact with dopamine and serotonin receptors within the central nervous system (CNS) [12]. However, predicting its BBB permeability and systemic responsiveness remains a significant challenge in drug development and therapeutic optimization [13,14]. The role of the BBB is to control the transportation of various metabolites, which consist of amino acids, lipids, peptides, nucleic acids, carbohydrates, vitamins, and minerals, along with drugs used for CNS-related disease treatment, between the blood and brain compartments. In this way, the BBB acts as both a safeguard and preserves the interstitial conditions of the CNS [15]. It remains unclear to what extent OLZ, particularly at dosages near the upper limit of its therapeutic range, influences the function of BBB cells. Such effects may have significant implications for the transport of OLZ across the BBB, its pharmacokinetics, and its overall impact on CNS function and clinical outcomes. Notably, OLZ concentrations comparable to those observed in the serum of patients receiving high doses have been shown to induce substantial changes in BBB permeability [16,17]. Psychostimulant drugs of abuse alter BBB function and increase permeability, likely contributing to their associated neurotoxicities. For instance, acute administration of 5 mM chlorpromazine increases the permeability of the BBB to $^{59}Fe^{3+}$, mannitol, and inulin [18,19]. This observed effect suggests a breach in the BBB at a drug concentration comparable to that occasionally detected in human serum (3 mM, 1000 ng/mL). Detecting OLZ in Transwell systems seeded with cells is crucial for understanding its pharmacokinetics and pharmacodynamics at the cellular level [20]. Accurate measurement of OLZ concentrations in this in vitro model helps to study its cellular uptake, distribution, metabolism, and potential cytotoxic effects. This information is vital for optimizing dosing regimens and improving therapeutic outcomes. Wang et al. investigated the penetration of OLZ into mice brains [16]. The results indicated that the expression of p-glycoprotien in the BBB significantly limits the penetration of olanzapine into the central nervous system. Moreover, the OLZ concentration was analyzed using the HPLC technique, which is a very expensive tool and time-consuming analysis. Traditional methods to assess BBB permeability, such as in vivo models and in vitro cell-based assays, often involve labor-intensive and time-consuming protocols [21]. Among in vitro techniques, the measurement of transepithelial electrical resistance (TEER) has been a cornerstone for evaluating BBB integrity, yet it lacks the precision to quantify drug penetration rates [22–24]. Developing sensitive, cost effective, and reproducible analytical platforms is essential to overcome these limitations. Numerous analytical techniques have been reported for the quantification of OLZ in pure as well as in dosage form. These include high-performance liquid chromatography (HPLC) [9,25–27], mass spectroscopy (MS) [26], gas chromatography (GC) [28], GC-MS and liquid chromatography coupled with MS [26,29], spectrophotometry [30], and electroanalytical techniques [31] and in combination with these techniques [32]. However, traditional chromatographic, spectroscopic, and analytical methods exhibit several limitations, including low sensitivity, high sample volumes, expensive equipment, complex sample pretreatment and dilution requirements, and time-intensive extraction and separation procedures. In contrast, electroanalytical methods offer notable advantages, such as rapid response, simplicity, cost-effectiveness, broad applicability in both qualitative and quantitative analyses, high accuracy, and the potential for miniaturization. These distinctive attributes have led to the widespread adoption of electrochemical methods for the analysis of environmental and pharmaceutical samples in recent years.

There are different materials used for olanzapine estimation by electrochemical techniques. Azab et al. used polyethylene glycol and silver nanoparticles to improve the sensitivity of the carbon paste electrode for electrochemical estimation of OLZ [33]. Ahmed et al. used the voltammetric technique to determine olanzapine in tablets and human urine samples with modified carbon paste electrodes by incorporating gold nanoparticles and glutamine in a micellar medium [34]. Platinum-black-modified electrodes, in particular, offer enhanced surface area and catalytic activity, making them highly suitable for detecting low-concentration analytes [35–39]. Previous studies have demonstrated the utility of such sensors in monitoring small-molecule drugs, but their application in antipsychotic permeability assessments, especially for OLZ, remains unexplored.

In our previous study, the OLZ was analyzed by an electrochemical technique using platinum-black-modified microelectrodes [35]. The OLZ content in undiluted serum with a limit of detection (LOD) of 28.6 ± 1.3 nM and a sensitivity of 0.14 ± 0.02 $\mu A/cm^2$ nM was analyzed. In this study, platinum-black-modified electrodes were utilized to monitor OLZ concentrations across an in vitro BBB model. The sensor-based data were correlated with TEER measurements to provide insights into OLZ permeability dynamics. The present work on the OLZ sample diluted with organic solvent achieved an LOD of $9.96 \pm 7.35 \times 10^{-6}$ nM and a sensitivity of 0.027 ± 0.001 $\mu A/cm^2$ nM. In this study, the permeability of OLZ through the BBB was investigated. The OLZ concentration was analyzed using an electrochemical technique using the platinum-black-modified electrode.

## 2. Materials and Methods

The following chemicals and materials are used in the experiment: OLZ (CAS number: 132539-06-1, Sigma-Aldrich, College park, MD, USA), sodium chloride (CAS number: 7647-14-5, Merck, Nantong, China), 2-propanol (CAS number: 67-63-0, Bio-Lab, Ltd., Jerusalem, Israel), potassium hexacyanoferrate (II) trihydrate ('Ferrocyanide,' CAS number: 14459-95-1, Merck, Darmstadt, Germany), Dihydrogen hexachloroplatinate (IV) hexahydrate (chloroplatinic acid; CAS number: 26023-84-7, Alfa Aesar, Petach Tikva, Israel), potassium hexacyanoferrate(III) ('Ferricyanide,' CAS number: 13746-66-2, Merck, Darmstadt, Germany), 99% Lead(II) acetate trihydrate (lead acetate; CAS number: 6080-56-4, Alfa Aesar, Petach Tikva, Israel), hydrochloric acid 32% (CAS number: 7647-01-0, Bio-Lab, Ltd, Jerusalem, Israel), di-sodium hydrogen phosphate dehydrate (CAS number: 10028-24-7, Merck, Raanana, Israel), sodium dihydrogen phosphate dihydrate (CAS number: 13472-35-0, Merck, Darmstadt, Germany), acetone (CAS number 67-64-1, Sigma-Aldrich, Darmstadt, Germany), and deuterium- depleted water (>18 MΩ) from Millipore (Millipore system, Thermo Scientific, Waltham, MA, USA). The solutions were diluted using PBS solution (10 mM, pH 7.4). The electrochemical experiments were conducted using the three-electrode system, a platinum wire counter electrode (catalog number: 012961, ALS Co., Ltd, Tokyo Japan), an Ag/AgCl reference electrode (catalog number: 011464, BAS, Inc., Kent Avenue, West Lafayette, USA), and an in-house fabricated gold electrodes on a glass substrate as the working electrode.

The OLZ solutions were prepared in 2-propanol [35] and stored at −20 °C. The calibration was performed by mixing the OLZ solution with PBS (10 mM, pH 7.4). The preparation of the platinum black solution [40], fabrication of the electrodes, and electrodeposition of the platinum black solution are elaborated on in the Supplementary File. The effective surface area of the platinum-black-modified electrodes was calculated using the Randles–Sevick relationship [41,42] is $5.23 \times 10^{-2} \pm 2.3 \times 10^{-3}$ $cm^2$, which is higher than the bare Au electrode, $3.81 \times 10^{-2} \pm 1.2 \times 10^{-3}$ $cm^2$. The higher surface area is due to non-identical and uneven roughness on the surface of the platinum black [43].

The induced pluripotent stem cells are collected from five different cell lines (Labelled as Line 1, 3, 5, 7, and 9) and used to prepare brain microvascular endothelial cells. These cells were transferred to the Transwell to identify the movement of OLZ through them to analyze the effect of the OLZ drug on different individual lines. The preparation of the Transwell is explained schematically in Figure 1. The samples were collected at five different times (0, 3, 6, 9, and 12 min) from the basolateral side. We collected a drug sample immediately after introducing it in the upper chamber to determine its initial concentration. Detecting the concentration of drugs that permeate from top to bottom was the objective. Hence, we collected and detected drug concentrations in the lower chamber. The electrochemical behavior of the OLZ was measured using chronopotentiometry before and after inducing the drug for all the individual lines at different concentrations with different potentials.

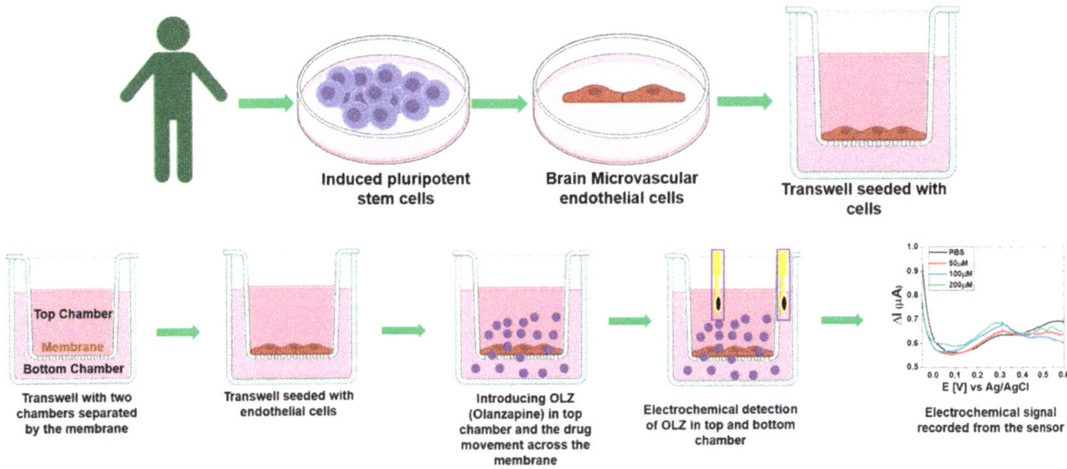


**Figure 1.** Schematic of preparing the BBB membrane from human induced pluripotent stem cells in a dual Transwell separated by a membrane and the introduction of OLZ with electrochemical measurements.

TEER measurements on Transwells were taken using an STX2 electrode and an EVOM2 Voltammeter (World Precision Instruments, Sarasota, FL, USA). This epithelial voltammeter uses two pairs of electrodes, one applying current and the other measuring the output voltage, on each side of a monolayer, for example, a Transwell membrane [23]. After 24 h of plating iBMECs onto Transwells, TEER measurements were taken every 24 h for three subsequence days. The STX2 electrode was positioned within the well and the resistance (Ω) was taken once equilibrated, before repeating in two more locations on each Transwell, to calculate the mean resistance. The resistance of the BBB is measured to identify the dose requirement for different individuals. The correlation between the TEER value and permeability was investigated to indicate the effect of OLZ on different individuals.

The platinum-black-modified electrodes were used to detect OLZ, and their electrochemical signals were recorded for the stock solution and solutions obtained from the Transwell. A chronoamperometry (CA) technique with a potential of 0.45 V was used to calibrate the response of OLZ. From the calibration curve, the LOD and sensitivity were analyzed. The response and the OLZ concentration curve were plotted to find the slope of the linear regression, which shows the sensitivity, and three standard deviations give the LOD. The modified electrodes were used to test different concentrations of OLZ ranging from 20 to 200 nM in steps of 20 nM with three modified electrodes. These three platinum-black-modified electrodes were used for measuring the OLZ signals with 10 repeats. The background signal was taken from 10 mM PBS solution with 40 repeats. After every measurement, the electrodes were cleaned with deionized water and dried using nitrogen gas.

## 3. Results and Discussion

### 3.1. Olanzapine Sensing

The sensing performance of the platinum-black-modified electrode for olanzapine was characterized with cyclic voltammetry (CV) of 100 nM OLZ in PBS solution and compared with CV of PBS solution (Figure 2A). The plot showed the peak around 0.4 V. To further characterize the OLZ, the chronopotentiometry technique was used by measuring current at three different potentials 0.35, 0.4, and 0.45 (Figure 2B). The response at 0.45 V shows a linear correspondence with the increase in OLZ concentration from 20 nM to 200 nM. So, the response at 0.45 V is considered for further analysis and sensitivity, and LOD calculations (Figure 2C) from Equations (1) and (2).

$$\text{LOD} = (3 \times \text{Std. Dev of PBS})/\text{Slope of the analyte from the calibration curve} \quad (1)$$

$$\text{Error of LOD} = (3 \times \text{Std. Dev} \times \text{Error of slope})/(\text{slope})^2 \quad (2)$$

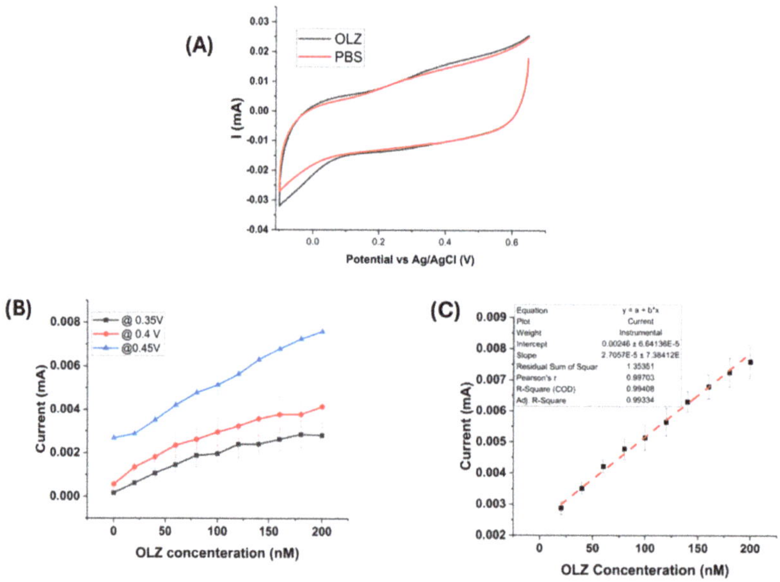


**Figure 2.** (**A**) Cyclic voltammetry of 100 nM OLZ in PBS solution and PBS solution, (**B**) Chronoamperometric current measured at 20 s for different concentrations of OLZ with three different potentials (0.35, 0.4, and 0.45 V), (**C**) Dose–response plot for platinum-black-modified electrode at 0.45 V.

The sensitivity of the measured response was $0.027 \pm 0.001\ \mu\text{A/cm}^2$, and LOD was $9.96 \pm 7.35 \times 10^{-6}$ nM.

### 3.2. Rate Constant for Precision Dosing of OLZ

The platinum-black-modified electrodes were used to test the response from the samples obtained (Transwell). The samples are obtained from schizophrenia patients. This approach to SCZ patients can contribute to understanding the molecular mechanism of SCZ and the pathophysiology of treatment resistance. Human-induced pluripotent stem cells (hiPSCs) give an interesting new road to investigate the function of BBB disruption [44]. We aimed to generate iPSCs from five SCZ patients. We utilized these iPSCs to induce pluripotent stem cell-derived brain microvascular endothelial cells (iBMECs) and developed in vitro models for the BBB to study the function in SCZ compared to HC. iBMECs were seeded on Transwell and barrier properties were evaluated by measuring TEER and paracellular permeability. The TEER values commonly examined the permeability of the epithelial and endothelial cell monolayers [45]. The values of TEER state the resistance to an electrical current passed through the cell monolayer to assess permeability to small inorganic ions [45]. Table 1 illustrates the variation in TEER values for different lines and also the effect of OLZ on the TEER value for all the membranes. Reduced TEER values indicate higher permeability and impaired barrier integrity. The permeability increased with the increase in OLZ concentration, which is the reason for the higher current observed in CA plots.

**Table 1.** TEER value of each line taken before the permeability of OLZ and after the permeability of the OLZ.

| Lines | Line 1 (Ω cm$^2$) | Line 3 (Ω cm$^2$) | Line 5 (Ω cm$^2$) | Line 7 (Ω cm$^2$) | Line 9 (Ω cm$^2$) |
|---|---|---|---|---|---|
| Before Permeability | 1879 | 2689 | 2877 | 3145 | 2851 |
| After Permeability | 1496 | 2371 | 2613 | 2773 | 2579 |

The media were collected from five different patients from the bottom of the Transwell after adding OLZ drug at 100 nM concentration. The samples were collected at five different times (0, 3, 6, 9, and 12 min) from the time of OLZ addition. These samples were tested for OLZ response using platinum-black-modified electrodes for all the patients. Figure 3 shows the results of the measured current for different samples at 20 s of chronoamperometry (CA). Lower TEER values exhibit a high transfer of OLZ through the membrane, which is reflected in the higher current in CA. Line 1 shows a higher current, which has a lower TEER value compared to other cell lines [46].

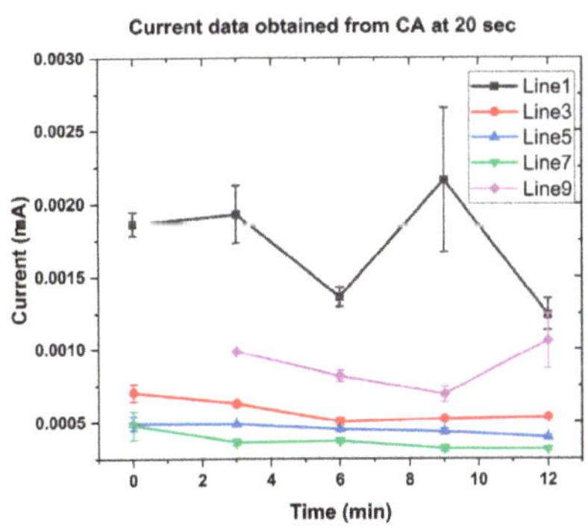


**Figure 3.** Current data from chronoamperometry measured at 20 s for different lines at 0.45 V.

The CA data obtained for all the cell lines at different times (0, 3, 6, 9, and 12 min) were normalized using the following equation. After normalization, the current at 20 s was used in Figure 4.

$$I_{normalized} = \frac{\mathbf{I}_{ij}}{\sqrt{\Sigma_j \mathbf{I}_{ij}^2}} \tag{3}$$

where 'i' is the time [min] and 'j' represents current [A].

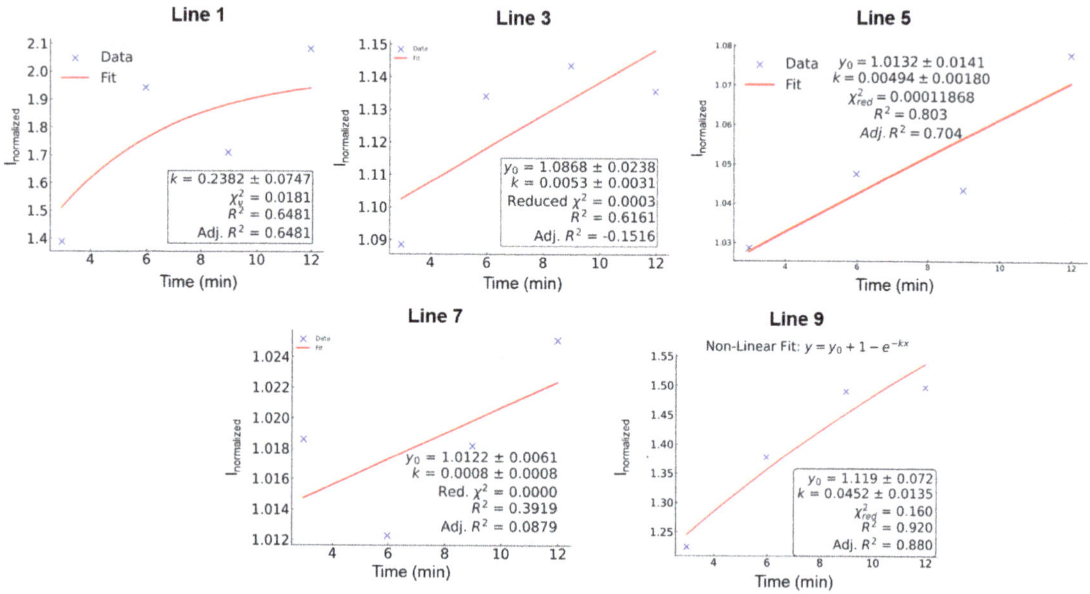


**Figure 4.** Permeability rate constant calculation of 5 SCZ patient's cell lines (1, 3, 5, 7, and 9) from the normalized current data obtained after the drug transportation through the membrane.

The permeability rate constant is calculated using the normalized response plot of the platinum-black-modified electrode (Equation (5)) [47,48].

$$\text{Normalized concentration} = \frac{C_t}{C_0} = 1 - e^{-kt} \tag{4}$$

Here, $\frac{C}{C_0}(t \rightarrow \infty) = 1 \frac{C}{C_0}(t = 0) = 0$

$$\mathbf{as C(t)} \propto \mathbf{I(t)},\ I_{normalized} = 1 - \mathbf{e}^{-\mathbf{kt}} \tag{5}$$

where $C_t$ is the concentration of OLZ that passed the barrier (bottom area), $C_0$ is the initial concentration of OLZ that was in the top area, I is the response current [A], k is the rate constant, and t is the time [min].

The above equation can qualitatively describe drug transport through a membrane, particularly in scenarios where drug accumulation follows first-order kinetics. The equation assumes a single-compartment model and may not fully capture complex active transport mechanisms or saturation kinetics (e.g., facilitated diffusion or carrier-mediated transport). The rate constant describes how fast the drug permeates through the membrane. It is directly influenced by factors such as membrane thickness, drug lipophilicity, surface area available for diffusion, and drug concentration gradient [49,50]. If passive diffusion is the primary transport mechanism, this equation provides an exponential growth model for drug accumulation. The model is plotted with an orthogonal distance regression iteration algorithm to fit the normalized current data from each line. Figure 4 shows the fitting of normalized current data with Equation (5) to determine the permeability rate constant. The nonlinear fitting of the data (Figure 4) for all the cell lines is not the best fit; however, to calculate the permeability rate constant, we considered the drug accumulation as first-order kinetics and calculated the rate constant. The rate at which drug transport through the membrane approaches equilibrium is determined by the permeability rate constant. A higher k means faster transport. The k values determined from the nonlinear fitting (Figure 4) are plotted along with TEER values in Figure 5 for the five SCZ patients (cell lines 1, 3, 5, 7, and 9). The data show that SCZ patients of lines 1 and 5 have faster transportation of the OLZ drug through the membrane, whereas 3, 7, and 9 cell line patients have slower transportation of the OLZ drug. The TEER value is a measure of ionic permeability through intercellular clefts [45]. Estimation of TEER guarantees noninvasive determination, which can also evaluate the barrier integrity of epithelial or endothelial cells at various stages of differentiation and growth [23]. The TEER is influenced by other parameters, including medium composition, temperature, and use of the equipment [51]. These parameters may also cause changes in the TEER values. The correlation between the TEER values and the permeability rate constant for the transport of the OLZ drug across the membrane is illustrated in Figure 5. Based on the TEER data and the predicted permeability rate constants, the membrane permeability of OLZ across the cell lines derived from five SCZ patients follows the order: Line 7 < Line 9 < Line 3 < Line 5 < Line 1. However, it is important to note that the TEER values and permeability rate constants do not exhibit a consistent trend across all cell lines; moreover, they are from five different SCZ patients. This discrepancy can be attributed to differences in the experimental methods and instrumentation used for data collection.

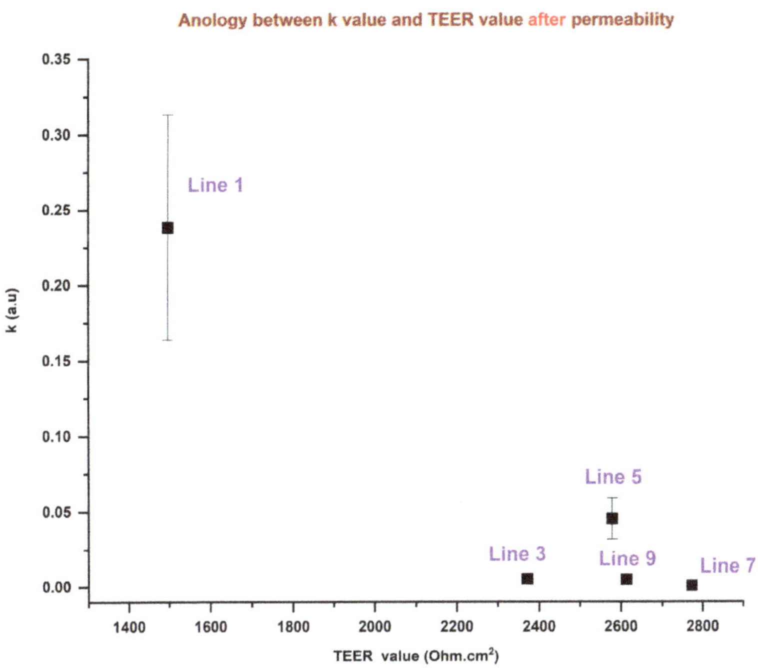


**Figure 5.** Correlation between the TEER value and permeability rate constant of OLZ.

For TEER measurement, the positioning of the chopstick electrodes is critical and requires careful attention, as improper placement can lead to inaccurate resistance readings when compared to measurements obtained through other techniques, such as electrochemical analysis using a potentiostat. In some cases, it may be difficult to accurately assess high drug permeability using TEER measurements, due to errors introduced by inaccurate resistivity values. Furthermore, the integrity of the cell monolayer may be compromised during transfer procedures or due to fluctuations in environmental conditions, such as temperature and pH, which can lead to significant variations in TEER measurements.

Additionally, the reproducibility of TEER measurements using chopstick electrodes presents a notable challenge, as variations in electrode positioning, the limited surface area of the electrodes, and their geometry relative to the membrane can all contribute to inconsistencies in the data [52]. These factors complicate the comparison of TEER values and permeability rate constants across different patients. In this study, we attempted to correlate TEER with the permeability rate constant, which was estimated by fitting current data to Equation (5); however, the fit was not optimal. This poor fitting further explains why TEER values and permeability rate constants did not exhibit a strong correlation in this analysis. Based on the current data, a decrease in the TEER value shows that the barrier has become less intact, which means higher permeability and the same trend was observed in Figure 5, where Line 1 has a higher rate constant and lower TEER value.

## 4. Conclusions

This study highlights the potential of electrochemical sensors for predicting human responsiveness to antipsychotic drugs. Platinum-black-modified electrodes were employed to detect the concentration of OLZ after their penetration through the BBB in various cell line models. The effective surface area of the platinum-black-modified electrodes is $5.23 \times 10^{-2} \pm 2.3 \times 10^{-3}$ cm$^2$, which is four times higher than the bare gold electrode, $3.81 \times 10^{-2} \pm 1.2 \times 10^{-3}$ cm$^2$. The dose–response of OLZ with platinum-black-modified electrodes was characterized using chronoamperometric electrochemical signals, which showed an increasing current at 0.45 V vs. Ag/AgCl with an increasing OLZ concentration. The detection limit of OLZ was calculated as $9.96 \pm 7.35 \times 10^{-6}$ nM. The OLZ concentration data were utilized to predict its permeability rate, which was further compared with the TEER values. Based on the TEER values and predicted permeability rate constant data, the decreasing order of the membrane permeability of the OLZ drug tested with five SCZ patients is Line 1, Line 5, Line 3, Line 9, and Line 7. Here, we were able to monitor the olanzapine permeability across the membrane, which demonstrates the efficacy of platinum-black-modified electrodes as a robust analytical tool for evaluating antipsychotic drug transport.

**Supplementary Materials:** The following supporting information can be downloaded at: https://www.mdpi.com/article/10.3390/s25072266/s1, Figure S1: Schematic representation of the fabrication of gold electrodes. Figure S2: (A) Electrodeposition of platinum black using chronopotentiometry; (B) CV measured in 5 mM ferrocyanide/ferricyanide at different scan rates (0.05 to 0.5 V/s) using bare gold electrode; (C) CV measured at different scan rates for the platinum-black-modified electrode; plot of oxidation peak current vs. square root of the scan rate of (D) bare gold and (E) platinum-black-modified electrode; and (F) Optical microscopic image of platinum-black-modified electrode. Reference [53] is cited in the Supplementary Materials.

**Author Contributions:** Conceptualization, H.B.-Y.; methodology, H.B.-Y., D.D. and M.S.I.; software, M.S.I.; validation, D.D., M.S.I., S.J. and A.S.; formal analysis, M.S.I. and S.J.; investigation, H.B.-Y., D.D., M.S.I. and G.V.; resources, H.B.-Y. and G.V.; data curation, D.D., M.S.I. and H.B.-Y.; writing—M.S.I. and D.D.; writing—review and editing, D.D., M.S.I., S.J., G.V. and H.B.-Y.; visualization, M.S.I., D.D. and H.B.-Y.; supervision, H.B.-Y., D.D. and G.V.; project administration, H.B.-Y.; funding acquisition, H.B.-Y. and G.V. All authors have read and agreed to the published version of the manuscript.

**Funding:** This research was funded by the Israel Science Foundation (ISF), grant number 1052/19.

**Institutional Review Board Statement:** Not applicable.

**Informed Consent Statement:** Not applicable.

**Data Availability Statement:** The data used to develop all the figures are not available to share publicly.

**Acknowledgments:** This work was supported by the Kreitman School at Ben-Gurion University of the Negev Postdoctoral Fellowship. We thank the Ilse Katz Institute for Nanoscale Science & Technology and the Nano-Fabrication Center, for providing the microfabrication facility.

**Conflicts of Interest:** There are no conflicts of interest to declare.

## References

1. Meftah, A.M.; Deckler, E.; Citrome, L.; Kantrowitz, J.T. New Discoveries for an Old Drug: A Review of Recent Olanzapine Research. *Postgrad. Med.* **2020**, *132*, 80–90. [CrossRef] [PubMed]
2. Jin, W.; Chen, S.; Li, D.; Chen, Q.; Zhu, M.; Wang, M.; Fu, X.; Lin, P. Metabolic Effects and Clinical Outcomes of Olanzapine in Schizophrenia: A Systematic Review and Meta-Analysis. *Heliyon* **2024**, *10*, e40424. [CrossRef] [PubMed]
3. Carli, M.; Kolachalam, S.; Longoni, B.; Pintaudi, A.; Baldini, M.; Aringhieri, S.; Fasciani, I.; Annibale, P.; Maggio, R.; Scarselli, M. Atypical Antipsychotics and Metabolic Syndrome: From Molecular Mechanisms to Clinical Differences. *Pharmaceuticals* **2021**, *14*, 238. [CrossRef] [PubMed]
4. Meltzer, H.Y. Update on Typical and Atypical Antipsychotic Drugs. *Annu. Rev. Med.* **2013**, *64*, 393–406. [CrossRef]
5. Lieberman, J.A.; Stroup, T.S.; McEvoy, J.P.; Swartz, M.S.; Rosenheck, R.A.; Perkins, D.O.; Keefe, R.S.E.; Davis, S.M.; Davis, C.E.; Lebowitz, B.D.; et al. Effectiveness of Antipsychotic Drugs in Patients with Chronic Schizophrenia. *N. Engl. J. Med.* **2005**, *353*, 1209–1223. [CrossRef]
6. Tollefson, G.D.; Beasley, C.M.; Tran, P.V.; Street, J.S.; Krueger, J.A.; Tamura, R.N.; Graffeo, K.A.; Thieme, M.E. Olanzapine versus Haloperidol in the Treatment of Schizophrenia and Schizoaffective and Schizophreniform Disorders: Results of an International Collaborative Trial. *Am. J. Psychiatry* **1997**, *154*, 457–465. [CrossRef]
7. Andreasen, N.C. Symptoms, Signs, and Diagnosis of Schizophrenia. *Lancet* **1995**, *346*, 477–481. [CrossRef]
8. Bymaster, F.P.; Hemrick-Luecke, S.K.; Perry, K.W.; Fuller, R.W. Neurochemical Evidence for Antagonism by Olanzapine of Dopamine, Serotonin, Alpha 1-Adrenergic and Muscarinic Receptors in Vivo in Rats. *Psychopharmacology* **1996**, *124*, 87–94. [CrossRef]
9. Catlow, J.T.; Barton, R.D.; Clemens, M.; Gillespie, T.A.; Goodwin, M.; Swanson, S.P. Analysis of Olanzapine in Human Plasma Utilizing Reversed-Phase High-Performance Liquid Chromatography with Electrochemical Detection. *J. Chromatogr. B Biomed. Appl.* **1995**, *668*, 85–90. [CrossRef]
10. Xu, H.; Yang, F. The Interplay of Dopamine Metabolism Abnormalities and Mitochondrial Defects in the Pathogenesis of Schizophrenia. *Transl. Psychiatry* **2022**, *12*, 464. [CrossRef]
11. Yun, S.; Yang, B.; Anair, J.D.; Martin, M.M.; Fleps, S.W.; Pamukcu, A.; Yeh, N.-H.; Contractor, A.; Kennedy, A.; Parker, J.G. Antipsychotic Drug Efficacy Correlates with the Modulation of D1 Rather than D2 Receptor-Expressing Striatal Projection Neurons. *Nat. Neurosci.* **2023**, *26*, 1417–1428. [CrossRef] [PubMed]
12. Pardridge, W.M. Blood–Brain Barrier Delivery. *Drug Discov. Today* **2007**, *12*, 54–61. [CrossRef] [PubMed]
13. Tu, K.-H.; Yu, L.-S.; Sie, Z.-H.; Hsu, H.-Y.; Al-Jamal, K.T.; Wang, J.T.-W.; Chiang, Y.-Y. Development of Real-Time Transendothelial Electrical Resistance Monitoring for an In Vitro Blood-Brain Barrier System. *Micromachines* **2021**, *12*, 37. [CrossRef]
14. Kincses, A.; Vigh, J.P.; Petrovszki, D.; Valkai, S.; Kocsis, A.E.; Walter, F.R.; Lin, H.-Y.; Jan, J.-S.; Deli, M.A.; Dér, A. The Use of Sensors in Blood-Brain Barrier-on-a-Chip Devices: Current Practice and Future Directions. *Biosensors* **2023**, *13*, 357. [CrossRef]
15. Paradis, A.; Leblanc, D.; Dumais, N. Optimization of an in Vitro Human Blood–Brain Barrier Model: Application to Blood Monocyte Transmigration Assays. *MethodsX* **2016**, *3*, 25–34. [CrossRef]
16. Wang, J.-S.; Taylor, R.; Ruan, Y.; Donovan, J.L.; Markowitz, J.S.; Lindsay De Vane, C. Olanzapine Penetration into Brain Is Greater in Transgenic Abcb1a P-Glycoprotein-Deficient Mice than FVB1 (Wild-Type) Animals. *Neuropsychopharmacology* **2004**, *29*, 551–557. [CrossRef]
17. Luptáková, D.; Vallianatou, T.; Nilsson, A.; Shariatgorji, R.; Hammarlund-Udenaes, M.; Loryan, I.; Andrén, P.E. Neuropharmacokinetic Visualization of Regional and Subregional Unbound Antipsychotic Drug Transport across the Blood–Brain Barrier. *Mol. Psychiatry* **2021**, *26*, 7732–7745. [CrossRef]
18. Ben-Shachar, D.; Livne, E.; Spanier, I.; Leenders, K.L.; Youdim, M.B. Typical and Atypical Neuroleptics Induce Alteration in Blood-Brain Barrier and Brain 59FeCl3 Uptake. *J. Neurochem.* **1994**, *62*, 1112–1118. [CrossRef]
19. Pardridge, W.M.; Crawford, I.L.; Connor, J.D. Permeability Changes in the Blood-Brain Barrier Induced by Nortriptyline and Chlorpromazine. *Toxicol. Appl. Pharmacol.* **1973**, *26*, 49–57. [CrossRef]
20. Dipta, P.; Sarsenbayeva, A.; Shmuel, M.; Forno, F.; Eriksson, J.W.; Pereira, M.J.; Abalo, X.M.; Wabitsch, M.; Thaysen-Andersen, M.; Tirosh, B. Macrophage-Derived Secretome Is Sufficient to Confer Olanzapine-Mediated Insulin Resistance in Human Adipocytes. *Compr. Psychoneuroendocrinol.* **2021**, *7*, 100073. [CrossRef]

21. Abbott, N.J.; Patabendige, A.A.K.; Dolman, D.E.M.; Yusof, S.R.; Begley, D.J. Structure and Function of the Blood-Brain Barrier. *Neurobiol. Dis.* **2010**, *37*, 13–25. [CrossRef] [PubMed]
22. Cecchelli, R.; Berezowski, V.; Lundquist, S.; Culot, M.; Renftel, M.; Dehouck, M.-P.; Fenart, L. Modelling of the Blood–Brain Barrier in Drug Discovery and Development. *Nat. Rev. Drug Discov.* **2007**, *6*, 650–661. [CrossRef] [PubMed]
23. Srinivasan, B.; Kolli, A.R.; Esch, M.B.; Abaci, H.E.; Shuler, M.L.; Hickman, J.J. TEER Measurement Techniques for In Vitro Barrier Model Systems. *SLAS Technol.* **2015**, *20*, 107–126. [CrossRef] [PubMed]
24. Holzreuter, M.A.; Segerink, L.I. Innovative Electrode and Chip Designs for Transendothelial Electrical Resistance Measurements in Organs-on-Chips. *Lab. Chip* **2024**, *24*, 1121–1134. [CrossRef]
25. Dusci, L.J.; Peter Hackett, L.; Fellows, L.M.; Ilett, K.F. Determination of Olanzapine in Plasma by High-Performance Liquid Chromatography Using Ultraviolet Absorbance Detection. *J. Chromatogr. B* **2002**, *773*, 191–197. [CrossRef]
26. Bonde, S.L.; Bhadane, R.P.; Gaikwad, A.; Gavali, S.R.; Katale, D.U.; Narendiran, A.S. Simultaneous Determination of Olanzapine and Fluoxetine in Human Plasma by LC-MS/MS: Its Pharmacokinetic Application. *J. Pharm. Biomed. Anal.* **2014**, *90*, 64–71. [CrossRef]
27. Basavaiah, K.; Rajendraprasad, N.; Vinay, K.B. Isocratic High-Performance Liquid Chromatographic Assay of Olanzapine: Method Development and Validation. *Int. Sch. Res. Not.* **2014**, *2014*, 616941. [CrossRef]
28. Jenkins, A.J.; Sarconi, K.M.; Raaf, H.N. Determination of Olanzapine in a Postmortem Case. *J. Anal. Toxicol.* **1998**, *22*, 605–609. [CrossRef]
29. Patel, D.S.; Sharma, N.; Patel, M.C.; Patel, B.N.; Shrivastav, P.S.; Sanyal, M. LC–MS/MS Assay for Olanzapine in Human Plasma and Its Application to a Bioequivalence Study. *Acta Pharm. Sin. B* **2012**, *2*, 481–494. [CrossRef]
30. Hussien, L.A.E.A.; Ghani, M.F.A.; El-Alamein, A.M.A.; Mohamed, E.H. Stability-Indicating Methods for the Determination of Olanzapine in Presence of Its Degradation Products. *Eur. J. Chem.* **2014**, *5*, 311–320. [CrossRef]
31. Arvand, M.; Pourhabib, A. Surfactant-Assisted Voltammetric Determination of Olanzapine at Amine Functionalized TiO2/Multi-Walled Carbon Nanotubes Nanocomposite. *J. Anal. Chem.* **2019**, *74*, 1096–1103. [CrossRef]
32. Heres, S.; Kraemer, S.; Bergstrom, R.F.; Detke, H.C. Pharmacokinetics of Olanzapine Long-Acting Injection: The Clinical Perspective. *Int. Clin. Psychopharmacol.* **2014**, *29*, 299–312. [CrossRef] [PubMed]
33. Azab, S.M.; Fekry, A.M. Role of Green Chemistry in Antipsychotics' Electrochemical Investigations Using a Nontoxic Modified Sensor in McIlvaine Buffer Solution. *ACS Omega* **2019**, *4*, 25–30. [CrossRef] [PubMed]
34. Ahmed, H.M.; Mohamed, M.A.; Salem, W.M. New Voltammetric Analysis of Olanzapine in Tablets and Human Urine Samples Using a Modified Carbon Paste Sensor Electrode Incorporating Gold Nanoparticles and Glutamine in a Micellar Medium. *Anal. Methods* **2015**, *7*, 581–589. [CrossRef]
35. Shukla, R.P.; Belmaker, R.H.; Bersudsky, Y.; Ben-Yoav, H. A Platinum Black-Modified Microelectrode for in Situ Olanzapine Detection in Microliter Volumes of Undiluted Serum. *J. Neural. Transm.* **2020**, *127*, 291–299. [CrossRef]
36. Qiang, L.; Vaddiraju, S.; Rusling, J.F.; Papadimitrakopoulos, F. Highly Sensitive and Reusable Pt-Black Microfluidic Electrodes for Long-Term Electrochemical Sensing. *Biosens. Bioelectron.* **2010**, *26*, 682–688. [CrossRef]
37. Seo, H.-K.; Park, D.-J.; Park, J.-Y. Fabrication and Characterization of Platinum Black and Mesoporous Platinum Electrodes for *in-Vivo* and Continuously Monitoring Electrochemical Sensor Applications. *Thin Solid. Films* **2008**, *516*, 5227–5230. [CrossRef]
38. Goh, A.; Roberts, D.; Wainright, J.; Bhadra, N.; Kilgore, K.; Bhadra, N.; Vrabec, T. Evaluation of Activated Carbon and Platinum Black as High-Capacitance Materials for Platinum Electrodes. *Sensors* **2022**, *22*, 4278. [CrossRef]
39. Li, Y.; Sella, C.; Lemaître, F.; Guille Collignon, M.; Thouin, L.; Amatore, C. Highly Sensitive Platinum-Black Coated Platinum Electrodes for Electrochemical Detection of Hydrogen Peroxide and Nitrite in Microchannel. *Electroanalysis* **2013**, *25*, 895–902. [CrossRef]
40. Stanca, S.E.; Hänschke, F.; Ihring, A.; Zieger, G.; Dellith, J.; Kessler, E.; Meyer, H.-G. Chemical and Electrochemical Synthesis of Platinum Black. *Sci. Rep.* **2017**, *7*, 1074. [CrossRef]
41. Konopka, S.J.; McDuffie, B. Diffusion Coefficients of Ferri- and Ferrocyanide Ions in Aqueous Media, Using Twin-Electrode Thin-Layer Electrochemistry. *Anal. Chem.* **1970**, *42*, 1741–1746. [CrossRef]
42. Bard, A.J.; Faulkner, L.R. *Electrochemical Methods: Fundamentals and Applications*, 2nd ed.; Wiley: New York, NY, USA; Weinheim, Germany, 2001; ISBN 978-0-471-04372-0.
43. Wayu, M.B.; Pannell, M.J.; Leopold, M.C. Layered Xerogel Films Incorporating Monolayer-Protected Cluster Networks on Platinum-Black-Modified Electrodes for Enhanced Sensitivity in First-Generation Uric Acid Biosensing. *ChemElectroChem* **2016**, *3*, 1245–1252. [CrossRef]
44. Matthiesen, I.; Voulgaris, D.; Nikolakopoulou, P.; Winkler, T.E.; Herland, A. Continuous Monitoring Reveals Protective Effects of N-Acetylcysteine Amide on an Isogenic Microphysiological Model of the Neurovascular Unit. *Small* **2021**, *17*, e2101785. [CrossRef] [PubMed]
45. Bednarek, R. In Vitro Methods for Measuring the Permeability of Cell Monolayers. *Methods Protoc.* **2022**, *5*, 17. [CrossRef]

46. Salama, H.A.; Mahmoud, A.A.; Kamel, A.O.; Abdel Hady, M.; Awad, G.A.S. Phospholipid Based Colloidal Poloxamer–Nanocubic Vesicles for Brain Targeting via the Nasal Route. *Colloids Surf. B Biointerfaces* **2012**, *100*, 146–154. [CrossRef]
47. McMartin, K.E.; Sebastian, C.S.; Dies, D.; Jacobsen, D. Kinetics and Metabolism of Fomepizole in Healthy Humans. *Clin. Toxicol.* **2012**, *50*, 375–383. [CrossRef]
48. Currie, G.M. Pharmacology, Part 2: Introduction to Pharmacokinetics. *J. Nucl. Med. Technol.* **2018**, *46*, 221–230. [CrossRef]
49. Menichetti, R.; Kanekal, K.H.; Bereau, T. Drug–Membrane Permeability across Chemical Space. *ACS Cent. Sci.* **2019**, *5*, 290–298. [CrossRef]
50. Dutta, A.; Vreeken, J.; Ghiringhelli, L.M.; Bereau, T. Data-Driven Equation for Drug–Membrane Permeability across Drugs and Membranes. *J. Chem. Phys.* **2021**, *154*, 244114. [CrossRef]
51. Blume, L.-F.; Denker, M.; Gieseler, F.; Kunze, T. Temperature Corrected Transepithelial Electrical Resistance (TEER) Measurement to Quantify Rapid Changes in Paracellular Permeability. *Pharmazie* **2010**, *65*, 19–24. [PubMed]
52. Vigh, J.P.; Kincses, A.; Ozgür, B.; Walter, F.R.; Santa-Maria, A.R.; Valkai, S.; Vastag, M.; Neuhaus, W.; Brodin, B.; Dér, A.; et al. Transendothelial Electrical Resistance Measurement across the Blood–Brain Barrier: A Critical Review of Methods. *Micromachines* **2021**, *12*, 685. [CrossRef] [PubMed]
53. Bard, A.J.; Faulkner, L.R. Basic Potential Step Methods. In *Electrochemical Methods: Fundamentals and Applications*; Wiley: New York, NY, USA; Weinheim, Germany, 2001; pp. 156–225.

*sensors*

MDPI

*Article*

# Research on Binary Mixed VOCs Gas Identification Method Based on Multi-Task Learning

**Haixia Mei [1,2,*], Ruiming Yang [1], Jingyi Peng [1], Keyu Meng [1], Tao Wang [3] and Lijie Wang [2,*]**

[1] Key Lab Intelligent Rehabil & Barrier Free Disable (Ministry of Education), Changchun University, Changchun 130022, China; 220401159@mails.ccu.edu.cn (R.Y.); 230401162@mails.ccu.edu.cn (J.P.); mengky@ccu.edu.cn (K.M.)

[2] State Key Laboratory of Integrated Optoelectronics, College of Electronic Science and Engineering, Jilin University, Changchun 130012, China

[3] Shanghai Key Laboratory of Intelligent Sensing and Detection Technology, School of Mechanical and Power Engineering, East China University of Science and Technology, Shanghai 200237, China; wangtao@ecust.edu.cn

* Correspondence: meihx@ccu.edu.cn (H.M.); wang_lij@jlu.edu.cn (L.W.)

**Highlights:**

**What are the main findings?**

- A multi-task residual network (MRCA) which generates dynamic feature depending on the cross-fusion module was invented to perform VOCs gas component identification and concentration prediction.
- The dynamic weighted loss function, which can dynamically adjust the weight according to the training progress of each task.

**What is the implication of the main finding?**

- The MRCA model showed a high classification accuracy of 94.86%, as well as achieving an $R^2$ score up to 0.95.
- Using only 35% of the total data length as input data leads to excellent identification performance.

**Abstract:** Traditional volatile organic compounds (VOCs) detection models separate component identification and concentration prediction, leading to low feature utilization and limited learning in small-sample scenarios. Here, we realize a Residual Fusion Network based on multi-task learning (MTL-RCANet) to implement component identification and concentration prediction of VOCs. The model integrates channel attention mechanisms and cross-fusion modules to enhance feature extraction capabilities and task synergy. To further balance the tasks, a dynamic weighted loss function is incorporated to adjust weights dynamically according to the training progress of each task, thereby enhancing the overall performance of the model. The proposed network achieves an accuracy of 94.86% and an $R^2$ score of 0.95. Comparative experiments reveal that using only 35% of the total data length as input data yields excellent identification performance. Moreover, multi-task learning effectively integrates feature information across tasks, significantly improving model efficiency compared to single-task learning.

**Keywords:** gas sensor; multi-task learning; mixed gases; feature fusion

## 1. Introduction

Volatile organic compounds (VOCs) are widely present in industrial production, vehicle emissions, and building materials, posing both short-term and long-term health risks [1–5]. Therefore, effective monitoring and control of VOCs are crucial for environmental protection and human health [6–12]. Compared to spectroscopy [13–18] and mass spectrometry [19–22], which have high equipment and environmental requirements [23], artificial olfaction technology [24] has gained widespread attention in recent years due to its miniaturization, intelligence, and low cost [25].

However, judging from the recent research trends of artificial olfaction technology, the reason why it has stagnated is due to the poor selectivity [26,27] of gas sensors, which leads to cross-sensitivity issues [28,29]. Recently, advanced artificial intelligence has been accelerating the development of artificial olfactory systems [30–34]. For example, Xia et al. [35] used Principal Component Analysis for feature extraction in a mixed gas classification task, achieving 96.88% accuracy with K-Nearest Neighbors (KNNs). Li et al. [36] improved gas concentration prediction by combining Variational Mode Decomposition with Extreme Learning Machine. Martono et al. [37] evaluated multiple algorithms for blood alcohol concentration prediction, with LightGBM achieving the best performance (accuracy: 0.908, F1 score: 0.617), highlighting the effectiveness of blood gas analysis for alcohol concentration estimation.

While traditional machine learning methods achieve gas detection, their reliance on manual feature extraction limits generalizability. Deep learning approaches like Convolutional Neural Network (CNN) [38] and Recurrent Neural Network (RNN) [39] have gained traction for automatic feature extraction, excelling in gas identification and concentration prediction. Chu et al. [40] transformed gas response data into grayscale images, effectively distinguishing CO and $NO_2$ under varying humidity. Song et al. [41] used LASSO-RNN for mine gas concentration prediction, reducing mean squared error (MSE) (0.0029) and mean absolute error (MAE) (0.0084). Zeng et al. [42] improved mixed gas concentration regression with a dual-channel Temporal Convolutional Network (TCN), surpassing Long Short-Term Memory (LSTM), Gated Recurrent Unit (GRU), and standard TCN in prediction accuracy.

With deep learning achieving remarkable success, researchers have shifted focus toward faster and more efficient detection methods. Li et al. [43] proposed a low-cost method, which uses only 60% of rising-phase response data, enabling mixed gas concentration prediction within 10 s. However, traditional qualitative and quantitative gas analysis relies on separate systems, rendering them impractical for resource-constrained portable devices. Multi-task learning (MTL) has emerged as a promising solution. Wang et al. [44] introduced an MTL-CNN that simultaneously identifies gas types, concentrations, and states of 12 VOCs. Wang et al. [45] proposed LSTM-Attention combined with MTL (MTL-LSTMA), which achieved the classification accuracy and concentration prediction up to 98% within 30 s. Fu et al. [46] developed a real-time Progressive Prediction Algorithm integrating TCN and GRU, enabling early detection of harmful gases with enhanced speed and accuracy. Kang et al. [47] applied a multi-task CNN with a 10 s time window to classify and predict concentrations of five gases.

However, until now, there is still the challenge of improving task collaboration and feature utilization in multi-task learning. In this study, we proposed the Residual Fusion Network based on multi-task learning (MTL-RCANet) method, which aims to simultaneously perform gas identification and concentration prediction while facilitating efficient information exchange between different tasks. The method dynamically extracts local peak features from gas response data through time windows, which enhances the model's ability to focus on key features. It introduces a channel attention mechanism and a cross-fusion module, which strengthen task collaboration by sharing information between tasks.

Additionally, a dynamic weighted loss function is used to adjust the weight of each task based on its specific requirements, further improving the overall performance of the model. This approach not only improves the utilization of multi-task parameters but also provides an efficient and reliable solution for resource-constrained scenarios that require rapid identification.

## 2. Gas Experiment

The sensors used in this experiment were provided by Henan Weisheng Technology, with $SnO_2$ as the sensing material. Table 1 presents the gas response characteristics of each sensor. During the experiment, ethanol and n-propanol were mixed at different concentrations. As shown in Figure 1, a total of seven tests were conducted, including two single-gas response tests and five mixed-gas response tests with varying concentration ratios. Each test yielded five gas sensor response data. In the mixed-gas experiments, the concentration of one gas remained constant, while the other increased gradually from 0 to 100 parts per million (ppm) in increments of 20 ppm. G1 represents the experimental scheme for pure ethanol gas, G2 represents the experimental scheme for pure 1-propanol gas, and G3–G7 represent the experimental schemes for mixtures of ethanol and 1-propanol gases. Each point in the figure corresponds to a specific ethanol-to-1-propanol concentration ratio. Throughout the experiment, fixed time intervals were used to define the response and recovery phases in gas response testing. In the gas concentration gradient cycling response test, the sensor was exposed to the target gas for 5 min, followed by a 5 min recovery period in air. The gas sampling frequency was set to 2 Hz, and each test was repeated at least five times, resulting in a total of 175 gas sensor responses.

**Table 1.** Sensor model and corresponding response gases.

| Model Number | Response Gas |
|---|---|
| MQ-2 | Liquefied Gas, $C_3H_8$, $H_2$ |
| MQ-3 | $C_2H_5OH$ |
| MQ-4 | $CH_4$ |
| MQ-5 | $C_4H_{10}$, $C_3H_8$, $CH_4$ |
| MQ-6 | $C_3H_8$, $C_4H_{10}$ |
| MQ-7 | CO |
| MQ-8 | $H_2$ |
| MQ-9 | CO |

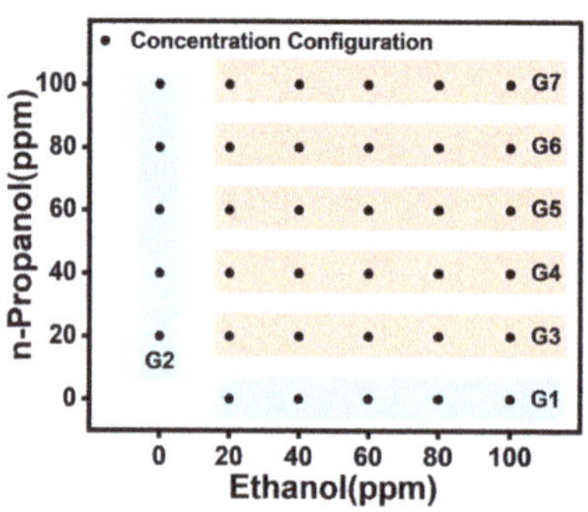


**Figure 1.** Gas experimental setup.

Figure 2a–g presents the time-resistance/response curves of the gas sensor array for each experimental group. Specifically, Figure 2a and 2b depict the sensor array's response to ethanol and n-propanol, respectively, while Figure 2c–g illustrate the responses to their mixed gases. The response is calculated as follows:

$$s = \frac{R_0}{R} \quad (1)$$

where $R_0$ is the resistance of the gas sensor in air, and $R$ is the resistance of the gas sensor in the measured gas.

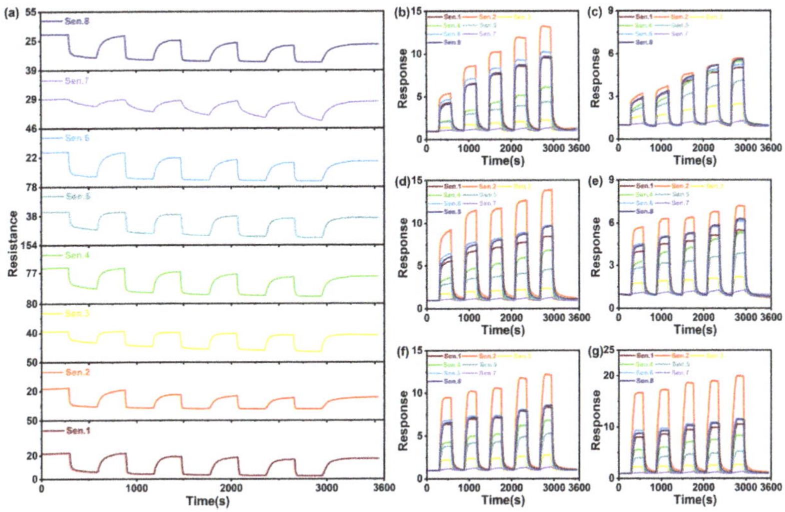


**Figure 2.** (**a**) Sensor output for single ethanol gas; (**b**) response curve for single n-propanol gas; (**c–g**) response curves for the mixed gases of both.

## 3. Method

### *3.1. Data Preprocessing*

As shown in Figure 3, the data preprocessing in this study involves three key stages: segmenting the response signals, selecting relevant features, and normalizing and reshaping the feature matrix.

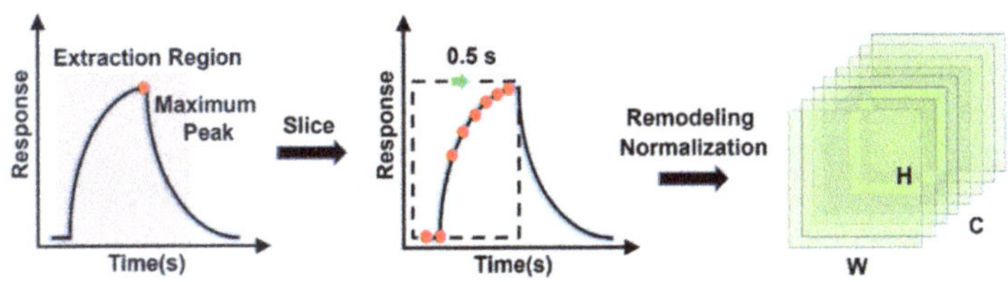


**Figure 3.** Data preprocessing flow.

#### 3.1.1. Response Fragment Segmentation

The gas response data undergo peak identification, where local peaks are classified based on a predefined threshold, which groups peaks within the threshold into the same category. The maximum peak from each category is then selected, which serves as a reference to segment the complete response signal. The calculation process is as follows:

The peak detection algorithm (find_peaks) is used to identify the index set $P = \{p_1, p_2, \cdots, p_n\}$ of all local peak points in the gas response data $y_i$, where each $p_j$ satisfies the following:

$$y_{p_j} > y_{p_{j-1}} \bigwedge y_{p_j} > y_{p_{j+1}}, j \in [1, n] \tag{2}$$

Calculate the distance set between adjacent peak indices, $D = \{d_1, d_2, \cdots, d_{n-1}\}$, where

$$d = p_{j+1} - p_j, j \in [1, n-1] \tag{3}$$

Set the threshold $d_{\text{threshold}}$, which is determined by the gas sampling frequency and the response recovery time of the gas sensor (i.e., the descending phase of the response curve). In the gas experiments of this study, the response recovery time of the gas sensor is 5 min (corresponding to 600 sample points). Therefore, the distance between peak point indices in each gas sensor response should be less than 600. Peak points satisfying $d < d_{\text{threshold}}$ are grouped into the same cluster, forming the final grouping set $G$.

$$G = \{g_1, g_2, \cdots, g_m\}, g = \{p_{k1}, p_{k2}, \cdots, p_{kl}\}, k \in [1, m] \tag{4}$$

For each group $g$, identify its corresponding maximum peak index $p_{max}$:

$$p_{max} = \underset{p \in g}{\text{argmax}}\, y[p] \tag{5}$$

The set of maximum peaks is $P_{max} = \{p_{max_1}, p_{max_2}, \cdots, p_{max_m}\}$.

Using each maximum peak index $p_{max}$ as a reference, set a left offset $l_{\text{offset}}$ and a right offset $r_{\text{offset}}$ to determine the index range of each response segment. This results in the response segment set $S = \{s_1, s_2, \cdots, s_m\}$. The determination strategy for $l_{\text{offset}}$ and $r_{\text{offset}}$ is as follows: After identifying the maximum peak point, the left side corresponds to the gas sensor's response phase, while the right side corresponds to the recovery phase. In this study, the gas sensor's response phase lasts for 5 min (i.e., 600 sample points). The data from the response phase will be fully utilized for feature extraction. To ensure data sufficiency, $l_{\text{offset}}$ is set to 700. In the initial part of the recovery phase, we assume the presence of "numerical features" since the final response values at the end of the response vary for mixed gases with different concentration ratios. Therefore, $r_{\text{offset}}$ is set to 400.

### 3.1.2. Feature Selection

We apply a sliding time window with a step size of 0.5 s along the time axis to the segmented response fragments, which allows calculating the ratio between data variation within the window and the window width. Then, extracting the segment with the maximum ratio as the feature data. Additionally, we set the window width to a square value (e.g., $n^2$) to facilitate subsequent convolution operations. The calculation process is as follows:

$$R_t = \frac{\Delta y_t}{w}, \Delta y_t = \sum_{i=t}^{t+w-1} (y_{i+1} - y_i) \tag{6}$$

Here, $t$ is the starting position of the time window, $\Delta y_t$ represents the total variation within the current time window, $w$ is the width of the time window, and $R_t$ is the variation rate of the window starting at position $t$. Among all the sliding time windows, the time segment with the maximum variation rate is selected as the feature data, and the starting position of the corresponding time window is denoted as $t_{max}$. The final feature data segment is as follows:

$$y_{t_{max}} = \{y_i \mid t_{max} \le i < t_{max} + w\} \tag{7}$$

### 3.1.3. Feature Matrix Normalization and Reshaping

The response segment $y_{t_{max}}$ selected is normalized using the min-max normalization method, with the formula:

$$y_{norm} = \frac{y_{t_{max}} - min(y_{t_{max}})}{max(y_{t_{max}}) - min(y_{t_{max}})} \tag{8}$$

For each gas sensor response, the response data from individual gas sensors are first reshaped into independent square feature layers using the reshape function. These layers are then stacked along the channel dimension to form multi-layer feature maps, enhancing feature extraction and analysis. As shown in Figure 3, where C represents the number of channels, H denotes the feature layer height, and W represents the feature layer width. The final gas compositions and their corresponding labels after preprocessing are shown in Table 2.

**Table 2.** Gas composition.

| Gas Type | Single n-Propanol | Single Ethanol | n-Propanol and Ethanol |
|---|---|---|---|
| Label | 01 | 10 | 11 |

Finally, each gas sensor response was divided into five folds based on acquisition time and different test groups. One fold was used as the test set, while the remaining four folds were used as the training set, maintaining a 4:1 ratio between the training and test sets for model training and validation.

### 3.2. Multi-Task Learning Model

The proposed MTL-RCANet (hereafter referred to as MRCA) model, shown in Figure 4, consists of a multi-task residual network, a channel attention mechanism module, and a cross-fusion module.

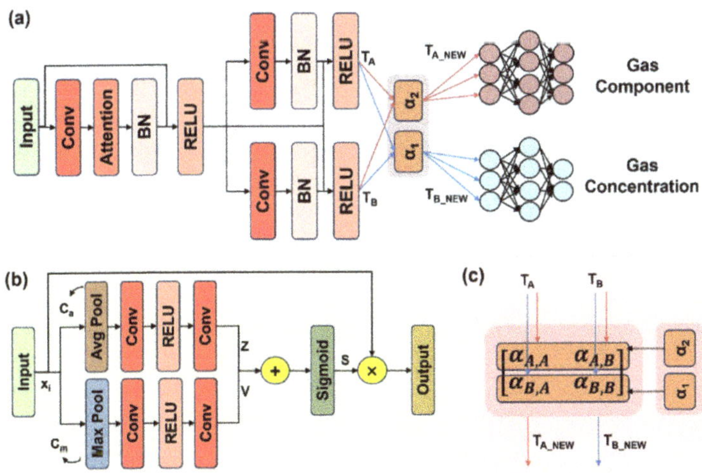


**Figure 4.** (**a**) Overall model architecture. (**b**) Channel attention mechanism module. (**c**) Cross-fusion module.

### 3.2.1. Channel Attention Mechanism

In residual networks (ResNet), the convolution operation primarily focuses on extracting deep features in spatial dimensions ($H \times W$) but pays less attention to inter-channel dependencies ($C$), which may limit the network's ability to capture channel relationships and impact feature representation [48]. To address this, we incorporate a channel attention mechanism that adaptively assigns weights to each channel based on the data's inherent characteristics, emphasizing key channel features while suppressing secondary or irrelevant ones. The details of the channel attention mechanism are as follows:

The input to the channel attention mechanism is $X_i$, with a shape of $C \times H \times W$, and it consists of two branches. The first branch is retained and will be weighted after the second branch has been computed, which highlights the key channel features. In the second branch, the global average value $C_a$ and the maximum value $C_m$ are computed separately for each channel.

$$C_a(c) = \frac{1}{H \times W}\sum_{i=1}^{H}\sum_{j=1}^{W} X_i(c,i,j), \forall c \in [1,C] \tag{9}$$

$$C_m(c) = max_{i=1,j=1}^{H,W} X_i(c,i,j), \forall c \in [1,C] \tag{10}$$

After global average pooling and global max pooling, two $1 \times 1$ convolutions are applied, with a $ReLU$ activation layer in between, to compress and aggregate features at the channel level, learning the correlation between channels. The final outputs are $Z$ and $V$.

$$Z = Conv_2(ReLU(Conv_1(C_a))) \tag{11}$$

$$V = Conv_2(ReLU(Conv_1(C_m))) \tag{12}$$

The two sub-branches, $Z$ and $V$, are summed and passed through a $Sigmoid$ activation function to output the final weight $S$. The weight $S$ is then element-wise multiplied with $X_i$ to produce the output.

$$S = Sigmoid(Z + V) \tag{13}$$

$$Output = S \odot X_i \tag{14}$$

### 3.2.2. Cross-Fusion Module

A key challenge in multi-task learning is how to effectively sharing beneficial weight information between tasks to enhance model performance. A typical hard parameter sharing strategy directly shares network weights. However, this approach can be unstable, with interfering weights dominating between tasks. In contrast, the cross-fusion module employs a soft parameter sharing strategy, which facilitates dynamic weight sharing that enables more flexible coordination of information flow between tasks [49]. Its core concept is to dynamically adjust the degree of feature sharing between tasks using a learnable weighting mechanism, allowing efficient fusion of task features. The fusion formula is as follows:

$$\begin{bmatrix} T'_A \\ T'_B \end{bmatrix} = \begin{bmatrix} \alpha_{A,A} & \alpha_{A,B} \\ \alpha_{B,A} & \alpha_{B,B} \end{bmatrix} \begin{bmatrix} T_A \\ T_B \end{bmatrix} \tag{15}$$

Here, $T_A$ and $T_B$ are the feature maps after the second convolution for the gas component identification and concentration prediction tasks (as shown in Figure 3). $T'_A$ and $T'_B$ are the new feature maps after the fusion of the two tasks. The learnable matrix $\alpha$ contains $\alpha_{A,A}$ and $\alpha_{B,B}$, which represent the self-preserved feature weights for tasks A and B, respectively. $\alpha_{A,B}$ and $\alpha_{B,A}$ are the contribution weights from task B to task A and from task A to task B, respectively. The matrix $\alpha$ is a learnable parameter that is dynamically adjusted based on the data, with its initial state as the identity matrix.

3.2.3. Multi-Task Residual Network

Residual neural networks have inherent advantages in parallel execution of multiple outputs and tasks, enabling them to handle various task requirements based on shared features. Additionally, through residual connections, they effectively preserve and propagate feature information, improving the model's training efficiency. In the backbone network (excluding the channel attention mechanism and cross-fusion modules), all convolution operations use a $3 \times 3$ kernel size, a stride of 1, and a padding size of 2, with *ReLU* as the activation function. The design process for each layer is as follows:

1. Input Layer: After preprocessing, the shape of the gas response data is $8 \times H \times W$ (where $H = W$). The number eight represents the number of input feature maps corresponding to the number of gas sensors in the array, and $H \times W$ refers to the height and width of the feature maps.
2. Convolutional Layer: In the backbone structure of the multi-task residual network, the convolution kernel is set to the common $3 \times 3$ size. To avoid information loss at the edges of the feature map due to convolution, the padding size is set to two, ensuring that edge regions fully participate in feature extraction. The main purpose of convolution is to extract deeper features, so after each convolution operation, the number of channels doubles compared to the previous layer. For example, after the second convolution, the number of channels increases to 32, gradually enhancing the network's expressive power.
3. Batch Normalization and Activation Function: To accelerate model convergence, batch normalization is applied after each convolution operation to standardize intermediate feature distributions. Since the length of the gas response data samples is relatively short, pooling and dropout operations are omitted, but batch normalization helps reduce overfitting. The *ReLU* activation function is chosen to improve the model's non-linear representation and reduce computational complexity.
4. Fully Connected Layer: After completing feature extraction and fusion for tasks A and B, the feature maps are flattened and passed through three fully connected layers for transformation. These layers gradually compress and map the high-dimensional feature space, enhancing the model's ability to represent the target task. Finally, task A outputs gas component recognition results using the Softmax function to calculate the probability distribution for each category, while task B predicts the concentrations of the two gases.

## 4. Experimental Results and Analysis

### *4.1. Hyperparameter Settings*

Table 3 presents the hyperparameters of the backbone network, including a batch size of 5. Specifically, the table defines several key parameters: Map represents the number of channels in the output feature maps generated by the convolutional operations; K denotes the kernel size used in the convolution; S refers to the stride applied during convolution; P indicates the padding around the feature maps during convolution; D corresponds to the channel dimension in batch normalization. In addition, $T_A$ and $T_B$ represent the gas component recognition and concentration prediction tasks, respectively. The term "1st Convolutional" refers to the convolutional layer within the first residual block, whereas "2.1st–2.2nd Convolutional" refers to the convolutional layers within the second residual block, which are dedicated to handling tasks $T_A$ and $T_B$. This setup ensures that each task is processed with specific configurations to optimize their individual performances within the network.

**Table 3.** Hyperparameter list of the backbone network.

| Layer | Configuration | Input Shape |
|---|---|---|
| 1st Convolutional | Map: 16, K: 3, S: 1, P: 2 | $8 \times H \times W$ |
| CA Module | / | $16 \times H \times W$ |
| BN, Activation | $D: 16,\ ReLU$ | / |
| 2.1st–2.2st Convolutional | Map: 32, K: 3, S: 1, P: 2 | / |
| BN, Activation | $D: 32,\ ReLU$ | $32 \times H \times W$ |
| FC1 | $32 \times H \times W, 128$ | / |
| FC2 | 128, 64 | 128 |
| FC3 | 64, 3 ($T_A$) \| \| 2 ($T_B$) | 64 |
| Output | $T_A$: 5 × 3, $T_B$: 5 × 2 | / |

### *4.2. Model Training and Validation*

All experimental results are the averages from 5-fold cross-validation. The gas component recognition task is a three-class classification, and the gas concentration prediction task is a two-variable regression, each with distinct loss functions: cross-entropy loss for recognition and MSE for prediction. Since the values of cross-entropy loss lie in the range [0, 1], whereas MSE has no upper limit, the numerical difference between the two loss functions is large. A simple addition would cause the cross-entropy loss to have little impact on the final loss. Therefore, this study proposed a learnable dynamic weighting loss function to balance the numerical differences between the various losses. Three learnable parameters are defined: $\sigma_p$ for the propane concentration prediction, $\sigma_e$ for the ethanol concentration prediction, and $\sigma_{ci}$ for the gas component recognition. These parameters can be dynamically optimized through neural network gradient descent. The total loss function is defined as follows:

$$TotalLoss = \frac{1}{2\sigma_p^2} \cdot Loss_p + \frac{1}{2\sigma_e^2} \cdot Loss_e + \frac{1}{\sigma_{ci}^2} \cdot Loss_{ci} + \log_{10}(\sigma_p \sigma_e \sigma_{ci}) \tag{16}$$

In the above formula, $TotalLoss$ is the final total loss, $Loss_p$ is the training loss for propane concentration prediction, $Loss_e$ is the training loss for ethanol concentration prediction, and $Loss_{ci}$ is the training loss for gas component recognition. This adaptive mechanism addresses the large numerical differences between the loss values, allowing for effective optimization of all tasks during training. The term $\log_{10}(\sigma_p \sigma_e \sigma_{ci})$ acts as a regularizer, preventing $\sigma_p$, $\sigma_e$, and $\sigma_e$ from growing or shrinking excessively, thus ensuring model stability. Figure 5 shows the loss curves and evaluation metrics during the training and validation processes, comparing the dynamic weighted loss function with the direct sum of individual losses.

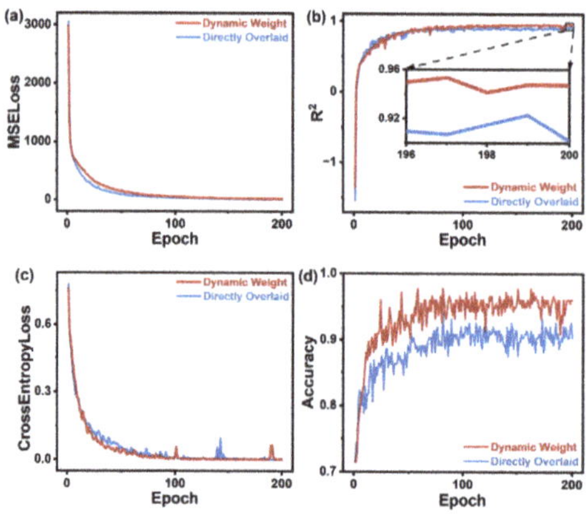


**Figure 5.** (**a**) Overall MSE loss for both gases during the training process; (**b**) overall $R^2$ value for both gases during the validation process; (**c**) cross-entropy loss for gas component recognition during the training process; (**d**) accuracy of gas component recognition during the validation process.

Figure 5a,b show that while directly summing individual losses leads to faster convergence, the $R^2$ score on the validation set is about four percentage points lower than with the dynamic weighted loss function. The dynamic weighted loss function accelerates convergence and improves accuracy by approximately five percentage points compared to direct summation, as shown in Figure 5c,d.

### *4.3. Model Performance*

The experimental results for gas component identification and concentration prediction are shown in Figure 6. Through 5-fold cross-validation, the model achieved 94.86% classification accuracy and an $R^2$ score of 0.95 for the regression task. Table 4 presents the key evaluation metrics for both tasks, along with their standard errors.

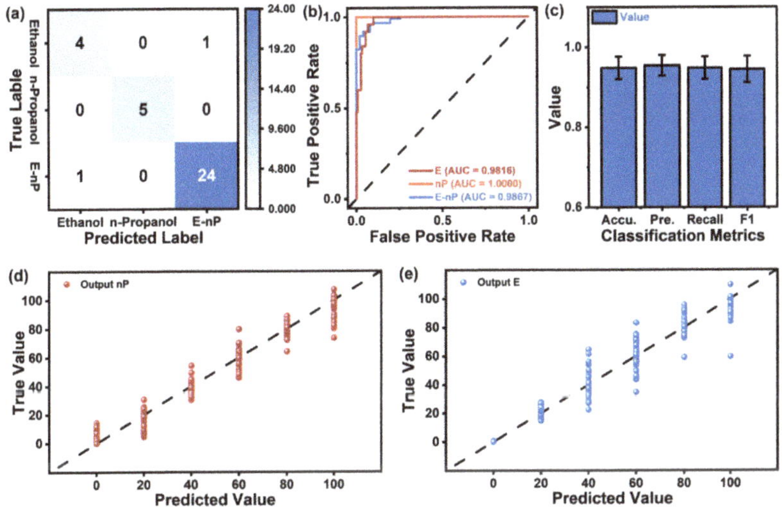


**Figure 6.** (**a**) Confusion matrix for the component recognition task; (**b**) ROC curve for each gas component classification; (**c**) accuracy, precision, recall, and F1 score for the classification task; (**d**) comparison between the true and predicted values for propanol; (**e**) comparison between the true and predicted values for ethanol.

**Table 4.** Model evaluation metrics.

| Accuracy | Std. | F1 | Std. | MAE | Std. | $R^2$ | Std. |
|---|---|---|---|---|---|---|---|
| 94.86% | 0.03 | 0.94 | 0.03 | 5.40 | 1.26 | 0.95 | 0.03 |

Figure 6a presents the confusion matrix for the classification task. On average, only 2 out of 35 validation samples per fold are misclassified, while the remaining samples are correctly identified. Figure 6b displays the Receiver Operating Characteristic (ROC) curves for each gas component, with all gas types achieving an area under the curve exceeding 0.98, indicating excellent classification performance. Figure 6c summarizes the accuracy, precision, recall, and F1 score of the classification task, which are 94.86%, 95.45%, 94.86%, and 0.94, respectively. Figure 6d,e illustrate scatter plots comparing the true and predicted values for propanol and ethanol under five-fold cross-validation. Some data points exhibit noticeable deviations, which we attribute to data drift occurring during long-term sample collection.

To evaluate the performance differences between the proposed model and baseline models, we conducted a systematic comparison with various mainstream machine learning and deep learning models. The baseline models include traditional machine learning methods (such as KNN, Support Vector Machine (SVM), and Random Forest (RF)) as well as deep neural networks (such as RNN, LSTM, CNN, and ResNet). Experimental results indicate that the proposed model surpasses all baseline models in both classification accuracy and regression prediction precision. The corresponding results are presented in Figure 7, while Table 5 provides a detailed comparison of each model's performance.

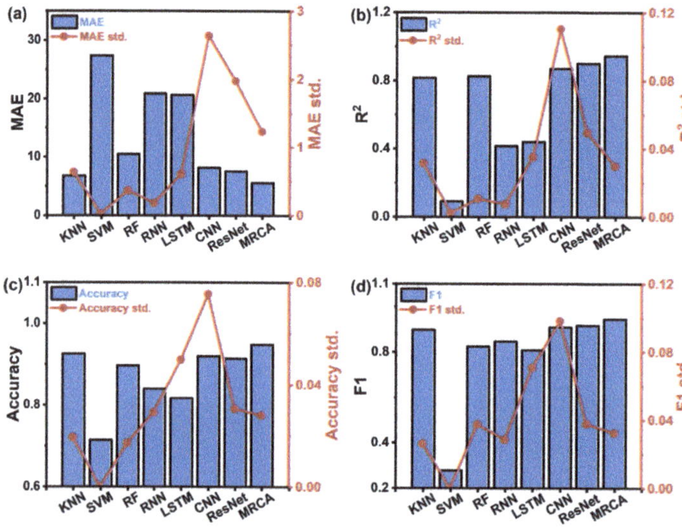

**Figure 7.** Performance evaluation metrics of eight algorithms in 5-fold cross-validation: (**a**) MAE, (**b**) $R^2$ score, (**c**) accuracy, and (**d**) F1 score, with the mean values and standard errors.

**Table 5.** Classification and regression performance of different models.

| Algo. | MAE | $R^2$ | Accuracy | F1 |
|---|---|---|---|---|
| KNN | 6.8000 | 0.8164 | 0.9257 | 0.8975 |
| SVM | 27.3700 | 0.0914 | 0.7143 | 0.2778 |
| RF | 10.4900 | 0.8263 | 0.8971 | 0.8249 |
| RNN | 20.8944 | 0.4151 | 0.8400 | 0.8463 |
| LSTM | 20.6333 | 0.4397 | 0.8171 | 0.8086 |
| CNN | 8.1681 | 0.8712 | 0.9200 | 0.9092 |
| ResNet | 7.3927 | 0.8882 | 0.8914 | 0.8984 |
| MRCA | 5.3961 | 0.9471 | 0.9486 | 0.9449 |

To validate the effectiveness of multi-task learning in jointly performing gas component identification and concentration prediction, we removed the cross-fusion module from the MRCA model and divided it into two separate networks: MRCA-C for gas component identification and MRCA-R for concentration prediction. Each network was trained independently on its respective task. The performance comparison among the three models is presented in Table 6.

**Table 6.** Performance comparison between multi-task and single-task models.

| Algo. | MAE | $R^2$ | Accuracy | F1 |
|---|---|---|---|---|
| MRCA-C | / | / | 0.9142 | 0.9068 |
| MRCA-R | 6.7737 | 0.9182 | / | / |
| MRCA | 5.3961 | 0.9471 | 0.9486 | 0.9449 |

To determine the optimal training sample length, we set different time window widths, 11, 12, 13, 14, 15, 16, 17, 19, 21, 23, 25, and 27 squared, and conducted comparison experiments using CNN, ResNet, and the MRCA network. The results show that when the sample length is $16^2$ (i.e., 256 data points), the model performance is comparable to that of the data with a sample length of $27^2$ (i.e., 729 data points). This indicates that appropriately shortening the sample length not only does not significantly reduce the model's prediction accuracy but can actually improve the model's running efficiency. The experimental results are shown in Figure 8. The pink area represents the evaluation metrics corresponding to the optimized sample length.

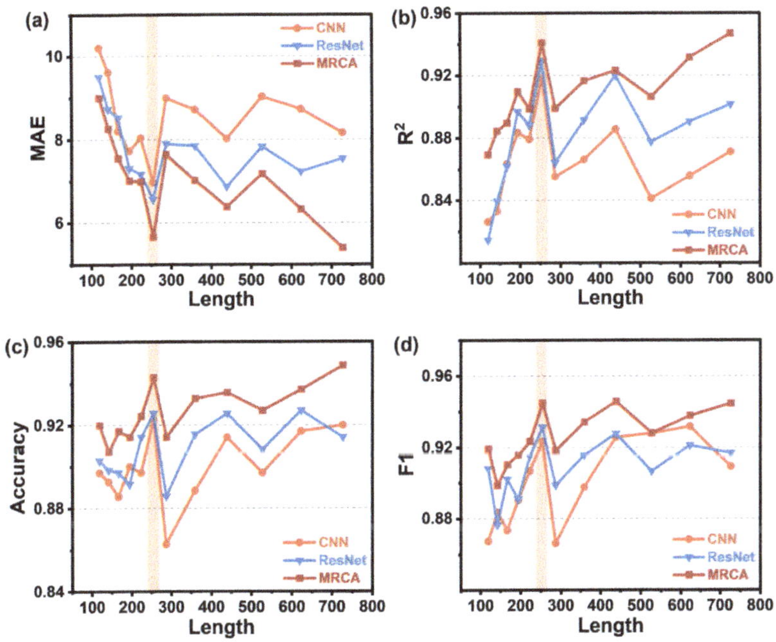


**Figure 8.** (**a**–**d**) MAE, $R^2$ score, classification accuracy, and F1 score of the three algorithms: CNN, ResNet, and MRCA at different sample lengths.

*4.4. Ablation Experiment*

To evaluate the contribution of each module to the model's performance, we conducted ablation experiments by sequentially removing or modifying key modules. The three core modules examined were the dynamic weighted loss function, the channel attention mechanism, and the cross-fusion module. We hypothesize that the channel attention mechanism enhances the model's focus on critical features, the cross-fusion module improves feature layer interaction and fusion, and the dynamic weighted loss function facilitates gradient descent optimization. To verify these hypotheses, we designed the following experiments:

1. MRCA-1: The dynamic weighted loss function's weight parameter $\sigma$ is initialized based on experience to evaluate the impact of weight initialization on model performance.
2. MRCA-2: The dynamic weighted loss function's weight parameter $\sigma$ is not initialized, aiming to evaluate the impact of not initializing the weights on model performance.
3. MRCA-3: The total loss is calculated by directly adding the individual losses to evaluate the impact of the dynamic weighted loss function on model performance.
4. NO Attention: The channel attention mechanism module is removed to evaluate its impact on model performance.
5. NO Cross: The cross-fusion module is removed to evaluate its contribution.

6. BaseLine: The baseline model, which removes both the channel attention mechanism and the cross-fusion module.

The experimental results are shown in Figure 9, and specific experimental performances are provided in Table 7.

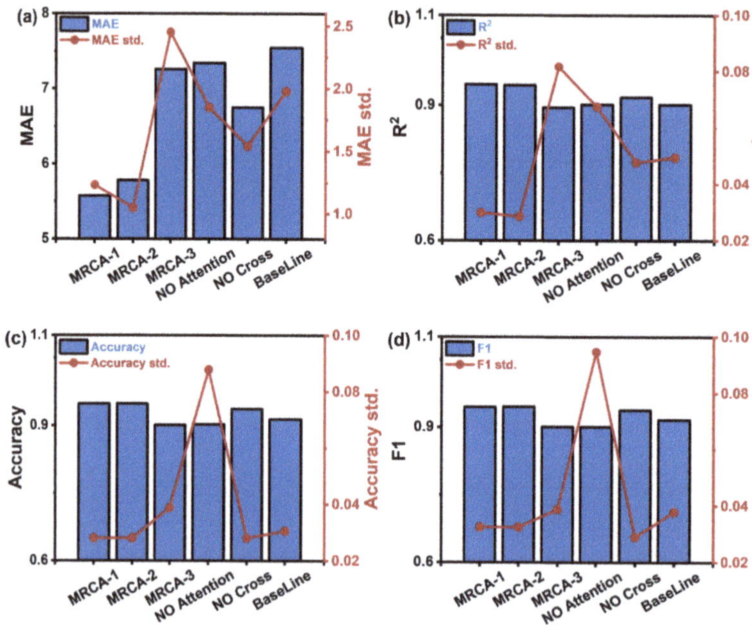

**Figure 9.** Ablation experiment performance evaluation metrics: (**a**) MAE, (**b**) $R^2$ score, (**c**) accuracy, (**d**) average F1 score, and standard error.

**Table 7.** Ablation experiment.

| Algo. | MAE | $R^2$ | Accuracy | F1 |
|---|---|---|---|---|
| MRCA-1 | 5.3961 | 0.9471 | 0.9486 | 0.9449 |
| MRCA-2 | 5.7812 | 0.9445 | 0.9486 | 0.9456 |
| MRCA-3 | 7.2567 | 0.8953 | 0.9014 | 0.9011 |
| NO Attention | 7.3411 | 0.9017 | 0.9029 | 0.9009 |
| NO Cross | 6.7503 | 0.9179 | 0.9371 | 0.9378 |
| BaseLine | 7.3927 | 0.8882 | 0.8914 | 0.8984 |

The experimental results show that the performance difference between using experience-initialized weight parameters ($\sigma$) and uninitialized weight parameters ($\sigma$) in the dynamic weighted loss function is minimal. This suggests that the dynamic weighted loss function is less sensitive to initial weights and allows the model to adaptively optimize loss weights, thereby simplifying the initialization process. In contrast, the MRCA-3 model, which lacks the dynamic weighted loss function, performs poorly in both classification and regression tasks.

When either the channel attention mechanism or the cross-fusion module is removed individually, model performance declines. However, when both modules are removed, performance deteriorates significantly, with the $R^2$ score dropping by approximately 6% points and accuracy decreasing by 5% points.

These ablation results highlight the positive impact of the channel attention mechanism and the cross-fusion module on model performance. Additionally, the dynamic weighted loss function not only optimizes performance but also improves training efficiency.

## 5. Conclusions

This study proposed the MTL-RCANet, which simultaneously achieves a classification accuracy of 94.86% and an $R^2$ score of 0.95 for concentration prediction. Compared to single-task models, MTL-RCANet significantly improves the performance of both classification and regression tasks by efficiently extracting and sharing key feature information across tasks. Additionally, a dynamic weighted loss function was introduced to address the varying loss requirements of different tasks during training. By dynamically adjusting the loss weight for each task based on its importance and difficulty, the model's overall performance is further enhanced. Therefore, our approach not only accelerates training convergence but also effectively balances the training processes of different tasks. Further research indicates that the model can still perform detection tasks efficiently, even with just 35% of the gas response data.

In summary, the proposed method offers a new solution for gas detection tasks in fast detection and low-resource consumption scenarios, which shows great application potential. Future work can further optimize the network structure to enhance task collaboration, particularly in more complex gas mixtures or dynamic response scenarios.

**Author Contributions:** Conceptualization, H.M. and R.Y.; methodology, J.P. and R.Y.; software, R.Y.; validation, H.M. and R.Y.; data curation, K.M.; writing—original draft preparation, H.M. and R.Y.; writing—review and editing, T.W. and L.W.; supervision, T.W.; project administration, L.W. and T.W. All authors have read and agreed to the published version of the manuscript.

**Funding:** This work was supported by the Jilin Provincial Science and Technology Development Program Project (YDZJ202501ZYTS591) and the National Natural Science Foundation of China (52205586 and 62301314).

**Institutional Review Board Statement:** Not applicable.

**Informed Consent Statement:** Not applicable.

**Data Availability Statement:** Data are contained within the article.

**Conflicts of Interest:** The authors declare no conflicts of interest.

## References

1. Li, P.; Wang, C.; Li, L.; Zheng, T. Bioaerosols and VOC emissions from landfill leachate treatment processes: Regional differences and health risks. *J. Hazard. Mater.* **2024**, *480*, 136232. [CrossRef] [PubMed]
2. Kim, S.J.; Lee, S.J.; Hong, Y.; Choi, S.D. Investigation of priority anthropogenic VOCs in the large industrial city of ulsan, south korea, focusing on their levels, risks, and secondary formation potential. *Atmos. Environ.* **2024**, *343*, 120982. [CrossRef]
3. Miao, G.; Wang, Y.; Wang, B.; Yu, H.; Liu, J.; Pan, R.; Zhou, C.; Ning, J.; Zheng, Y.; Zhang, R.; et al. Multi-omics analysis reveals hepatic lipid metabolism profiles and serum lipid biomarkers upon indoor relevant VOC exposure. *Environ. Int.* **2023**, *180*, 108221. [CrossRef]
4. Wang, H.; Zhang, R.; Kong, H.; Wang, K.; Sun, L.; Yu, X.; Zhao, J.; Xiong, J.; Tran, P.T.M. Balasubramanian. long-term emission characteristics of VOCs from building materials. *J. Hazard. Mater.* **2024**, *480*, 136337. [CrossRef]
5. Zhou, L.; Huang, C.; Lu, R.; Wang, X.; Sun, C.; Zou, Z. Associations between VOCs and childhood asthma in Shanghai, China: Impacts of daily behaviors. *Atmos. Pollut. Res.* **2024**, *16*, 102359. [CrossRef]
6. Hussain, M.S.; Gupta, G.; Mishra, R.; Patel, N.; Gupta, S.; Alzarea, S.I.; Kazmi, I.; Kumbhar, P.; Disouza, J.; Dureja, H.; et al. Unlocking the secrets: Volatile organic compounds (VOCs) and their devastating effects on lung cancer. *Pathol.-Res. Pract.* **2024**, *255*, 155157. [CrossRef]
7. Peng, J.; Mei, H.; Yang, R.; Meng, K.; Shi, L.; Zhao, J.; Zhang, B.; Xuan, F.; Wang, T.; Zhang, T. Olfactory diagnosis model for lung health evaluation based on pyramid pooling and shap-based dual encoders. *ACS Sens.* **2024**, *9*, 4934–4946. [CrossRef]

8. Song, J.; Li, R.; Yu, R.; Zhu, Q.; Li, C.; He, W.; Liu, J. Detection of VOCs in exhaled breath for lung cancer diagnosis. *Microchem. J.* **2024**, *199*, 110051. [CrossRef]
9. Liu, H.; Fang, C.; Zhao, J.; Zhou, Q.; Dong, Y.; Lin, L. The detection of acetone in exhaled breath using gas pre-concentrator by modified metal-organic framework nanoparticles. *Chem. Eng. J.* **2024**, *498*, 155309. [CrossRef]
10. Lv, E.; Wang, T.; Yue, X.; Wang, H.; Zeng, J.; Shu, X.; Wang, J. Wearable SERS sensor based on bionic sea urchin-cavity structure for dual in-situ detection of metabolites and VOCs gas. *Chem. Eng. J.* **2024**, *499*, 156020. [CrossRef]
11. Mei, H.; Peng, J.; Wang, T.; Zhou, T.; Zhao, H.; Zhang, T.; Yang, Z. Overcoming the limits of cross-sensitivity:pattern recognition methods for chemiresistive gas sensor array. *Nano-Micro Lett.* **2024**, *16*, 285–341.
12. Raina, S.; Bharti, A.; Singh, H.M.; Kothari, R.; Tyagi, V.V.; Pathania, D.; Buddhi, D. Chapter 1—Applications of gas and VOC sensors for industry and environmental monitoring: Current trends and future implications. In *Complex and Composite Metal Oxides for Gas, VOC, and Humidity Sensors*; Yadav, B.C., Kumar, P., Eds.; Elsevier: Amsterdam, The Netherlands, 2024; pp. 3–26. [CrossRef]
13. Fan, H.; Wang, P.; Zhang, H.; Hu, M.; Zhu, C.; Wang, Q. Zero-absorption-assisted multitask learning for simultaneous measurement of acetylene concentration and gas pressure from overlap-deformed spectra. *Opt. Laser Technol.* **2024**, *176*, 110968. [CrossRef]
14. Gong, Z.; Fan, Y.; Guan, Y.; Wu, G.; Mei, L. Empirical Modal Decomposition Combined with Deep Learning for Photoacoustic Spectroscopy Detection of Mixture Gas Concentrations. *Anal. Chem.* **2024**, *96*, 18528–18536.
15. Hou, J.; Liu, X.; Sun, H.; He, Y.; Qiao, S.; Zhao, W.; Zhou, S.; Ma, Y. Dual-Component Gas Sensor Based on Light-Induced Thermoelastic Spectroscopy and Deep Learning. *Anal. Chem.* **2025**, *97*, 5200–5208.
16. Kan, Z.; Zhang, Y.; Luo, L.; Cao, Y. Ultraviolet absorption spectrometry with symmetrized dot patterns and deep learning for quantitative analysis of SO2, H2S, CS2 mixed gases. *Eng. Appl. Artif. Intell.* **2024**, *133*, 108366. [CrossRef]
17. Kistenev, Y.V.; Skiba, V.E.; Prischepa, V.V.; Borisov, A.V.; Vrazhnov, D.A. Gas-mixture IR absorption spectra denoising using deep learning. *J. Quant. Spectrosc. Radiat. Transf.* **2024**, *313*, 108825. [CrossRef]
18. Zhou, Y.; Jiang, M.; Dou, W.; Meng, D.; Wang, C.; Wang, J.; Wang, X.; Sun, L.; Jiang, S.; Chen, F.; et al. Narrow-band multi-component gas analysis based on photothermal spectroscopy and partial least squares regression method. *Sens. Actuators B Chem.* **2023**, *377*, 133029.
19. Chakraborty, P.; Borras, E.; Rajapakse, M.Y.; McCartney, M.M.; Bustamante, M.; Mitcham, E.J.; Davis, C.E. Non-destructive method to classify walnut kernel freshness from volatile organic compound (VOC) emissions using gas chromatography-differential mobility spectrometry (GC-DMS) and machine learning analysis. *Appl. Food Res.* **2023**, *3*, 100308. [CrossRef]
20. Dwyer, D.B.; Niedziela, J.L.; Miskowiec, A. Tandem pyrolysis evolved gas–gas chromatography–mass spectrometry. *J. Anal. Appl. Pyrolysis* **2025**, *186*, 106904.
21. Rahmani, N.; Mani-Varnosfaderani, A. Profiling volatile organic compounds of different grape seed oil genotypes using gas chromatography-mass spectrometry and chemometric methods. *Ind. Crops Prod.* **2024**, *222*, 119928.
22. Zhang, Z.; Zhang, Q.; Xi, Y.; Zhou, Y.; Zhan, M. Establishment of a headspace-thermal desorption and gas chromatography-mass spectrometry method (HS-TD-GC-MS) for simultaneous detection of 51 volatile organic compounds in human urine: Application in occupational exposure assessment. *J. Chromatogr. A* **2024**, *1722*, 464863. [CrossRef] [PubMed]
23. Zhao, Y.; Liu, Y.; Han, B.; Wang, M.; Wang, Q.; Zhang, Y.-n. Fiber optic volatile organic compound gas sensors: A review. *Coord. Chem. Rev.* **2023**, *493*, 215297. [CrossRef]
24. Bing, Y.; Zhang, F.; Han, J.; Zhou, T.; Mei, H.; Zhang, T. A method of ultra-low power consumption implementation for MEMS gas sensors. *Chemosensors* **2023**, *11*, 236. [CrossRef]
25. Chen, H.; Huo, D.; Zhang, J. Gas Recognition in E-Nose System: A Review. *IEEE Trans. Biomed. Circuits Syst.* **2022**, *16*, 169–184. [CrossRef] [PubMed]
26. Li, J.; Zhao, H.; Wang, Y.; Zhou, Y. Approaches for selectivity improvement of conductometric gas sensors: An overview. *Sens. Diagn.* **2024**, *3*, 336–353. [CrossRef]
27. Yin, X.-T.; Dastan, D.; Gity, F.; Li, J.; Shi, Z.; Alharbi, N.D.; Liu, Y.; Tan, X.-M.; Gao, X.-C.; Ma, X.-G.; et al. Gas sensing selectivity of SnO2-xNiO sensors for homogeneous gases and its selectivity mechanism: Experimental and theoretical studies. *Sens. Actuators A Phys.* **2023**, *354*, 114273. [CrossRef]
28. Panda, S.; Mehlawat, S.; Dhariwal, N.; Kumar, A.; Sanger, A. Comprehensive review on gas sensors: Unveiling recent developments and addressing challenges. *Mater. Sci. Eng. B* **2024**, *308*, 117616. [CrossRef]
29. Wang, Z.; Li, Y.; He, X.; Yan, R.; Li, Z.; Jiang, Y.; Li, X. Improved deep bidirectional recurrent neural network for learning the cross-sensitivity rules of gas sensor array. *Sens. Actuators B Chem.* **2024**, *401*, 134996. [CrossRef]
30. Pan, X.; Chen, J.; Wen, X.; Hao, J.; Xu, W.; Ye, W.; Zhao, X. A comprehensive gas recognition algorithm with label-free drift compensation based on domain adversarial network. *Sens. Actuators B Chem.* **2023**, *387*, 133709. [CrossRef]
31. Wei, G.; Xu, Y.; Lv, X.; Jiao, S.; He, A. An adaptive drift compensation method based on integrated dual-channel feature fusion for electronic noses. *IEEE Sens. J.* **2024**, *24*, 26814–26824. [CrossRef]

32. Yao, Y.; Chen, B.; Liu, C.; Qu, C. Investigation on the combined model of sensor drift compensation and open-set gas recognition based on electronic nose datasets. *Chemom. Intell. Lab. Syst.* **2023**, *242*, 105003. [CrossRef]
33. Laref, R.; Losson, E.; Sava, A.; Adjallah, K.; Siadat, M. A comparison between SVM and PLS for E-nose based gas concentration monitoring. In Proceedings of the 2018 IEEE International Conference on Industrial Technology (ICIT), Lyon, France, 20–22 February 2018; pp. 1335–1339.
34. Xu, W.; Tang, J.; Xia, H.; Sun, Z. Prediction method of dioxin emission concentration based on PCA and deep forest regression. In Proceedings of the 2021 40th Chinese Control Conference (CCC), Shanghai, China, 26–28 July 2021; pp. 1212–1217.
35. Xia, W.; Song, T.; Yan, Z.; Song, K.; Chen, D.; Chen, Y. A method for recognition of mixed gas composition based on PCA and KNN. In Proceedings of the 2021 19th International Conference on Optical Communications and Networks (ICOCN), Qufu, China, 23–27 August 2021; pp. 1–3.
36. Li, K.; Yang, G.; Wang, K.; Lu, B.; Jia, J.; Sun, T. Prediction of dissolved gases concentration in transformer oil based on VMD and ELM. In Proceedings of the 2023 IEEE 7th Conference on Energy Internet and Energy System Integration (EI2), Hangzhou, China, 15–18 December 2023; pp. 3749–3753.
37. Martono, N.P.; Kuramaru, S.; Igarashi, Y.; Yokobori, S.; Ohwada, H. Blood alcohol concentration screening at emergency room: Designing a classification model using machine learning. In Proceedings of the 2023 14th International Conference on Information & Communication Technology and System (ICTS), Surabaya, Indonesia, 4–5 October 2023; pp. 255–260.
38. Zhu, R.; Gao, J.; Li, M.; Gao, Q.; Wu, X.; Zhang, Y. A ppb-level online detection system for gas concentrations in CS2/SO2 mixtures based on UV-DOAS combined with VMD-CNN-TL model. *Sens. Actuators B Chem.* **2023**, *394*, 134440. [CrossRef]
39. Mao, G.; Zhang, Y.; Xu, Y.; Li, X.; Xu, M.; Zhang, Y.; Jia, P. An electronic nose for harmful gas early detection based on a hybrid deep learning method H-CRNN. *Microchem. J.* **2023**, *195*, 109464. [CrossRef]
40. Chu, J.; Li, W.; Yang, X.; Wu, Y.; Wang, D.; Yang, A.; Yuan, H.; Wang, X.; Li, Y.; Rong, M. Identification of gas mixtures via sensor array combining with neural networks. *Sens. Actuators B Chem.* **2021**, *329*, 129090. [CrossRef]
41. Song, S.; Chen, J.; Ma, L.; Zhang, L.; He, S.; Du, G.; Wang, J. Research on a working face gas concentration prediction model based on LASSO-RNN time series data. *Heliyon* **2023**, *9*, e14864. [CrossRef]
42. Zeng, L.; Xu, Y.; Ni, S.; Xu, M.; Jia, P. A mixed gas concentration regression prediction method for electronic nose based on two-channel TCN. *Sens. Actuators B Chem.* **2023**, *382*, 133528. [CrossRef]
43. Li, X.; Guo, J.; Xu, W.; Cao, J. Optimization of the mixed gas detection method based on neural network algorithm. *ACS Sens.* **2023**, *8*, 822–828. [CrossRef]
44. Wang, T.; Zhang, H.; Wu, Y.; Jiang, W.; Chen, X.; Zeng, M.; Yang, J.; Su, Y.; Hu, N.; Yang, Z. Target discrimination, Concentration prediction of binary mixed gases based on random forest algorithm in the electronic nose system mixed gases, and status judgment of electronic nose system based on large-scale measurement and multi-task deep learning. *Sens. Actuators B Chem.* **2022**, *351*, 130915. [CrossRef]
45. Wang, X.; Zhao, W.; Ma, R.; Zhuo, J.; Zeng, Y.; Wu, P.; Chu, J. A novel high accuracy fast gas detection algorithm based on multi-task learning. *Measurement* **2024**, *228*, 114383. [CrossRef]
46. Fu, C.; Zhang, K.; Guan, H.; Deng, S.; Sun, Y.; Ding, Y.; Wang, J.; Liu, J. Progressive prediction algorithm by multi-interval data sampling in multi-task learning for real-time gas identification. *Sens. Actuators B Chem.* **2024**, *418*, 136271. [CrossRef]
47. Kang, M.; Cho, I.; Park, J.; Jeong, J.; Lee, K.; Lee, B.; Del Orbe Henriquez, D.; Yoon, K.; Park, I. High accuracy real-time multi-gas identification by a batch-uniform gas sensor array and deep learning algorithm. *ACS Sens.* **2022**, *7*, 430–440. [PubMed]
48. Zhang, S.; Cheng, Y.; Luo, D.; He, J.; Wong, A.K.Y.; Hung, K. Channel attention convolutional neural network for chinese baijiu detection with E-Nose. *IEEE Sens. J.* **2021**, *21*, 16170–16182.
49. Misra, I.; Shrivastava, A.; Gupta, A.; Hebert, M. Cross-stitch networks for multi-task learning. In Proceedings of the 2016 IEEE Conference on Computer Vision and Pattern Recognition (CVPR), Las Vegas, NV, USA, 27–30 June 2016; pp. 3994–4003.

*sensors*

MDPI

*Article*

# Electrospun (La,Ba)$FeO_3$ Nanofibers as Materials for Highly Sensitive VOC Gas Sensors

**Vadim Platonov [1], Nikolai Malinin [2], Darya Filatova [1], Ivan Sapkov [3] and Marina Rumyantseva [1,*]**

[1] Chemistry Department, Moscow State University, 119991 Moscow, Russia
[2] Faculty of Materials Science, Moscow State University, 119991 Moscow, Russia
[3] Physics Department, Moscow State University, 119991 Moscow, Russia
* Correspondence: roum@inorg.chem.msu.ru; Tel.: +7-495-939-5471

**Abstract:** In this work, we report the synthesis of perovskite-type Ba-doped $LaFeO_3$ ($La_{1-x}Ba_xFeO_3$, x = 0.00, 0.02, 0.04, and 0.06) nanofibers (NFs) using the electrospinning method. The synthesized $La_{1-x}Ba_xFeO_3$ materials have a fibrous structure with an average fiber diameter of 250 nm. The fibers, in turn, consist of smaller crystalline particles of 20–50 nm in size. The sensor properties of $La_{1-x}Ba_xFeO_3$ nanofibers were studied when detecting 20 ppm CO, $CH_4$, methanol, and acetone in dry air in the temperature range of 50–350 °C. Doping with barium leads to a significant increase in sensor response and a decrease in operating temperature when detecting volatile organic compounds (VOCs). The process of acetone oxidation on the surface of the most sensitive $La_{0.98}Ba_{0.02}FeO_3$ material was studied using in situ diffuse reflectance infrared Fourier transform spectroscopy (DRIFTS) and temperature-programmed desorption in combination with mass spectrometry (TPD-MS). A mechanism for the sensor signal formation is proposed.

**Keywords:** Ba-doped $LaFeO_3$; perovskites; semiconductor gas sensor; VOCs; DRIFTS; TPD-MS

## 1. Introduction

Intensive development of chemical and metallurgical industries, use of power plants operating on biofuel, and high-temperature waste processing and combustion processes lead to an increase in environmental pollution. To control air quality in residential and industrial areas, it is necessary to create selective, stable, and highly sensitive gas sensors. Traditional semiconductor gas sensors based on nanocrystalline oxides ($MO_x$ = ZnO, $SnO_2$, $WO_3$, $In_2O_3$) have a number of serious drawbacks for the quantitative determination of the concentration of toxic impurities in the air, primarily due to low selectivity and drift of parameters caused by low stability of the microstructure during long-term operation at a temperature of 300–500 °C, which is necessary for the detection of reducing gases, including volatile organic compounds (VOCs) [1–4]. The factors responsible for the instability of the parameters of a semiconductor gas sensor can be classified as follows [5]: (i) structural changes in microheaters and electrodes; (ii) changes caused by processes in the layer of the sensitive material; (iii) changes caused by the processes occurring at the boundaries between the sensitive layer, substrate, and electrodes; and (iv) changes caused by the interaction of the sensitive layer with the components of the surrounding gas phase. The processes occurring in the layer of the sensitive material include structural and phase transformations, as well as the diffusion of oxygen vacancies and other point defects in the volume of the semiconductor oxide. These processes occur with greater intensity under high working temperatures. So, the problem of the stability of the sensor's characteristics can be solved (at least partially) by creating materials that provide high sensor response at low operating temperatures. Metal oxide perovskites offer tunable electronic properties and the capability for low-temperature operation [6]. Several perovskite materials have been studied as gas sensors due to their stability [7]. Of particular interest as a gas-sensitive material is lanthanum ferrite ($LaFeO_3$), a compound belonging to the class of orthoferrites with a distorted orthorhombic perovskite structure. Lanthanum ferrite is a *p*-type semiconductor with a band gap of 2.1–2.6 eV.

Due to its crystal structure, $LaFeO_3$ has the ability to substitute atoms in A and B positions within one structural type that provides ample opportunities for creating materials with controlled chemical, electrophysical, sensor, and catalytic properties [8]. The nature of the cations occupying the A and B positions has a very important effect on the electronic structure of materials. The electrostatic potential created by a given cation and the hybridization of its orbitals with the orbitals of the nearest ions affect the width of the energy bands and their arrangement on the energy diagram. In addition, the size of the A cation affects the degree of distortion of the perovskite structure [9]. For this reason, the cationic composition of the A-position also has an important effect on the main properties of perovskite materials (electrophysical, chemical, sensor, and catalytic properties). Based on the available literature data, heterosubstitution of $La^{3+}$ with alkaline earth metal cations ($Ba^{2+}$, $Sr^{2+}$, $Ca^{2+}$) as well as $Mg^{2+}$ and $Pb^{2+}$ cations can increase the sensitivity of the sensor material to volatile organic compounds (VOCs) [10–17] when detecting carbon monoxide [18], sulfur dioxide ($SO_2$) [19], carbon dioxide ($CO_2$) [20–23]. Thus, changing the chemical composition of the A-position of perovskite can improve the sensitivity of materials to many analyte gases due to a change in the base resistance of the compounds and the appearance of various point defects (oxygen vacancies, a change in the population of A-positions, a change in the oxidation state of the B-cation) capable of entering into physicochemical interactions with gases in the environment.

The morphology of the $LaFeO_3$-based material is another key factor influencing sensor characteristics. Of considerable interest is the production of perovskite oxides with unique morphology and improved properties using new synthesis methods [24]. Gas-sensitive materials based on nanofibers/nanorods/nanowires/nanotubes/etc. can provide high gas permeability even with a small crystallite size and a high surface-to-volume ratio, making them promising for creating gas sensors [25]. The most widely used methods for obtaining nanocrystalline semiconductor oxides are co-precipitation or separate chemical precipitation from aqueous solutions with subsequent annealing in the temperature range of 300–500 °C. The main disadvantage of such methods is the formation of a surface hydrate-hydroxyl layer, which affects the sintering process during the formation of sensitive porous layers and the sensor properties of materials. An alternative is the electrospinning (electroforming) method—the process of forming nanofibers from an electrically charged polymer solution or melt under the action of electrostatic forces [26]. This method allows for a smaller number of steps from non-aqueous polymer solutions to obtain highly dispersed materials of complex composition in the form of fibers, from which a porous structure is formed during synthesis and thermal annealing. Detection of analyte gases by chemoresistive sensors is associated with processes occurring on the surface of a solid; thus, the availability of the surface for the adsorption of detected gases is an important parameter that directly affects the efficiency of the material. The bimodal pore distribution characteristic for materials synthesized by electrospinning improves their sensor properties due to the formation of numerous branched channels in the sample mass, through which the diffusion of molecules of the analyzed gas and reaction products occurs faster than, for example, in sintered powders. In this regard, electrospinning has become widespread in the synthesis of semiconductor oxide materials [27–37]. Successful synthesis of $LaFeO_3$-based nanofibers by electrospinning has been reported [38–40]. However, the sensor properties of electrospun lanthanum ferrite have been poorly studied to date [41,42].

In this work, $La_{1-x}Ba_xFeO_3$ (x = 0.02, 0.04, and 0.06) nanofibers were synthesized by electrospinning precursor-filled polymer solutions with subsequent heat treatment. The effect of Ba content on the structure, morphology, surface properties, and electrophysical and sensor properties in detecting gases of various chemical natures (CO, $CH_4$, methanol, and acetone) was studied. Based on the results obtained by in situ diffuse reflectance infrared Fourier transform spectroscopy (DRIFTS) and temperature-programmed desorption coupled with mass spectrometry (TPD-MS), a mechanism for the multi-stage decomposition of acetone on the surface of the sensor material is proposed.

## 2. Materials and Methods

### *2.1. Materials Synthesis*

The following reagents were used in the synthesis: polyvinylpyrrolidone (PVP) (Mw = 1,300,000), lanthanum nitrate ($La(NO_3)_3 \times 6\ H_2O$), iron acetylacetonate ($Fe(acac)_3$), barium nitrate ($Ba(NO_3)_2$), *N,N*-dimethylformamide ($C_3H_7NO$, DMF), and ethanol ($C_2H_5OH$) (all from Sigma-Aldrich (Saint Louis, MO, USA), analytical pure grade). Nanofibers of pristine $LaFeO_3$ and $La_{1-x}Ba_xFeO_3$ (x = 0.02, 0.04, and 0.06) were synthesized by electrospinning precursor-filled polymer solutions with subsequent heat treatment.

To obtain pure $LaFeO_3$ nanofibers, 1.2987 g of lanthanum nitrate and 1.0595 g of iron acetylacetonate were dissolved in a mixture of ethyl alcohol (25 mL) and dimethylformamide (25 mL). When preparing barium-doped lanthanum ferrite nanofibers, the molar ratio of [La]:[Ba]:[Fe] = (1 − x):x:1 (x = 0.02, 0.04, and 0.06) was selected. After complete dissolution of the cation's precursors, 9 g of polyvinylpyrrolidone was added and stirred until complete dissolution of the polymer and homogenization of the solution. To effectuate the electrospinning process, the polymer solution was placed in a medical syringe, which was fixed in a syringe infusion pump. The solution was fed at a rate of 1 mL/h through a G21 gauge metal needle with an internal diameter of 510 μm. The polymer fibers were formed on a metal collector under a potential difference of 10–11 kV and at a distance of 120–130 mm between the needle and the collector. The resulting polymer fibers were annealed at 600 °C for 5 h, at a heating rate of 1 K/min. This heat treatment mode was selected based on the data obtained for a series of pure $LaFeO_3$ samples annealed at different temperatures [43].

### *2.2. Materials Characterization*

The elemental composition of the synthesized materials was studied by X-ray fluorescence microanalysis with full external reflection using a S2 PICOFOX spectrometer (Bruker Nano GmbH, Berlin, Germany). Mo Kα radiation was used to excite X-ray fluorescence. The spectrum set time was 250 s. A 5 μL aliquot of the sample solution was applied to a quartz substrate using a dispenser, dried, and analyzed.

The phase composition of the samples was determined by powder X-ray diffraction (XRD). The diffractograms were recorded at room temperature using a DRON-4-0.7 diffractometer (Burevestnik, St. Petersburg, Russia; λ = 1.54051 Å, $CuK_{\alpha1+\alpha2}$ radiation) in the 20–70° 2θ range with 0.1° step and 1°/min speed. Phase identification was effectuated with the use of PDF-1 and PDF-2 databases. The diffraction patterns were processed using the STOE "WinXPOW" software package (Version 1.06). The size of the coherent scattering region ($d_{XRD}$) was calculated from the reflection broadening using the Scherrer formula:

$$d_{XRD} = \frac{k\lambda}{\beta cos\theta}, \quad (1)$$

where $d_{XRD}$ is the average size of the coherent scattering region, $\beta$ is the full width at half maximum (FWHM) of the corresponding diffraction peak, $\lambda$ is the wavelength of the radiation used, $\theta$ is the diffraction angle, and $k$ is the shape coefficient for spherical particles ($k$ = 0.9). $\beta$ was determined as follows:

$$\beta^2 = \beta^2_{sample} + \beta^2_{standard}, \quad (2)$$

where $\beta_{sample}$ is the FWHM of the diffraction peak of the sample, and $\beta_{standard}$ = 0.085 is the FWHM of the diffraction peak of the standard with a crystallite size over 200 nm.

The specific surface area was determined by low-temperature nitrogen adsorption using a Chemisorb 2750 apparatus (Micromeritics, Norcross, GA, USA) with a thermal conductivity detector. The BET (Brunauer–Emmett–Teller) model was used to calculate the surface area available for gas adsorption. The powder (~100 mg) was placed in a flowing quartz reactor and annealed in 99.999% He flow at 50 mL/min and 300 °C for 1 h to remove adsorbed impurities. Then, a flow of $N_2$/He mixture (30 vol.% $N_2$) was passed through the reactor at a rate of 12 mL/min, and the sample was cooled to 77 K.

The microstructure and morphology of the nanofibers were studied by scanning electron microscopy (SEM) using a Zeiss Supra 40 FE-SEM electron microscope (Carl Zeiss, Inc., Oberkochen, Germany) with an intralens secondary electron detector at an accelerating voltage of 10 kV and an aperture of 30 µm.

The sample designation, elemental and phase compositions, as well as the microstructure parameters of the synthesized nanofibers are summarized in Table 1.

**Table 1.** Sample designations, elemental and phase compositions, and microstructure parameters of synthesized nanofibers.

| Sample | Material | [Ba]/([La]+[Ba]) at.%, RSD [1] = 4% (XRF) | Phase Composition | $d_{XRD}$, nm |
|---|---|---|---|---|
| LFO | $LaFeO_3$ | - | lanthanum ferrite $LaFeO_3$ ICDD [37-1493] | 17 ± 2 |
| LBFO-2 | $La_{0.98}Ba_{0.02}FeO_3$ | 2 | | 14 ± 1 |
| LBFO-4 | $La_{0.96}Ba_{0.04}FeO_3$ | 4 | | 14 ± 1 |
| LBFO-6 | $La_{0.94}Ba_{0.06}FeO_3$ | 6 | | 14 ± 1 |

[1] Relative standard deviation.

The surface composition and the charge state of elements were studied by infrared (IR) and X-ray photoelectron (XPS) spectroscopies. IR-spectra were recorded in the transmission mode in the wavenumbers range of 4000–400 $cm^{-1}$ with a step of 4 $cm^{-1}$ using a Perkin Elmer Frontier spectrometer (Perkin Elmer, Waltham, MA, USA). The survey was carried out with potassium bromide tablets (7 mm diameter) containing 1 mass% of the test sample. The tablets were prepared by carefully grinding KBr together with the sample, followed by pressing into tablets at a pressure of 50 bar.

In situ surface composition studies were carried out by diffuse reflectance IR Fourier transform spectroscopy (DRIFTS) using a DiffusIR set-top box and a heated flow chamber HC900 (Pike Technologies, Cottonwood, WI, USA) enclosed by a KBr window. DRIFT spectra were recorded in the region of 4000–1000 $cm^{-1}$ with 4 $cm^{-1}$ resolution. Powder samples weighing 30 mg were placed in aluminum oxide crucibles (5 mm diameter). Before measurement, the samples were heated in a stream of purified air at 450 °C to remove adsorbed impurities.

Studies of the charge state of the elements were carried out using an Axis Ultra DLD spectrometer (Kratos Analytical, Manchester, UK) with monochromatic Al Kα radiation (hν = 1486.7 eV, 150 W) in a vacuum not lower than $10^{-9}$ Torr. The charge shift was compensated by the C1s ground state peak with a binding energy of 285 eV. Survey spectra in the range of 1300–0 eV with 0.5 eV step were obtained for all samples. The spectra of the Fe2p, La3d, Ba3d, O1s, and C1s regions were recorded with increments of 0.05 eV. The Unifit 2014 program was used to process the spectra, the background was described using the Shirley method, and the spectra were approximated by mixed Gauss–Lorentz functions.

The adsorption/desorption of acetone and its oxidation products on the surface was analyzed using an MS7–200 mass spectrometer (Atomtyazhmash, St. Petersburg, Russia) equipped with an RGA–200 analyzer (Stanford Research Systems, Sunnyvale, CA, USA) during stepwise heating of a sample placed in a quartz tube with an inner diameter of 10 mm. Before the experiment, the sample was kept at 450 °C for 1 h in a flow of $He/O_2$ mixture (99:1 vol.%, 30 mL/min) to remove molecules adsorbed from the air on the surface. After cooling the sample to room temperature, a helium–oxygen mixture containing 200 ppm acetone was passed through the sample, and the temperature was gradually raised to 500 °C.

### 2.3. Gas Sensor Measurements

The electrophysical and sensor properties of the synthesized materials were studied in situ by measuring the electrical conductivity of thick films. Materials in the form of a terpineol-based paste were deposited on $Al_2O_3$ micro hotplates with platinum contacts. After applying the paste, the thick films were dried for 5 h at 50 °C, then heated to 550 °C (2 K/min) and kept at this temperature for 3 h to completely remove the binder. The temperature of the films was controlled via resistance heating of the hotplates, and a heating rate of 2 K/min was necessary to prevent destruction of the films due to the difference in the thermal expansion coefficients of the ceramic substrate and deposited material. Three sensors were made from each material.

Measurements of the electrical conductivity of the fibers were carried out in DC stabilized voltage mode using an automated flow cell (PTFE, 100 mL). An RRG-12 electronic gas flowmeter (Eltochpribor, Zelenograd, Russia) was used to create stable flows and obtain gas mixtures with specified concentrations of analyte gases. The IVTM-7 humidity and temperature meter (EXIS, Zelenograd, Russia) was used to control the humidity. The attested gas mixtures were used as sources of analyte gases: 2540 ppm CO in $N_2$, 4060 ppm $CH_4$ in $N_2$, 1020 ppm methanol in $N_2$, and 1460 ppm acetone in $N_2$. To create gas mixtures containing analyte gases at a given concentration (20 ppm), the attested gas mixtures (from cylinders) were diluted with dry air obtained from a GCHV-2.0-3.5 pure air generator (NPP Chemelectronics, Moscow, Russia). The same dry air was used as the reference gas. The selected concentration (20 ppm) was close to the maximum permissible concentration of the working area for CO (20 ppm) and methanol (15 ppm). In order to be able to compare the sensor signal during the detection of acetone and methane, it was decided to use the same concentration of 20 ppm.

In all experiments, the gas flow through the cell was 100 mL/min. During measurement, the resistance of the sensors was recorded depending on the temperature of the sensor material and the composition of the gas mixture. The sensor response $S$ was calculated as follows:

$$S = \frac{R_{gas} - R_{air}}{R_{air}} = \frac{R_{gas}}{R_{air}} - 1, \tag{3}$$

where $R_{air}$ is the resistance of the material in the background air, and $R_{gas}$ is the resistance of the material in the presence of the target gas.

## 3. Results and Discussion

### 3.1. Morphology, Microstructure, Elemental and Phase Composition

Using X-ray powder diffraction, it was shown that all materials from this series were single phase. The XRD patterns corresponded to $LaFeO_3$ with an orthorhombic crystal structure (ICDD card [37-1493]) (Figure 1). The absence of any extraneous reflections indicated the successful incorporation of $Ba^{2+}$ into the $La^{3+}$ sites, which was enabled by the similarity of the effective ion radii between $Ba^{2+}$ (1.61 Å) and $La^{3+}$ (1.36 Å), all of which were in a cuboctahedral environment (coordination number CN = 12). In turn, the introduction of $Ba^{2+}$ into the perovskite B-sublattice was excluded from geometric considerations due to significant discrepancies between the effective radii of these ions (1.35 Å for $Ba^{2+}$) compared to HS $Fe^{3+}$ (0.645 Å) in an octahedral site (CN = 6).

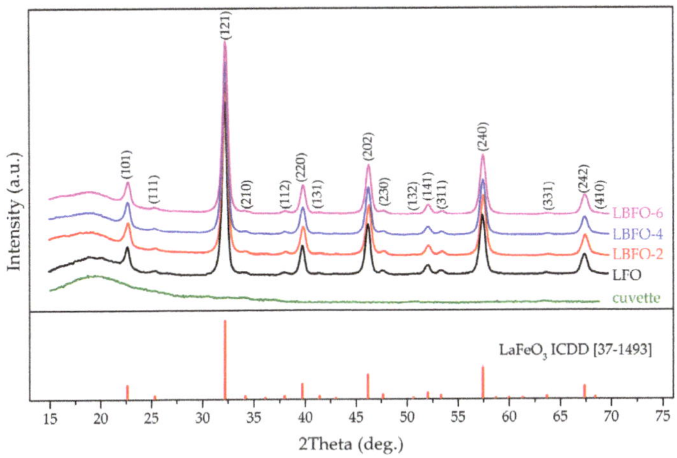


**Figure 1.** XRD patterns of pure $LaFeO_3$ and $La_{1-x}Ba_xFeO_3$ series with different Ba contents.

The calculation of the Goldschmidt tolerance factor was carried out according to the formula:

$$t = \frac{r_A + r_O}{\sqrt{2} * (r_B + r_O)}, \quad (4)$$

where $r_A$, $r_B$, and $r_O$ are the effective ionic radii of atoms A, B, and O, respectively, which are part of the perovskite composition $ABO_3$. For oxygen in an octahedral environment, $r_O = 1.4$.

In the case of pure $LaFeO_3$, the tolerance factor was $t = 0.95$. For the limiting composition of $BaFeO_3$, $t$ was approximately 1.04, which confirmed the possibility of heterovalent substitution of $La^{3+}$ cations by $Ba^{2+}$ ions at the A-position within the same structural type (perovskite $0.85 < t < 1.11$) [44]. At the same time, according to the XRD data, the perovskite lattice was distorted and had orthorhombic symmetry with the *Pnma* space group, while ideal perovskites had a primitive cubic structure ($Pm\bar{3}m$ space group). This may be due to oxygen vacancies, the concentration of which may increase because of heterovalent substitution to maintain electroneutrality [14,18,21,22]. Lattice distortion can occur due to another factors, such as changes in the relative positions of the octahedral units relative to each other (e.g., swing-rotation), as well as compression and stretching of these octahedrons [45,46]. The presence of distortions in the perovskite structure after doping was also indicated by a slight shift in the diffraction peaks toward larger angles, despite the larger ionic radius of $Ba^{2+}$ compared to $La^{3+}$. As the dopant concentration increased, the distance between planes tended to decrease, which can be explained by nonstoichiometric oxygen and the inclination of octahedra $BO_6$ [47].

The specific surface area of the materials, determined by the low-temperature nitrogen adsorption method using the BET model, was 2–5 $m^2/g$. In turn, the introduction of barium into lanthanum ferrite led to a decrease in the crystallite size (Table 1).

The morphology of the $La_{1-x}Ba_xFeO_3$ samples was investigated using scanning electron microscopy (Figure 2). The samples exhibited a fibrous structure, with an average fiber diameter of 250 ± 20 nm. The fibers, in turn, consisted of smaller crystalline particles with a size of 20–50 nm. Therefore, it could be inferred that the doping process did not significantly alter the macroscopic structure of the materials, which was mostly determined by the synthesis parameters (needle diameter, distance between the needle and collector, voltage, etc.).

**Figure 2.** SEM images of pure $LaFeO_3$ and $La_{1-x}Ba_xFeO_3$ series with different Ba contents.

### 3.2. Surface Properties

The analysis of the surface composition of $La_{1-x}Ba_xFeO_3$ fibers was carried out using the IR spectroscopy method (Figure 3). The IR spectra of all of the studied samples contained the same set of vibration bands. The broad band at 3700–3050 $cm^{-1}$ and the band at 1635 $cm^{-1}$ were associated with the stretching vibrations of various hydroxyl groups and the deformation vibrations of water molecules adsorbed from atmospheric air. The vibration bands in the range of 2920–2840 $cm^{-1}$ corresponded to C–H bonds, the presence of which was associated with the remains of organic molecules used in the synthesis of $LaFeO_3$ nanofibers. The presence of bands at 1480 $cm^{-1}$ and 1400 $cm^{-1}$ was associated with the formation of surface carbonates during annealing, as well as during the adsorption of $CO_2$ from air [48]. The band at 844 $cm^{-1}$ was attributed to vibrations of the La–O bond [48]. For an ideal cubic perovskite, the optically active internal vibrations can be classified as $\Gamma_{vib}$ = 3 F1u + F2u; of which the F1u modes are IR active, while the F2u modes are inactive. These vibrations can be roughly described as follows: the v1(F1u) mode is the B–O stretching vibration of $BO_6$ octahedra; the v2(F1u) mode is the deformation vibration of the OBO angle, weakly coupled to the vibrations of the A–O bond; the v3(F1u) mode is associated with the vibration of the A sublattice relative to the $BO_6$ octahedra;

and the inactive v4(F2u) mode is also associated with deformation vibrations of the OBO angle [49,50]. The expected order of the bands was v1 > v3 > v4 > v2. Since the v4 mode was inactive and the v2 mode was at very low frequencies, the infrared spectrum of cubic perovskite should have two bands. In the case of distorted and low-symmetry perovskites, the v1(F1u) mode may split. Thus, in the recorded spectra (Figure 3), the bands at 594 and 550 $cm^{-1}$ were attributed to vibrations of the Fe–O bonds in the octahedral $FeO_6$ groups, corresponding to the v1(F1u) mode [51]. The observed splitting, which appeared as an additional band in the spectrum of the pure lanthanum ferrite sample and as a shoulder in the spectra of barium-doped samples, was apparently due to the distorted crystalline structure. This statement was supported by the X-ray diffraction results, according to which the materials under study had an orthorhombic crystal structure.

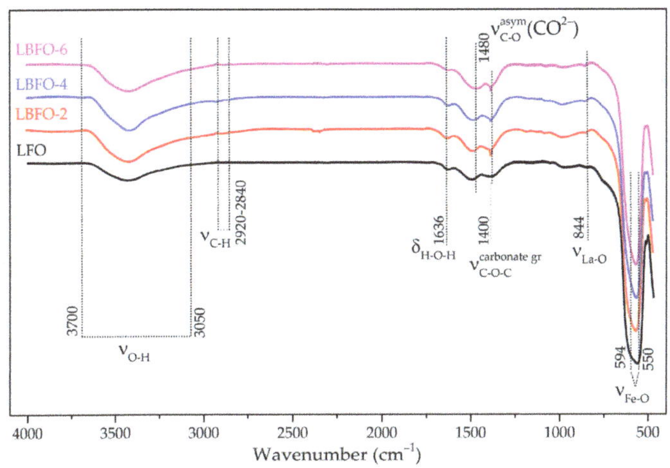


**Figure 3.** FTIR spectra of pure $LaFeO_3$ and $La_{1-x}Ba_xFeO_3$ series with different Ba contents.

Using the method of X-ray photoelectron spectroscopy, we determined the charge state of the elements in the samples. Among them, the spectra of O1s, Fe2p, and Ba3d were the most informative. In the oxygen spectra (Figure 4a), three main components could be distinguished, corresponding to lattice oxygen, adsorbed oxygen, and oxygen in the surface hydroxyl groups. The calculation of the proportions of each component in the spectra showed that the proportion of adsorbed components increased for Ba-doped materials. This indirectly indicated an increase in the number of oxygen vacancies in the $LaFeO_3$ structure with the introduction of $Ba^{2+}$. Due to the fact that the oxidation of reducing gases by the Mars–van Krevelen mechanism [52] is unlikely at low temperatures, an increase in adsorbed oxygen may contribute to improving the sensor properties of the doped materials through reactions by Langmuir–Hinshelwood [53] or Eley–Rideal [54] mechanisms.

Because of the complexity of describing the Fe2p region resulting from the various Fe charge states, it was decided to describe the experimental spectrum with the minimum possible number of components (Figure 4b). The most intense components were attributed to $Fe^{3+}$, located in different coordination environments, while the remaining component, located at high energies, was attributed to $Fe^{4+}$ [17,18]. Based on this, it could be concluded that, as the amount of dopant increased, the ratio of $Fe^{4+}$ to $Fe^{3+}$ also increased.

The Ba3d region (Figure 4c) showed that barium had a single charge state of $Ba^{2+}$ on the surface. This, together with the XRD data, allowed us to exclude the possibility of a separate barium-containing phase forming on the grain surface of lanthanum ferrite's main phase.

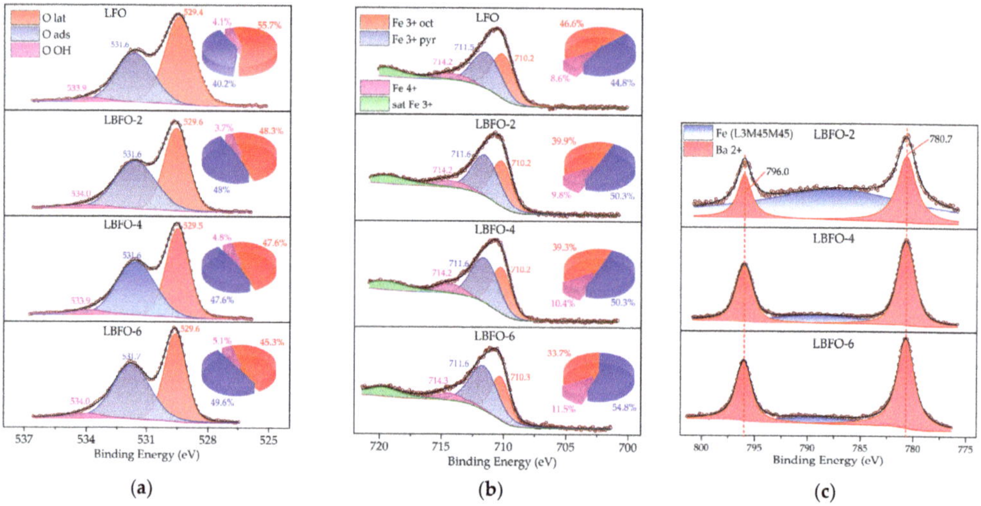

Figure 4. XP-spectra in O1s (a), Fe2p (b), and Ba3d (c) regions of pure $LaFeO_3$ and $La_{1-x}Ba_xFeO_3$ series with different Ba contents.

Based on an analysis of the literature data, it is known that charge compensation during the doping of $LaFeO_3$ with $Ba^{2+}$ ions can occur through the following mechanisms [14,18,21,22]:

(a) Formation of $Fe^{4+}$ ions, $La^{3+}_{1-x}Ba^{2+}_xFe^{3+}_{1-x}Fe^{4+}_xO_3$;
(b) Oxygen nonstoichiometry, $La^{3+}_{1-x}Ba^{2+}_xFeO_{3-x/2}$;
(c) Combination of mechanism (a) and (b), $La^{3+}_{1-x}Ba^{2+}_xFe^{3+}_{1-y}Fe^{4+}_yO_{3-(x-y)/2}$.

An increase in the amount of oxygen vacancies and of Fe in a higher charge state indicated the possibility of charge compensation through the combination of oxygen nonstoichiometry and $Fe^{4+}$ formation.

### 3.3. Electrical and Gas Sensor Properties

Measuring the temperature dependence of the materials' resistance in dry air showed that, with an increase in the dopant content, the base resistance decreased (Figure 5). This can be explained by an increase in the concentration of free charge carriers in a *p*-type semiconductor during heterovalent substitution, according to the quasi-chemical equation below (Wink–Kröger notation):

$$Ba^{\times}_{La} \rightarrow Ba'_{La} + h^{\cdot} \quad (5)$$

The temperature growth leads to a decrease in the base resistance, which is typical for semiconductors. It follows from Figure 5 that the temperature dependence of conductivity, plotted in Arrhenius coordinates (lnG vs. 1/T), was divided into two regions with different activation energies (Table 2). The low-temperature region (region II in Figure 5) could be attributed to the conductivity associated with the ionization of acceptor levels formed with Ba doping. Conductivity in the high-temperature region (region I in Figure 5) can be described using the small polaron hopping (SPH) mechanism. The introduction of barium into lanthanum ferrite can be considered as the dissolution of $BaFeO_3$ in $LaFeO_3$ [55]:

$$BaFeO_3 \overset{LaFeO_3}{\rightarrow} Ba'_{La} + Fe^{\cdot}_{Fe} + 3O^{\times}_{O}, \quad (6)$$

This leads to the formation of $Fe^{\cdot}_{Fe}$. The defect equilibrium is ensured by the incorporation of oxygen into the perovskite matrix, leading to the disappearance of oxygen vacancies and the formation of $Fe^{4+}$ cations:

$$V_O^{\cdot\cdot} + 2Fe^{\times}_{Fe} + \frac{1}{2}O_2 \leftrightarrow O^{\times}_O + 2Fe^{\cdot}_{Fe}. \quad (7)$$

Charge disproportionation is also possible:

$$2Fe^{\times}_{Fe} \leftrightarrow Fe^{\cdot}_{Fe} + Fe'_{Fe}. \quad (8)$$

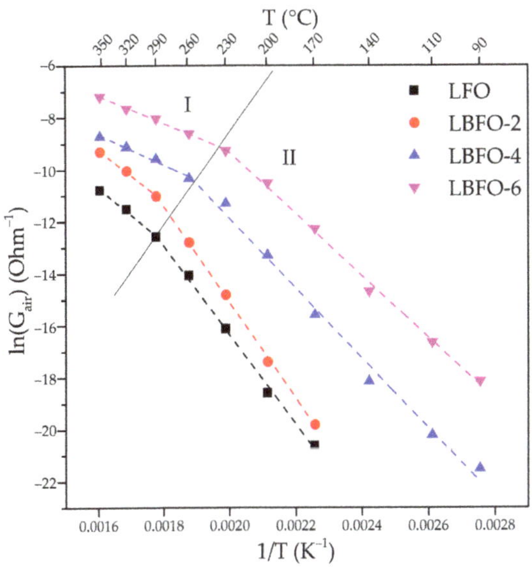


**Figure 5.** Temperature dependence of the conductivity of pure $LaFeO_3$ and materials with different Ba contents.

**Table 2.** Activation energies of the conductivity of pure $LaFeO_3$ and materials with different Ba contents in high- ($E_{aI}$) and low-temperature ($E_{aII}$) regions (Figure 5). $T_k$ denotes the kink temperature on the lnG vs. 1/T dependences (Figure 5).

| Sample | $E_{aI}$, eV | $E_{aII}$, eV | $T_k$, °C |
|---|---|---|---|
| LFO | 0.91 | 1.48 | 290 |
| LBFO-2 | 0.87 | 1.59 | 290 |
| LBFO-4 | 0.50 | 1.16 | 260 |
| LBFO-6 | 0.46 | 1.02 | 230 |

Within the SPH mechanism, it is assumed that the electronic conductivity in ferrites with a perovskite structure occurs via the path $\cdots Fe^{4+} - O - Fe^{3+} \cdots$ [56]. The degree of overlap of the O2p and the Fe3d orbitals depends on the distance and angle between the iron and oxygen atoms, that is, on the local symmetry of the iron cations. Thus, the distortion of the perovskite crystal structure affects the electronic conductivity realized by the SPH mechanism. At low concentrations of acceptor impurity ($x \leq 0.2$ in the $La_{1-x}Sr_xFeO_{3-\delta}$ system), electron charge compensation, which is realized by changing the degree of oxidation of iron from 3+ to 4+, occurs predominantly [57]. Since the ionic radii of $Fe^{4+}$ (0.585 Å) and $Fe^{3+}$ (0.645 Å) differ significantly, distortion of the crystal structure is very likely. In this regard, the binding energy of the hole with the iron cation in the polaron may decrease, resulting in a lower activation energy of SPH conductivity and a decrease in the kink temperature ($T_k$) on the lnG vs. 1/T dependences (Table 2).

The sensor properties of $LaFeO_3$ nanofibers were studied when detecting CO, $CH_4$, methanol, and acetone. The concentration of all analytes in the gas phase was 20 ppm. The measurements were performed in the mode of periodic change of the gas phase composition (15 min of clean air, 15 min of air with a pollutant gas) in the temperature range of 50–350 °C. Figure 6 shows the dynamic change in the resistance of pure $LaFeO_3$ and materials with different Ba contents when detecting 20 ppm acetone in dry air at different operating temperatures. The figure shows five stages of measurements effectuated sequentially at temperatures of 320, 290, 260, 230, and 200 °C. At each stage, three injections of a gas mixture containing acetone were carried out in the sensor cell, followed by purging with clean air at the same temperature. In the presence of the reducing gas, the resistance of the samples increased, corresponding to the *p*-type conductivity of $LaFeO_3$. When the sensor chamber was blown with pure air, the resistance of the materials decreased and reproducibly reached the initial value. As the detection temperature decreased, the base resistance of the sensors increased in pure air, reflecting the semiconductor nature of the materials.

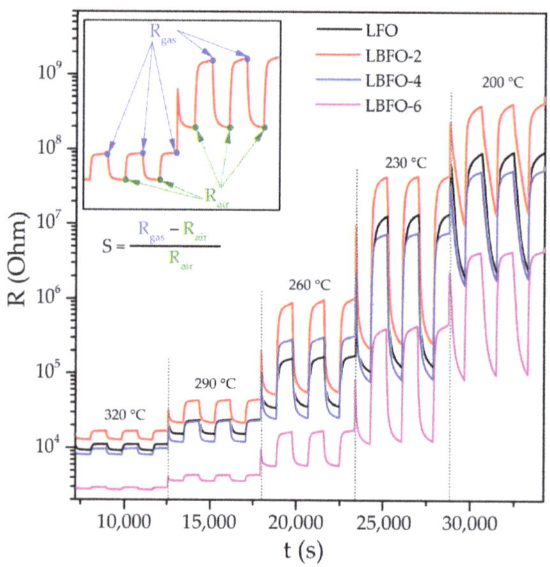


**Figure 6.** Dynamic change in the resistance of pure $LaFeO_3$ and materials with different Ba contents when detecting 20 ppm acetone in dry air at different working temperatures.

The temperature dependences of the sensor signal toward 20 ppm CO, $CH_4$, methanol, and acetone are shown in Figure 7. The formation of the sensor response of semiconductor oxide-based materials was due to the oxidation reaction of the analyte gases by oxygen chemisorbed on the surface:

$$CO_{(gas)} + \frac{1}{m}O^{n-}_{m\,(ads)} \rightarrow CO_{2\,(gas)} + \frac{n}{m}e^-,. \quad (9)$$

$$CH_{4\,(gas)} + \frac{4}{m}O^{n-}_{m\,(ads)} \rightarrow 2H_2O_{(gas)} + CO_{2\,(gas)} + \frac{4n}{m}e^-, \quad (10)$$

$$CH_3OH_{(gas)} + \frac{3}{m}O^{n-}_{m\,(ads)} \rightarrow CO_{2\,(gas)} + 2H_2O_{(gas)} + \frac{3n}{m}e^-, \quad (11)$$

$$CH_3C(O)CH_{3\ (gas)} + \frac{8}{m}O^{n-}_{m\ (ads)} \rightarrow 3CO_{2\ (gas)} + 2H_2O_{(gas)} + \frac{8n}{m}e^-. \quad (12)$$

where $CO_{(gas)}$, $CH_{4(gas)}$, $CH_3OH_{(gas)}$, and $CH_3C(O)CH_{3(gas)}$ are molecules in the gas phase; $O^{n-}_{m(ads)}$ is a particle of chemisorbed oxygen (m = 1 or 2, n = 1 or 2, possible options: $O_2^-, O^-, O^{2-}$); $e^-$ is an electron that is injected into the conduction band as a result of the reaction; and $CO_{2(gas)}$ and $H_2O_{(gas)}$ are molecules of reaction products desorbed from the surface of the material into the gas phase.

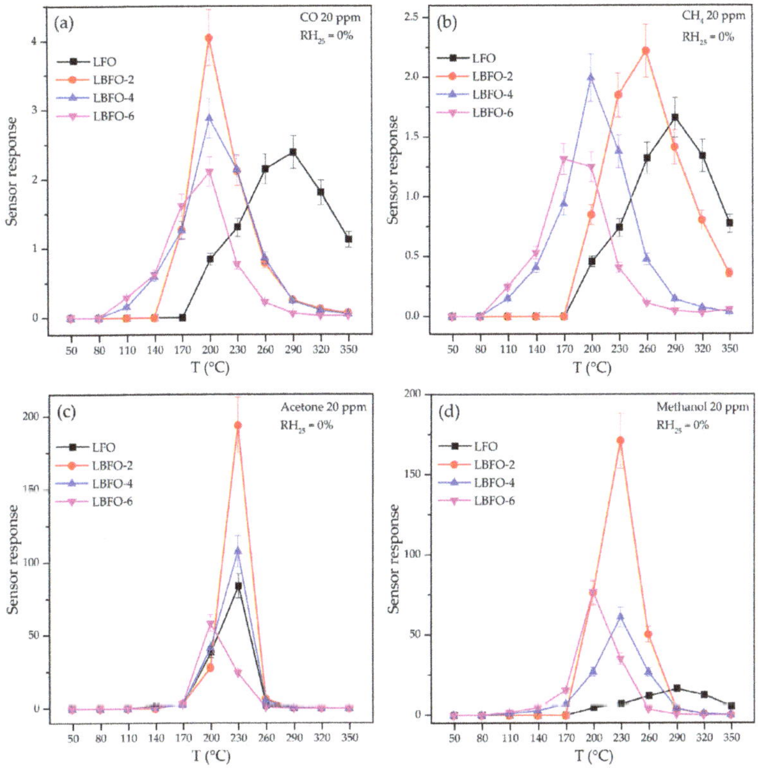


**Figure 7.** Temperature dependence of the sensor response (S) of $La_{1-x}Ba_xFeO_3$ series with different Ba contents toward 20 ppm CO (**a**), $CH_4$ (**b**), acetone (**c**), and methanol (**d**).

The measurement error of the sensor response (based on the reproducibility of data obtained from three sensors made of each material) was no more than 10%. It is worth noting the significantly greater sensor responses of barium-doped and pure $LaFeO_3$ when detecting volatile organic compounds compared to CO and $CH_4$ (Figure 7). On the one hand, this effect can be explained by the "multi-electron" nature of the oxidation process of VOC molecules on the surface of the sensor material by chemisorbed oxygen (Equations (9)–(12)). The more chemisorbed oxygen molecules are involved in the oxidation of the analyte gas molecule, the more electrons released during the reaction are injected into the conduction band, leading to a greater change in the conductivity of the material and, accordingly, a greater sensor response. On the other hand, this may be due to the high catalytic activity of coordinatively unsaturated iron cations on the $La_{1-x}Ba_xFeO_3$ surface. As is known, iron oxides are used as a catalyst for the oxidation of methanol and acetone [58–61]. In addition, the high sensor responses to acetone and methanol may be due to both the lower dissociation energy of bonds in VOC molecules relative to carbon monoxide and methane [62]

(Table 3) and the lower adsorption energy of VOC molecules due to the formation of a bond between the coordinatively unsaturated $Fe^{n+}$ cations and the oxygen of the organic molecule. From the obtained results, it followed that the introduction of barium led to some decrease in the operating temperature (Figure 7) and, in the case of VOCs (methanol and acetone), to a significant increase in the sensor response (Figure 8). The above discussed combined mechanism of charge compensation when replacing $La^{3+}$ cations with $Ba^{2+}$ cations provided an increase in the amount of chemisorbed oxygen and the concentration of $Fe^{4+}$, which, in turn, facilitated the oxidation of VOCs on the surface of barium-doped materials. A comparison of the sensor responses of $LaFeO_3$-based materials described in the literature and obtained in this work (Table 4) led to the conclusion that the combination of nanofiber morphology and barium doping made it possible to form materials that provide a high sensor response to acetone at a reduced operating temperature.

**Table 3.** Bond dissociation energies in selected gas molecules [62].

| Molecule | Bond | $\Delta H_{dis}$, kJ/mol |
|---|---|---|
| CO | C≡O | 1075 |
| $CH_4$ | H–$CH_3$ | 431 |
| Acetone | C=O<br>C–H<br>C–C | 745<br>414<br>347 |
| Methanol | C–O<br>C–H<br>O–H | 384<br>337<br>428 |

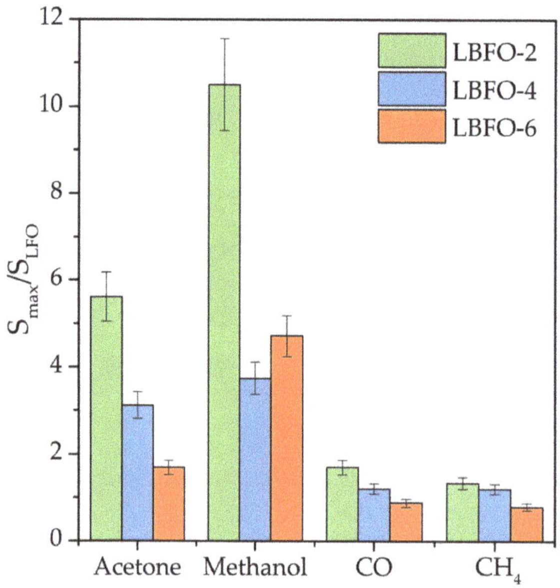


**Figure 8.** The ratio of the maximum sensor response of barium-doped materials to the response of $LaFeO_3$ nanofibers when detecting VOCs, CO, and $CH_4$ at appropriate operating temperatures.

**Table 4.** Summary of the sensor response values toward acetone of $LaFeO_3$-based materials described in the literature and obtained in this work.

| Material | Synthesis Method | Working Temperature (°C) | Acetone Concentration, ppm | Sensor Response | Ref. |
|---|---|---|---|---|---|
| $La_{0.7}Sr_{0.3}FeO_3$ | Sol-Gel | 275 | 500 | 0.7 | [10] |
| $(La,Ba)(Fe,Ti)O_3$ | Sol-Gel | 132 | 100 | 19 | [11] |
| $La_{0.75}Ba_{0.25}FeO_3$ | Sol-Gel | 275 | 20 | 2.1 | [12] |
| $La_{0.75}Ba_{0.25}FeO_3$ | Sol-Gel | 240 | 500 | 172 | [14] |
| $La_{0.98}Ba_{0.02}FeO_3$ | hydrothermal | 200 | 100 | 8 | [15] |
| $La_{0.9}Sr_{0.1}FeO_3$ | PLD | 303 | 4 | 0.77 | [16] |
| $La_{0.98}Mg_{0.02}FeO_3$ | PMMA template method | 190 | 100 | 50 | [17] |
| $La_{0.8}FeO_3$ | electrospinning | 180 | 100 | 6 | [63] |
| $La_{0.98}Ba_{0.02}FeO_3$ | electrospinning | 230 | 20 | 194 | This work |

To assess the prospects of practical use of sensors based on $La_{1-x}Ba_xFeO_3$ nanofibers, we estimated the response ($\tau_{90}^{res}$) and recovery ($\tau_{90}^{rec}$) times when detecting acetone (Figure 9). Even though the absolute $\tau_{90}^{res}$ and $\tau_{90}^{rec}$ values are strongly dependent on the parameters of the testing system, they are useful for comparing the characteristics of materials if the measurements are performed under identical conditions. It was noted that the measurement temperature had a major influence on the $\tau_{90}^{res}$ and $\tau_{90}^{rec}$ values, while the composition of the material was of minor importance for the dynamic characteristics. This is not surprising since the morphology of nanofibers (which may be an important parameter determining the transport of gases to the surface of crystallites of a sensitive material) did not change when $La^{3+}$ cations were replaced by $Ba^{2+}$ cations. It should be noted that, at a temperature of 230 °C, which corresponded to the maximum sensor response toward acetone, a minimum response time was observed, while the recovery time was significantly longer. To speed up sensor recovery when operating under real conditions, a two-temperature operation mode can be used, in which the sensor is shortly heated (for several seconds) to a higher temperature (for example, up to 300 °C). This operation mode will not lead to a significant increase in power consumption; however, it will ensure effective desorption of the products of oxidation of the analyte gas from the surface of the sensitive layer.

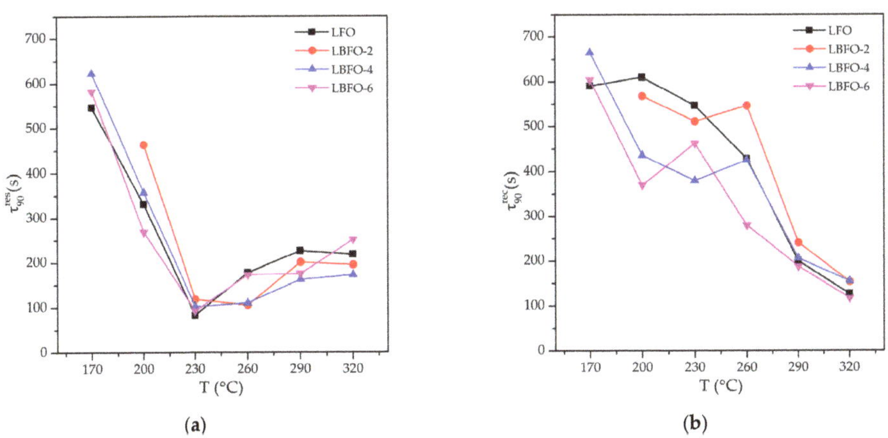


**Figure 9.** The temperature dependences of response time ($\tau_{90}^{res}$) (**a**) and recovery time ($\tau_{90}^{rec}$) (**b**) of $La_{1-x}Ba_xFeO_3$ nanofibers when detecting acetone.

### 3.4. Model of Forming $La_{1-x}Ba_xFeO_3$ Nanofibers' Sensor Response

It is noteworthy that the temperature range at which the sensor response was observed during VOC detection was quite narrow (170–290 °C), which may be due to a non-trivial mechanism of the oxidation reaction of the VOC molecules on the surface of the sensor material. To establish the reaction pathway during acetone oxidation on the $La_{1-x}Ba_xFeO_3$ surface, additional studies were carried out, including in situ IR spectroscopy in the diffuse reflectance mode (DRIFTS) and temperature-programmed desorption of acetone in combination with mass-spectral analysis of desorption products (TPD-MS). The LBFO-2 sample characterized by the maximum sensor response in acetone detection was selected for this study.

Figure 10 shows the evolution of the LBFO-2 DRIFT spectra in the presence of 200 ppm acetone with increasing temperature. After keeping the sample in a flow of acetone at a temperature of 50 °C, bands appeared in the spectra at 1094, 1235, 1370, 1706, 1740, 2925, 2970, 3652, and 3686 $cm^{-1}$, corresponding to vibrations of $\omega(CH_2)$, ν(C–C), $\delta_s(CH_3)$, ν(C=O), ν(C–O), $\nu_{as}(CH_3)$, $\nu_s(CH_3)$, and ν(O–H) of the acetone molecule adsorbed on the surface, respectively [64–67]. The spectra also contained low-intensity bands at 1425, 1575, and 2850 $cm^{-1}$, the intensity of which increased significantly with increasing temperature; the intensity of the bands at 1740 and 1370 $cm^{-1}$ did not change; and the intensity of the bands corresponding to acetone decreased sharply. The band at 1425 $cm^{-1}$ may have corresponded to $\nu_s$(C–O–C); the bands at 1575 and 1370 $cm^{-1}$ were related to $\nu_{as}(COO^-)$ and $\nu_s(COO^-)$, respectively; and the band at 2850 $cm^{-1}$ was related to $\nu_s(CH_2)$ vibrations in formate groups formed on the surface during acetone oxidation [64,65,68–70]. The retention of the intensity of the band at 1740 $cm^{-1}$ may have been related to the formation of carbonates on the surface of the material. At temperatures above 300 °C, the bands corresponding to the vibrations of formate groups disappeared. The signals at 1740 and 1437 $cm^{-1}$ corresponded to vibrations of ν(C–O) and $\nu_s$(O–C–O) of carbonates formed during the oxidation of formates, respectively.

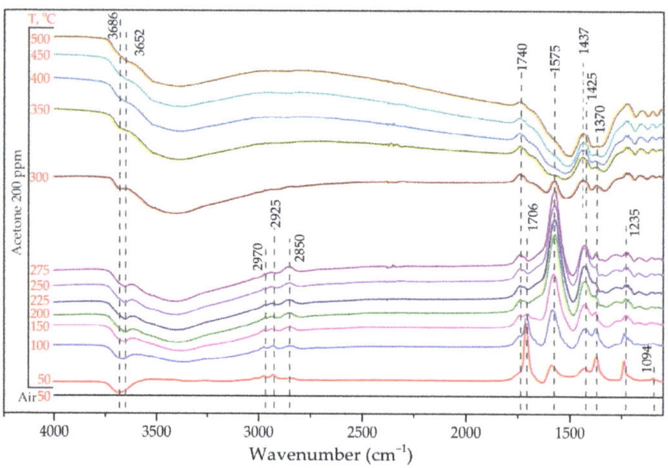


**Figure 10.** DRIFT spectra of $La_{0.98}Ba_{0.02}FeO_3$ nanofibers (sample LBFO-2) under acetone adsorption (200 ppm in dry air) in the temperature range of 50–500 °C.

At the same time, according to the mass spectra, starting from a temperature of 300 °C, significant increases in the amounts of water and $CO_2$ occurred when the sample was purged with acetone (Figure 11). It was assumed that, under these conditions, acetone combustion occurred, which was probably facilitated by Fe ions in the perovskite composition acting as a catalyst.

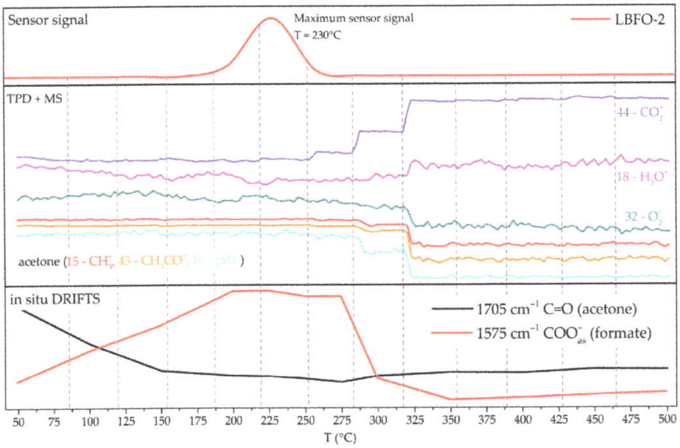


**Figure 11.** Acetone oxidation on the surface of $La_{0.98}Ba_{0.02}FeO_3$ nanofibers (sample LBFO-2).

If we consider the average temperature interval, increasing the temperature up to 250 °C led to the disappearance of carbonyl groups and the appearance of formate groups. The greatest sensor signal was observed in the same temperature interval, indicating that it was the process of acetone oxidation to formate groups and their further decomposition to $CO_2$ and $H_2O$ that included the largest amount of chemisorbed oxygen, which greatly changed the concentration of free charge carriers in the semiconductor.

Based on this, we can propose the following mechanism for the reaction of acetone with the surface of the sensing material and the formation of the sensor response.

At low temperatures, acetone adsorbs on the active centers of the lanthanum ferrite surface and forms formates (region 50–190 °C in Figure 11).

$$CH_3COCH_{3(gas)} \rightarrow CH_3COCH_{3(ads)}, . \tag{13}$$

$$CH_3COCH_{3(ads)} + \frac{2}{\beta}O^{\alpha-}_{\beta(ads)} \rightarrow 3HCOH_{(ads)} + 2\alpha e^-. \tag{14}$$

At temperatures above 250 °C, acetone oxidation occurs mostly due to oxygen from the surrounding air and not due to oxygen chemisorbed on the surface. This assumption was supported by a decrease in the ion current corresponding to oxygen in the mass spectra. It appears that acetone combustion occurs at the surface of the material, forming carbon dioxide and water, without intermediate stages (region 250–500 °C in Figure 11).

$$CH_3COCH_{3(gas)} + O_{2(gas)} \overset{Fe_xO_y}{\rightarrow} 3CO_{2(gas)} + H_2O_{(gas)}. \tag{15}$$

The processes occurring in the temperature range of 190–250 °C (Figure 11) can be described by the following reactions of multi-stage oxidation of formate to $CO_2$ and $H_2O$, accompanied by the release of a large number of localized electrons from chemisorbed oxygen [71]. Schematically, the multi-stage process of acetone oxidation on the surface of the sensor material is shown in Figure 12.

$$\beta{\cdot}HCOH_{(ads)} + O^{\alpha-}_{\beta(ads)} \rightarrow \beta{\cdot}H_2COO^-_{(ads)} + (\alpha-1){\cdot}e^-, \tag{16}$$

$$\beta \cdot H_2COO^{-}_{(ads)} + O^{\alpha-}_{\beta(ads)} \rightarrow \beta \cdot HCOO^{2-}_{(ads)} + \beta \cdot OH^{(\alpha-1)-}_{(ads)}, \quad (17)$$

$$2\beta \cdot HCOO^{2-}_{(ads)} + O^{\alpha-}_{\beta(ads)} \rightarrow 2\beta \cdot CO_{2(gas)} + \beta \cdot H_2O_{(gas)} + (4\beta + \alpha) \cdot e^{-} \quad (18)$$

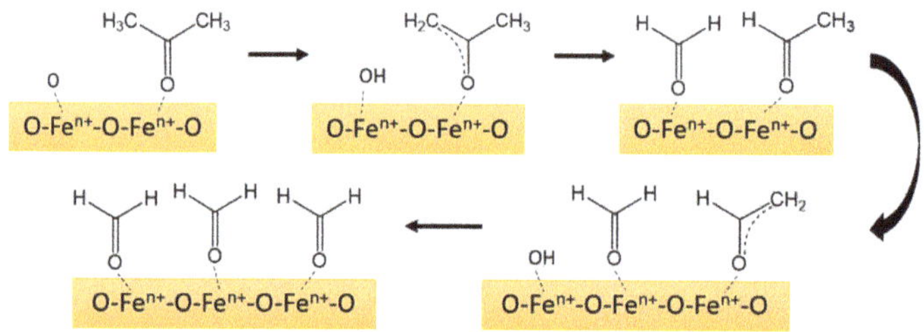

**Figure 12.** Scheme of step-by-step oxidation of acetone to formate on the $La_{1-x}Ba_xFeO_3$ surface.

## 4. Conclusions

Ba-doped $LaFeO_3$ nanofibers ($La_{1-x}Ba_xFeO_3$, x = 0.00, 0.02, 0.04, and 0.06) with an average diameter of about 200 nm were obtained by electrospinning. The obtained materials were single phase, had an orthorhombic structure, and consisted of nanocrystallites with a size of about 14–16 nm. The introduction of barium led to the inhibition of crystallite growth during isothermal annealing and promoted an increase in the sensor response of the $LaFeO_3$ nanofiber-based sensors toward VOCs. The $La_{0.98}Ba_{0.02}FeO_3$ sample demonstrated the highest sensor response and a decrease in the operating temperature. The improvement in the gas-sensitive properties of the doped materials can be explained by the high catalytic activity of the surface of synthesized materials associated with the formation of oxygen vacancies, highly active iron cations ($Fe^{4+}$), and coordinatively unsaturated cations ($Fe^{3+}$). The mechanism of acetone oxidation on the sensor surface, studied by DRIFTS and TPD-MS methods, is assumed to have a multi-stage nature.

**Author Contributions:** Conceptualization, V.P. and M.R.; methodology, V.P., D.F. and I.S.; formal analysis, N.M. and V.P.; investigation, N.M., V.P., D.F. and I.S.; data curation, N.M. and V.P.; writing—original draft preparation, N.M., V.P. and M.R.; writing—review and editing, V.P. and M.R. All authors have read and agreed to the published version of the manuscript.

**Funding:** This research was funded by Russian Science Foundation, grant number 23-73-01093, https://rscf.ru/en/project/23-73-01093/ (accessed on 25 April 2025).

**Institutional Review Board Statement:** Not applicable.

**Informed Consent Statement:** Not applicable.

**Data Availability Statement:** The original contributions presented in this study are included in the article. Further inquiries can be directed to the corresponding author.

**Acknowledgments:** The spectral experiments (Perkin Elmer Frontier FT-IR and Axis Ultra DLD XPS) were performed using the equipment purchased through the Lomonosov Moscow State University Program of Development. The SEM experiments were carried out using the research infrastructure of the "Educational and Methodical Center of Lithography and Microscopy" of Lomonosov Moscow State University supported by the Interdisciplinary Scientific and Educational School of Moscow State University "Photonic and Quantum Technologies. Digital Medicine".

**Conflicts of Interest:** The authors declare no conflicts of interest.

## References


1. Galstyan, V.; Moumen, A.; Kumarage, G.W.C.; Comini, E. Progress towards Chemical Gas Sensors: Nanowires and 2D Semiconductors. *Sens. Actuators B* **2022**, *357*, 131466. [CrossRef]
2. Staerz, A.; Weimar, U.; Barsan, N. Current State of Knowledge on the Metal Oxide Based Gas Sensing Mechanism. *Sens. Actuators B* **2022**, *358*, 131531. [CrossRef]
3. Yamazoe, N.; Shimanoe, K. Fundamentals of Semiconductor Gas Sensors. In *Semiconductor Gas Sensors*, 1st ed.; Jaaniso, R., Tan, O.K., Eds.; Elsevier: Amsterdam, The Netherlands, 2020; pp. 3–38. [CrossRef]
4. Mirzaei, A.; Leonardi, S.G.; Neri, G. Detection of Hazardous Volatile Organic Compounds (VOCs) by Metal Oxide Nanostructures-Based Gas Sensors: A Review. *Ceram. Int.* **2016**, *42*, 15119–15141. [CrossRef]
5. Korotcenkov, G.; Cho, B.K. Instability of Metal Oxide-Based Conductometric Gas Sensors and Approaches to Stability Improvement (Short Survey). *Sens. Actuators B* **2011**, *156*, 527–538. [CrossRef]
6. Wei, C.; Guo, Z.; Wang, H.; Zhang, S.; Hao, D.; Huang, J. Recent Progress of Gas Sensors Based on Perovskites. *Mater. Horiz.* **2025**, *12*, 317–342. [CrossRef]
7. Souri, M.; Salar Amoli, H. Gas Sensing Mechanisms in $ABO_3$ Perovskite Materials at Room Temperature: A Review. *Mater. Sci. Semicond. Process.* **2023**, *156*, 107271. [CrossRef]
8. Gibin, G.; Ede, S.R.; Luo, Z. *Fundamentals of Perovskite Oxides: Synthesis, Structure, Properties and Applications*, 1st ed.; CRC Press: Boca Raton, FL, USA, 2021; pp. 1–384. [CrossRef]
9. Wolfram, T.; Ellialtioglu, S. *Electronic and Optical Properties of d-Band Perovskites*, 1st ed.; Cambridge University Press: Cambridge, UK, 2006; pp. 1–315. [CrossRef]
10. Murade, P.A.; Sangawar, V.S.; Chaudhari, G.N.; Kapse, V.D.; Bajpeyee, A.U. Acetone Gas-Sensing Performance of Sr-Doped Nanostructured $LaFeO_3$ Semiconductor Prepared by Citrate Sol–Gel Route. *Curr. Appl. Phys.* **2011**, *11*, 451–456. [CrossRef]
11. Cao, E.; Feng, Y.; Guo, Z.; Wang, H.; Song, G.; Zhang, Y.; Hao, W.; Sun, L.; Nie, Z. Ethanol Sensing Characteristics of $(La,Ba)(Fe,Ti)O_3$ Nanoparticles with Impurity Phases of $BaTiO_3$ and $BaCO_3$. *J. Sol-Gel Sci. Technol.* **2020**, *96*, 431–440. [CrossRef]
12. Fan, K.; Qin, H.; Zhang, Z.; Sun, L.; Sun, L.; Hu, J. Gas Sensing Properties of Nanocrystalline $La_{0.75}Ba_{0.25}FeO_3$ Thick-Film Sensors. *Sens. Actuators B* **2012**, *171–172*, 302–308. [CrossRef]
13. Benali, A.; Azizi, S.; Bejar, M.; Dhahri, E.; Graça, M.F.P. Structural, Electrical and Ethanol Sensing Properties of Double-Doping $LaFeO_3$ Perovskite Oxides. *Ceram. Int.* **2014**, *40*, 14367–14373. [CrossRef]
14. Sun, L.; Qin, H.; Wang, K.; Zhao, M.; Hu, J. Structure and Electrical Properties of Nanocrystalline $La_{1-x}Ba_xFeO_3$ for Gas Sensing Application. *Mater. Chem. Phys.* **2011**, *125*, 305–308. [CrossRef]
15. Hao, P.; Qu, G.M.; Song, P.; Yang, Z.X.; Wang, Q. Synthesis of Ba-Doped Porous $LaFeO_3$ Microspheres with Perovskite Structure for Rapid Detection of Ethanol Gas. *Rare Met.* **2021**, *40*, 1651–1661. [CrossRef]
16. Cyza, A.; Cieniek, L.; Moskalewicz, T.; Maziarz, W.; Kusinski, J.; Kowalski, K.; Kopia, A. The Effect of Strontium Doping on $LaFeO_3$ Thin Films Deposited by the PLD Method. *Catalysts* **2020**, *10*, 954. [CrossRef]
17. Qin, J.; Cui, Z.; Yang, X.; Zhu, S.; Li, Z.; Liang, Y. Three-Dimensionally Ordered Macroporous $La_{1-x}Mg_xFeO_3$ as High Performance Gas Sensor to Methanol. *J. Alloys Compd.* **2015**, *635*, 194–202. [CrossRef]
18. Shi, C.; Qin, H.; Zhao, M.; Wang, X.; Li, L.; Hu, J. Investigation on Electrical Transport, CO Sensing Characteristics and Mechanism for Nanocrystalline $La_{1-x}Ca_xFeO_3$ Sensors. *Sens. Actuators B* **2014**, *190*, 25–31. [CrossRef]
19. Aranthady, C.; Jangid, T.; Gupta, K.; Mishra, A.K.; Kaushik, S.D.; Siruguri, V.; Rao, G.M.; Shanbhag, G.V.; Sundaram, N.G. Selective $SO_2$ Detection at Low Concentration by Ca Substituted $LaFeO_3$ Chemiresistive Gas Sensor: A Comparative Study of $LaFeO_3$ Pellet vs Thin Film. *Sens. Actuators B* **2021**, *329*, 129211. [CrossRef]
20. Wang, X.; Chen, Y.; Qin, H.; Li, L.; Shi, C.; Liu, L.; Hu, J. $CO_2$ Sensing of $La_{0.875}Ca_{0.125}FeO_3$ in Wet Vapor: A Comparison of Experimental Results and First-Principles Calculations. *Phys. Chem. Chem. Phys.* **2015**, *17*, 13733–13742. [CrossRef]
21. Li, L.; Qin, H.; Shi, C.; Zhang, L.; Chen, Y.; Hu, J. $CO_2$ Sensing Properties of $La_{1-x}Ba_xFeO_3$ Thick Film and Packed Powder Sensors. *RSC Adv.* **2015**, *5*, 103073–103081. [CrossRef]
22. Fan, K.; Qin, H.; Wang, L.; Ju, L.; Hu, J. $CO_2$ Gas Sensors Based on $La_{1-x}Sr_xFeO_3$ Nanocrystalline Powders. *Sens. Actuators B* **2013**, *177*, 265–269. [CrossRef]
23. Duan, X.; Jiang, Y.; Liu, B.; Duan, Z.; Zhang, Y.; Yuan, Z.; Tai, H. Enhancing the Carbon Dioxide Sensing Performance of $LaFeO_3$ by Co Doping. *Sens. Actuators B* **2024**, *402*, 135136. [CrossRef]
24. Thirumalairajan, S.; Girija, K.; Mastelaro, V.R.; Ponpandian, N. Surface Morphology-Dependent Room-Temperature $LaFeO_3$ Nanostructure Thin Films as Selective $NO_2$ Gas Sensor Prepared by Radio Frequency Magnetron Sputtering. *ACS Appl. Mater. Interfaces* **2014**, *6*, 13917–13927. [CrossRef] [PubMed]
25. Singh, P.; Bansal, N.K.; Dey, S.; Singh, R.; Singh, T. Recent Progress on Perovskite Materials for VOC Gas Sensing. *Langmuir* **2024**, *40*, 21931–21956. [CrossRef] [PubMed]
26. Doshi, J.; Reneker, D.H. Electrospinning Process and Applications of Electrospun Fibers. *J. Electrost.* **1995**, *3*, 151–160. [CrossRef]

27. Bohr, C.; Pfeiffer, M.; Öz, S.; von Toperczer, F.; Lepcha, A.; Fischer, T.; Schütz, M.; Lindfors, K.; Mathur, S. Electrospun Hybrid Perovskite Fibers—Flexible Networks of One-Dimensional Semiconductors for Light-Harvesting Applications. *ACS Appl. Mater. Interfaces* **2019**, *11*, 25163–25169. [CrossRef]
28. Huang, B.; Zhang, Z.; Zhao, C.; Cairang, L.; Bai, J.; Zhang, Y.; Mu, X.; Du, J.; Wang, H.; Pan, X.; et al. Enhanced Gas-Sensing Performance of ZnO@$In_2O_3$ Core@shell Nanofibers Prepared by Coaxial Electrospinning. *Sens. Actuators B* **2018**, *255*, 2248–2257. [CrossRef]
29. Rumyantseva, M.N.; Vladimirova, S.A.; Platonov, V.B.; Chizhov, A.S.; Batuk, M.; Hadermann, J.; Khmelevsky, N.O.; Gaskov, A.M. Sub-Ppm $H_2S$ Sensing by Tubular ZnO-$Co_3O_4$ Nanofibers. *Sens. Actuators B* **2020**, *307*, 127624. [CrossRef]
30. Ichangi, A.; Shvartsman, V.V.; Lupascu, D.C.; Lê, K.; Grosch, M.; Kathrin Schmidt-Verma, A.; Bohr, C.; Verma, A.; Fischer, T.; Mathur, S. Li and Ta-Modified KNN Piezoceramic Fibers for Vibrational Energy Harvesters. *J. Eur. Ceram. Soc.* **2021**, *41*, 7662–7669. [CrossRef]
31. Lê, K.; von Toperczer, F.; Ünlü, F.; Paramasivam, G.; Mathies, F.; Nandayapa, E.; List-Kratochvil, E.J.W.; Fischer, T.; Lindfors, K.; Mathur, S. Electrospun Electroluminescent $CsPbBr_3$ Fibers as Flexible Perovskite Networks for Light-Emitting Application. *Adv. Eng. Mater.* **2023**, *25*, 2201651. [CrossRef]
32. Li, H.; Chu, S.; Ma, Q.; Li, H.; Che, Q.; Wang, J.; Wang, G.; Yang, P. Multilevel Effective Heterojunctions Based on $SnO_2$/ZnO 1D Fibrous Hierarchical Structure with Unique Interface Electronic Effects. *ACS Appl. Mater. Interfaces* **2019**, *11*, 31551–31561. [CrossRef]
33. Platonov, V.B.; Rumyantseva, M.N.; Frolov, A.S.; Yapryntsev, A.D.; Gaskov, A.M. High-Temperature Resistive Gas Sensors Based on ZnO/SiC Nanocomposites. *Beilstein J. Nanotechnol.* **2019**, *10*, 1537–1547. [CrossRef]
34. Platonov, V.; Nasriddinov, A.; Rumyantseva, M. Electrospun ZnO/Pd Nanofibers as Extremely Sensitive Material for Hydrogen Detection in Oxygen Free Gas Phase. *Polymers* **2022**, *14*, 3481. [CrossRef] [PubMed]
35. Platonov, V.; Rumyantseva, M. Electrospun ZnO/$MO_x$ Nanocomposites as Sensitive Materials for Biomarker Gas Sensors: Role of $MO_x$ in C1–C4 Short-Chain Fatty Acids Detection. *Sens. Actuators B* **2025**, *433*, 137535. [CrossRef]
36. Platonov, V.; Rumyantseva, M.; Khmelevsky, N.; Gaskov, A. Electrospun ZnO/Pd Nanofibers: CO Sensing and Humidity Effect. *Sensors* **2020**, *20*, 7333. [CrossRef] [PubMed]
37. Platonov, V.; Sinyashin, O.; Rumyantseva, M. ZnO/$MO_x$ Nanofiber Heterostructures: $MO_x$ Receptor's Role in Gas Detection. *Sensors* **2025**, *25*, 376. [CrossRef]
38. Fan, H.; Zhang, T.; Xu, X.; Lv, N. Fabrication of N-Type $Fe_2O_3$ and P-Type $LaFeO_3$ Nanobelts by Electrospinning and Determination of Gas-Sensing Properties. *Sens. Actuators B* **2011**, *153*, 83–88. [CrossRef]
39. Fang, F.; Feng, N.; Wang, L.; Meng, J.; Liu, G.; Zhao, P.; Gao, P.; Ding, J.; Wan, H.; Guan, G. Fabrication of Perovskite-Type Macro/Mesoporous $La_{1-x}K_xFeO_{3-\delta}$ Nanotubes as an Efficient Catalyst for Soot Combustion. *Appl. Catal. B* **2018**, *236*, 184–194. [CrossRef]
40. Lee, W.Y.; Yun, H.J.; Yoon, J.W. Characterization and Magnetic Properties of $LaFeO_3$ Nanofibers Synthesized by Electrospinning. *J. Alloys Compd.* **2014**, *583*, 320–324. [CrossRef]
41. Alharbi, A.A.; Sackmann, A.; Weimar, U.; Bârsan, N. A Highly Selective Sensor to Acetylene and Ethylene Based on $LaFeO_3$. *Sens. Actuators B* **2020**, *303*, 127204. [CrossRef]
42. Hu, J.; Chen, X.; Zhang, Y. Batch Fabrication of Formaldehyde Sensors Based on $LaFeO_3$ Thin Film with Ppb-Level Detection Limit. *Sens. Actuators B* **2021**, *349*, 130738. [CrossRef]
43. Platonov, V.; Malinin, N.; Sapkov, I.; Rumyantseva, M. $LaFeO_3$ Nanofibers as Materials for Gas Sensors. *Russ. J. Gen. Chem.* **2025**, *94*, 1121–1132. [CrossRef]
44. Mitchell, R.H. *Perovskites: Modern and Ancient*, 1st ed.; Almaz Press: Thunder Bay, ON, Canada, 2002; pp. 1–318.
45. Alonso, J.A.; Martinez-Lope, M.J.; Casais, M.T.; Fernandez-Diaz, M.T. Evolution of the Jahn−Teller Distortion of $MnO_6$ Octahedra in $RMnO_3$ Perovskites (R = Pr, Nd, Dy, Tb, Ho, Er, Y): A Neutron Diffraction Study. *Inorg. Chem.* **2000**, *39*, 917–923. [CrossRef] [PubMed]
46. Glazer, A.M. The Classification of Tilted Octahedra in Perovskites. *Acta Crystallogr. B* **1972**, *28*, 3384–3392. [CrossRef]
47. Howard, C.J.; Stokes, H.T. Structures and Phase Transitions in Perovskites—A Group-Theoretical Approach. *Acta Crystallogr. A* **2005**, *61*, 93–111. [CrossRef]
48. Mocwana, M.L.; Mokoena, P.P.; Mbule, P.S.; Beas, I.N.; Kabongo, G.L.; Ogugua, S.N.; Tshabalala, T.E. Photocatalytic Degradation of Methylene Blue and Ortho-Toluidine Blue: Activity of Lanthanum Composites $La_xMO_y$ (M: Fe, Co, Ni). *Catalysts* **2022**, *12*, 1313. [CrossRef]
49. Lavat, A.E.; Baran, E.J. IR-Spectroscopic Characterization of $A_2BB'O_6$ Perovskites. *Vib. Spectrosc.* **2003**, *32*, 167–174. [CrossRef]
50. Baran, E.J. Structural Chemistry and Physicochemical Properties of Perovskite-like Materials. *Catal. Today* **1990**, *8*, 133–151. [CrossRef]
51. Barbero, B.P.; Gamboa, J.A.; Cadús, L.E. Synthesis and Characterisation of $La_{1-x}Ca_xFeO_3$ Perovskite-Type Oxide Catalysts for Total Oxidation of Volatile Organic Compounds. *Appl. Catal. B* **2006**, *65*, 21–30. [CrossRef]

52. Mars, P.; van Krevelen, D.W. Oxidations Carried out by Means of Vanadium Oxide Catalysts. *Chem. Eng. Sci.* **1954**, *3*, 41–59. [CrossRef]
53. Laidler, K.J.; Meiser, J.H. *Physical Chemistry*, 1st ed.; Benjamin/Cummings Publishing Company: San Francisco, CA, USA, 1982; pp. 1–919.
54. Rideal, E.K. A Note on a Simple Molecular Mechanism for Heterogeneous Catalytic Reactions. *Math. Proc. Camb. Philos. Soc.* **1939**, *35*, 130–132. [CrossRef]
55. Sogaard, M.; Vang Hendriksen, P.; Mogensen, M. Oxygen Nonstoichiometry and Transport Properties of Strontium Substituted Lanthanum Ferrite. *J. Solid State Chem.* **2007**, *180*, 1489–1503. [CrossRef]
56. Ren, Y.; Küngas, R.; Gorte, R.J.; Deng, C. The Effect of A-Site Cation (Ln=La, Pr, Sm) on the Crystal Structure, Conductivity and Oxygen Reduction Properties of Sr-Doped Ferrite Perovskites. *Solid State Ion.* **2012**, *212*, 47–54. [CrossRef]
57. Patrakeev, M.V.; Leonidov, I.A.; Kozhevnikov, V.L.; Poeppelmeier, K.R. P-Type Electron Transport in $La_{1-x}Sr_xFeO_{3-\delta}$ at High Temperatures. *J. Solid State Chem.* **2005**, *178*, 921–927. [CrossRef]
58. Nunes, P.; Strapasson, G.; Bafero, G.; Zanchet, D. Exploring Iron Oxide Catalysts for Acetone Hydrodeoxygenation: Making Use of Earth-Abundant Resources. *J. Braz. Chem. Soc.* **2024**, *35*, e20240126. [CrossRef]
59. Hellier, P.; Wells, P.P.; Bowker, M. Methanol Oxidation over Shell-Core MO/$Fe_2O_3$ (M = Mo, V, Nb) Catalysts. *Chin. J. Catal.* **2019**, *40*, 1686–1692. [CrossRef]
60. Malińnski, R.; Gallus-Olender, J.; Kubicka, T. Phase Composition of a Selective $V_2O_5-Fe_2O_3$ Catalyst of Methanol Oxidation to Formaldehyde. *React. Kinet. Catal. Lett.* **1979**, *10*, 219–223. [CrossRef]
61. Bowker, M.; Gibson, E.K.; Silverwood, I.P.; Brookes, C. Methanol Oxidation on $Fe_2O_3$ Catalysts and the Effects of Surface Mo. *Faraday Discuss.* **2016**, *188*, 387–398. [CrossRef]
62. Dean, J.A.; Lange, N.A. *Lange's Handbook of Chemistry*, 15th ed.; McGraw-Hill: New York, NY, USA, 1998; pp. 1–1424.
63. Zhu, L.; Wang, J.; Liu, J.; Chen, X.; Xu, Z.; Ma, Q.; Wang, Z.; Liang, J.; Li, S.; Yan, W. Designing Highly Sensitive Formaldehyde Sensors via A-Site Cation Deficiency in $LaFeO_3$ Hollow Nanofibers. *Appl. Surf. Sci.* **2022**, *590*, 153085. [CrossRef]
64. Würger, T.; Heckel, W.; Sellschopp, K.; Müller, S.; Stierle, A.; Wang, Y.; Noei, H.; Feldbauer, G. Adsorption of Acetone on Rutile $TiO_2$: A DFT and FTIRS Study. *J. Phys. Chem. C* **2018**, *122*, 19481–19490. [CrossRef]
65. Marins, A.A.L.; Banhos, S.G.; Cruz, P.C.M.; Rodrigues, R.V.; Marciniak, L.; Muri, E.J.B.; Strek, W.; de Freitas, M.B.J.G. Synthesis of Ni and Rare Earth Metal (La, Pr, and Nd) Oxides from Spent Ni–MH Batteries by Selective Precipitation with Formic Acid an Investigation of Photoluminescence Properties. *Ionics* **2020**, *26*, 311–321. [CrossRef]
66. Alminshid, A.H.; Abbas, M.N.; Alalwan, H.A.; Sultan, A.J.; Kadhom, M.A. Aldol Condensation Reaction of Acetone on MgO Nanoparticles Surface: An in-Situ Drift Investigation. *Mol. Catal.* **2021**, *501*, 111333. [CrossRef]
67. Wang, X.; Barakat, C.; Jia, Z.; Romanias, M.N.; Thévenet, F.; Rousseau, A. Adsorption of VOCs Is a Key Step in Plasma-Catalyst Coupling: The Case of Acetone onto $TiO_2$ vs. $CeO_2$. *Catalysts* **2021**, *11*, 350. [CrossRef]
68. Lucks, C.; Rossberg, A.; Tsushima, S.; Foerstendorf, H.; Fahmy, K.; Bernhard, G. Formic Acid Interaction with the Uranyl(vi) Ion: Structural and Photochemical Characterization. *Dalt. Trans.* **2013**, *42*, 13584. [CrossRef] [PubMed]
69. Calatayud, M.; Collins, S.E.; Baltanas, M.A.; Bonivardi, A.L. Stability of Formate Species on $\beta$-$Ga_2O_3$. *Phys. Chem. Chem. Phys.* **2009**, *11*, 1397. [CrossRef] [PubMed]
70. Alalwan, H.; Alminshid, A. An In-Situ DRIFTS Study of Acetone Adsorption Mechanism on $TiO_2$ Nanoparticles. *Spectrochim. Acta Part A Mol. Biomol. Spectrosc.* **2020**, *229*, 117990. [CrossRef]
71. Nasriddinov, A.; Platonov, V.; Garshev, A.; Rumyantseva, M. Low Temperature HCHO Detection by $SnO_2/TiO_2$@Au and $SnO_2/TiO_2$@Pt: Understanding by In-Situ DRIFT Spectroscopy. *Nanomaterials* **2021**, *11*, 2049. [CrossRef]

sensors

MDPI

*Article*

# The Role of Convection and Size Effects in Microhotplate Heat Exchange: Semiconductor and Thermomagnetic Gas Sensors †

**Alexey Vasiliev [1,2,*], Alexey Shaposhnik [3], Oleg Kul [2] and Artem Mokrushin [4]**

1 Laboratory of Sensor Systems, Dubna State University, Universitetskaya Strt., 19, 141980 Dubna, Russia
2 LLC "C-Component", Tushinskaya Strt., 17, 125362 Moscow, Russia; okul@c-component.ru
3 Department of Chemistry, Voronezh State Agrarian University, Michurin Strt., 1, 394087 Voronezh, Russia; a.v.shaposhnik@gmail.com
4 Kurnakov Institute of General and Inorganic Chemistry, Russian Academy of Sciences, Leninsky Av. 31, 119991 Moscow, Russia; artyom.nano@gmail.com
* Correspondence: a-a-vasiliev@yandex.ru; Tel.: +7-(903)-194-95-51
† This paper is an extended version of Alexey Vasiliev, Alexey Shaposhnik, and Oleg Kul. The Role of Convection and Size Effects in Sensor Microhotplate Heat Exchange. In proceedings of the Eurosensors XXXIV, Lecce, Italy, 10–13 September 2023 (MDPI Proceedings, 2024, 97(1), 150).

**Abstract:** The analysis of the influence of microhotplate size on the convective heat exchange of gas sensors is presented. Usually, the role of convection in the heat exchange of gas sensors is not considered in thermal simulation models because of the complexity of the convection process. As a result, the contribution of this process to the overall heat loss of sensors remains without detailed analysis. We analyzed convection issues in two groups of gas sensors: semiconductor and thermocatalytic (calorimetric) sensors and, on the other hand, in the oxygen sensors of the thermomagnetic type. It is demonstrated that there is a critical size leading to the formation of convective heat exchange flow. Below this critical value, only thermal conductivity of ambient air, IR (infrared) radiation from the heated microhotplate surface, and thermal conductivity of the microhotplate-supporting elements should be considered as channels for heat dissipation by the microhotplate, and the contribution of free convection can be neglected. The expression for the critical size contains only fundamental constants of air, $d_{cr} \sim 4 \cdot \sqrt[3]{\nu \cdot D / g}$, where ν is the kinematic viscosity of air, D is the diffusion coefficient, and g is the acceleration of free fall, $d_{cr} \sim 0.5$ cm. Therefore, if the size of the microhotplate $d \ll d_{cr}$, the influence of convection heat exchange can be neglected. Similar results were obtained in the analysis of the behavior of thermal magnetic sensors of oxygen, which use paramagnetic properties of molecular oxygen for the determination of $O_2$ concentration. In this case, the critical size of the sensor is also of significance; if the size of the magnetic sensor is much below this value, the oxygen concentration value measured with such a device is independent of the orientation of the sensor element. The results of the simulation were compared with the measurement of heat loss in micromachined gas sensors. The optimal dimensions of the sensor microhotplate are given as a result of these simulations and measurements.

**Keywords:** microheater; convective heat losses; thermal conductivity; Grashof number

## 1. Introduction

Sensors of different physical and chemical parameters of air (or of a gas, in general) have recently offered a variety of important online information about the systems that control the environment, technological processes, and atmosphere in residential and office buildings. Sensors of airflow, pressure, chemical composition and impurities in the air are


https://doi.org/10.3390/s25092830

based on different physical and chemical principles. However, many such sensors rely on the temperature difference between elements of the sensor as a main measurement parameter, ultimately providing information about the target value characterizing the ambient environment.

Among these sensors are the following: thermal anemometers, which measure gas flow velocity using the temperature difference between hotplates placed in a series within the gas flow [1]; thermocatalytic (calorimetric) sensors, which measure temperature difference between two hot elements coated and uncoated with a catalyst, active in the process of surface oxidation of target combustible gas [2]; and magnetic oxygen sensors, which measure gas flow rate of oxygen-containing gases using hot elements placed in a magnetic field. This flow arises due to the paramagnetic properties of oxygen, the only paramagnetic gas commonly present in the atmosphere (aside from relatively rare NO) [3].

Another group of sensors includes semiconductor gas sensors [4,5], which use a single hotplate to maintain a desirable high temperature required for fast catalytic oxidation of the target gas on the surface of the semiconducting catalytic nanomaterial, and Pirani pressure gauges [6,7]. In these sensors a microhotplate serves as a heating element with either stabilized (in semiconductor gas sensors) or measured (in Pirani gauges) temperature.

A current trend in the fabrication of all these sensors is the application of microhotplates made using one of the possible microelectronic technologies, such as silicon MEMS, thin ceramic alumina or LTCC (low temperature co-fired ceramic) membranes [8], screen-printed glass-ceramic membranes or cantilevers [9], and others. The main reason for using microfabrication technologies, aside from the optimization of sensor cost, is the minimization of the power consumption of the sensor's hotplate.

In order to minimize the power consumption of these microhotplates, it is necessary to analyze the paths of heat loss. This analysis is relatively simple in the case of losses due to radiation or thermal conduction, but an analysis of convection is always problematic because of the complex characteristics of this process. In this paper, we formulate criteria enabling the contribution of convection to the heat losses of a sensor operating at high temperature to be neglected.

## 2. Heat Dissipation by Microhotplates

A microhotplate dissipates heat in four ways (Figure 1):

(1) Conduction through supporting structures, such as wires suspending the hotplate, or a membrane supporting the microhotplate in the case of MEMS-based sensor elements;

(2) Conduction through ambient gas, resulting from the thermal conductivity of the surrounding medium;

(3) Natural thermal convection occurring in the air gap between the microhotplate and the sensor housing;

(4) Thermal radiation emitted by the surface of the microhotplate.

Sensor radiation losses are negligible under typical sensor dimensions and operating temperatures. Indeed, the radiation loss from the sensor microhotplate obeys the Stefan–Boltzmann law, $W = \sigma \cdot T^4 \cdot s$, where W—power dissipated due to radiation; σ—Stefan–Boltzmann constant; $\sigma = 5.67 \times 10^{-8}$ W·m$^{-2}$·K$^{-4}$; T—absolute temperature of the sensor; and s—the area of the sensor. Under typical conditions, s is of the order of 1 mm$^2$ and T = 700 K. Taking into account these values, radiation heat losses can be estimated as W~2 mW. This value is significantly lower than the power consumption of microhotplates commonly used in, for example, methane sensors operating at 450 °C, where the power consumption is approximately 200 mW [8].

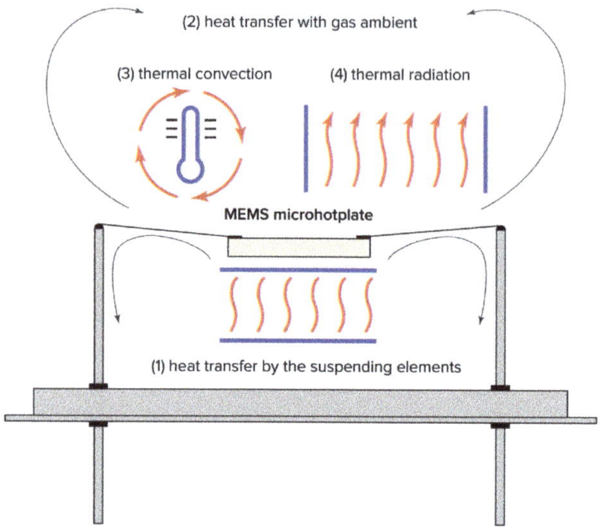


**Figure 1.** The paths of heat loss in a MEMS microhotplate, assembled in housing: (1) heat transfer by the supporting elements; (2) heat transfer due to heat conductivity of ambient gas; (3) free thermal convection; (4) thermal radiation.

In these estimations, the microhotplate is assumed to behave as a black-body emitter. In reality, its emissivity is lower, so the actual radiation losses are even smaller than this maximum theoretical estimate.

Understandings of the thermal conductivity of air and of the sensor elements are more or less clear. The heat losses related to these effects are proportional to the temperature difference between the microhotplate and the housing of the sensor. Heat loss due to the thermal conductivity of the supporting elements is independent of the microhotplate area, while loss due to the thermal conductivity of air is proportional to the area.

For example, as demonstrated in [8], in membrane-type microhotplates consisting of a platinum microheater on 1.2 μm thick $SiO_2/Si_3N_4$ dielectric membrane, heat losses exhibit a monotonic dependence on membrane size. With a fixed size of the microheater itself equal to 250 × 250 μm, the thermal losses decrease as the membrane size increases, eventually reaching a minimum constant value. This saturation occurs when the ratio of membrane size to microheater size reaches approximately 8. Further increase in membrane size doesn't reduce heat losses but significantly increases the membrane's fragility.

Based on both simulation results and experimental data, the optimal size of the membrane supporting a 250 × 250 μm microheater was determined to be approximately 1.5 × 1.5 mm.

In contrast to heat conduction and radiation losses, the role of convection in the heat dissipation of the microheater has, to the best of our knowledge, never been thoroughly investigated. This remains a common and unresolved issue in the design of low power-consuming semiconductor and thermocatalytic (calorimetric) gas sensors based on microhotplates [10,11], as well as other sensors working at elevated temperatures, such as thermoconductometric sensors of gas concentration [12], airflow sensors, magnetic sensors of oxygen concentration, etc.

In principle, convection could be important, because it may not only increase total heat loss and, consequently, power consumption, but can also introduce orientation-dependent effects in the sensor's response. In most thermal simulations, however, only heat conduction through the air and through the supporting structures of the microhotplate is considered, while radiation is typically neglected, as previously discussed.

Yet, clear estimates of the minimum microhotplate size required for the onset of natural convection are rarely found in the literature.

From an intuitive standpoint, if the microhotplate is extremely small, essentially acting as a hot point, then natural convection is unlikely to be initiated. In such a case, the buoyant (Archimedean) force acting on the heated air above the microhotplate is insufficient to generate convective flow. On the other hand, it is also clear that, if the temperature difference between the microhotplate and housing is too small, the convection also cannot start. Therefore, in our consideration, we should take into account these two boundary conditions.

In this work, we aim to estimate the critical size below which the contribution of convection to total heat loss can be neglected. This threshold is essential for optimizing the design of gas sensors that utilize microhotplates and operate under high-temperature conditions.

Another issue, which should be taken into account, is the optimization of sensor microhotplate size from both technological and physical points of view. Currently, the majority of metal oxide semiconductor and thermocatalytic gas sensors are fabricated using sensing layers prepared by sol–gel, plasma, or other technologies, giving this material in powder form. This approach offers higher stability and higher responses of such materials compared to sputtered ones. Therefore, the sensing material should be deposited onto the microhotplate in the form of ink or paste, and this restricts minimum size of the microhotplate by a value of about 200 μm.

On the other hand, the size of the total membrane supporting the sensing layer should be chosen by taking into account the minimization of power consumption at the appropriate working temperature, for example, at 450 °C for methane detection. This balance between minimizing power consumption and ensuring effective sensor performance is crucial.

The most comprehensive classical consideration of all aspects of the heat and mass transfer processes and their influence onto the kinetics of chemical reactions was presented many years ago in the book [13]. While this work provides a broad overview of thermal and mass transfer considerations, the discussion of convection is relatively brief. Unfortunately, this analysis does not provide significant insight into the specific role of convection in heat exchange within microhotplates. In fact, the author only mentions a dimensionless parameter, which can be related to convection, the so called Grashof number:

$$Gr = \frac{g \cdot L^3 \cdot \beta \cdot (T - T_r)}{v^2}$$

where $g$ is the gravitational acceleration, $L$ is the typical geometrical size of the surface, $T$ is the temperature of the hot surface, $T_r$ is the ambient temperature, ν is the kinematic viscosity of air, and β is the coefficient taking into account the heat exchange process on the surface.

This number was constructed to form a dimensionless value, taking into account all parameters, which could control the process of convection.

Unfortunately, there is no recommendation on how to use this number for the analysis of heat transfer from microhotplates.

In the sensor community, there are some legends, which are not confirmed by reference to any published documents. For example, we found an opinion that the authors of the book [14] analyzed this problem and found the criteria (the critical size of the hot element), which permits the neglecting of free convection. In fact, the authors of this book analyzed the operation of thermocatalytic sensors based on platinum spirals and supposed that the small dimension of the spiral, equal to ~100 μm, permitted them not to consider the influence of the free convection of air in housing as a mechanism of cooling the microsensor.

## 3. Convection in Thermomagnetic Sensors of Oxygen

A similar problem arises in the design and analysis of the operation of magnetic sensors of oxygen.

The operation principle of magnetic sensors of oxygen is based on paramagnetic properties of molecular oxygen [15]. There are two different approaches to the application of magnetic properties of oxygen for the measurement of its concentration.

The first one is the use of bulk properties of oxygen as paramagnetic gas and the use of its attraction by the magnetic field [15]. The second and, perhaps more interesting, is the application of the so-called Senftleben effect [16–18]. The Senftleben effect consists in a change in the diffusion cross-section of the triplet-state oxygen molecules in the magnetic field. The reason for this increase in the cross-section is the precession of oxygen molecules in the magnetic field. Therefore, in magnetic field, the coefficients of temperature conductivity and thermal conductivity decrease, and this change can be detected by measuring the variation in microhotplate temperature in the pulsing magnetic field.

This effect depends on the ratio of magnetic-field strength to gas pressure. Therefore, this effect is valuable at a low pressure of 0.01–2 Torr. The threshold of $O_2$ detection is very low and is about $10^{-5}$ Torr. Due to its high sensitivity, this effect is particularly valuable for leak detection in vacuum systems.

However, in the present work, we focus primarily on more conventional devices that exploit the bulk magnetic properties of gases containing paramagnetic components.

The contemporary state of the art in the field of magnetic sensors of oxygen concentration is presented, for example, in a review article [15].

The paramagnetic properties of oxygen are related to the electronic configuration of the oxygen molecule. It is known that according to Hund's law, electrons in oxygen molecules occupying π-orbital first occupy the state with parallel spins. Therefore, the total magnetic moment of the oxygen molecule is unity, and, as a result, it has a magnetic moment in contrast to almost all common gases in the atmosphere, except nitrogen oxide. The magnetic moment of oxygen exceeds the moment of all other gases by a factor of 100, and this property enables selective detection of oxygen in the presence of other gases.

On the other hand, this property imposes limitations on the minimum detectable concentration of oxygen. As a result, magnetic oxygen sensors are predominantly employed for industrial monitoring and medical diagnostics, where oxygen concentrations are relatively high and comparable to those found in ambient air.

The magnetic susceptibility of oxygen follows Curie's law, being inversely proportional to absolute temperature. Therefore, the interaction of oxygen with the magnetic field is much stronger at a low temperature; this property enables the fabrication of the most simple and the most widely usable instrument [15]. The scheme of this instrument is presented in Figure 2, and in our work we will analyze mostly the instruments of this type.

The operating principle of the sensor based on the so-called "magnetic wind" effect is illustrated in the schematic diagram and description provided in Figure 2. The sample gas passes through two symmetrical sections of the ring-shaped chamber, designed to balance the pressure and velocity at the inlet and outlet of a glass (diamagnetic) tube connecting these sections. A wire coil is wound around this tube and forms two arms of a Wheatstone bridge used both for gas heating and for the measurement of its velocity; these coils work as a thermal anemometer.

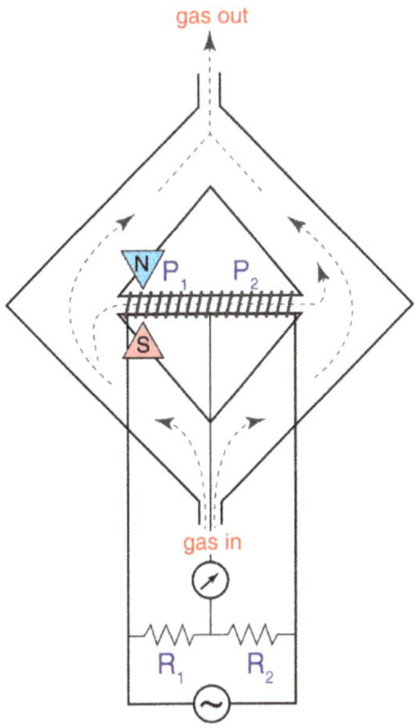


**Figure 2.** The scheme of the magnetic oxygen sensor based on the "magnetic wind" effect. The instrument consists of two branches (the left and the right ones). The gas that is to be analyzed enters from the "gas in" tube and is divided into two flows. As a result, gas pressures and velocities are the same in the left and right ends of the horizontal tube (point $P_1$ and $P_2$, respectively). The wire coil is made of platinum and is wound around a horizontal glass (diamagnetic) tube. This coil is used at the same time as a heater and as a thermoanemometer measuring gas velocity in the horizontal tube. This velocity is measured using a Wheatstone bridge composed of two parts of the coil and resistors $R_1$ and $R_2$. If the gas flows in the tube from left to right, the temperature of the left part of the coil is lower than the temperature of the right part of the coil (temperature in the circular branches is equal to room temperature). The left part of the horizontal tube is placed between the poles of the magnet, and paramagnetic oxygen is attracted by the magnetic field. Gas moving from left to right is heated gradually by the coil and loses paramagnetic susceptibility due to Curie's law. Therefore, gas from the left side is attracted by the magnetic field more strongly than gas from the right side. This force difference assures permanent gas flow in the horizontal tube, and that the velocity of gas is proportional to the oxygen content in the gas to be analyzed.

The working principle of this instrument is based on a decrease in magnetic susceptibility of oxygen with an increase in temperature (Curie's law). Consequently, the magnetic susceptibility of gas on the left end of the glass tube is higher than the susceptibility on the right side, and this creates the macroscopic gas motion from left to right. The velocity of this flow is directly related to the oxygen concentration in gas.

The key problem restricting the limit of detection of such a sensor is the competing influence of the free convection of gas. If the sensor tube is not perfectly horizontal, gas can move in the same (or opposite) direction due to unforced convection. The analysis presented in [19] demonstrates that the limit of detection of such sensors can be of the order of 100 ppm of oxygen in a diamagnetic background gas.

Therefore, it is crucial to determine the optimal geometry of thermal magnetic oxygen sensors that would eliminate or minimize the influence of free thermal convection, ensuring that measurement results are independent of sensor orientation. Achieving such immunity to orientation effects would significantly improve the sensor's applicability in portable devices. Moreover, miniaturization of the sensor design would reduce the characteristic diffusion time and make the sensor applicable for the monitoring of oxygen in real-time breath analysis, even in sport medicine, where the quantification frequency should be up to 100 Hz.

The primary objective of this study is to investigate the influence of microhotplate and magnetic sensor geometry on the effects of free thermal convection, and to establish design criteria that allow its contribution to be reliably neglected.

## 4. Methods and Analysis

### *4.1. Convection Effects in Semiconductor and Thermocatalytic Sensors*

As noted previously, there is a common consensus that, if a microhotplate with a size of a few tens of hundreds of microns is used as a heater of a semiconductor or thermocatalytic gas sensor, it is possible not to take into account the influence of free thermal convection on the heat exchange processes of gas sensors. This conclusion stems both from available experimental observations and from the inherent complexity of modeling convective processes. However, it remains difficult to find rigorous justification for the assumption that convection is negligible in systems involving micro-scale heating elements.

For the evaluation of the minimum size of microhotplate capable of inducing an ascending (buoyant or flowing up in gravitation field) air flow from the hotplate to, for instance, the colder wall of the sensor housing, we formulated a model based on the following assumptions.

(1) There is competition between two processes: the ascending air flow due to free thermal convection and the back-diffusion of air molecules. If the convection of gas plays an important role in heat transfer, the characteristic time of convective transfer should be shorter (or, preferably, much shorter) than that of diffusion. If the diffusion (it is known that the diffusion coefficient in ideal gas is equal to the temperature conductance coefficient) of gas is faster than the convection flow, the role of convection is negligible, and all heat exchange processes are only due to the process of heat conductivity.

(2) The convective flow in a virtual convection tube is laminar, because the velocity of this flow is very small, and the Reynolds number $Re = v \cdot L/\nu$, where v—average velocity of gas flow, $v < 10$ cm/s, L—characteristic size of the sensor, $L < 1$ cm, and $\nu$—kinematic viscosity of air, $\nu \sim 1$ cm$^2$/s. Therefore, $Re < 10$, and the flow is in the deep laminar zone. A virtual convection tube is a tube placed over the hot surface with gas flowing up due to free convection.

(3) Air in this ascending flow (convection tube), heated by a microheater, is then cooled by surrounding room-temperature air due to the heat conductance of air along the sides of the virtual convection tube.

(4) The result should meet the criteria of reasonability: under isothermal conditions, the convection is suppressed, and, for a point-like heat source, the Archimedes' force is insufficient to establish a stable upward flow.

Let us consider a microhotplate with diameter d and, respectively, radius $r = d/2$ (Figure 3). We suppose that there is a convection tube over this microhotplate, the temperature of lower end of this tube is equal to the temperature of the microhotplate T, and the temperature of the upper end is equal to the room temperature $T_r$. The height of the tube is equal to h. The convection velocity of gas in the tube is v. The condition, which should be met to neglect convection is $t > \tau$, or, better, $t >> \tau$ (t—is convection time, $\tau$—diffusion time) or

$$h/v > h^2/D, \quad (1)$$

where D is diffusion coefficient.

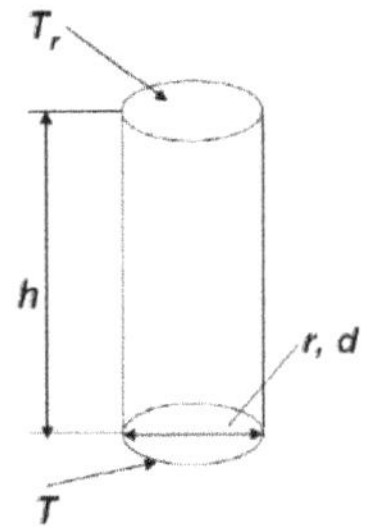


**Figure 3.** Microhotplate and convection tube over it. The microhotplateis is located in the bottom of the convection tube, the diameter and radius of the microhotplate are equal to d and r, respectively, the temperature of the microhotplate is T, and the ambient temperature is equal to $T_r$. The temperature of gas surrounding the cylindrical part of convection tube is also equal to $T_r$.

The gas flow in the convection tube is laminar; therefore, there is no gas mixing, and the cooling time of the gas in the tube is equal to $\tau_1 = r^2/D$. As a result, the height of the tube, necessary for the cooling of gas in it, is equal to $h = v \cdot \tau_1 = v \cdot r^2/D$. This is the height of the diffusion tube in Figure 3. Formula (1) can be rewritten as

$$\frac{v \cdot r}{D} < 1 \quad (2)$$

The velocity of gas in the convection tube can be evaluated using the Poiseuille formula

$$Q = \frac{\pi \cdot r^4}{8 \cdot \eta \cdot h} \cdot \Delta P,\ \eta = \nu \cdot \rho,\ Q = \pi \cdot r^2 \cdot v,$$

where ν—kinematic viscosity of air, η—dynamic viscosity, ρ—air density, Q—gas flow, v—linear velocity of gas, and ΔP—pressure difference on the ends of the convection tube.

Therefore, gas velocity in the convection tube is equal to

$$v = \frac{r^2}{8 \cdot \eta \cdot h} \cdot \Delta P. \quad (3)$$

Let us try to estimate the Archimedes' force acting on the gas in the convection tube. The mass of the gas over the microheater at room temperature $T_r$ and at high temperature $T$ are equal to, respectively,

$$m_r = \frac{\mu \cdot P \cdot V}{R \cdot T_r} = \frac{\mu \cdot P \cdot \pi \cdot r^2 \cdot h}{R \cdot T_r}$$

and

$$m_T = \frac{\mu \cdot P \cdot V}{R \cdot T_r} = \frac{\mu \cdot P \cdot \pi \cdot r^2 \cdot h}{R \cdot T}$$

The Archimedes' force acting on the air in the convection tube is equal to the difference between the weights of hot and cold air in the convection tube:

$$F = g \cdot (m_r - m_T) = \frac{g \cdot \mu \cdot P \cdot \pi \cdot r^2 \cdot h}{R} (\frac{1}{T_r} - \frac{1}{T}),$$

where μ—mass of mole of air, P—ambient pressure, and R—universal gas constant. Therefore, the Archimedes' pressure is of

$$\Delta P = \frac{F}{\pi \cdot r^2} = \frac{g \cdot \mu \cdot P \cdot h}{R} (\frac{1}{T_r} - \frac{1}{T})$$

We can rewrite this expression (3) as

$$v = \frac{r^2}{8\eta \cdot h} \cdot \Delta P = \frac{g \cdot \mu \cdot P \cdot h \cdot r^2}{8\eta \cdot h \cdot R} \cdot \left(\frac{1}{T_r} - \frac{1}{T}\right) = \frac{g \cdot \mu \cdot P \cdot r^2}{8\eta \cdot R} \cdot \left(\frac{1}{T_r} - \frac{1}{T}\right) \quad (4)$$

We should account for the ideal gas laws and relationship between kinematic and dynamic viscosity. As a result, the expression (4) can be written as

$$v = \frac{g \cdot r^2 \cdot T_{av}}{8\nu} \left(\frac{1}{T_r} - \frac{1}{T}\right), \quad (5)$$

where $T_{av}$ is a certain average temperature in the convection tube.

Substituting this expression into Formula (2), it is possible to obtain the following limit for the microhotplate.

$$\frac{g \cdot r^3}{8\nu \cdot D} \cdot \left(\frac{T_{av}}{T_r} - \frac{T_{av}}{T}\right) < 1 \text{ or } r < \sqrt[3]{\frac{8 \cdot \nu \cdot D}{g \cdot \left(\frac{T_{av}}{T_r} - \frac{T_{av}}{T}\right)}}. \quad (6)$$

For usual methane sensors, for example [8], the value of temperatures used in expression (6) are $T_r$ = 300 K, T = 750 K, and $T_{av}$ = 525 K. Therefore, the term in brackets is equal to 1.05. However, if the temperature of the microhotplate is very low, close to room temperature, the critical size of the microhotplate is very big, and convection is not observed more or less at any size of hotplate according to the reasonable boundary condition of this problem.

However, for usual conditions and usual gas sensor microhotplates, the cubic route of the term in brackets is close to unity; therefore, generally, the convection part can be neglected, if the following condition is met:

$$d < 4 \cdot \sqrt[3]{\frac{\nu \cdot D}{g}} \approx 0.5\text{cm}$$

Of course, the condition of the negligible convection role in the heat exchange process of the microhotplate is better met, if the size of the microhotplate is

$$d << 4 \cdot \sqrt[3]{\frac{\nu \cdot D}{g}} \approx 0.5\text{cm} \quad (7)$$

This means that the majority of microhotplates fabricated using microelectronic technology satisfy this condition. In practice, the cooling of the microhotplate occurs primarily through the thermal conductivity of the surrounding air and the thermal conductivity of the supporting elements that suspend the microhotplate.

This expression defines the conditions under which only thermal conductivity and radiation should be taken into account for the estimation of heat loss of a microhotplate.

Formula (7) can be compared with the aforementioned Grashof number Gr. It is known that, for ideal gas, the values of the diffusion coefficient and kinematic viscosity coincide with each other. The analysis presented here shows that, to neglect the convective cooling of the sensor microhotplate, the value of the Grashof number of the system, including the microhotplate and surrounding gas, should be Gr << 64 or, to be more realistic, Gr << 50.

### 4.2. Convection Effects in Thermal Magnetic Oxygen Sensors

A similar analysis can be applied to the consideration of the requirements used in the design of thermal magnetic gas sensors. As previously mentioned, a very important problem is how to minimize the influence of free thermal convection in the readings of oxygen sensors.

The magnetic sensor of oxygen consists of a tube made of a diamagnetic material placed within a magnetic field of a permanent magnet. This tube is equipped with a heater and thermoanemometer for measuring gas flow through the tube. Oxygen containing gas is attracted by the magnetic field; on the other hand, magnetic susceptibility of this paramagnetic gas is inversely proportional to its temperature. Therefore, if the gas is heated within the tube, the gradient of the temperature leads to asymmetry in the system, causing a continuous flow of gas from the cooler part to the hotter part of the system.

A very important advantage of this type of gas sensor is its ability to function even in an aggressive atmosphere. However, a significant drawback is the need to maintain the orientation of the device because of the competition between the magnetic and natural convection, due to the gradient of the temperature.

To minimize this effect, it is crucial to examine the interplay between thermomagnetic and natural convection. The ideal scenario occurs when the natural convection could be entirely neglected, ensuring that the thermomagnetic convection remains independent of the orientation of the tube placed in the magnetic field. In addition, to achieve a response time sufficient for the measurement of oxygen concentration with a characteristic time of less than 10 ms, the dimension of the tube should be below a certain value defined by the diffusion processes.

Here, we will utilize the results of the previous section of this paper, which considered the role of convection in the heat exchange of the microhotplate. In that analysis, we suggested that over a microhotplate exists a virtual tube of upstreaming gas. In the case of "magnetic wind" oxygen sensors, this corresponds to a real tube with diameter d and radius r heated up to temperature T. It is evident that the most critical case, when the natural convection is most important compared to the "magnetic wind", occurs, when the tube is oriented vertically.

Natural convection can be neglected if the time required for the gas to move through the tube due to convection (with length h) exceeds the time for the gas to diffuse back.

$$\frac{h}{v} >> \frac{h^2}{D}, \text{ or } v << \frac{D}{h};$$

substituting Formula (5) for the last one, we obtain $\frac{g \cdot r^2 \cdot T_{av}}{8 \cdot v}\left(\frac{1}{T_r} - \frac{1}{T}\right) < \frac{D}{h}$, and $r^2 h << \frac{8D \cdot \nu}{g \cdot \left(\frac{T_{av}}{T_r} - \frac{T_{av}}{T}\right)}$, in our case $T_{av} = T$, therefore

$$r^2 h << \frac{8D \cdot \nu}{g \cdot \left(\frac{T}{T_r} - 1\right)} \tag{8}$$

Taking into account that both the values of the diffusion coefficient and the kinematic viscosity of air at working temperature of the magnetic oxygen sensor are equal to about 2 $cm^2/s$, we can evaluate the dimension of the sensing element that would render the gas sensor insensitive to its orientation in the gravitation field and, therefore, suitable for use in portable instruments.

Let us consider the usual macroscopic gas sensor that is a glass tube equipped with a flow meter and put to a magnetic field. If the diameter of this tube is, for example, 0.1 cm, $T = 2 \cdot T_r$, the length of the tube should be by the order of magnitude $h << 3 \times 10^{-2}/r^2 \sim 3$ cm, that is about, or less than, 0.1 cm; the total size of the sensor is, in this case, a small cylinder with a diameter and length of less than 1 mm.

It is clear that the fabrication of such a small sensor using conventional macroscopic tools is rather complicated. Therefore, a thermomagnetic oxygen sensor free of the influence of the sensor orientation should be fabricated using alternative methods, such as microfabrication, which requires extremely small and precise machinery or similar instrumentation.

## 5. Discussion

### 5.1. Microhotplates

As discussed in the Introduction, the heat exchange processes of microhotplates have been extensively described in the literature by various research groups. One of the recent examples of such a detailed consideration is presented in [20]. In this comprehensive review, the authors overviewed the application of thermoconductometric gas sensors for the analysis of gases. However, even in this article, there is no well pronounced criterion enabling the neglect of the influence of free thermal convection on the heat exchange processes in gas sensors based on thin wires or microhotplates as heating sources.

This situation is similarly observed in other publications describing the heat exchange in microhotplates used for thermocatalytic and semiconductor gas sensors. For example, in our early publications [8], we supposed that, for the rather small microhotplates with 250 × 250 μm size, the influence of convection could be disregarded. This conclusion was confirmed by the good agreement between the results obtained by the simulation of the heat exchange process and the experimental results. In both cases, we had approximately 25 mW heating power necessary to heat the sensor up to a working temperature of 450 °C used for methane detection with MOS (metal oxide semiconductor) and thermocatalytic gas sensors.

The same assumptions were applied in the design of thermocatalytic gas sensors based on microspirals and microhotplates [21]. In this study, the authors fabricated a small microhotplate based on thin anodic alumina film (thickness is of about 30 μm). The size of the hot area was of about 200 × 200 μm. Power consumption at a working temperature of 450 °C was of about 35 mW for continuous heating to this temperature.

On the other hand, today there is another tendency in gas sensor development. It consists of the analysis of multisensory responses, and one of the possible solutions is the application of multi-heater and multi-electrode sensor chips [22,23]. In cases where the sensor array is fabricated as a single microhotplate, the size of the hotplate becomes significantly larger, typically around 2 × 2 cm. In such instances, convection heat exchange processes must be taken into account.

Another scenario, in which the application of relatively big hotplates is reasonable, is the investigation of chemical, gas sensing, and other properties of sensing materials [24]. The authors of this work used a hotplate fabricated by screen printing, and the size of the hot area of this hotplate is about 6 × 3 mm (Figure 4). It consumes ~3 W at a working temperature of 350 °C.

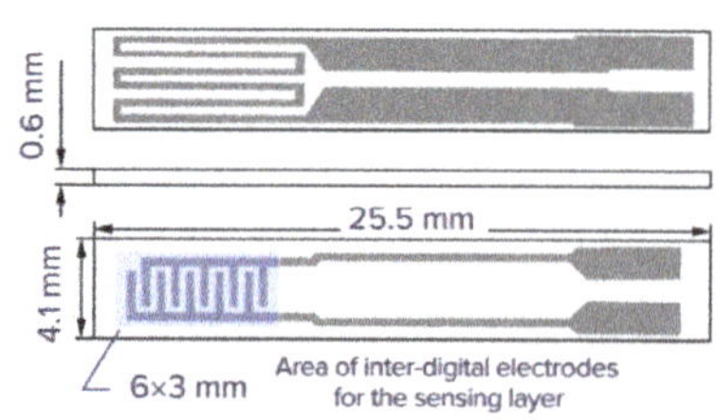


**Figure 4.** The hotplate with similar dimensions was used in [24] for the physical and chemical investigation of sensing materials.

The main reasons for using a large hotplate are to ensure uniformity of the temperature field over the hotplate and the possibility to easily deposit the sensing layer using both a screen-printing process and inkjet printing and to easily control the uniformity of the deposited sensing layer. In both last cases, the convection heat loss should be taken into account. The same situation also occurs with the application of early gas sensors manufactured by Figaro Inc., such as the TGS 812 and related models.

The application of different types of microfabricated hotplates is now widespread. In addition to the microhotplates mentioned above, it is possible to discuss commercial microhotplates fabricated by companies such as Figaro Inc. (Osaka, Japan) [25], Sensirion (Stäfa, Switzerland) [26,27], and SGX (Corcelles-Cormondreche, Switzerland) [28], and the microhotplates developed by a number of research groups. In the most part, for these microhotplates, condition (7) is met and, therefore, the free convection as a channel for heat losses can be neglected.

### 5.2. Thermomagnetic Sensors

Our colleagues [29] made an attempt to fabricate an oxygen sensor, which can be applied in portable devices (including medical instruments) using microfabrication. The sensor was fabricated as a small spiral made of 10 μm Pt glass-coated wire and the thickness of glass was of about 1 μm. The spiral diameter was about 100 μm and the length was about 150 μm. The spiral was made by winding the Pt wire around hot NiCr (Nickel-Chromium) wire followed by etching away the NiCr wire. The high temperature during the winding process facilitated the sintering of the glass insulation around the Pt wire, resulting in a structure resembling a very small glass tube containing the Pt wire spiral. The spiral was suspended on two platinum wires within a TO-18- or TO-46-type housing and placed in a magnetic field gradient. The orientation of the spiral in the magnetic field was more or less random. A schematic representation of this sensor is presented in Figure 5.

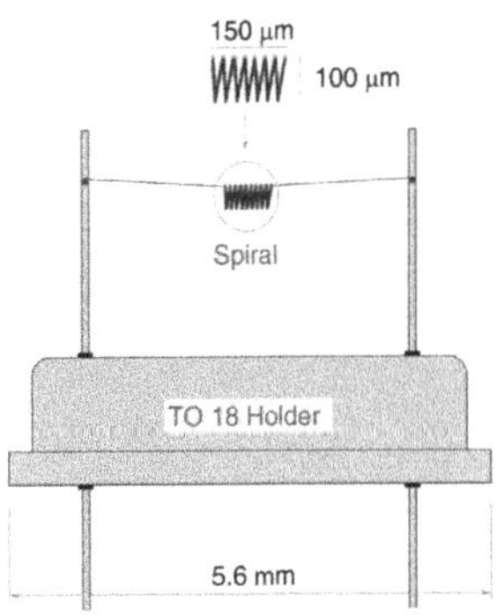


**Figure 5.** The scheme of the thermomagnetic oxygen sensor, which meets the requirements of Equation (8). The spiral is made of 10 μm platinum coated wire with a spiral size is of ~150 × 100 μm.

Figure 6 presents a photograph of the thermomagnetic sensor with a magnetic system consisting of two neodymium (Nd) magnets with iron magnetic conductors. The Pt wire coil is placed within the gap, and another arm with a reference coil is placed within the gap between two details made of diamagnetic Al alloy. The results of the measurement of oxygen concentration are presented in Figure 7. (This figure presents results obtained in [29]). The plot shows the potential difference between two branches of a Wheatstone bridge similar to those presented in Figure 2. The configuration of the particular microsensor is given in Figure 6. This potential difference is due to gas flow through the tubes formed by the Pt wire spiral. As previously discussed, there is a permanent gas flow of oxygen-containing gas in the gap of the permanent magnet, whereas no flow occurs in the tube positioned between the poles made of aluminum. Therefore, the spiral within the magnetic field is cooled by the gas flow, and this leads to the different resistance of the two spirals. This difference results in a potential difference between the two branches of the

Wheatstone bridge, as shown in Figure 7. This potential difference is proportional to the concentration of the paramagnetic component (oxygen) in the gas-analyte.

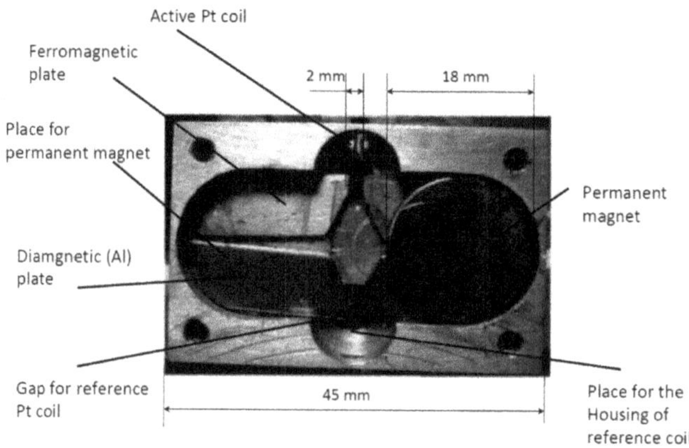


**Figure 6.** Photo of the thermomagnetic sensor [29].

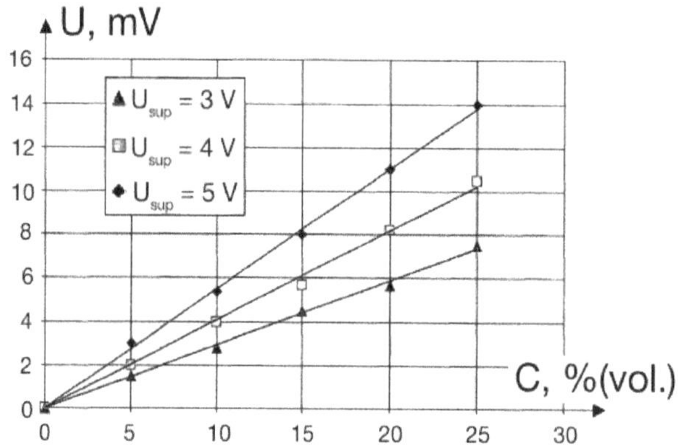


**Figure 7.** Thermomagnetic sensor response as a function of oxygen concentration in diamagnetic gas. This plot presents the potential difference between two branches of a Wheatstone bridge presented in Figure 2; the configuration of the particular sensor is given in Figure 6. The voltage indicated in the plot is a heating voltage supplied to the Wheatstone bridge heating two spirals (Figure 5) connected in series. The potential difference (Y axis) is a misbalance of two branches due to additional cooling of the spiral into the magnetic gap between the two poles of the magnet.

The results presented in [29] show that the thermomagnetic sensor is capable of determining oxygen concentrations as low as approximately 0.1 vol.%. The response time of the sensor is below 0.1 s. The small size of the sensor (~$10^{-2}$ cm, diffusion time of ~1 ms), together with the low thermal response time of the miniaturized Pt coil (~50 ms), results in a total response time of around 0.05 s. Consequently, the sensor is suitable for medical monitoring at a normal breath rate (12 $min^{-1}$) and can even function at a high rate of up to 100 $min^{-1}$.

Additionally, it was demonstrated that the measurement of oxygen concentration is less influenced by the orientation of the sensor compared to the macro sensor shown in Figure 1 (where the tube length is 12 mm). These findings support our conclusion regarding methods to mitigate the influence of free convection on the response of thermomagnetic oxygen sensors.

However, the sensor described in [29] requires further optimization. This concerns, first of all, the strong fixation and orientation of the microcoil with respect to the magnetic field. The suspension with two 10 μm Pt wires is insufficient for ensuring stable orientation. This issue will be addressed in future developments of the sensor, particularly in the advanced microelectronic versions of both thermomagnetic sensors and sensors based on the Senftleben effect.

## 6. Conclusions

We investigated the influence of convection on the heat exchange processes of microhotplates used in the fabrication of semiconductor and thermocatalytic (calorimetric) gas sensors, as well as the heat exchange of thermal magnetic sensors of oxygen. The analysis was based on the consideration of the competition of convection flow and back diffusion. It was shown that there is a certain critical size of the microhotplate equal to $d_{cr} \sim 4 \cdot \sqrt[3]{\frac{\nu \cdot D}{g}}$, where ν is kinematic viscosity of air, D is the diffusion coefficient, g is the acceleration of free fall, and d~0.5 cm. If the size of the microhotplate $d << d_{cr}$, the influence of convection heat exchange can be neglected and only the thermal conductivity of air and of the elements of the sensor together with the IR (infrared) radiation (if the sensor is heated up to a very high temperature) should be taken into account as channels of heat losses. This expression involves only the fundamental constants of ambient gas (air, for example). It can be compared with the Grashof number $Gr = gL^3/\nu^2$, where $g$ is the gravitational acceleration, $L$ is the typical geometrical size of the surface, and ν is the kinematic viscosity of air, designed to form a dimensionless value, taking into account all parameters, which could control the process of convection.

Similarly, for the thermal magnetic oxygen sensor consisting of a hot dielectric tube placed within a magnetic field gradient and used to measure the concentration of paramagnetic gas oxygen, certain conditions must be met to avoid the influence of free convection on the measurement results. In this case, the convection can be neglected and, respectively, the results of the measurements of oxygen concentration should not depend on the sensor orientation, if the size of the sensor size satisfies the following condition: $r^2 h << \frac{8D \cdot \nu}{g \cdot (\frac{T}{T_r} - 1)}$, where r is the radius of the dielectric tube of the sensor and h is the length of the tube.

**Author Contributions:** Concept formulation and evaluation of the hotspot size sufficient for air convection formation, A.V. and O.K.; proof of the concept, A.S. and A.V.; editing and preparation of figures, A.M. All authors have read and agreed to the published version of the manuscript.

**Funding:** The research was supported by the Russian Science Foundation grant №. 24-13-00254, https://rscf.ru/project/24-13-00254/ (accessed on 24 April 2025).

**Institutional Review Board Statement:** Not applicable.

**Informed Consent Statement:** Not applicable.

**Data Availability Statement:** All available data are presented in this paper.

**Conflicts of Interest:** Authors Oleg Kul and Alexey Vasiliev were employed by the LLC "C-Component" company. The remaining authors declare that the research was conducted in the absence of any commercial or financial relationships that could be construed as a potential conflict of interest.

## References

1. Available online: https://smks.fbk.eu/wp-content/uploads/2018/08/Mass-Flow-Sensor.pdf (accessed on 24 April 2025).
2. Bíró, F.; Dücso, C.; Radnóczi, G.Z.; Baji, Z.; Takács, M.; Bársony, I. ALD nano-catalyst for micro-calorimetric detection of hydrocarbons. *Sens. Actuators B* **2017**, *247*, 617–625. [CrossRef]

3. Schmid, U.; Seidel, H.; Mueller, G.; Becker, T. Theoretical considerations on the design of a miniaturized paramagnetic oxygen sensor. *Sens. Actuators B* **2006**, *116*, 213–220. [CrossRef]
4. Santra, S.; Sinha, A.K.; De Luca, A.; Ali, S.Z.; Udrea, F.; Guha, P.K.; Ray, S.K.; Gardner, J.W. Mask-less deposition of Au–$SnO_2$ nanocomposites on CMOS MEMS platform for ethanol detection. *Nanotechnology* **2016**, *27*, 125502. [CrossRef]
5. Cui, Z.; Wu, H.; Yang, K.; Wang, X.; Lv, Y. Adsorption of gas molecules on intrinsic and defective $MoSi_2N_4$ monolayer: Gas sensing and functionalization. *Sens. Actuators A Phys.* **2024**, *366*, 114954. [CrossRef]
6. Wang, X.; Liu, C.; Zhang, Z.; Liu, S.; Luo, X. A micro-machined Pirani gauge for vacuum measurement of ultra-small sized vacuum packaging. *Sens. Actuators A* **2010**, *161*, 108–113. [CrossRef]
7. Schelcher, G.; Lefeuvre, E.; Brault, S.; Parrainn, F.; Martincic, E.; Dufour-Gergam, E.; Bosseboeuf, A. Micro Pirani vacuum gauges manufactured by a film transfer process. *Procedia Eng.* **2010**, *5*, 1136–1139. [CrossRef]
8. Vasiliev, A.; Pavelko, R.; Gogish-Klushin, S.; Kharitonov, D.; Gogish-Klushina, O.; Pisliakov, A.; Sokolov, A.; Samotaev, N.; Guarnieri, V.; Zen, M.; et al. Sensors based on technology "nano-on-micro" for wireless instruments preventing ecological and industrial catastrophes. In *Sensors for Environment, Health and Security*; Baraton, M.-I., Ed.; Springer: Berlin, Germany, 2009; pp. 205–228.
9. Kul, O.; Vasiliev, A.; Shaposhnik, A.; Nikitin, A.; Dmitrieva, A.; Bolshakov, A.; Liu, Z.; Ma, M.; Mokrushin, A.; Simonenko, N.; et al. Novel screen-printed ceramic MEMS microhotplate for MOS sensors. *Sens. Actuators A Phys.* **2024**, *379*, 115907. [CrossRef]
10. Fau, P.; Sauvan, M.; Trautweiler, S.; Nayral, C.; Erades, L.; Maisonnat, A.; Chaudret, B. Nanosized tin oxide sensitive layer on a silicon platform for domestic gas applications. *Sens. Actuators B Chem.* **2001**, *78*, 83–88. [CrossRef]
11. Vincenzi, D.; Butturi, M.A.; Guidi, V.; Carotta, M.C.; Martinelli, G.; Guarnieri, V.; Brida, S.; Margesin, B.; Giacomozzi, F.; Zen, M.; et al. Development of a low-power thick-film gas sensor deposited by screen-printing technique onto a micromachined hotplate. *Sens. Actuators B* **2001**, *77*, 95–99. [CrossRef]
12. Berndt, D.; Muggli, J.; Wittwer, F.; Langer, C.; Heinrich, S.; Knittel, T.; Schreiner, R. MEMS-based thermal conductivity sensor for hydrogen gas detection inautomotive applications. *Sens. Actuators A Phys.* **2020**, *305*, 111670. [CrossRef]
13. Frank-Kamenetskii, D.A. *Diffusion and Heat Transfer in Chemical Kinetics*; Plenum Press: New York, NK, USA, 1969.
14. Karpov, E.F. *Physical and Chemical Basis of Automated Protection from Methane Emission*; Nauka: Moscow, Russia, 1981; 185p. (In Russian)
15. Jasek, K.; Pasternak, M.; Grabka, M. Paramagnetic Sensors for the Determination of Oxygen Concentration in Gas Mixtures. *ACS Sens.* **2022**, *7*, 3228–3242. [CrossRef]
16. Senftleben, H.; Pitzner-Breslau, J. Die Eiwirkung magnetischer Felder auf das Waermeleitvermoegen von Gasen. *Ann. Der Phys.* **1933**, *16*, 907. [CrossRef]
17. Senftleben, H.; Pitzner-Breslau, J. Die Einwirkung maynetischer Felder auf Waermeleitvermoegen van gasen. *IV Untersuchungen Mischungen Van Sauerst. Mit Diamagnetischen Gasen.* **1937**, *30*, 541.
18. Babichev, A.P.; Brayko, V.D.; Gorshunov, N.M.; Muromkin, Y.A.; Pashkovsky, V.G. Determination of oxygen admixture in fluorine. *Russ. J. Appl. Phys.* **2012**, *4*, 37–39.
19. Zhang, X.; Zhang, J.; Song, K.; Wang, L. Numerical study of the sensing mechanism of the oxygen concentration sensor based on thermal magnet convection. *Int. J. Therm. Sci.* **2016**, *99*, 71–84. [CrossRef]
20. Gardner, E.L.W.; Gardner, J.W.; Udrea, F. Micromachined Thermal Gas Sensors—A Review. *Sensors* **2023**, *23*, 681. [CrossRef] [PubMed]
21. Karpov, E.E.; Karpov, E.F.; Suchkov, A.; Mironov, S.; Baranov, A.; Calliari, L. Energy efficient planar catalytic sensor for methane measurement. *Sens. Actuators A* **2013**, *194*, 176–180. [CrossRef]
22. Pronin, I.A.; Plugin, I.A.; Kolosov, D.A.; Karmanov, A.A.; Yakushova, N.D.; Varezhnikov, A.S.; Komolov, A.S.; Lazneva, E.F.; Koroleva, A.V.; Moshnikov, V.A.; et al. Sol-gel derived ZnO film as a gas sensor: Influence of UV processing versus a thermal annealing. *Sens. Actuators A Phys.* **2024**, *377*, 115707. [CrossRef]
23. Musatova, V.Y.; Sysoeva, V.V.; Sommer, M.; Kiselev, I. Assessment of meat freshness with metal oxide sensor microarray electronic nose: A practical approach. *Sens. Actuators B* **2010**, *144*, 99–103. [CrossRef]
24. Mokrushin, A.S.; Fisenko, N.A.; Gorobtsov, P.Y.; Simonenko, T.L.; Glumov, O.V.; Melnikova, N.A.; Simonenko, N.P.; Bukunov, K.A.; Simonenko, E.P.; Sevastyanov, V.G.; et al. Pen plotter printing of ITO thin film as a highly CO sensitive component of a resistive gas sensor. *Talanta* **2021**, *221*, 121455. [CrossRef]
25. Figaro Engineering Inc., n.d. Available online: https://www.figaro.co.jp/en/ (accessed on 24 April 2025).
26. Sensirion Inc., n.d. Available online: www.sensirion.com (accessed on 24 April 2025).
27. Weller, F.; Keller, D.; Wettstein, S.; Graf, M. Sensing Characteristics of Smart Microsensor Systems for Measuring Relative Humidity in Pressurized Air. *Proceedings* **2018**, *2*, 871. [CrossRef]

28. SGX Sensortech, (n.d.). Available online: http://Sgxsensortech.com (accessed on 24 April 2025).
29. Krupin, M.V. Early Detection of Fire on Nuclear Power Station Using Thermomagnetic Oxygen Sensor. Ph.D. Thesis, State Academy of the Fire Protection Service of the Ministry of Emergency Situations of Russian Federation, Moscow, Russia, 2015. (In Russian)

sensors

MDPI

*Article*

# Optical Ammonia Sensors Based on Spray-Coated Polyaniline Complexes with Polysulfonic Acids

**O. L. Gribkova [1,*], V. A. Kabanova [1], E. I. Rodina [1], M. A. Teplonogova [2], L. I. Demina [1] and A. A. Nekrasov [1,*]**

[1] Frumkin Institute of Physical Chemistry and Electrochemistry of the Russian Academy of Sciences, Leninskii prospect 31, 119071 Moscow, Russia
[2] Kurnakov Institute of General and Inorganic Chemistry of the Russian Academy of Sciences, Leninskii prospect 31, 119071 Moscow, Russia
* Correspondence: gribkova@elchem.ac.ru (O.L.G.); secp@elchem.ac.ru (A.A.N.)

**Abstract:** The optical ammonia-sensing properties of water-dispersible polyaniline (PANI) complexes chemically synthesized in the presence of polysulfonic acids of different structure and chain flexibility were compared for the first time. Flexible-chain poly(styrene-4-sulfonic acid) and poly-(2-acrylamido-2-methyl-1-propanesulfonic acid), as well as semi-rigid-chain poly-4,4′-(2,2′-disulfonic acid)diphenylene-iso-phthalamide and rigid-chain poly-4,4′-(2,2′-disulfonic acid)diphenylene-tere-phthalamide (t-PASA) were used. The sensor films were prepared by a convenient and scalable method—spray coating of aqueous solutions on glass substrates. The optical response time and amplitude of the sensor films in the range of ammonia concentrations from 5 to 200 ppm were investigated. To overcome the influence of humidity and presence of over-stoichiometric protons of the polyacid on the accuracy of ammonia determination treatments of the films in aqueous solutions of NaCl, $CaCl_2$ and $BaCl_2$ were tested. The treatment in 1 M $CaCl_2$ solution for all of the PANI complexes results in a significant improvement in the response time, amplitude and reproducibility. The films of PANI complexes with the flexible-chain polyacids have the highest response amplitude in the range of ammonia concentrations 5–25 ppm. PANI-t-PASA film demonstrated the best sensory properties at ammonia concentrations more than 50 ppm. FTIR spectroscopy showed that $CaCl_2$ treatment results in cross-linking of sulfoacid groups from adjacent polyacid chains by $Ca^{2+}$ ions. Thus, such a treatment results both in the neutralization of excessive protons and a significant reduction in the films' swelling at high humidity.

**Keywords:** polyaniline complexes; polyacid; spray coating; ammonia sensors; optical gas sensors

## 1. Introduction

One of the current trends in the progress of materials science is the development and production of new functional materials with predetermined properties. From this point of view, electrically conductive polymers (ECP) are of great interest. The advantages of these polymers (polyaniline (PANI), poly(3,4-ethylenedioxythiophene) (PEDOT) and polypyrrole (PPy)) show high conductivity and optical transparency in the conducting state, and high stability in the doped state. They have a unique combination of physicochemical, electrical and optical characteristics that makes these materials promising for use in chemical sensors [1–3].

PANI is distinguished by the capability of detecting both oxidation-reduction and pH active gases, such as ammonia, amines, etc. [4]. Most of these gases are dangerous to humans and are among the most common industrial pollutants. Also, the detection of amines is one of the methods for controlling freshness, quality and safety of food products [5].

Chemical sensors can be classified based on their operating principle: electrochemical [6], thermochemical [7], resistive [8–10], mass-sensitive [11] and optical [12]. The majority of ammonia and amines sensors based on conductive polymers operate using the resistive detection mechanism [1,2]. For example, PANI obtained by chemical synthesis in the presence of hydrochloric acid and indium oxide nanotubes was used for detecting sprayed pesticides [13]. The resistive method of ammonia determination is simple to implement and is currently well studied. However, the resistance value can be affected by external factors. At the same time, optical sensors can provide high sensitivity and short response time, while they are less affected by external factors such as electromagnetic interference, humidity and temperature. Also, the optical analytical signal can be transmitted over long distances without distortion [1,14].

Upon exposition to ammonia, PANI film undergoes deprotonation, which results in a transition from the salt form to the base one [15]. This change in the PANI electronic structure is reflected in the change in the absorption spectrum of the film.

PANI films can be obtained in various ways, for example, by electrochemical deposition on a conductive substrate during electropolymerization. Chemically synthesized PANI can be applied on a substrate using various methods: casting onto a horizontal substrate, spin coating, spray coating, inkjet printing, etc. Many studies have shown that the method of film fabrication significantly affects both the mechanical properties and morphology of the resulting film (adhesion to the substrate, thickness uniformity, roughness), as well as the sensory properties [12,15,16].

The comparative studies of the optical ammonia sensing properties of PANI, PEDOT and polypyrrole obtained electrochemically in inorganic electrolytes [17] and in the presence of polyelectrolytes of different structure [12] were presented, and the prospects of such films for ammonia detection in air were shown. Also, double layer films of PANI/PEDOT electrodeposited in inorganic electrolytes were used as optical ammonia sensors [18,19].

The literature contains enough work on the development of optical sensors for ammonia based on chemically synthesized PANI. PANI films obtained in the presence of dodecylbenzenesulfonic and laurylsulfonic acids and $TiO_2$ nanoparticles can be used to detect ammonia in solution [20]. A significant influence of the film deposition method (sedimentation from the synthesis solution or spin coating) and the nature of dopant (hydrochloric acid, camphorsulfonic acid or iodine) on the sensitivity of PANI to ammonia was demonstrated [15]. A more developed surface area of the film contributes to the improvement of sensing characteristics. An optical sensor for detecting ammonia and amines that can be connected to a smartphone was developed [5]. It is intended to be used to determine the freshness of fish. It was shown that a chemically synthesized complex of PANI with polystyrene sulfonic acid can be applied to filter paper and used as a colorimetric sensor for amines [21].

The synthesis of PANI in the presence of polysulfonic acids of various structures makes it possible to control the optical and electrical properties of the resulting PANI complexes, the morphology of their films [22,23] and, consequently, their sensory characteristics. The use of water-soluble polysulfonic acids in PANI synthesis allows the preparation of water-soluble polymer compositions, films of which can be applied to various substrates, including flexible ones, using scalable methods (spray coating, inkjet printing, etc.).

The deposition of PANI layers by inkjet printing has been mostly used to create biological and chemical sensors based on the resistive sensing mechanism [24,25]. In this case, solutions of PANI synthesized in the presence of camphorsulfonic, dodecylbenzenesulfonic and polystyrene sulfonic acids were used for printing. High-boiling organic solvents (dimethyl sulfoxide, N-methylpyrrolidone) or water with the addition of sodium dodecyl sulfonate served as solvents in these works. Resistive sensing properties of films prepared by spray coating and inkjet printing of aqueous solutions of PANI synthesized chemically in the presence of polyvinylpyrrolidone were compared in [16]. The studies showed that the spray-coated film has 2–3 times greater roughness and demonstrates twice the conductivity change. PANI obtained in the presence of dodecylbenzenesulfonic acid in water can be used to determine the level of ammonia in the blood by spectrophotometry or electrochemical impedance spectroscopy [6]. A composition of PANI with $CuCl_2$ applied by spray coating onto paper was used as a hydrogen sulfide sensor [26]. Also, PANI compositions with 2-D carbides of transition metals deposited on cotton fabrics were developed for detection of ammonia [14]. Most of the above mentioned PANI compositions were prepared based on conventional water-insoluble PANI prepared in HCl, which was then processed and/or mixed with various additives to obtain water-dispersible PANI compositions. Only in the work [24], aqueous solutions of PANI were prepared in the presence of an organic sulfonic acid for designing printed flexible humidity sensors and in [27] a composite of cellulose/PANI–poly-(2-acrylamido-2-methyl-1-propanesulfonic acid) was used as resistive humidity sensor and biomedical sensor for heart rate or respiration activity. Spray-coated films of PANI complexes with sulfonated polysulfone demonstrated good sensing properties in optical sensors for ammonia in air [28].

Earlier, we proposed treating the films of PANI complexes with 1M aqueous solution of $CaCl_2$. Such treatment did not influence the electronic structure or morphology of PANI films and was shown to decrease the solubility of drop-cast films of chemically synthesized PANI complexes with polyacids [22]. Such treatment has also led to an improvement in the ammonia-sensing properties of electrochemically deposited films of ECP complexes with sulfonated polyelectrolytes [12,29].

A more or less similar approach was used in [30]: the treatment of inkjet-printed PEDOT:PSS film on paper in $FeCl_3$ solution improved the ammonia-sensing properties. The authors hypothesized that the sensor's high humidity tolerance was achieved by the molecular interaction between the sulfonate groups in PSS and the iron(III) ions which suppressed the swelling of the PSS shell surrounding PEDOT in high humidity, resulting in the maintenance of electronic coupling between PEDOT chains. Also, calcium salts are commonly used for crosslinking of biopolymers such as alginate [31,32] and pectin [33] to prepare hydrogels with the purpose of improving their mechanical properties (including swelling) and/or regulate their gas and vapor permeability.

In this work, we have first performed a comparative study of the optical ammonia-sensing properties of water-dispersible PANI complexes chemically synthesized in the presence of polysulfonic acids of different structures and chain flexibility. The films were prepared by spray coating onto transparent glass substrates. The effect of the structure and flexibility of the polyacid dopants on the morphology spray-coated films and ammonia-sensing properties is considered. Also, the influence of treatment of PANI films in aqueous solutions of different cations ($Na^+$, $Ca^{2+}$, $Ba^{2+}$) on the ammonia-sensing properties is discussed.

## 2. Materials and Methods

### 2.1. Materials

PANI was synthesized by oxidative chemical polymerization of aniline in aqueous solutions of polymeric sulfonic acids of various structures (Figure 1): flexible-chain poly(styrene-4-sulfonic acid) (PSSA) and poly-(2-acrylamido-2-methyl-1-propanesulfonic acid) (PAMPSA), as well as semi-rigid-chain poly-4,4′-(2,2′-disulfonic acid)diphenylene-iso-phthalamide (i-PASA) and rigid-chain poly-4,4′-(2,2′-disulfonic acid)diphenylene-terephthalamide (t-PASA). The polymerization was carried out at room temperature (~24 °C) according to the method described in [22,34]. Prior to the synthesis, aniline (Sigma-Aldrich, St. Louis, MO, USA, reagent grade) was distilled under reduced pressure with nitrogen bubbling. PAMPSA (MW 2,000,000, 15% aqueous solution) was purchased from Sigma-Aldrich. Sodium salts of i-PASA, t-PASA were synthesized as described in [34,35]. PSSNa (Sigma-Aldrich, MW 1,000,000, 25% aqueous solution), i-PASNa and t-PASNa were converted into $H^+$-forms using ion-exchange column. All polyacid solutions were purified via dialysis (cellulose membrane ZelluTrans MWCO 8000–10,000, Roth, Karlsruhe, Germany) against 18 MOhm deionized water for 3 days and diluted by 18 MOhm deionized water to achieve necessary concentration. We have determined viscosity-average molar weight of polysulfonic acids using the method described in [34]: PSSA—600,000, PAMPSA—1,100,000, i-PASA—17,000, t-PASA—59,000.

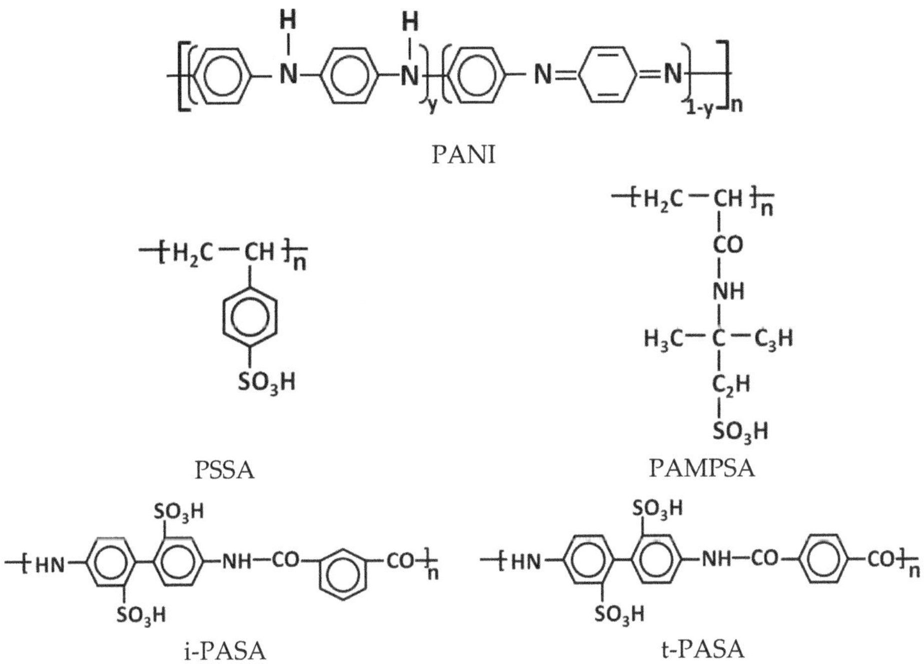


**Figure 1.** Chemical structure of PANI and polysulfonic acids used in the paper.

The ratio of the concentrations of aniline to sulfoacid groups of the polyacids was 0.5 mol/g-eq. of sulfogroups: for the single base polyacids (PAMPSA, PSSA) one aniline molecule corresponded to two monomer units of the polyacid, while for the double base polyacids (t-PASA and i-PASA) (Figure 1) this ratio was 1:1. Ammonium persulfate (APS, Sigma-Aldrich, reagent grade) was used as the oxidizing agent, and the ratio of the concentrations of aniline to APS was 1:1 mol/mol. The concentration of aniline was 0.01 M for the synthesis in the presence of PAMPSA (0.02 g-eq. $SO_3$), PSSA (0.02 g-eq. $SO_3$) and i-PASA (0.01 g-eq. $SO_3$). In the case of the synthesis in t-PASA, in order to prevent sedimentation of the resulting solution and ensure it is suitable for spray coating, the aniline and the polyacid were taken in 0.005 M and 0.005 g-eq. $SO_3$ concentrations, respectively. The progress of the synthesis was controlled by in situ registration of electronic absorption spectra (AvaSpec 2048 spectrophotometer, Avantes BV, Apeldoorn, The Netherlands) of the reaction solution. After the polymerization was completed, the PANI complexes with polyacids were purified from oligomers, residues of unreacted oxidizer and aniline via dialysis (ZelluTrans MWCO 8000–10,000) against deionized water for 3 days. The mass concentrations of the aqueous solutions of PANI complexes thus obtained were: 3.7 g/L PANI-PSSA, 4.9 g/L PANI-PAMPSA, 3.2 g/L PANI-i-PASA, 2.9 g/L PANI-t-PASA.

The films of PANI complexes were obtained by spraying the aqueous solutions onto glass substrates. Before spraying, the solutions were treated in an ultrasonic bath for 10 min. The substrates were placed onto a horizontally leveled platform of an IKA MSC BASICS magnetic stirrer (IKA-Werke GmbH, Staufen, Germany), heated up to 70–80 °C. After thermal stabilization of the glass substrates during 3–4 min, spray coating was performed using a JAS 1147 aerograph (nozzle diameter 0.3 mm, JAS, Shanghai, China), fixed at a distance of 20 cm from the platform. The solutions were spray-coated step by step, with drying intermediate layers.

Samples for investigating the sensing properties were prepared as follows: for each PANI–polyacid complex, 5 samples were taken as-coated, 5 samples were treated during 30 min in 1 M aqueous solution of NaCl, 5 samples in $CaCl_2$ and 5 samples in $BaCl_2$. After the treatment excess of the salts' solutions was removed from the films by keeping them in deionized water for 5 min, followed by air drying (Figure S1, Supplementary materials, see the link after the Conclusions).

### 2.2. Characterization Techniques

The thickness of the films was measured by the MII-4 microinterferometer (LOMO, St. Petersburg, Russia). The thicknesses of PANI–polyacid films depending on the polyacid used were (nm) as follows: 275 ± 125 (PANI-PSSA), 375 ± 50 (PANI-PAMPSA), 185 ± 15 (PANI-i-PASA) and 425 ± 25 (PANI-t-PASA). The wide spread of thickness for PANI-PSSA is due to its rough surface (see AFM data below), which increases the possibility of error in thickness determination by interferometry. The differences in thickness are explained by (1) different molecular weights of the polyacids; (2) possibly different yield of aniline polymerization in the presence of different polyacids; (3) necessity to prepare films with comparable absorbance changes in the spectral area of optical sensor response to ensure accuracy of the determination.

The surface morphology of PANI films was recorded using Enviroscope atomic force microscope (AFM) with a Nanoscope V controller (Bruker, Billerica, MA, USA) in tapping mode. The roughness of PANI films was averaged from 5 different areas of 3 films. Scanning electron microscopy (SEM) images were taken using a Tescan Amber GMH scanning electron microscope. Images were obtained using Everhart-Thornley SE detector at ×3000–100,000 magnifications and at an accelerating voltage of 0.5–1.0 kV.

The ammonia-sensing properties of PANI films were studied similarly as described in [29]. The spray-coated films on glass substrates were placed into a closed 5 cm spectrophotometric quartz cell filled with ammonia vapors in equilibrium (at 22–25 °C), with a 5 mm layer of the aqueous solutions of different ammonia concentrations on the bottom of the cell (Figure S2). Using literature reference data on the equilibrium concentrations of ammonia in the aqueous and gas phases at 25 ± 0.1 °C [36], we have calculated the volume of the 30% $NH_3$ solution (analytical grade, Chimmed, Russia) necessary to obtain different $NH_3$ concentrations in 3 mL of water. Then, the concentrations of ammonia in water were recalculated to the concentrations in air expressed in ppm.

UV-visible–NIR (350–1000 nm) absorption spectra of PANI films in air and their evolution when exposited to the ammonia vapors was registered in situ using the AvaSpec 2048 spectrophotometer. The recording time of each spectrum was 2 s.

The sensor response ($\Delta A$) was calculated as the relative variation of the absorbance amplitude at characteristic wavelengths (different for different PANI complexes)

$$\Delta A = \frac{A_{NH_3vap} - A_{air}}{A_{air}} 100\% \quad (1)$$

where $A_{NH3vap}$ is the value of absorbance when the sample is exposed to $NH_3$, and $A_{air}$ is the value of absorbance when the sample is exposed to air.

The response time ($t_r$) was calculated as the time necessary to reach 90% of the response amplitude. The diffusion coefficient (D) was calculated as described in [12]

$$\frac{A_t - A_0}{A_k - A_0} = \frac{2}{l}\left(\frac{D_t}{\pi}\right)^{0.5} \quad (2)$$

where $A_0$ and $A_k$ are the optical absorbance at characteristic wavelength in the initial and final moments of sensing, respectively, $A_t$ is the optical absorbance at characteristic wavelength at the time t and l is the film thickness.

FTIR spectra were registered in the range from 4000 to 400 $cm^{-1}$ on a Nicolet NEXUS scanning single-beam Fourier transform IR spectrometer (CsI beam splitter, TGS–CsI detector, photometric accuracy of 0.1%, resolution of 2 $cm^{-1}$). The samples were spray-coated on pieces of Ge wafer. The spectra were recorded in transmission mode. The measurements were performed under standard conditions.

## 3. Results and Discussion

### 3.1. Spectral Changes in PANI Films During Exposition to Ammonia

Typical evolutions in time of the electronic absorption spectra of the films of PANI complexes with different polyacids measured in situ during exposition to 50 ppm $NH_3$ are presented in Figure 2. The blue arrows indicate simultaneous growth/drop of absorbance at characteristic wavelengths. One can see that PANI complexes have different electronic structures depending on the structure of polyacid. The flexibility of the polyacid chain, the distance between the sulfoacid groups on the polymer chain and the length and rigidity of the side chains containing sulfoacid groups affect the character of the synthesis and the spectral properties of the PANI complexes [??] The presence of rigid-chain polyacids (Figure 2c,d) in the complexes facilitates the electron exchange between neighboring PANI chains, which is manifested in the increased absorption in the NIR spectral range. In contrast, flexible-chain polyacids (Figure 2a,b) are able to adjust their conformation to the conformation of PANI and possibly isolate PANI chains from each other, which leads to increased absorption of localized polarons (800 nm) [37].

In the case of PANI complexes with flexible-chain PSSA and PAMPSA (Figure 2a,b), during exposition to ammonia we observe the decrease in absorption in the region of localized polarons and near 420 nm, which is related to the radical cations [37]. Simultaneously, an increase in absorption in the region of 500–650 nm corresponding to the deprotonated form of PANI [37,38] is observed. Such changes indicate the transition of PANI from the salt to the base form. Two isosbestic points confirm mutual transitions between the fragments of chemical structure of PANI. For PANI complexes with rigid-chain i-PASA and t-PASA (Figure 2c,d), one can see the noticeable absorption growth in the range of 600–750 nm.

For building the optical response transients, we have chosen the wavelength areas where response amplitudes were highest: for PANI-PSSA and PANI-PAMPSA—570 nm, for PANI-i-PASA—630 nm and for PANI-t-PASA—670 nm.

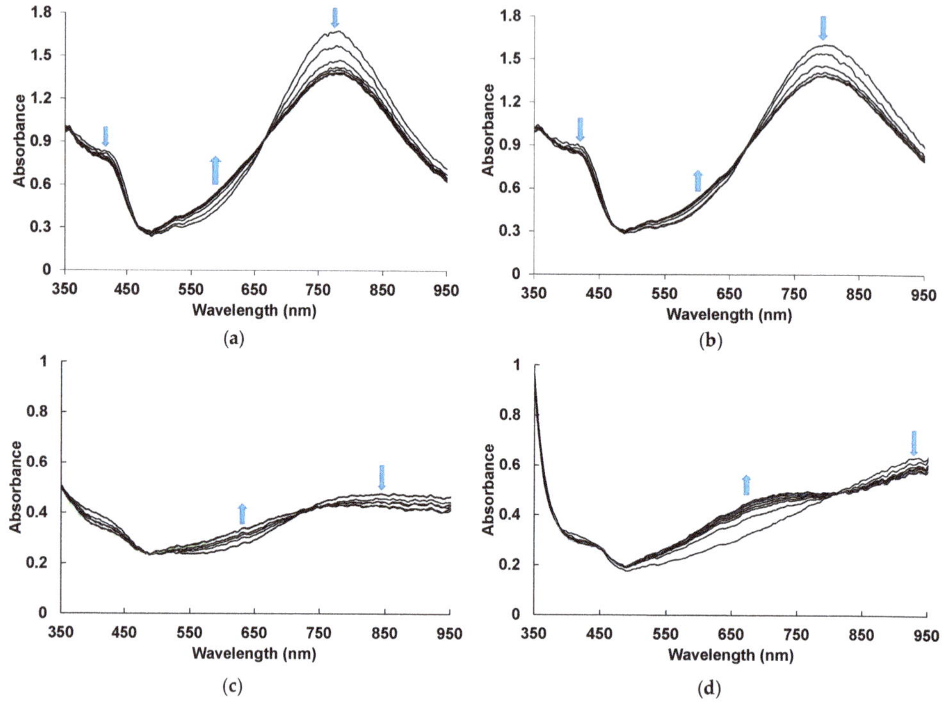


**Figure 2.** Time evolution of UV–visible–NIR electronic absorption spectra of PANI-PSSA (**a**), PANI-PAMPSA (**b**), PANI-i-PASA (**c**) and PANI-t-PASA (**d**) films during exposition to 50 ppm $NH_3$. The blue arrows indicate simultaneous growth/drop of absorbance at characteristic wavelengths.

### *3.2. Specific Features of Ammonia-Sensing Properties of PANI–Polyacids Films*

When studying the sensing properties of PANI complexes with polyacids, some specific features should be taken into account. First, hydrogen ions in the films of PANI complexes with polyacids can easily migrate through sulfonic acid centers to the film surface, where they can react with ammonia molecules. This can lead to a faster sensor response. The second important feature is possible neutralization of some ammonia molecules by excessive protons of the polyacids, which are not ionically linked with the positively charged fragments of PANI chain. This neutralization does not cause spectral changes in the PANI films, thus reducing the response amplitude. The over-stoichiometry excess of sulfoacid groups is needed to ensure reproducible chemical synthesis of PANI–polyacid complexes [34]. However, the presence of these excessive protons can reduce the sensitivity of the films.

To solve the latter problem, we have tried treatment of the films of PANI–polyacid complexes by aqueous solutions of chlorides of various metals ($Na^+$, $Ca^{2+}$, $Ba^{2+}$) and compared the influence of such treatment on the ammonia-sensing properties. The aim of using NaCl-treatment was to replace excessive protons of polyacids by cation exchange process. $Ca^{2+}$, $Ba^{2+}$ were chosen because they can form strong ionic bonds with sulfonic groups of polyacids by analogy with weakly soluble $CaSO_4$ and $BaSO_4$. In addition to the replacement of excessive protons $Ca^{2+}$, $Ba^{2+}$ decrease solubility of PANI complexes by creating bridges between sulfonic groups belonging to adjacent polyacid molecules (or different fragments of flexible polyacid chain).

Unfortunately, the treatment of all PANI complexes with NaCl results in unstable sensory properties (a large spread of response amplitudes, shown in Figure S3a,b), which may be due to the influence of humidity. In our opinion, humidity may cause the swelling of PANI films, as increasing their thickness resulted in a change in the balance of the absorption/refraction/reflection phenomena at the air/film/glass interfaces. Since the polymer film has submicron thickness, this influences optical properties of the films in the visible range of spectrum. Treatment with $Ba^{2+}$ ions gives a slightly lower response amplitude (Figure S3e,f) compared to that in the case of $Ca^{2+}$ treatment (Figure S3c,d). The reason for this may be the significantly lower solubility of Ba-$SO_3$ salt, resulting in crosslinking and neutralization of the sulfoacid groups only on the surface of the film, thus hindering further penetration of the neutralizing agent to the bulk. Therefore, further investigations in this work were performed for $CaCl_2$ treatment only.

According to the World Health Organization, workplace concentration limit of ammonia in the air at 8 h exposure is 25 ppm [39]. Figure 3 shows response transients at the exposition to ammonia concentration of 25 ppm for the PANI films untreated and treated by $Ca^{2+}$ ions. It is clearly seen in Figure 3c,d that the treatment of the PANI complexes with rigid-chain polyacids leads to a greater increase in the response amplitude than in the case of PANI complexes with flexible-chain polyacids (Figure 3a,b). At the same time, for all treated PANI complexes, the response amplitudes are comparable.

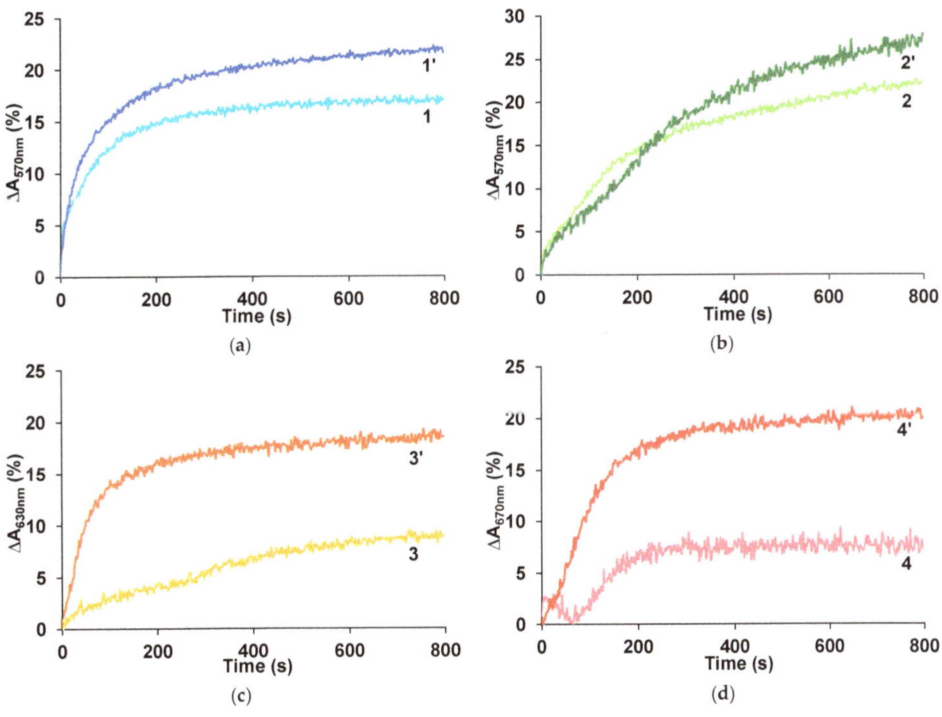


**Figure 3.** Response transients at the exposition to 25 ppm of ammonia for the films of PANI complexes with PSSA (**a**), PAMPSA (**b**), i-PASA (**c**) and t-PASA (**d**), untreated (1, 2, 3, 4) and treated with $CaCl_2$ (1′, 2′, 3′, 4′).

On the base of the time dependences of the optical absorbance at the chosen wavelengths, we have calculated the dependences of $(A_t - A_0)/(A_k - A_0)$ on square root of time (Figure S4) in accordance with the Equation (2). From the linear parts of these dependences, we have found the values of the ammonia diffusion coefficients (Table 1). The S-shaped graph indicates the deviation of ammonia diffusion in these films from Fick's second law. It may be due to structure relaxation (changes in the polymer structure and/or conformation upon deprotonation) accompanying the diffusion of ammonia in the PANI film. The rate of these processes may be comparable to the ammonia diffusion rate.

**Table 1.** Values of the sensor response amplitude (ΔA), response time ($t_r$) and diffusion coefficient (D) at 50 ppm of ammonia (smell detection limit [40]) of the spray-coated PANI–polyacid films.

| | ΔA at 50 ppm, % | $t_r$, s (50 ppm) | D, $10^{-12}$, $cm^2/s$ |
|---|---|---|---|
| PANI-PSSA | 10.3 | 171 | 2.5 |
| PANI-PSSA+$Ca^{2+}$ | 22.5 | 66 | 7.9 |
| PANI-PAMPSA | 12.4 | 378 | 3.4 |
| PANI-PAMPSA+$Ca^{2+}$ | 25.8 | 206 | 17.4 |
| PANI-i-PASA | 7.6 | 216 | 3.1 |
| PANI-i-PASA+$Ca^{2+}$ | 21.8 | 115 | 9.2 |
| PANI-t-PASA | 17.7 | 142 | 25.2 |
| PANI-t-PASA+$Ca^{2+}$ | 40.3 | 104 | 55.8 |

From Table 1, it is seen that after the treatment by $CaCl_2$, the diffusion coefficients increase for all PANI films: PANI-PAMPSA~5 times; PANI-PSSA, PANI-i-PASA~3 times; PANI-t-PASA~2 times. The response times of all films decrease by about two times. This value does not correlate with the change in diffusion coefficients due to different morphology of PANI complexes (see below). The treatment positively affected the response amplitudes: PANI-PAMPSA-growth~2 times; PANI-PSSA~2.2 times; PANI-t-PASA~2.3 times; PANI-i-PASA~2.8 times. Importantly, PANI-t-PASA exhibits the highest response amplitude, which directly influences the accuracy of ammonia detection.

FTIR spectroscopy was used to reveal possible interactions between the cations ($Na^+$, $Ca^{2+}$) and sulfoacid groups of polyacids in PANI complexes. Since PANI-PAMPSA is a composite of polyaniline and sulfonic acid, the FTIR spectrum contains bands that describe the vibrations of PANI (1608, 804 $cm^{-1}$) and bands that describe the vibrations of exclusively PAMPSA (1652, 1040, 720, 625 $cm^{-1}$), as well as combined bands that include both PANI and PAMPSA (3301, 3260, 3065, 2997, 2986, 2941, 2919, 2850, 1555, 1459, 1393, 1372, 1301, 1215, 1181, 1155 $cm^{-1}$ (Figure S5)). As we see, most bands are combined.

Since the treatment of PANI-PAMPSA with NaCl and $CaCl_2$ leads to deprotonation of the acid to form sulfonic acid salts, the most significant changes in the FTIR spectra are to be expected for the bands describing the sulfoacid fragment: asymmetric stretching $\nu_{as}$(O=S=O), symmetric stretching $\nu_s$(O=S=O) and stretching ν(S-O).

To determine the spectroscopic criteria of PAMPSA binding in the composition of PANI-PAMPSA, a preliminary experiment was carried out for PAMPSA+$Na^+$ and PAMPSA+$Ca^{2+}$ films (Figure S5). It is shown that in the PAMPSA+$Na^+$ and PAMPSA+$Ca^{2+}$ spectra, a shift of the $SO_3^-$ anion vibration bands relative to the position of the corresponding bands in the sulfonic acid spectrum is recorded (Figure S5). These results allowed us to unambiguously define bands for the analysis of sulfogroup.

In the PANI-PAMPSA spectrum (Figure 4, curve 1), the sulfonic acid group is characterized by the bands: $\nu_{as}$(O=S=O) 1215 $cm^{-1}$, $\nu_s$(O=S=O) 1039 $cm^{-1}$ and ν(S-O) 626 $cm^{-1}$. It should be noted that the band at 1215 $cm^{-1}$ is combined of stretching ν(C-N•+) in the polaron lattice of PANI [41] and $\nu_{as}$(O=S=O), but the main contribution to the position and intensity of this band is made by an asymmetric stretching vibration.

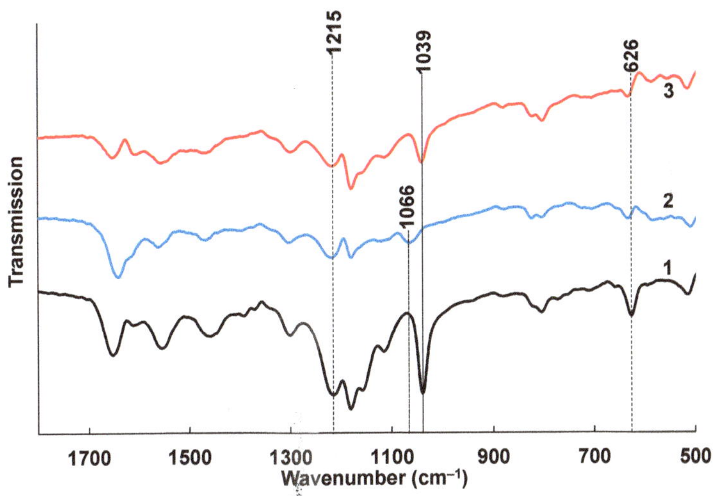


**Figure 4.** The FTIR spectra of spray-coated PANI-PAMPSA film (1), treated with $CaCl_2$ (2) and NaCl (3).

As observed in the preliminary experiment (Figure S5), the deprotonation of the acid with the formation of PAMPSA$^{-}$+$Na^+$ does not lead to significant changes in the bands characterizing ν(O=S=O) (Figure 4, curve 3). A shift to the high-frequency region at 7 $cm^{-1}$ is recorded for the ν(S-O) band, which is certainly explained by deprotonation of $O_2SOH$ group. However, the formation of a calcium salt linker chain in PAMPSA+$Ca^{2+}$ leads, in addition to a similar (as for PAMPSA+$Na^+$) shift of the ν(S-O) band at 7 $cm^{-1}$, to a significant shift of the symmetric stretching band $\nu_s$(O=S=O) at 27 $cm^{-1}$ to the high-frequency region (Figure 4, curve 2). The shift of this band can be used as a criterion for the formation of a salt with $Ca^{2+}$, leading to the formation of a chain structure rather than simple deprotonation (as in the case of $Na^+$).

Thus, the treatment with $CaCl_2$ leads to cross-linking of the polyacid matrix by strong ionic bonds of $Ca^{2+}$ ions with two sulfogroups belonging to neighboring polyacid chains. This leads to an effective substitution of the excessive protons of the polyacids.

### 3.3. Morphology

Our studies of the morphology of spray-coated films revealed that they are more dense and uniform and have roughness lower in small scale (5 μm × 5 μm) than the electrochemically synthesized films [12]. This is due to the coating procedure, which includes drying at each step of spraying of nano-sized aqueous dispersions. AFM and SEM images of the PANI films are shown in Figures 5 and 6. One can see the presence of macro-roughness with abrupt height peculiarities on the scale of 40 μm × 40 μm and the intrinsic roughness on the scale of 5 μm × 5 μm. PANI−PSSA film has a very uniform and smooth morphology on the small scale (Figures 5a and 6a). At the same time, the macro-roughness of PANI-PSSA is the highest—150 nm, and oval droplets from 14 to 25 μm are clearly visible, with a height difference of up to 400 nm on the 40 μm × 40 μm scale (Figure 5b). PANI-PAMPSA film has fibrous morphology (Figures 5c and 6c) and consists of round, dried-up droplets, with the diameter from 15 to 20 μm; the height of the edges of droplets is 100 nm (Figure 5d). The cross-section of PANI-PAMPSA and PANI-PSSA revealed a very dense and homogeneous structure, with lower nanorelief in the case of PANI-PSSA (Figure 6d,b).

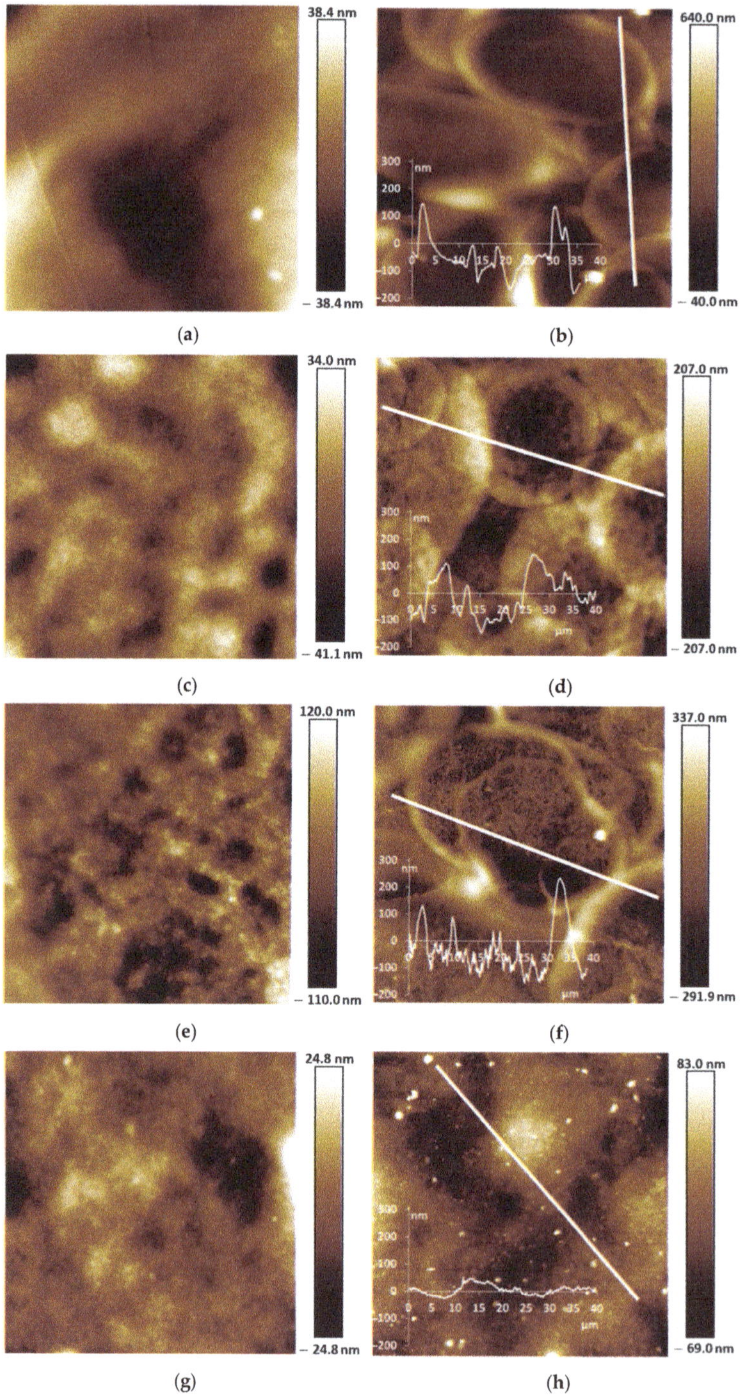


**Figure 5.** AFM images of PANI-PSSA (**a**,**b**), PANI-PAMPSA (**c**,**d**), PANI-i-PASA (**e**,**f**) and PANI-t-PASA (**g**,**h**) films applied by the spray coating method in 5 μm × 5 μm (**a**,**c**,**e**,**g**) and 40 μm × 40 μm (**b**,**d**,**f**,**h**) scales.

For PANI-i-PASA, particles with a diameter of 50–60 nm are clearly visible (Figures 5e and 6e), dried-up droplets, with a diameter from of 20 to 30 μm; the edges of the drops are thin—4 μm; the height of the edges is up to 200 nm (Figure 5f). One can observe an even finer internal morphology of the films with barely noticeable layers along the substrate (Figure 6f).

In the case of PANI-t-PASA, the droplets are practically not pronounced and the height fluctuations do not exceed 50 nm (Figure 5h). The roughness of PANI-t-PASA films is the smallest—22 nm. PANI-t-PASA film has uniform, fibrous morphology on a small scale (Figures 5g and 6g). At the same time, on a cross-section, one can see the layered-like structure (Figure 6h). It is seen that the PANI-t-PASA layers are located in parallel to the substrate, and their number possibly corresponds to the number of spraying steps.

Such differences can relate to molecular structures of the complexes. A more pronounced relief may be associated with the higher molecular weight of PAMPSA. In PANI-PAMPSA complex, PANI macromolecules, which are rigid in the conjugated emeraldine form, are surrounded by flexible-chain PAMPSA and formed coiled chains uniformly distributed on the substrate. Such complex presumably should have a double-strand structure [22,42]. PANI-PSSA has a fine nanorelief in spite of the similar flexible backbone of PSSA. However, the phenyl containing side chains bearing sulfonic groups sterically hinder free bending of the PSSA chain. At the same time, large height fluctuations and clearly defined dried-up droplets can be caused by the greater hydrophobicity of the PSSA.

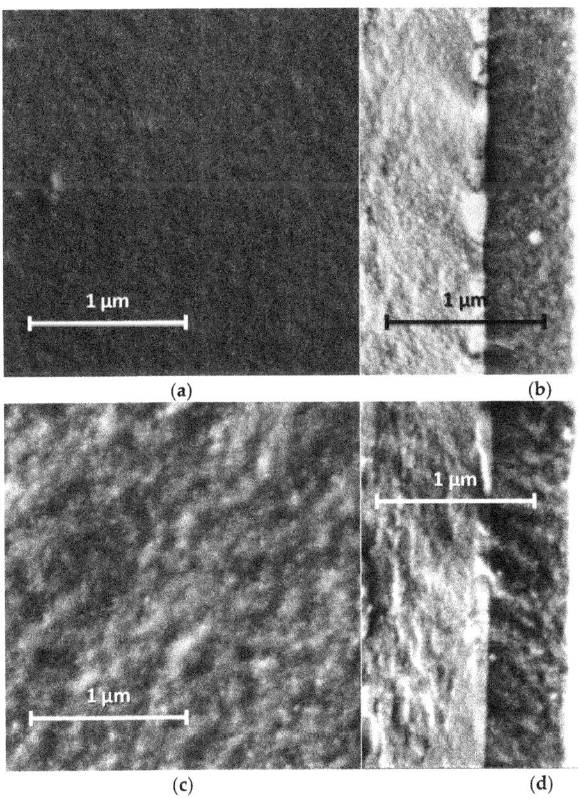


Figure 6. *Cont.*

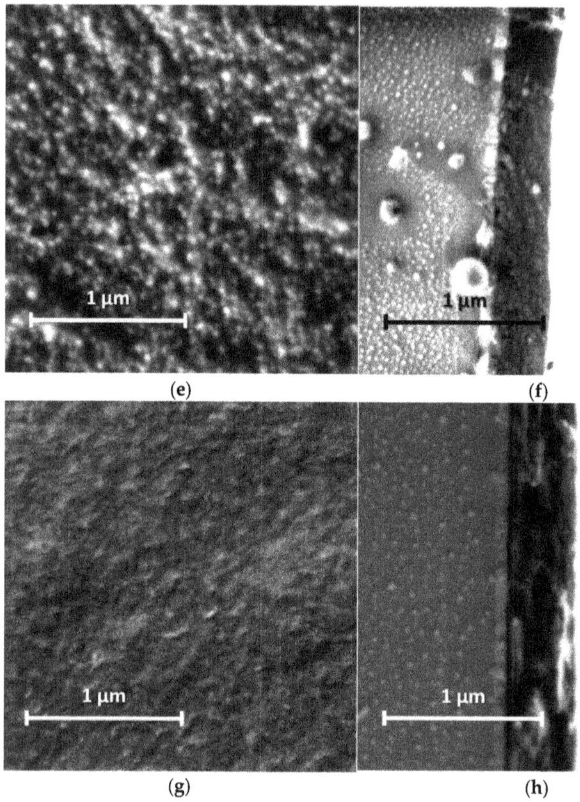


**Figure 6.** SEM images of PANI-PSSA (**a**,**b**), PANI-PAMPSA (**c**,**d**), PANI-i-PASA (**e**,**f**) and PANI-t-PASA (**g**,**h**) films applied by the spray coating method: planar (**a**,**c**,**e**,**g**) and cross-sectional (**b**,**d**,**f**,**h**) views. Cross-sections from left to right: glass, ITO, PANI.

In the case of rigid-chain t-PASA, PANI macromolecules are presumably located perpendicular to several macromolecules of the rigid-chain polyacid [22] and form less mobile rigid structures which can be arranged in parallel layers during spraying. The complex with semi-rigid PANI-i-PASA has a mixed structure [22] in which some of the fragments of PANI macromolecule units are surrounded by the polyacid chains, while others are linked by the polyacid with the neighboring ones. This may lead to the imperfect mutual packing of the macromolecules and, accordingly, to aggregation and sedimentation. Such an imperfect structure results in a slight deterioration in sensory properties.

Based on the morphology investigation and differences in the complexes' structure, we suppose that PANI-t-PASA film with more uniform, self-organized morphology and layered structure after the treatment in $CaCl_2$ can effectively detect ammonia at concentrations more than 25 ppm with high sensitivity and low response time.

### *3.4. Sensing Properties*

From the dependences of the maximum response amplitude ($\Delta A$) at the characteristic wavelengths on the concentration of ammonia in air (Figure 7), it is evident that PANI complexes with the flexible-chain polyacids (PSSA and PAMPSA) after the treatment in $CaCl_2$ can detect ammonia at very low concentrations (5 and 10 ppm) and demonstrate high response amplitude at 25 ppm (Figure 7b,d). The untreated PANI-PAMPSA film demonstrates the highest response amplitude among all untreated films (Figure 7c). PANI complexes with the rigid-chain polyacids, especially t-PASA, after the treatment determine ammonia in the air with high response amplitude in the range of concentrations above 25 ppm (Figure 7h). The saturation of ammonia-sensing response for PANI-PSSA and PANI-i-PASA appears after 50 ppm (Figure 7b,f). In the case of PANI-PAMPSA and PANI-t-PASA the growth of ΔA continues up to higher ammonia concentrations (Figure 7d,h). Importantly, standard deviation of the determination is the lowest for PANI-t-PASA film (Figure 7h).

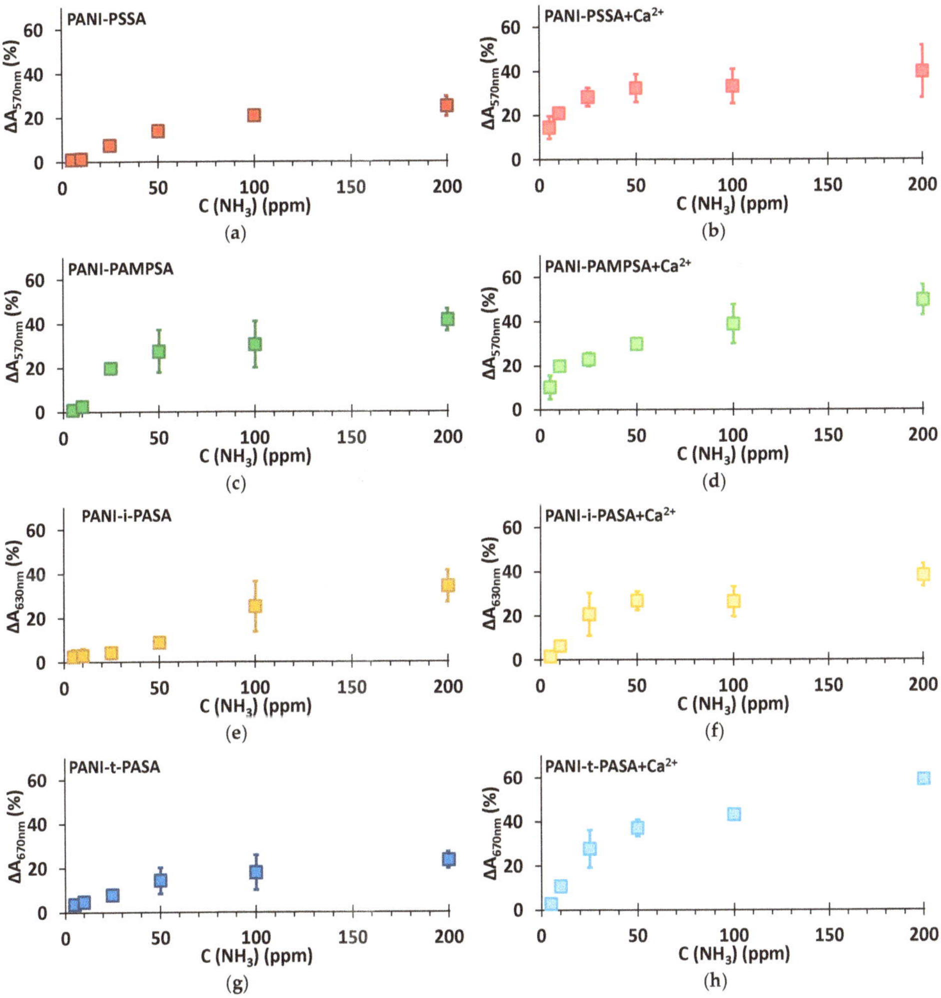


**Figure 7.** Dependence of response amplitude on the different concentrations of ammonia for the films of PANI complexes with PSSA (a), PSSA+$Ca^{2+}$ (**b**), PAMPSA (c), PAMPSA+$Ca^{2+}$ (**d**), i-PASA (**e**), i-PASA+$Ca^{2+}$ (**f**), t-PASA (**g**) and t-PASA+$Ca^{2+}$ (**h**).

It should be noted that the films of PANI complexes are reusable 2–4 times, the reversibility decreasing at high ammonia concentration. These films are therefore preferably used as alarm detectors.

If we compare the sensory properties of $Ca^{2+}$-treated spray-coated PANI–polyacid films with those of electrodeposited PANI–polyacid films [12] (Table 2), one can see the influence of method of films preparation. The differences may relate to specific features of formation of PANI films by electrodeposition and spray coating: in the first case, the polyacid content in the film is mostly determined by the electrosynthesis conditions, while in the second case the polyacid content is predetermined by the composition of the synthesis solution. At the same time, there are several general patterns. The response amplitudes are higher for all PANI films prepared by spray coating, but the response times (except in PANI-t-PASA) are longer. The response amplitude of PANI-PAMPSA at 50 ppm is higher for both methods of preparation. Diffusion coefficient of spray-coated PANI-t-PASA film is the highest [12]. Electrochemically prepared PANI-PSSA film demonstrated the worst sensing properties due to its lower doping level [12]. On the contrary, the spray-coated PANI-PSSA complex has a response amplitude similar to other PANI complexes. Just as in the case of electrochemically obtained PANI films, treatment by $CaCl_2$ leads to a significant increase in the sensitivity, especially for PANI-t-PASA film.

**Table 2.** Comparison of the sensor response amplitude (ΔA) and response time ($t_r$), at 50 ppm of ammonia of the spray-coated (this work) and the electrodeposited PANI–polyacid films [12] treated with 1 M $CaCl_2$ solution.

| | Spray Coating | | Electrodeposition | |
|---|---|---|---|---|
| | ΔA, % | $t_r$, s | ΔA, % | $t_r$, s |
| PANI-PAMPSA | 25.8 | 206 | 18.0 | 70 |
| PANI-t-PASA | 40.7 | 104 | 19.7 | 194 |

Thus, the electrodeposition method results in the sensor films with shorter response time, but it requires conducting substrates and expensive electrochemical equipment. The spray coating allows us to obtain films with higher sensitivity to ammonia in air on various substrates, particularly on flexible ones.

## 4. Conclusions

In the present work, we have first performed a comparative study of the optical ammonia-sensing properties of water-dispersible PANI complexes chemically synthesized in the presence of polysulfonic acids of different structures and chain flexibility. The sensor films were obtained by a convenient and scalable method—spray coating. The range of ammonia vapor concentrations from 5 to 200 ppm was studied. It was shown that the as-coated films of PANI complexes with the flexible-chain polyacids have the highest response amplitude in the range of ammonia concentrations 5–25 ppm, but it is subject to bigger fluctuation under the influence of humidity.

To overcome the influence of humidity on the accuracy of ammonia determination, treatments of the films in aqueous solutions of NaCl, $CaCl_2$ and $BaCl_2$ were tested. All three variants of treatment produced no influence on the electron absorption spectra of all films. These treatments were chosen to solve two problems—(1) to remove excessive protons of the polyacid, which can partly neutralize ammonia molecules penetrated into the films thus reducing the optical response amplitude and, therefore, the accuracy of determination; (2) to reduce the solubility and swelling of the spay-coated films of PANI water-dispersible complexes, which induce fluctuations in their optical absorbance. It was shown that the treatment of all of the films of PANI complexes in 1 M NaCl solution effectively neutralizes the excessive protons but does not solve the problem of swelling. The treatment in 1 M $CaCl_2$ solution for all of the PANI complexes results in a significant improvement in the response time, amplitude and reproducibility. The treatment in 1 M $BaCl_2$ solution also improved the ammonia-sensing properties of all the films, but the response amplitudes in this case were lower in general than those after the $CaCl_2$ treatment. FTIR spectroscopy showed that $CaCl_2$ treatment results in cross-linking of sulfoacid groups from adjacent polyacid chains by $Ca^{2+}$ ions. Thus, such treatment results both in the neutralization of excessive protons and significant reduction in the films' swelling.

AFM and SEM studies of the films revealed no changes in their morphology after all of the three treatments. From the analysis of AFM cross-sectional profiles of the films and their ammonia-sensing properties, it was concluded that macro-relief (on the scale of $40 \times 40$ mm) induced by spray-coating procedure has a smaller influence on the ammonia-sensing properties than the intrinsic nano-relief predetermined by the peculiarities of molecular structure of the PANI complexes.

Among the films of the PANI complexes treated with $CaCl_2$ the best sensory properties demonstrated PANI-t-PASA film at ammonia concentrations more than 50 ppm. The films of PANI complexes give reproducible results at reusing 2–4 times, the reversibility decreasing at high ammonia concentration. So, at this stage of the investigations, these films are preferably to be used as alarm detectors.

**Supplementary Materials:** The following supporting information "Supplementary materials. Optical ammonia sensors based on spray-coated polyaniline complexes with polysulfonic acids" can be downloaded at https://www.mdpi.com/article/10.3390/s25113348/s1.

**Author Contributions:** Conceptualization: O.L.G. and A.A.N.; experimental investigation: O.L.G., M.A.T., L.I.D., V.A.K. and E.I.R.; data analysis and interpretation: O.L.G., L.I.D. and A.A.N.; writing—original draft preparation: O.L.G.; writing—review and editing: A.A.N.; supervision: A.A.N.; project administration: A.A.N.; funding acquisition: A.A.N. All authors have read and agreed to the published version of the manuscript.

**Funding:** This study was supported by the Ministry of Science and Higher Education of Russia (theme No. 125012200626-9 for the Frumkin Institute of Physical Chemistry and Electrochemistry of the Russian Academy of Sciences).

**Institutional Review Board Statement:** Not applicable.

**Informed Consent Statement:** Not applicable.

**Data Availability Statement:** Data are contained within the article and Supplementary Materials.

**Acknowledgments:** The UV–Vis–NIR spectroscopy and AFM measurements were performed using the equipment of the CKP FMI IPCE RAS. The SEM measurements were performed using the equipment of the JRC PMR IGIC RAS.

**Conflicts of Interest:** The authors declare no conflicts of interest. The funders had no role in the design of the study; in the collection, analyses, or interpretation of the data; in the writing of the manuscript; or in the decision to publish the results.

## References

1. Bai, H.; Shi, G. Gas Sensors Based on Conducting Polymers. *Sensors* **2007**, *7*, 267–307. [CrossRef]
2. Wong, Y.C.; Ang, B.C.; Haseeb, A.S.M.A.; Baharuddin, A.A.; Wong, Y.H. Review—Conducting Polymers as Chemiresistive Gas Sensing Materials: A Review. *J. Electrochem. Soc.* **2020**, *167*, 37503. [CrossRef]
3. Lange, U.; Roznyatovskaya, N.V.; Mirsky, V.M. Conducting Polymers in Chemical Sensors and Arrays. *Anal. Chim. Acta* **2008**, *614*, 1–26. [CrossRef]
4. Mishra, P.K.; Sharma, H.K.; Gupta, R.; Manglik, M.; Brajpuriya, R. A Critical Review on Recent Progress on Nanostructured Polyaniline (PANI) Based Sensors for Various Toxic Gases: Challenges, Applications, and Future Prospects. *Microchem. J.* **2025**, *208*, 112369. [CrossRef]
5. Yumnam, M.; Hatiboruah, D.; Mishra, R.; Sathyaseelan, K.; Nath, P.; Mishra, P. A Smartphone-Based Optical Sensor with Polyaniline Label for Quantitative Determination of Freshness of Freshwater Fish Fillets. *Sens. Actuators A Phys.* **2023**, *361*, 114557. [CrossRef]
6. Brannelly, N.T.; Killard, A.J. An Electrochemical Sensor Device for Measuring Blood Ammonia at the Point of Care. *Talanta* **2017**, *167*, 296–301. [CrossRef] [PubMed]

7. Kim, S.; Lee, Y.-I.; Choi, Y.-M.; Lim, H.-R.; Lim, J.-H.; Myung, N.V.; Choa, Y.-H. Thermochemical Hydrogen Sensor Based on Chalcogenide Nanowire Arrays. *Nanotechnology* **2015**, *26*, 145503. [CrossRef]
8. Huang, H.; Pan, Z.; Wang, J.; Wang, T.; Yang, W.; Yu, H.; Li, F.; Dong, X.; Yang, Y. Ag Nanoparticles Sensitized ZnO/$MoS_2$ Composites to Detect Ppb-Level $NO_2$ and Automobile Exhaust Gas. *Sens. Actuators B Chem.* **2025**, *433*, 137569. [CrossRef]
9. Yang, W.; Huo, Y.; Wang, T.; Liu, X.; Li, D.; Yu, H.; Dong, X.; Yang, Y. RGO@$In_2O_3$ Based Flexible Gas Sensor: Efficient Monitoring of Trace $NO_2$ Gas at Room Temperature. *Sens. Actuators B Chem.* **2025**, *430*, 137359. [CrossRef]
10. Qiu, L.; Huo, Y.; Pan, Z.; Wang, T.; Yu, H.; Liu, X.; Tong, X.; Yang, Y. Resister-Type Sensors Based on $Ti_3C_2Tx$ MXene Decorated $In_2O_3$ p-n Heterojunction for Ppb-Level $NO_2$ Detection at Room Temperature. *J. Environ. Chem. Eng.* **2025**, *13*, 115249. [CrossRef]
11. Ayad, M.M.; Torad, N.L. Quartz Crystal Microbalance Sensor for Detection of Aliphatic Amines Vapours. *Sens. Actuators B Chem.* **2010**, *147*, 481–487. [CrossRef]
12. Gribkova, O.; Kabanova, V.; Tverskoy, V.; Nekrasov, A. Comparison of Optical Ammonia-Sensing Properties of Conducting Polymer Complexes with Polysulfonic Acids. *Chemosensors* **2021**, *9*, 206. [CrossRef]
13. Yang, L.; Huang, H.; Zhou, Q.; Kan, Z.; Qi, Y.; Dong, B.; Song, H.; Ren, L.; Xu, L.; Yu, S.; et al. A Multimodal Sensing Network Based on Synergistically Sensitized Polyaniline Composites Strategy for Safety Monitoring in Pesticide Spraying. *Sens. Actuators B Chem.* **2024**, *420*, 136486. [CrossRef]
14. Li, D.Y.; Liu, L.X.; Wang, Q.W.; Zhang, H.B.; Chen, W.; Yin, G.; Yu, Z.Z. Functional Polyaniline/MXene/Cotton Fabrics with Acid/Alkali-Responsive and Tunable Electromagnetic Interference Shielding Performances. *ACS Appl. Mater. Interfaces* **2022**, *14*, 12703–12712. [CrossRef]
15. Kebiche, H.; Debarnot, D.; Merzouki, A.; Poncin-Epaillard, F.; Haddaoui, N. Relationship between Ammonia Sensing Properties of Polyaniline Nanostructures and Their Deposition and Synthesis Methods. *Anal. Chim. Acta* **2012**, *737*, 64–71. [CrossRef]
16. Peřinka, N.; Držková, M.; Randjelović, D.V.; Bondavalli, P.; Hajná, M.; Bober, P.; Syrový, T.; Bonnassieaux, Y.; Stejskal, J. Characterization of Polyaniline-Based Ammonia Gas Sensors Prepared by Means of Spray Coating and Ink-Jet Printing. *Sens. Lett.* **2014**, *12*, 1620–1627. [CrossRef]
17. Ismail, A.H.; Mohd Yahya, N.A.; Yaacob, M.H.; Mahdi, M.A.; Sulaiman, Y. Optical Ammonia Gas Sensor of Poly(3,4-Polyethylenedioxythiophene), Polyaniline and Polypyrrole: A Comparative Study. *Synth. Met.* **2020**, *260*, 116294. [CrossRef]
18. Tsizh, B.; Aksimentyeva, O. Ways to Improve the Parameters of Optical Gas Sensors of Ammonia Based on Polyaniline. *Sens. Actuators A Phys.* **2020**, *315*, 112273. [CrossRef]
19. Ismail, A.H.; Mohd Yahya, N.A.; Mahdi, M.A.; Yaacob, M.H.; Sulaiman, Y. Gasochromic Response of Optical Sensing Platform Integrated with Polyaniline and Poly(3,4-Ethylenedioxythiophene) Exposed to NH3 Gas. *Polymer* **2020**, *192*, 122313. [CrossRef]
20. Mikhaylov, S.; Ogurtsov, N.; Noskov, Y.; Redon, N.; Coddeville, P.; Wojkiewicz, J.-L.; Pud, A. Ammonia/Amine Electronic Gas Sensors Based on Hybrid Polyaniline–$TiO_2$ Nanocomposites. The Effects of Titania and the Surface Active Doping Acid. *RSC Adv.* **2015**, *5*, 20218–20226. [CrossRef]
21. Li, L.; Ferng, L.; Wei, Y.; Yang, C.; Ji, H.-F. Effects of Acidity on the Size of Polyaniline-Poly(Sodium 4-Styrenesulfonate) Composite Particles and the Stability of Corresponding Colloids in Water. *J. Colloid. Interface Sci.* **2012**, *381*, 11–16. [CrossRef] [PubMed]
22. Gribkova, O.L.; Nekrasov, A.A.; Trchova, M.; Ivanov, V.F.; Sazikov, V.I.; Razova, A.B.; Tverskoy, V.A.; Vannikov, A.V. Chemical Synthesis of Polyaniline in the Presence of Poly(Amidosulfonic Acids) with Different Rigidity of the Polymer Chain. *Polymer* **2011**, *52*, 2474–2484. [CrossRef]
23. Sezen-Edmonds, M.; Loo, Y.L. Beyond Doping and Charge Balancing: How Polymer Acid Templates Impact the Properties of Conducting Polymer Complexes. *J. Phys. Chem. Lett.* **2017**, *8*, 4530–4539. [CrossRef] [PubMed]
24. Kulkarni, M.V.; Apte, S.K.; Naik, S.D.; Ambekar, J.D.; Kale, B.B. Ink-Jet Printed Conducting Polyaniline Based Flexible Humidity Sensor. *Sens. Actuators B Chem.* **2013**, *178*, 140–143. [CrossRef]
25. Crowley, K.; Smyth, M.; Killard, A.; Morrin, A. Printing Polyaniline for Sensor Applications. *Chem. Pap.* **2013**, *67*, 771–780. [CrossRef]
26. Sarfraz, J.; Ihalainen, P.; Määttänen, A.; Peltonen, J.; Lindén, M. Printed Hydrogen Sulfide Gas Sensor on Paper Substrate Based on Polyaniline Composite. *Thin Solid. Films* **2013**, *534*, 621–628. [CrossRef]
27. Ragazzini, I.; Gualandi, I.; D'Altri, G.; Di Matteo, V.; Yeasmin, L.; Cassani, M.C.; Scavetta, E.; Bernardi, E.; Ballarin, B. Polyaniline/Poly (2-Acrylamido-2-Methyl-1-Propanesulfonic Acid) Modified Cellulose as Promising Material for Sensors Design. *Carbohydr. Polym.* **2023**, *316*, 121079. [CrossRef]
28. Kabanova, V.A.; Gribkova, O.L.; Pozin, S.I.; Tverskoy, V.A.; Nekrasov, A.A. Complexes of Polyaniline with Sulfonated Polysulfone. Their Structure and Sensory Properties. *Prot. Met. Phys. Chem. Surf.* **2024**, *60*, 148–157. [CrossRef]
29. Gribkova, O.L.L.; Kabanova, V.A.A.; Nekrasov, A.A.A. Electrodeposition of Thin Films of Polypyrrole-Polyelectrolyte Complexes and Their Ammonia-Sensing Properties. *J. Solid. State Electrochem.* **2020**, *24*, 3091–3103. [CrossRef]
30. Fujita, H.; Hao, M.; Takeoka, S.; Miyahara, Y.; Goda, T.; Fujie, T. Paper-Based Wearable Ammonia Gas Sensor Using Organic–Inorganic Composite PEDOT:PSS with Iron(III) Compounds. *Adv. Mater. Technol.* **2022**, *7*, 2101486. [CrossRef]

31. Jost, V.; Reinelt, M. Effect of $Ca^{2+}$ Induced Crosslinking on the Mechanical and Barrier Properties of Cast Alginate Films. *J. Appl. Polym. Sci.* **2018**, *135*, 45754. [CrossRef]
32. Mikula, K.; Skrzypczak, D.; Ligas, B.; Witek-Krowiak, A. Preparation of Hydrogel Composites Using $Ca^{2+}$ and $Cu^{2+}$ Ions as Crosslinking Agents. *SN Appl. Sci.* **2019**, *1*, 643. [CrossRef]
33. Silva, M.A.d.; Bierhalz, A.C.K.; Kieckbusch, T.G. Alginate and Pectin Composite Films Crosslinked with $Ca^{2+}$ Ions: Effect of the Plasticizer Concentration. *Carbohydr. Polym.* **2009**, *77*, 736–742. [CrossRef]
34. Gribkova, O.L.; Melnikova, E.I.; Teplonogova, M.A.; Tverskoy, V.A.; Nekrasov, A.A. Conductive Polyaniline-Based Composition for Spray Coating. *Colloids Surf. A Physicochem. Eng. Asp.* **2024**, *692*, 133903. [CrossRef]
35. Kirsh, Y.E.; Fedotov, Y.A.; Iudina, N.A.; Artemov, D.Y.; Yanul', N.A.; Nekrasova, T.N. Polyelectrolyte Properties of Sulphur-Containing Polyamides Based on Isophthalic and Terephthalic Acids in Aqueous Solutions. *Polym. Sci. USSR* **1991**, *33*, 1040–1047. [CrossRef]
36. Kosandrovich, E.G.; Shachenkova, L.N.; Soldatov, V.S. Distribution of Ammonia, Mono-, Di- and Triethylamine in the Water Solution–Gas Phase System. *Dokl. Natl. Acad. Sci. Belarus.* **2019**, *63*, 164–168. [CrossRef]
37. Nekrasov, A.A.; Ivanov, V.F.; Vannikov, A. Analysis of the Structure of Polyaniline Absorption Spectra Based on Spectroelectrochemical Data. *J. Electroanal. Chem.* **2000**, *482*, 11–17. [CrossRef]
38. Stejskal, J.; Kratochvíl, P.; Radhakrishnan, N. Polyaniline Dispersions 2. UV—Vis Absorption Spectra. *Synth. Met.* **1993**, *61*, 225–231. [CrossRef]
39. Health and Safety Executive. *EH40/2005 Workplace Exposure Limits for Use with the Control of Substances*, 4th ed.; TSO: Hobart, Australia, 2020; Volume 61.
40. Timmer, B.; Olthuis, W.; Berg, A. van den Ammonia Sensors and Their Applications—A Review. *Sens. Actuators B Chem.* **2005**, *107*, 666–677. [CrossRef]
41. Trchová, M.; Stejskal, J. Polyaniline: The Infrared Spectroscopy of Conducting Polymer Nanotubes (IUPAC Technical Report). *Pure Appl. Chem.* **2011**, *83*, 1803–1817. [CrossRef]
42. Sun, L.; Liu, H.; Clark, R.; Yang, S.C. Double-Strand Polyaniline. *Synth. Met.* **1997**, *84*, 67–68. [CrossRef]

sensors

MDPI

*Article*

# Wearable Device for Continuous and Real-Time Monitoring of Human Sweat Sodium

**Anas Mohd Noor [1,2,*], Muhammad Salman Al Farisi [3,*], Mazlee Mazalan [1,4], Nur Fatin Adini Ibrahim [1], Asnida Abdul Wahab [5], Zulkarnay Zakaria [1,2], Nurul Izni Rusli [1], Norhayati Sabani [1,4] and Asrulnizam Abd Manaf [6]**

1 Faculty of Electronic Engineering & Technology, Universiti Malaysia Perlis, Arau 02600, Malaysia; mazleemazalan@unimap.edu.my (M.M.); fatinadini@studentmail.unimap.edu.my (N.F.A.I.); zulkarnay@unimap.edu.my (Z.Z.); nurulizni@unimap.edu.my (N.I.R.); hayatisabani@unimap.edu.my (N.S.)
2 Centre of Excellence for Sports Engineering Research Center, Universiti Malaysia Perlis, Arau 02600, Malaysia
3 Department of Biomedical Information Sciences, Hiroshima City University, Hiroshima 731-3194, Japan
4 Centre of Excellence for Micro System Technology, Universiti Malaysia Perlis, Arau 02600, Malaysia
5 Department of Biomedical Engineering and Health Sciences, Faculty of Electrical Engineering, Universiti Teknologi Malaysia, Johor Bahru 81310, Malaysia; asnida.aw@utm.my
6 Collaborative Microelectronic Design Excellence Center, Universiti Sains Malaysia, Bayan Lepas 11900, Malaysia; eeasrulnizam@usm.my
* Correspondence: anasnoor@unimap.edu.my (A.M.N.); alfarisi@hiroshima-cu.ac.jp (M.S.A.F.)

**Abstract:** Wearable sweat-sensing devices hold significant potential for non-invasive, continuous health monitoring. However, challenges such as ensuring data accuracy, sensor reliability, and measurement stability persist. This study presents the development of a wearable system for the real-time monitoring of human sweat sodium levels, addressing these challenges through the integration of a novel microfluidic chip and a compact potentiostat. The microfluidic chip, fabricated using hydrophilic materials and designed with vertical channels, optimizes sweat flow, prevents backflow, and minimizes sample contamination. The developed wearable potentiostat, as a measurement device, precisely measures electrical currents across a wide dynamic range, from nanoamperes to milliamperes. Validation results demonstrated accurate sodium concentration measurements ranging from 10 mM to 200 mM, with a coefficient of variation below 4% and excellent agreement with laboratory instruments (intraclass correlation = 0.998). During physical exercise, the device measured a decrease in sweat sodium levels, from 101 mM to 67 mM over 30 min, reflecting typical physiological responses to sweating. These findings confirm the system's reliability in providing continuous, real-time sweat sodium monitoring. This work advances wearable health-monitoring technologies and lays the groundwork for applications in fitness optimization and personalized hydration strategies. Future work will explore multi-biomarker integration and broader clinical trials to further validate the system's potential.

**Keywords:** microfluidic chip; sweat sodium measurement; wearable device

## 1. Introduction

Wearable sweat-sensing devices are transforming health monitoring by offering non-invasive, real-time analysis of physiological and biochemical parameters. Sweat, as a biofluid, contains various biomarkers, including electrolytes, metabolites, and hormones, which provide critical insights into hydration status, electrolyte balance, and overall health [1,2]. Unlike traditional blood tests, which are invasive and pose risks such as skin inflammation, wearable sweat sensors enable continuous monitoring, making them

 https://doi.org/10.3390/s25113467

more convenient and user-friendly [3]. These systems have broad applications, from helping athletes optimize performance and prevent dehydration by tracking electrolyte loss during exercise to assisting individuals with chronic medical conditions, such as diabetes, in monitoring biomarkers such as sweat glucose and reducing the need for frequent blood draws. Physicians also benefit from these devices, as the continuous data stream supports more informed diagnoses and treatment plans [4].

Wearable sweat-sensing systems typically consist of three main components: (1) a sweat collection module, (2) sensors for biomarker detection, and (3) an electrochemical interface, such as a potentiostat, which converts chemical signals into measurable electrical outputs for real-time monitoring [5,6]. However, effective sweat collection remains a challenge, as sweat secretion is limited to microliter volumes. Identifying optimal body locations for efficient sweat collection is critical to device performance. Sweat-mapping studies have revealed that the forehead produces the highest sweat rates, followed by the anterior and posterior torso, hands, and legs [7,8]. These findings guide the design of sweat collection devices capable of capturing limited sweat volumes effectively [9]. However, challenges such as power efficiency, data security, material stability, and mass production hinder widespread adoption. Future advancements aim to integrate artificial intelligence, enhance multifunctionality, and improve interoperability, paving the way for personalized, remote healthcare solutions [10,11].

Microfluidic chips or devices have emerged as promising solutions for sweat collection and analysis. These chips offer significant advantages, such as reducing evaporation, minimizing interference from skin debris, and enabling real-time biomarker analysis through integrated channels [12–15]. Despite these benefits, existing designs face limitations in controlling sweat flow rates. High flow rates can lead to rapid discharge of sweat before sensors can capture the data, while low flow rates may cause the mixing of old and new samples, compromising data accuracy [16–18]. Material properties also pose challenges. Many microfluidic chips are made from hydrophobic materials, which hinder efficient fluid flow. Although surface treatments can improve hydrophilicity, these modifications often degrade over time, reverting to hydrophobicity due to environmental exposure [19,20]. Consequently, robust hydrophilic materials that maintain consistent fluid flow over extended periods are needed to enhance device performance [21].

Microfluidic methods utilizing passive pumps, such as gravity and capillary forces, offer cost-effective and energy-efficient fluid manipulation. These methods rely on natural forces to move fluids through microchannels without requiring external power sources. Gravity-driven flow uses the force of gravity to drive fluid, while capillary forces exploit the fluid's tendency to move through narrow channels. These passive approaches are ideal for wearable devices, enabling continuous and real-time monitoring, such as sweat collection [22]. In contrast, active pumping methods use external pumps to control fluid flow, offering more precise flow regulation. While these systems are versatile and allow for accurate flow control, they require additional components, increasing system complexity and power consumption. For sweat collection, passive methods are advantageous because they simplify design and reduce energy requirements [23,24]. However, both passive and active methods face challenges. Variations in sweat rate, skin conditions, and environmental factors can impact accuracy. Active systems offer better flow control but are more complex and power-intensive, while passive systems may face limitations in sensitivity and consistent flow. The choice depends on balancing efficiency, complexity, and specific application needs.

Beyond fluid handling, wearable sweat-sensing devices must achieve high accuracy and selectivity. High accuracy is essential for detecting specific analytes, such as sodium, particularly at low concentrations [25–27]. Selectivity is equally important to differentiate target ions such as sodium from interfering ions such as potassium and calcium, ensuring reliable data in the complex composition of sweat [23,28]. Additionally, maintaining stability and consistency in electrical signal measurements over prolonged use remains a significant challenge, as external factors such as temperature changes and user movement can introduce variability [29,30].

This study addresses current limitations in wearable sweat sensing by developing a novel platform for real-time sodium monitoring with three key advancements: (1) a 3D-printed hydrophilic microfluidic chip featuring vertical inlets to ensure continuous sweat flow, prevent backflow, and minimize contamination—overcoming the challenges of flow control and material hydrophobicity in existing designs; (2) integration with our previously developed We-VoltamoStat [31], a compact, Bluetooth-enabled potentiostat that achieves lab-grade accuracy in a wearable form factor; and (3) a complete system enabling real-time wireless monitoring during physical exercise, demonstrated by tracking physiological sodium dynamics. By combining these innovations in fluid handling, sensor accuracy, and practical usability, the platform significantly advances wearable health-monitoring technologies for both fitness and clinical applications.

## 2. Materials and Methods

### 2.1. Fabrication of Microfluidic Chip

The design was chosen to leverage gravity-driven fluid dynamics, ensuring uninterrupted sweat flow while preventing backflow. This approach minimizes the mixing of old and new sweat samples, thereby enhancing the accuracy of real-time measurements. Additionally, the vertical channel design allows for efficient use of space within the wearable device, making it compact and user-friendly. Water-washable resin is chosen to enhance the wettability properties of microfluidic chips, which are crucial for fluid flow applications, especially in microfluidic chips. In contrast to standard non-washable resin, which typically exhibits low wettability and high contact angles (hydrophobic), water-washable resin offers improved hydrophilicity.

The microfluidic chip was fabricated using a Creality LD-002H SLA 3D printer (Micro Center, Hilliard, OH, USA). The device's geometry was designed in SolidWorks 2020, sliced with Chitubox software, and printed using a clear, UV-curable, water-washable resin (E-Sun, W100). Post-processing included cleaning, curing, and assembly. The microfluidic chip, shown in Figure 1, displays the schematic drawing of the chip and an example of the printed chip. Magnets were incorporated into the design to enable easy and secure attachment or removal of the sensor. The Horiba ISE sodium sensor (B722, Laqua Twin Sensor, Horiba, Kyoto, Japan) was selected for its compact size and reliable performance, as shown in Figure 1. The sensor casing was dismantled, leaving only the sensor part, where the electrode connection is used to read the current. The microfluidic chip collects sweat droplets from the inlet and directs them to the sensing area. The sensor holder part includes a sensor holder secured by magnets, a lid hinge for the sensor, holes for wires, and a band for holding the wearable microfluidic chip.

Eight inlets were incorporated into the microfluidic chip to maximize sweat collection within a short time. The inlets were designed with a channel size width of 1.8 mm to enhance capillary force and maintain a good flow while reducing contamination. The microchannel was designed for vertical fluid flow driven by gravity. To control the flow rate of sweat reaching the outlet and ensure the adequate filling of the sensing area, an outlet with a width and height of less than 1 mm was designed. This small outlet channel increases the time required for sweat to flow out due to the high-pressure difference between the large sensing area and the small outlet channel [32]. Additionally, the design is based on hydrostatic principles, requiring sweat to flow from a lower position to a higher position. This creates increased pressure at the outlet, which generates a back pressure that opposes the fluid outflow, reducing and controlling the sweat flow rate.

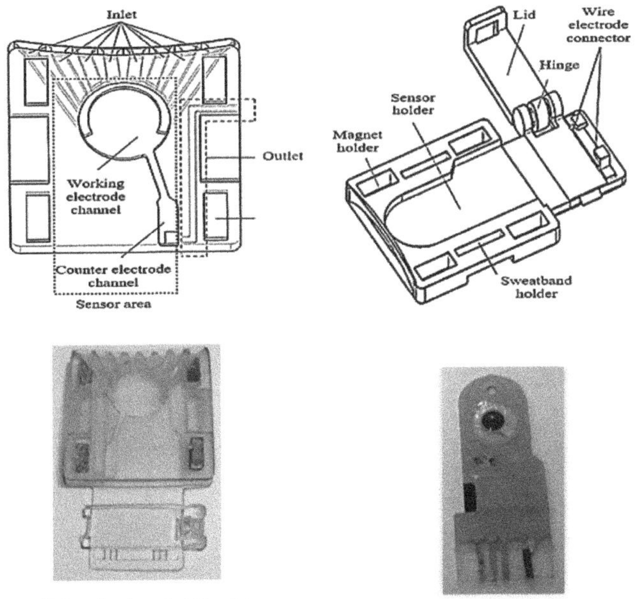


Printed microfluidic chip Sensor

**Figure 1.** Schematic drawing and 3D print of the microfluidic chip and sodium sensor.

To optimize the printing process and ensure high-quality fabrication, we carefully adjusted the printing recipe by conducting multiple trials and testing different parameter values, focusing on slicing settings. The bottom layer requires a time exposure of 12 s, while the normal layer needs only 2.5 s, optimizing the curing process for different layers. The lift distance for both bottom and normal layers is set to 5 mm, maintaining uniform detachment from the resin vat. The speed during operation is also consistent, with both the bottom and normal layers having a lift speed of 50 mm/min. The retraction speed is adjusted for efficiency, with the bottom layer retracting at 150 mm/min and the normal layer at 190 mm/min. Optimizing these parameters was crucial for achieving precise dimensions, smooth surfaces, and consistent device performance. By fine-tuning the slicing settings, we were able to minimize defects and improve the overall quality of the fabricated microfluidic chip. Table 1 shows the 3D printer settings used to achieve the chip dimensions.

**Table 1.** The 3D printer settings used to achieve the chip dimensions.

| Setting Parameter | Bottom Layer | Normal Layer | Bottom Retract Speed | Normal Retract Speed |
|---|---|---|---|---|
| Time Exposure (s) | 12 | 2.5 | - | - |
| Lift Distance (mm) | 5 | 5 | - | - |
| Speed (mm/min) | 50 | 50 | 150 | 190 |

## 2.2. *Microfluidic Chip Flow Test*

The performance of the printed microfluidic chip was assessed through two primary tests: a hydrophilicity test and an efficiency test for fluid delivery to the sensor channel. The hydrophilicity test focused on evaluating the wettability of the microfluidic chip, fabricated with a water-washable resin, by measuring the contact angle of its surface. A contact angle of less than 90°, ideally below 60°, is considered optimal for promoting efficient fluid flow [30]. Meanwhile, the efficiency test examined the device's ability to deliver sweat to the sensor channel. Several parameters were evaluated, including the volume of the channel, the time taken for sweat to flow out of the outlet, and the time required to completely fill the sensing area. The channel volumes were defined using SolidWorks and remained constant for the inlets and the sensing area, while the outlet channel sizes varied. The results indicated that the volume of the outlet channel significantly affected the time required for sweat to exit. To ensure the complete filling of the sensing area, the time for the fluid to exit the outlet was measured while monitoring the fluid level within the sensing area, as shown in Figure 2.

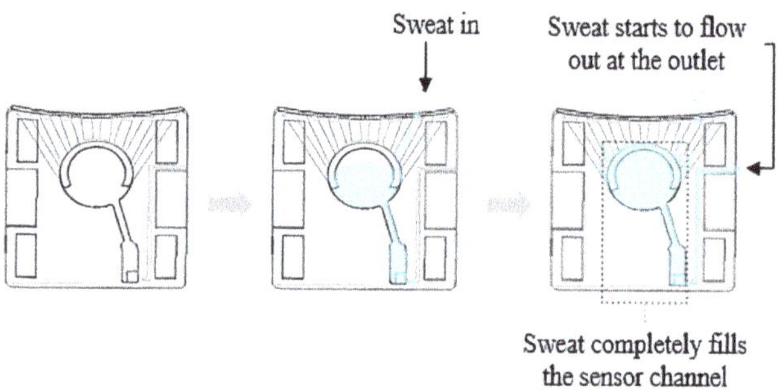


**Figure 2.** Illustration of sweat flow through the microfluidic chip. The sequential images show sweat entering the sensing channel, filling the sensing area, and flowing out through the outlet, demonstrating the device's efficiency in fluid transport.

*2.3. Performance Comparison of Developed We-Voltamostat Device with Standard Instrument*

The current measurement capabilities of the We-VoltamoStat were evaluated through a calibration test using a photodiode and a current source (Keithley 6221, Keithley Instruments, Inc., Cleveland, OH, USA). The photodiode was illuminated by a light source, and the current source generated a range of currents, as shown in Figure 3. Both the We-VoltamoStat and the Agilent multimeter (Agilent 34401) measured the generated currents. By comparing the readings from both instruments across the range of currents, the accuracy and precision of the We-VoltamoStat were assessed. The photodiode provided an independent and reliable reference, ensuring the accuracy of the calibration process.

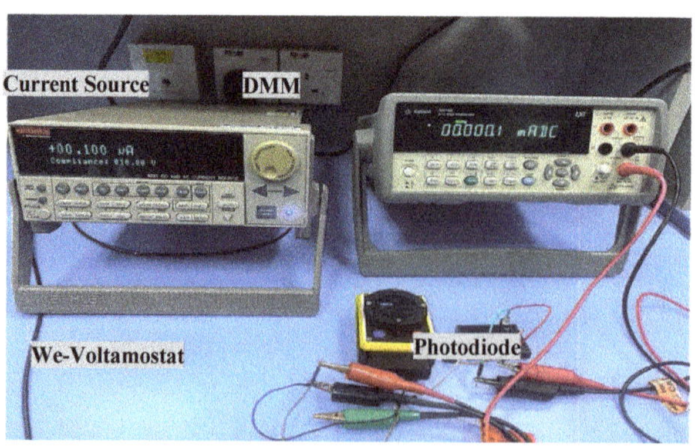


**Figure 3.** Setup for calibrating the We-VoltamoStat using a Keithley current source. The calibration ensures a reliable measurement by comparing the device's measurements to a standard instrument across a range of currents.

### 2.4. Dummy Cell Test

A dummy cell was used to test the electronic system of the We-VoltamoStat and verify its ability to accurately detect and resolve potentiostat issues. The dummy cell creates a stable electrochemical environment, essential for evaluating the device's performance [33]. Based on the approach outlined by Caux et al., a simple series circuit was used, consisting of a 1 kΩ resistor and a 1000 µF capacitor [34]. However, the specific configuration of the dummy cell can vary depending on the potentiostat setup. On the other hand, in a commercial potentiostat, such as the Metrohm, the dummy cell is connected to the working electrode (WE) through a resistor, while the counter electrode (CE) and reference electrode (RE) are connected to a capacitor, as shown in Figure 4. In contrast, the We-VoltamoStat uses a simplified two-electrode configuration, where the WE is connected to a resistor and the CE to a capacitor. While this configuration is adequate for current measurement, it is important to note that a three-electrode configuration can offer more precise results in certain applications.

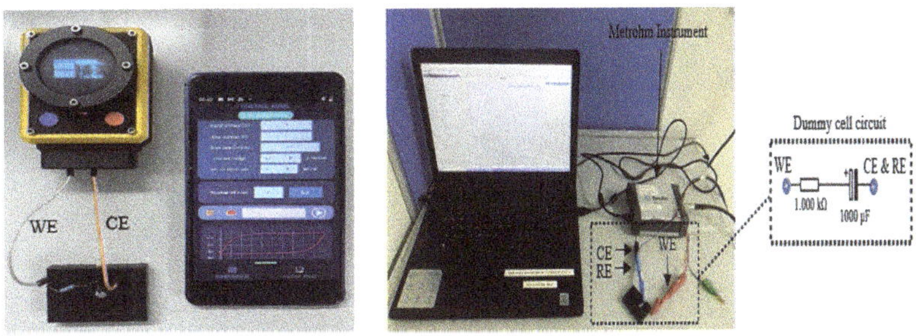


(**a**) (**b**)

**Figure 4.** Dummy cell test configurations for validating the We-VoltamoStat. (**a**) Simplified two-electrode setup with the We-VoltamoStat, demonstrating its reliability in measuring electrochemical currents. (**b**) Standard three-electrode setup using the Metrohm instrument.

### 2.5. Accuracy Test

Several tests were conducted to evaluate the accuracy, selectivity, and stability of the We-VoltamoStat and sensor in detecting sodium ions. These tests involved analyzing samples with known sodium ion concentrations as artificial sweat, ranging from 10 mM to 200 mM [35]. The artificial sweat is prepared using a standard process involving sodium chloride (NaCI), potassium chloride (KCI), and calcium chloride ($CaCI_2$), as these are the major components of sweat ions. Combinations of NaCl solution, KCl solution, and $CaCl_2$ solution were also mixed in a volume ratio of 2:1:1, respectively, aligning with the typical higher volume and dominance of NaCl compared to other interfering ions commonly found in sweat. To assess the accuracy of the We-VoltamoStat in measuring sodium ions, we conducted a series of amperometry tests using a sodium sensor. The tests involved comparing the electrical currents measured by the We-VoltamoStat to those measured by a commercial potentiostat, the Metrohm instrument (µStat-i 400s) (Metrohm, Herisau, Switzerland). For each sodium concentration in artificial sweat, ranging from 10 mM to 200 mM, we performed amperometry measurements for 300 s. The resulting current data were analyzed to evaluate the graphical similarities and the percentage error of average electrical currents between the two devices. To assess the statistical significance of the differences between the two devices, we conducted a Bland–Altman plot analysis, a paired t-test, and calculated the intraclass correlation coefficient (ICC) and coefficient of variation (CV). The Bland–Altman plot illustrates the agreement between the devices, identifying any systematic bias or outliers. The paired t-test determined if there are statistically significant differences in the mean currents measured by the two devices. The ICC and CV were used to evaluate the reliability and consistency of the measurements.

### 2.6. Selectivity and Stability Test

The selectivity for sodium ions over potassium and calcium was ensured using ion-selective electrodes (ISEs) optimized for sodium. These electrodes utilize a membrane composition specifically designed to preferentially interact with sodium ions, thereby minimizing interference from other ions commonly found in sweat. In the selectivity tests, solutions containing varying concentrations of sodium, potassium, and calcium were prepared to simulate real sweat samples. The electrode's response to each ion was measured and compared against a reference curve generated for sodium. The statistical analyses, including graphical representation and two-way ANOVA with Tukey's HSD test, demonstrated significant differences between the sodium responses and those of interference ions. The results indicated a mean difference exceeding 100 nA for potassium and calcium, which confirms the high specificity of the sensor for sodium ions. This differentiation ensures accurate sodium measurement even in the presence of interfering ions. The tests were repeated with a mixture of all three ions to evaluate real-world scenarios. The consistent performance of the sensor under these conditions validates its robustness and applicability for the continuous monitoring of sweat sodium levels in practical applications.

To assess the stability and reproducibility of the amperometry measurements, we conducted multiple repetitions for each sweat sample concentration and each 10 min exercise session. Five measurements were performed for each condition, allowing us to calculate the standard deviation of the measured currents. A standard deviation (SD) value less than 10 was considered indicative of high stability in the measurements [36]. By repeating the experiments, we were able to evaluate the consistency of the device's performance over time and under varying conditions.

### 2.7. Sweat Sodium Measurement During Physical Exercise

Human volunteer testing was conducted to evaluate the wearable device's performance in real-world conditions, focusing on continuous flow and real-time measurement of sweat sodium levels. This study, which involved a single volunteer, was approved by the institutional ethics procedure. The microfluidic chip was placed on the subject's forehead, a region known for its high sweat production, as shown in the experimental setup in Figure 5. This setup allowed for the wireless monitoring of sweat sodium levels, with real-time data transmitted to a smartphone app for convenient analysis. The volunteer followed a standardized exercise protocol, running on a treadmill at a speed of 15 km/h for 10 min, followed by a 3 min rest period. Sweat was collected at 10 min intervals using the microfluidic chip over a total duration of 30 min. Amperometry measurements were conducted at 0 V on the collected sweat samples. The testing protocol was repeated ten times over 10 consecutive days. Sweat sodium concentrations were estimated using the determined reference curve, providing reliable and repeatable measurements. The conversion from measured electrical current (I) to sodium ion concentration in an amperometric sensor is performed using the general empirical calibration formula, as shown in Equation (1). This curve represents a mathematical relationship derived from experimental measurements of current at known sodium concentrations.

$$[\mathrm{Na}^{+}] = \mathrm{f(I)} = \mathrm{aI}^{\mathrm{n}} + \mathrm{bI}^{\mathrm{n}-1} + \ldots + \mathrm{c} \quad (1)$$

where

- $[Na^+]$ is the sodium ion concentration (typically expressed in mmol/L);

- I is the measured current (in nA or μA);
- a, b, c are empirically determined calibration coefficients;
- n is the degree of the polynomial fit (commonly 1 for a linear model, or 2–3 for nonlinear models).

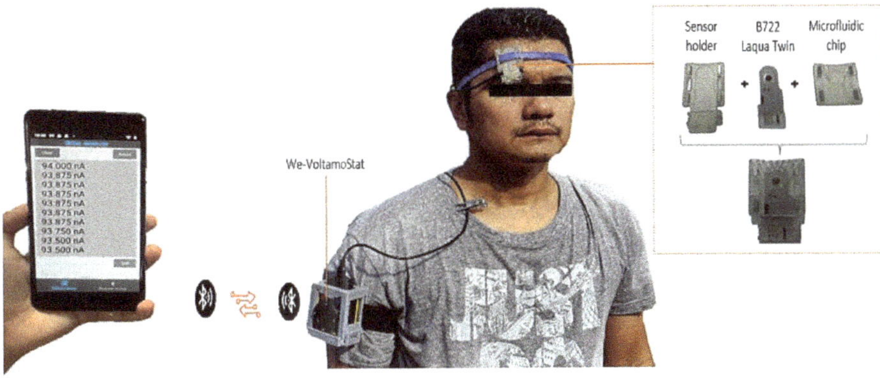


**Figure 5.** Wearable microfluidic sweat-sensing system during real-time monitoring. The microfluidic chip is worn on the forehead, with data transmitted wirelessly to a smartphone app for continuous sodium analysis.

## 3. Results and Discussion

### *3.1. Flow Test*

Figure 6 shows the surface of the printed microfluidic chip, which was fabricated using a commercial water-washable resin (eSUN, Shenzhen, China). To evaluate the hydrophilicity of the surface, a water droplet was placed on the chip, and its contact angle was measured. The chip demonstrated excellent hydrophilicity with a contact angle of less than 45°. In comparison, a chip made with commercial conventional photopolymer resin (eSUN, W100, China) exhibited a contact angle greater than 90°, indicating a hydrophobic surface. The superior hydrophilic properties of the water-washable resin promote better wetting and smoother fluid flow within the microfluidic channels, significantly enhancing the device's performance. Figure 7 shows the microfluidic chip fabricated in early 2023. A hydrophilicity test conducted more than two years later, using the same fabrication materials, shows that the contact angle remains below 45°. The exact angle measured is approximately 25°, determined using ImageJ 1.54f. This confirms that the device maintains its hydrophilic properties over time, demonstrating stable and long-lasting wettability.

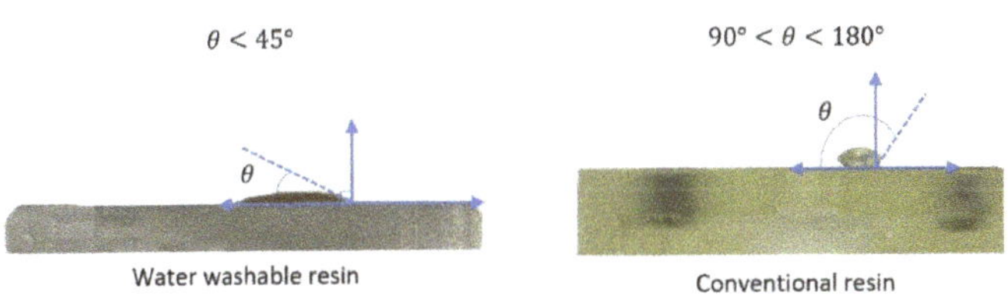


**Figure 6.** Contact angle measurements of water on microfluidic chip surfaces fabricated with water-washable resin (hydrophilic) and conventional resin (hydrophobic). The results demonstrate the superior hydrophilic properties of the water-washable resin, which enhances sweat flow within the microchannels.

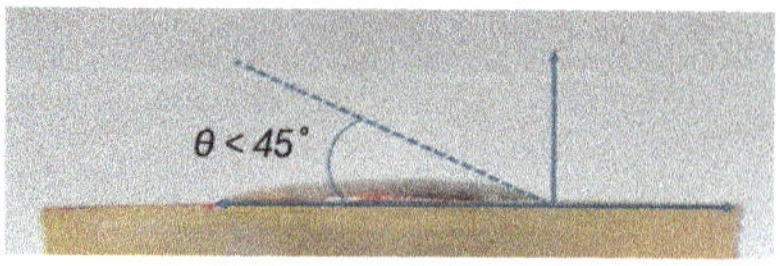


**Figure 7.** Wettability test of a microfluidic chip after more than two years, showing a contact angle below 45°, indicating that it remains hydrophilic.

### 3.2. Calibration Test

The We-VoltamoStat exhibited strong comparability to the Metrohm instrument in measuring electrical currents across varying voltage ranges. Both devices produced nearly identical current-voltage curves in cyclic voltammetry mode at a scan rate of 25 mV/s, as shown in Figure 8. This close alignment underscores the high accuracy and reliability of the We-VoltamoStat for electrochemical analysis. The resulting current-potential curve displayed a characteristic decay loop, which aligns with patterns reported in earlier studies [37,38]. This loop shape arises from the exponential decay of current spikes within the dummy cell, a well-documented phenomenon of resistor–capacitor (RC) circuits. A detailed comparative analysis of this wearable device against the bench-top instrument will be presented in the next sections.

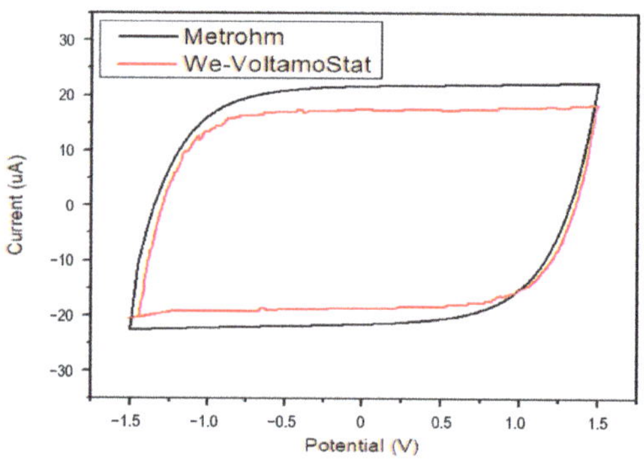


**Figure 8.** The current-voltage curves generated by the We-VoltamoStat and the Metrohm instrument using a scan rate of 25 mV/s in cyclic voltammetry mode. The nearly identical curves demonstrate the correlation of the We-VoltamoStat with electrochemical measurements.

### 3.3. Validation of the We-Voltamostat, Including Accuracy, Selectivity, and Stability

#### 3.3.1. Accuracy

The We-VoltamoStat consistently measured slightly lower current readings compared to the Metrohm instrument, as illustrated in Figure 9. This discrepancy is primarily attributed to the high current offset setting in the We-VoltamoStat, which can introduce noise and obscure certain current data points, resulting in a slight underestimation of the measured values. Despite this limitation, the We-VoltamoStat demonstrated remarkable accuracy in amperometry measurements, producing current-voltage graphs that closely aligned with those of the Metrohm instrument, as previously shown in Figure 9. The calculated percentage errors for average currents across various sodium chloride concentrations were minimal: 13.7% for 10 mM, 9.9% for 50 mM, 4.2% for 100 mM, 7.8% for 150 mM, and 8.8% for 200 mM. These findings highlight the We-VoltamoStat's exceptional precision and reliability in current measurements, even when accounting for the influence of the current offset.

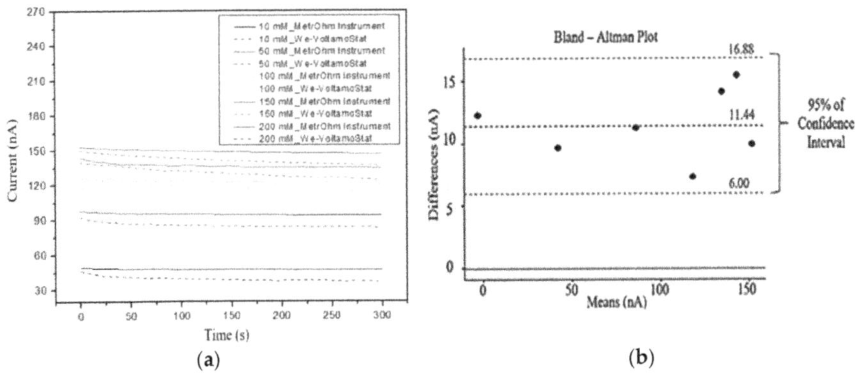


**Figure 9.** Measurement comparison between the We-VoltamoStat and the Metrohm instruments. (**a**) The graph compares the average currents recorded by the We-VoltamoStat and Metrohm instruments. (**b**) Bland–Altman plots comparing the sodium ion measurements of the across concentrations ranging from 1 mM to 250 mM (each black dot represents an individual sample).

Table 2 presents the results of the sodium analysis tests, which evaluated the accuracy of the We-VoltamoStat in measuring sodium ion concentrations ranging from 1 mM to 250 mM. The calculated percentage errors for average currents were generally low, with values ranging from 4.2% to 13.7%. While lower concentrations (10 mM and below) exhibited slightly higher percentage errors, the overall performance of the We-VoltamoStat was satisfactory. The Bland–Altman plot in Figure 9 reveals an acceptable level of agreement between the We-VoltamoStat and the reference device for all concentrations.

**Table 2.** Sensor output for NaCl concentrations ranging from 1–250 mM.

| mM | Current Measurement (nA) | | Percentage Error (%) | Mean Difference (nA) |
|---|---|---|---|---|
| | Metrohm | We-Voltamostat | | |
| | Average | | | |
| 1 | 3.252 | −9.054 | 37.8 | 12.306 |
| 10 | 46.881 | 37.203 | 14.6 | 9.678 |
| 50 | 91.825 | 80.586 | 12.2 | 11.239 |
| 100 | 121.465 | 114.126 | 6 | 7.339 |
| 150 | 141.76 | 127.624 | 9.9 | 14.136 |
| 200 | 151.221 | 135.78 | 10.2 | 15.441 |
| 250 | 156.882 | 146.95 | 6.3 | 9.932 |

The slight measurement discrepancies observed between the We-VoltamoStat and the Metrohm instruments are attributed to the high current offset setting in the We-VoltamoStat. This offset can mask subtle current data points, particularly at low concentrations, leading to an underestimation of the measured currents. While these differences were minor (mean difference of ~11.44 nA), they are within an acceptable range for wearable sensors and do not compromise practical applications. The We-VoltamoStat's strong agreement with the Metrohm instrument, as confirmed by a high intraclass correlation coefficient (ICC) value of 0.998, ensures its reliability for real-world use. For practical applications, this level of precision is sufficient for monitoring sweat sodium in dynamic conditions such as exercise, where relative changes are more critical than absolute values.

Although there are higher percentage errors in current measurements at 1 mM and 10 mM, the Bland–Altman plot (Figure 9) shows that the We-VoltamoStat agrees well with the reference device across all concentrations, from 1 mM to 250 mM. The We-VoltamoStat consistently shows a small bias, approximately 11.44 nA lower than the reference device. This bias is more noticeable at lower concentrations, where the percentage errors are larger. However, the mean difference between the two devices for low concentrations (1 mM and 10 mM) is consistent with this expected bias. The higher errors at low concentrations are due to challenges in measuring very small sensor currents, where background noise can also contribute to the accuracy. Overall, the Bland–Altman plot confirms that the We-VoltamoStat delivers reliable measurements of sodium ions over a wide concentration range, even at low current values.

The statistical analyses comparing the We-VoltamoStat with the Metrohm instrument are shown in Table 3. The small standard error of the mean (SEM) of 1.469 nA reflects minimal measurement uncertainty, underscoring the device's precision [39]. An ICC value of 0.998 indicates a high level of agreement between the two devices. The low coefficient of variation (CV) of 3.5% suggests consistent results across specific concentrations. Although the $p$-value of 0.0014 slightly exceeds the 0.001 significance threshold, it still points to no significant differences between the instruments, confirming the We-VoltamoStat's comparable performance.

**Table 3.** Validation of the We-VoltamoStat was performed using a paired sample t-test against the Metrohm instrument.

| | **Metrohm Instrument—We-VoltamoStat** |
|---|---|
| Mean difference (nA) | 11.566 |
| SD of differences (nA) | 3.285 |
| SEM | 1.469 |
| 95% CI of difference (nA) | 5.128–17.996 |
| ICC | 0.998 |
| Sign. (2-tailed) | 0.0014 |
| t | 7.873 |
| CV (%) | 3.518 |

SD, standard deviation; SEM, standard error of mean; CI, confident interval; ICC intraclass correlation coefficient; t-test statistic; CV, coefficient of variation; sign. (2-tailed); two-tailed probability, $p < 0.001$.

Figure 10 shows the relationship between the sensor current and sodium concentration during continuous sweat sample flow conditions. As the NaCl concentration decreased from 200 mM to 10 mM, the sensor current also decreased, reflecting the reduction in sweat sodium concentration. The average current readings from the We-VoltamoStat closely followed a polynomial curve, which can be used to estimate unknown sodium concentrations. The sensor's current ranged from 37 to 135 nA for sodium concentrations between 10 mM and 200 mM. The average currents recorded by the We-VoltamoStat and the Metrohm instruments over five repetitions of amperometry are shown. Both instruments showed a strong correlation between current readings and sodium concentration, with the curves closely aligning. The results demonstrated that the sensor can be effectively modeled with a polynomial curve for sodium concentrations ranging from 1 mM to 250 mM. This highlights the sensor's versatility and accuracy in measuring the sodium-related current across a broad range. Overall, these results demonstrate that the sensor is reliable, sensitive, and capable of accurately measuring sodium concentrations in sweat across a wide concentration range.

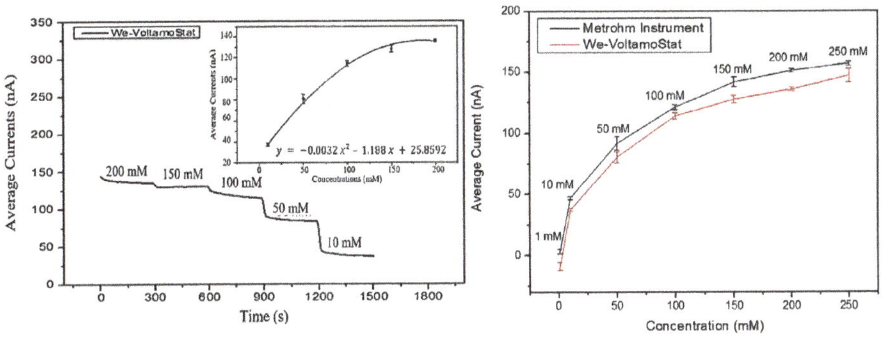


(a) (b)

**Figure 10.** Average current plots for (**a**) sensor response and (**b**) sodium concentration range of 1–250 mM.

3.3.2. Selectivity and Stability

Figure 11 shows the sensor's ability to distinguish sodium ions from interfering ions such as potassium chloride (KCl) and calcium chloride ($CaCl_2$) across a wide concentration range. This demonstrates the sensor's strong selectivity for sodium ions and its resistance to interference from other common ions in sweat. The close match between the measured currents and the reference curve indicates minimal interference from KCl and $CaCl_2$, even at higher concentrations. These findings confirm the sensor's reliability in measuring sodium ions in real sweat samples.

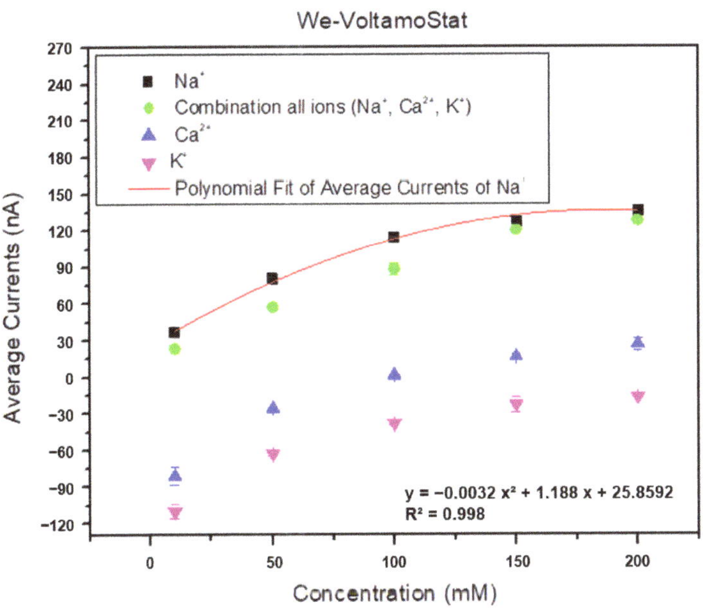


**Figure 11.** Graphical analysis of selectivity test results for the target ion in the presence of the interference ions.

The sodium-selective sensor evaluated in this study is based on a cation-selective membrane that operates according to the Nernstian principle. This membrane is engineered to respond specifically to monovalent and divalent cations (e.g., $Na^+$, $K^+$, $Ca^{2+}$) through the incorporation of selective ionophores, enabling it to differentiate sodium ions from other ionic species. Importantly, due to its anionic nature, $Cl^-$ does not participate in the ion-exchange process across the membrane, thereby reducing the likelihood of interference in sodium ion detection [40].

Table 4 presents the mean differences between sodium ions and mixed ions ($Na^+$, $K^+$, $Ca^{2+}$) for concentrations ranging from 10 mM to 200 mM. The results further highlight the sensor's high selectivity for sodium ions, as the mean differences for $K^+$ and $Ca^{2+}$ are significantly higher (over 100 nA) than those observed for sodium ions. While $K^+$ ions typically exhibit negative deviations from the reference curve, $Ca^{2+}$ ions contribute to positive current values, especially at concentrations exceeding 100 mM. However, considering the typical calcium concentration in sweat (0.07–12 mM), the degree of interference in sodium ion measurement remains minimal [41]. The amperometric measurements, repeated five times, exhibited consistently low standard deviations (SD) of less than 7 nA, indicating stable sensor performance. Moreover, the small standard error of the mean (SEM), typically less than 5 nA, further underscores the accuracy and consistency of the measurements.

**Table 4.** Mean differences between $Na^+$ sensor responses in the presence of interfering ions ($Ca^{2+}$, $K^+$, and a combination of ions) at varying concentrations (10–200 mM).

| Concentration (mM) | Target Ion (A) | Interference Ions (B) | Mean Differences (A − B) | SEM | SD A | SD B |
|---|---|---|---|---|---|---|
| 10 mM | $Na^+$ | Combination all ions | 14.305 | 1.7142 | 1.233 | 3.428 |
| | | $Ca^{2+}$ | 118.580 | 3.5405 | | 4.081 |
| | | $K^+$ | 147.548 | 2.9386 | | 5.877 |
| 50 mM | $Na^+$ | Combination all ions | 23.773 | 0.4624 | 4.339 | 2.813 |
| | | $Ca^{2+}$ | 106.9358 | 1.4066 | | 0.925 |
| | | $K^+$ | 143.838 | 2.1695 | | 1.352 |
| 100 mM | $Na^+$ | Combination all ions | 26.442 | 0.1688 | 2.613 | 4.746 |
| | | $Ca^{2+}$ | 112.956 | 2.3729 | | 0.338 |
| | | $K^+$ | 152.893 | 1.3065 | | 0.788 |
| 150 mM | $Na^+$ | Combination all ions | 7.294 | 1.0129 | 3.182 | 2.302 |
| | | $Ca^{2+}$ | 111.592 | 1.1509 | | 2.026 |
| | | $K^+$ | 150.999 | 3.3026 | | 6.605 |
| 200 mM | $Na^+$ | Combination all ions | 8.222 | 0.655 | 1.520 | 1.937 |
| | | $Ca^{2+}$ | 109.786 | 0.968 | | 4.648 |
| | | $K^+$ | 152.873 | 0.760 | | 1.309 |

### 3.4. Real-Time Sweat Sodium Monitoring

In this study, the developed wearable device was used to stream real-time current data from the $Na^+$-selective ISE sensor during human testing. While the conversion of current values to $Na^+$ concentration (in mmol/L) is performed offline using a predefined calibration curve, the measured current is directly correlated with sodium concentration and provides real-time insight into relative changes in $Na^+$ levels. Two related studies demonstrated the feasibility of wearable electrochemical sensors for real-time sweat monitoring, with sodium concentrations determined offline. Schazmann et al. developed a Sodium Sensor Belt (SSB) using a sodium-selective electrode and fabric-based sweat collection to stream real-time signals, later converted to concentration via calibration [42]. Similarly, Pirovano et al. introduced the SwEatch platform, combining solid-contact ISEs with a 3D-printed microfluidic system and Bluetooth transmission, with offline conversion of signals to sodium levels [43].

Figure 12 illustrates the changes in sweat current (measured in nA) over a 30 min exercise session, with data collected at three distinct 10 min intervals: 0–600 s, 1200–1800 s, and 2400–3000 s. During these exercise periods, the sweat current showed a gradual decline, indicating a reduction in sweat sodium concentration over time. Physiologically, increased sweat rates are often linked to reduced sodium reabsorption efficiency in the sweat ducts, which can lead to higher sodium concentrations in sweat [44]. However, our study observed a progressive decline in sodium levels during the exercise phases, as reflected by the decreasing current. This trend is consistent with previous findings by Xu et al., who reported that subject 1 exhibited a reduction in sweat sodium concentration during low-intensity exercise. In contrast, subject 2 in the same study demonstrated an opposite trend, underscoring the high degree of inter-individual variability in sweat composition [45]. Furthermore, Pirovano et al. observed that the sodium concentration in sweat rose to 2.97 mM at approximately 34 min, before decreasing to 2.21 mM at around 58 min and further to 0.61 mM by 78 min [43]. Such discrepancies in sweat sodium profiles can be attributed to multiple physiological and environmental factors, including individual fitness level, heat acclimatization status, hydration practices, dietary sodium intake, genetics, and the functional responsiveness of eccrine sweat glands. These findings reinforce the notion that sweat composition is highly individualized, and that the accurate interpretation of sweat biomarkers necessitates careful consideration of both intrinsic and extrinsic variables [46].

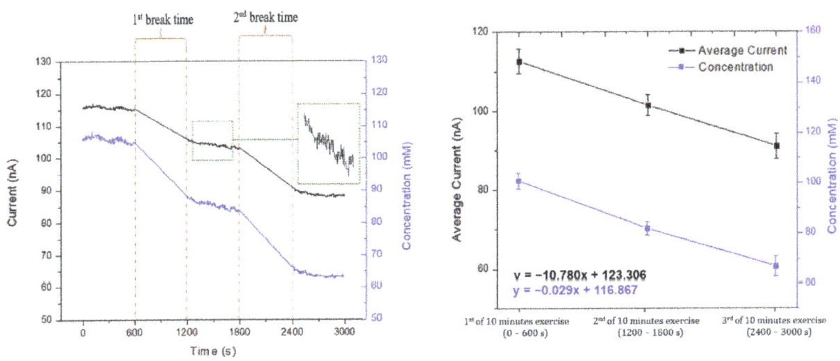


(a) (b)

**Figure 12.** (**a**) Real-time measurement of current (black line) and sodium ion concentration (blue line) during three sessions of 10-min exercise with two rest intervals. (**b**) Averaged values of current (black squares) and sodium concentration (blue circles) during each 10-min exercise session.

## 3.5. Discussion

This study presents a novel wearable sweat-sensing platform that addresses key limitations in existing sodium monitoring systems. The integrated platform combines a sodium-selective ion-selective electrode (ISE), a Bluetooth-enabled miniaturized potentiostat (We-VoltamoStat), and a microfluidic chip with vertically oriented, gravity-assisted channels. Together, these components enable the real-time streaming of current signals during exercise, with sodium concentrations estimated offline via a predefined calibration curve. In comparison to previous wearable systems such as those developed by Schazmann et al. [42] and Pirovano et al. [43], which employed similar offline data processing approaches, the current system advances the field by integrating a more compact, low-power, and accurate electrochemical acquisition module.

The microfluidic chip further distinguishes this work through its vertical fluid transport design, which enhances flow stability and minimizes sample mixing. Fabricated using a water-washable photopolymer resin, the chip maintains long-term hydrophilicity, as demonstrated by sustained low contact angles (<45°) over a year after fabrication. This eliminates the need for post-fabrication surface treatments commonly required for PDMS-based systems, which often suffer from hydrophobic recovery within a day [46]. The use of a passive, gravity-driven system with a controlled outlet geometry also prevents backflow and ensures that fresh sweat is consistently delivered to the sensing area, addressing key challenges noted in earlier microfluidic designs [22].

An important physiological observation from this study is the progressive decline in sweat sodium concentration during the 30 min exercise session, as measured by decreasing current values. This phenomenon is consistent with established physiological mechanisms, in which sweat glands become more efficient at sodium reabsorption over time under the influence of aldosterone, particularly during prolonged or repeated sweating episodes [44,45]. As the body attempts to conserve electrolytes, sodium is increasingly reabsorbed from the primary sweat fluid in the duct before reaching the skin surface. This leads to lower sodium levels in the collected sweat over time, a trend also reported by Xu et al. [45] in human sweat analysis studies during exercise. Additionally, inter-individual variability in sweat electrolyte profiles—driven by factors such as fitness level, acclimatization, and glandular response—may further explain observed differences in sodium dynamics, as supported by prior research [47].

## 4. Conclusions

This research demonstrates the successful development of a novel wearable sweat-sensing platform for real-time monitoring of sodium ions. The microfluidic chip, fabricated using a water-washable resin and 3D printing technology, offers high resolution, rapid fabrication, and excellent hydrophilicity. The We-VoltamoStat potentiostat exhibited high accuracy and precision in measuring electrical currents, with minimal deviations from the reference instrument. The sensor's current measurements exhibited less than 15% error compared to the Metrohm instrument across the 10 to 200 mM NaCl range. Bland–Altman analysis confirmed a 95% difference interval, while statistical analyses revealed low variation, with a SEM of 1.469 nA, an ICC of 0.998, and a CV of 3.5%. Moreover, the selectivity tests showed minor differences (less than 30 nA) between mixed ions ($Na^+$, $K^+$, $Ca^{2+}$) and target ions ($Na^+$), with a standard deviation below 7. Real-time monitoring during exercise indicated a decrease in sweat sodium levels, with average concentrations ranging from 101 mM to 67 mM over a 50 min exercise session. Future studies could explore the integration of additional sensors to measure other biomarkers in sweat, such as glucose or lactate. Moreover, we also consider integrating flow rate sensing function in the future to simultaneously measure the sweat rate, for example by incorporating the flow rate sensors [48,49]. Additionally, investigating the performance of the proposed device in larger-scale clinical trials could further validate its potential for real-world applications.

**Author Contributions:** Conceptualization, A.M.N.; Formal analysis, N.F.A.I.; Resources, N.S.; Writing—original draft, N.F.A.I.; Writing—review & editing, A.M.N., M.S.A.F., M.M., A.A.W., Z.Z., N.I.R., N.S. and A.A.M.; Supervision, A.M.N.; Funding acquisition, A.A.M. All authors have read and agreed to the published version of the manuscript.

**Funding:** This research was funded by the Fundamental Research Grant Scheme (FRGS), grant number FRGS/1/2019/TK05/UNIMAP/03/6.

**Institutional Review Board Statement:** Not applicable.

**Informed Consent Statement:** Not applicable.

**Data Availability Statement:** Data are contained within the article.

**Conflicts of Interest:** The authors declare no conflicts of interest.

## Abbreviations

The following abbreviations are used in this manuscript:

| | |
|---|---|
| WE | Working electrode |
| CE | Counter electrode |
| RE | Reference Electrode |
| NaCI | Sodium Chloride |
| KCI | Potassium Chloride |
| $CaCI_2$ | Calcium chloride |
| mM | Milimolar |
| ICC | Intraclass correlation coefficient |
| CV | Coefficient of variation |
| HSD | Honestly significant difference |
| SD | Standard deviation |
| RC | Resistor-capacitor |
| nA | Nano ampere |

# References


1. Ji, W.; Zhu, J.; Wu, W.; Wang, N.; Wang, J.; Wu, J.; Wu, Q.; Wang, X.; Yu, C.; Wei, G.; et al. Wearable sweat biosensors refresh personalized health/medical diagnostics. *Research* **2021**, *2021*, 9757126. [CrossRef] [PubMed]
2. Moonen, E.J.; Haakma, J.R.; Peri, E.; Pelssers, E.; Mischi, M.; den Toonder, J.M. Wearable sweat sensing for prolonged, semicontinuous, and nonobtrusive health monitoring. *View* **2020**, *1*, 20200077. [CrossRef]
3. Gao, F.; Liu, C.; Zhang, L.; Liu, T.; Wang, Z.; Song, Z.; Cai, H.; Fang, Z.; Chen, J.; Wang, J.; et al. Wearable and flexible electrochemical sensors for sweat analysis: A review. *Microsyst. Nanoeng.* **2023**, *9*, 1. [CrossRef] [PubMed]
4. Ibrahim, N.F.A.; Sabani, N.; Johari, S.; Manaf, A.A.; Wahab, A.A.; Zakaria, Z.; Noor, A.M. A comprehensive review of the recent developments in wearable sweat-sensing devices. *Sensors* **2022**, *22*, 7670. [CrossRef]
5. Smith, C.J.; Havenith, G. Body mapping of sweating patterns in male athletes in mild exercise-induced hyperthermia. *Eur. J. Appl. Physiol.* **2011**, *111*, 1391–1404. [CrossRef]
6. Xing, Z.; Hui, J.; Lin, B.; Wu, Z.; Mao, H. Recent advances in wearable sensors for the monitoring of sweat: A comprehensive tendency summary. *Chemosensors* **2023**, *11*, 470. [CrossRef]
7. Coull, N.A.; West, A.M.; Hodder, S.G.; Wheeler, P.; Havenith, G. Body mapping of regional sweat distribution in young and older males. *Eur. J. Appl. Physiol.* **2021**, *121*, 109–125. [CrossRef]
8. Liu, H.; Gu, Z.; Zhao, Q.; Li, S.; Ding, X.; Xiao, X.; Xiu, G. Printed circuit board integrated wearable ion-selective electrode with potential treatment for highly repeatable sweat monitoring. *Sens. Actuators B Chem.* **2022**, *355*, 131102. [CrossRef]
9. Chen, X.M.; Li, Y.J.; Han, D.; Zhu, H.C.; Xue, C.D.; Chui, H.C.; Cao, T.; Qin, K.R. A capillary-evaporation micropump for real-time sweat rate monitoring with an electrochemical sensor. *Micromachines* **2019**, *10*, 457. [CrossRef]
10. Kazanskiy, N.L.; Khonina, S.N.; Butt, M.A. A review on flexible wearables–Recent developments in non-invasive continuous health monitoring. *Sens. Actuators A Phys.* **2024**, *366*, 114993. [CrossRef]
11. Vaghasiya, J.V.; Mayorga-Martinez, C.C.; Pumera, M. Wearable sensors for telehealth based on emerging materials and nanoarchitectonics. *Npj Flex. Electron.* **2023**, *7*, 26. [CrossRef] [PubMed]
12. Zhang, Y.; Chen, Y.; Huang, J.; Liu, Y.; Peng, J.; Chen, S.; Song, K.; Ouyang, X.; Cheng, H.; Wang, X. Skin-interfaced microfluidic devices with one-opening chambers and hydrophobic valves for sweat collection and analysis. *Lab Chip* **2020**, *20*, 2635–2645. [CrossRef] [PubMed]
13. Escobedo, P.; Ramos-Lorente, C.E.; Martínez-Olmos, A.; Carvajal, M.A.; Ortega-Muñoz, M.; de Orbe-Payá, I.; Hernández-Mateo, F.; Santoyo-González, F.; Capitán-Vallvey, L.F.; Palma, A.J.; et al. Wireless wearable wristband for continuous sweat pH monitoring. *Sens. Actuators B Chem.* **2021**, *327*, 128948. [CrossRef]
14. Jain, V.; Ochoa, M.; Jiang, H.; Rahimi, R.; Ziaie, B. A mass-customizable dermal patch with discrete colorimetric indicators for personalized sweat rate quantification. *Microsyst. Nanoeng.* **2019**, *5*, 29. [CrossRef]
15. Wicaksono, I.; Tucker, C.I.; Sun, T.; Guerrero, C.A.; Liu, C.; Woo, W.M.; Pence, E.J.; Dagdeviren, C. A tailored, electronic textile conformable suit for large-scale spatiotemporal physiological sensing in vivo. *Npj Flex. Electron.* **2020**, *4*, 1–3. [CrossRef]
16. Ding, L.; Yang, X.; Gao, Z.; Effah, C.Y.; Zhang, X.; Wu, Y.; Qu, L. A holistic review of the state-of-the-art microfluidics for exosome separation: An overview of the current status, existing obstacles, and future outlook. *Small* **2021**, *17*, 2007174. [CrossRef]
17. Heikenfeld, J. Non-invasive analyte access and sensing through eccrine sweat: Challenges and outlook circa 2016. *Electroanalysis* **2016**, *28*, 1242–1249. [CrossRef]

18. Bandodkar, A.J.; Jeang, W.J.; Ghaffari, R.; Rogers, J.A. Wearable sensors for biochemical sweat analysis. *Annu. Rev. Anal. Chem.* **2019**, *12*, 1–22. [CrossRef]
19. Elvira, K.S.; Gielen, F.; Tsai, S.S.; Nightingale, A.M. Materials and methods for droplet microfluidic device fabrication. *Lab Chip* **2022**, *22*, 859–875. [CrossRef]
20. Hashimoto, Y.; Ishihara, T.; Kuwabara, K.; Amano, T.; Togo, H. Wearable microfluidic sensor for the simultaneous and continuous monitoring of local sweat rates and electrolyte concentrations. *Micromachines* **2022**, *13*, 575. [CrossRef]
21. Annabestani, M.; Esmaeili-Dokht, P.; Fardmanesh, M. A novel, low cost, and accessible method for rapid fabrication of the modifiable microfluidic devices. *Sci. Rep.* **2020**, *10*, 16513. [CrossRef] [PubMed]
22. Narayanamurthy, V.; Jeroish, Z.E.; Bhuvaneshwari, K.S.; Bayat, P.; Premkumar, R.; Samsuri, F.; Yusoff, M.M. Advances in passively driven microfluidics and lab-on-chip devices: A comprehensive literature review and patent analysis. *RSC Adv.* **2020**, *10*, 11652–11680. [CrossRef] [PubMed]
23. Steijlen, A.S.; Jansen, K.M.; Bastemeijer, J.; French, P.J.; Bossche, A. Low-cost wearable fluidic sweat collection patch for continuous analyte monitoring and offline analysis. *Anal. Chem.* **2022**, *94*, 6893–6901. [CrossRef] [PubMed]
24. Wang, S.; Wu, Y.; Gu, Y.; Li, T.; Luo, H.; Li, L.H.; Zhang, T. Wearable sweatband sensor platform based on gold nanodendrite array as efficient solid contact of ion-selective electrode. *Anal. Chem.* **2017**, *89*, 10224–10231. [CrossRef]
25. Wu, L.; Guo, Z.; Liu, W. Surface behaviors of droplet manipulation in microfluidics devices. *Adv. Colloid Interface Sci.* **2022**, *308*, 102770. [CrossRef]
26. Adams, S.D.; Doeven, E.H.; Quayle, K.; Kouzani, A.Z. MiniStat: Development and evaluation of a mini-potentiostat for electrochemical measurements. *IEEE Access* **2019**, *7*, 31903–31912. [CrossRef]
27. Anshori, I.; Mufiddin, G.F.; Ramadhan, I.F.; Ariasena, E.; Harimurti, S.; Yunkins, H.; Kurniawan, C. Design of smartphone-controlled low-cost potentiostat for cyclic voltammetry analysis based on ESP32 microcontroller. *Sens. Bio-Sens. Res.* **2022**, *36*, 100490. [CrossRef]
28. Pungjunun, K.; Yakoh, A.; Chaiyo, S.; Siangproh, W.; Praphairaksit, N.; Chailapakul, O. Smartphone-based electrochemical analysis integrated with NFC system for the voltammetric detection of heavy metals using a screen-printed graphene electrode. *Microchim. Acta* **2022**, *189*, 191. [CrossRef]
29. Chung, M.; Fortunato, G.; Radacsi, N. Wearable flexible sweat sensors for healthcare monitoring: A review. *J. R. Soc. Interface* **2019**, *16*, 20190217. [CrossRef]
30. Brothers, M.C.; DeBrosse, M.; Grigsby, C.C.; Naik, R.R.; Hussain, S.M.; Heikenfeld, J.; Kim, S.S. Achievements and challenges for real-time sensing of analytes in sweat within wearable platforms. *Acc. Chem. Res.* **2019**, *52*, 297–306. [CrossRef]
31. Ibrahim, N.F.; Noor, A.M.; Sabani, N.; Zakaria, Z.; Wahab, A.A.; Manaf, A.A.; Johari, S. We-VoltamoStat: A wearable potentiostat for voltammetry analysis with a smartphone interface. *HardwareX* **2023**, *15*, e00441. [CrossRef] [PubMed]
32. Goedecke, N.; Eijkel, J.; Manz, A. Evaporation driven pumping for chromatography application. *Lab Chip* **2002**, *2*, 219–223. [CrossRef] [PubMed]
33. Vamos, I.; Kertesz, V. HunStat–a simple and low-cost potentiostat for analytical and educational purposes. *Anal. Methods* **2024**, *16*, 4198–4204. [CrossRef] [PubMed]
34. Caux, M.; Achit, A.; Var, K.; Boitel-Aullen, G.; Rose, D.; Aubouy, A.; Argentieri, S.; Campagnolo, R.; Maisonhaute, E. PassStat, a simple but fast, precise and versatile open source potentiostat. *HardwareX* **2022**, *11*, e00290. [CrossRef]
35. Baker, L.B. Sweating rate and sweat sodium concentration in athletes: A review of methodology and intra/interindividual variability. *Sports Med.* **2017**, *47*, 111–128. [CrossRef]
36. Hu, W.; Shao, M.; Wang, Q.; Reichardt, K. Time stability of soil water storage measured by neutron probe and the effects of calibration procedures in a small watershed. *Catena* **2009**, *79*, 72–82. [CrossRef]
37. Setiyono, R.; Lestari, T.F.; Anggraeni, A.; Hartati, Y.W.; Bahti, H.H. UnpadStat design: Portable potentiostat for electrochemical sensing measurements using screen printed carbon electrode. *Micromachines* **2023**, *14*, 268. [CrossRef]
38. Dobbelaere, T.; Vereecken, P.M.; Detavernier, C. A USB-controlled potentiostat/galvanostat for thin-film battery characterization. *HardwareX* **2017**, *2*, 34–49. [CrossRef]
39. Lee, D.K.; In, J.; Lee, S. Standard deviation and standard error of the mean. *Korean J. Anesthesiol.* **2015**, *68*, 220–223. [CrossRef]
40. Bakker, E.; Pretsch, E.; Bu1hlmann, P. Selectivity of Potentiometric Ion Sensors. *Anal. Chem.* **2000**, *72*, 1127–1133. [CrossRef]
41. Maughan, R.J.; Watson, P.; Shirreffs, S.M. Implications of active lifestyles and environmental factors for water needs and consequences of failure to meet those needs. *Nutr. Rev.* **2015**, *73*, 130–140. [CrossRef] [PubMed]
42. Schazmann, B.; Morris, D.; Slater, C.; Beirne, S.; Fay, C.; Reuveny, R.; Moyna, N.; Diamond, D. A wearable electrochemical sensor for the real-time measurement of sweat sodium concentration. *Anal. Methods* **2010**, *2*, 342–348. [CrossRef]
43. Pirovano, P.; Dorrian, M.; Shinde, A.; Donohoe, A.; Brady, A.J.; Moyna, N.M.; Wallace, G.; Diamond, D.; McCaul, M. A wearable sensor for the detection of sodium and potassium in human sweat during exercise. *Talanta* **2020**, *219*, 121145. [CrossRef] [PubMed]
44. Buono, M.J.; Claros, R.; DeBoer, T.; Wong, J. Na+ secretion rate increases proportionally more than the Na+ reabsorption rate with increases in sweat rate. *J. Appl. Physiol.* **2008**, *105*, 1044–1048. [CrossRef]

45. Xu, G.; Cheng, C.; Liu, Z.; Yuan, W.; Wu, X.; Lu, Y.; Low, S.S.; Liu, J.; Zhu, L.; Ji, D.; et al. Battery-free and wireless epidermal electrochemical system with all-printed stretchable electrode array for multiplexed in situ sweat analysis. *Adv. Mater. Technol.* **2019**, *4*, 1800658. [CrossRef]
46. Neves, L.B.; Afonso, I.S.; Nobrega, G.; Barbosa, L.G.; Lima, R.A.; Ribeiro, J.E. A Review of Methods to Modify the PDMS Surface Wettability and Their Applications. *Micromachines* **2024**, *15*, 670. [CrossRef]
47. Presta, V.; Ambrosini, L.; Carubbi, C.; Masselli, E.; Mirandola, P.; Arcari, M.L.; Gobbi, G.; Vitale, M. Different waters for different performances: Can we imagine sport-related natural mineral spring waters? *Water* **2021**, *13*, 166. [CrossRef]
48. Al Farisi, M.S.; Wang, Y.; Hasegawa, Y.; Matsushima, M.; Kawabe, T.; Shikida, M. Facile In-Tube-Center Packaging of Flexible Airflow Rate Microsensor for Simultaneous Respiration and Heartbeat Measurement. *IEEE Sens. J.* **2023**, *23*, 12626–12633.
49. Al Farisi, M.S.; Okazaki, J.; Hasegawa, Y.; Matsushima, M.; Kawabe, T.; Shikida, M. Non-contact sensor module for rapid detection of extravasation during intravenous drug administration. *Biomed. Microdevices* **2024**, *26*, 44. [CrossRef]

*sensors*

MDPI

*Review*

# Perspectives on the Application of Biosensors for the Early Detection of Oral Cancer

**Sanket Naresh Nagdeve, Baviththira Suganthan and Ramaraja P. Ramasamy ***

Nano Electrochemistry Laboratory, School of Chemical, Materials and Biomedical Engineering, University of Georgia, Athens, GA 30602, USA; sanket.nagdeve@uga.edu (S.N.N.); baviththira.suganthan@uga.edu (B.S.)
* Correspondence: rama@uga.edu

**Abstract:** Oral cancer continues to cause profound suffering and is associated with high mortality rates. Early detection techniques are crucial in enhancing patient outcomes. This review paper thoroughly evaluates the significance of biomarkers and recent advancements in oral cancer detection, emphasizing cutting-edge electrochemical methods. The paper provides an epidemiological and etiological overview, outlining its clinical importance and reviewing the current state of the art in detection methods. Despite considerable progress, conventional methods exhibit limitations such as invasiveness, long wait times, and a lack of accuracy, creating a critical need for more robust technologies. This review emphasizes the significance of oral cancer biomarkers, which are considered promising cues for early detection, facilitating the development of innovative biosensing technologies. This review seeks to illuminate the recent advances in early detection and precision diagnostics, along with the usage of artificial intelligence strategies, ultimately contributing to significant progress in the battle against oral cancer.

**Keywords:** biomarkers; biofluids; electrochemical sensors; molecular analytical techniques; diagnostic tools; commercial test kits

## 1. Introduction: Understanding the Etiology, Prevalence, and Current Diagnostic Tools and Emphasizing the Importance of Early Detection

Oral cancer ranks sixth globally in terms of cancer incidence. Almost 90% of such cancers are oral squamous cell carcinomas (OSCCs) [1]. This is the most prevalent malignant tumor that develops in the oral cavity, starting at the lips and ending at the anterior surface, and is usually lined by squamous cell epithelia with minor salivary gland involvement [2]. The main factors contributing to OSCC are tobacco, betel quid, alcohol consumption, and a sexually transmitted virus called human papillomavirus (HPV) [3]. Typical symptoms include red or white patches on the tongue, ulcers, unusual bleeding, neck swelling, or a lump [4]. Oral cancer affects various areas within the oral cavity, including the lips, tongue (particularly the sides and dorsal surface), floor of the mouth, buccal mucosa (inner lining of cheeks), palate (hard and soft), gums and gingiva, oropharynx (base of the tongue, tonsils, and soft palate), and retromolar trigone (the area behind the last molar teeth on the lower jaw) [5], as shown in Figure 1.

The Global Cancer Observatory reported 476,000 new cases of oral cancer worldwide in 2020 [6]. Furthermore, data from the National Institute of Dental and Craniofacial Research suggests that the incidence of oral cancer tends to increase with advancing age, particularly among adults aged 65 and older [4]. The five-year survival rate for the period between 2013 and 2019 stood at 68.5%, with approximately 11,580 deaths recorded during this time [7]. Figure 2 showcases the reported new cases and deaths due to oral cancer across different age groups. One of the significant reasons that a large proportion of patients are affected by oral cancer is that it is diagnosed at the later stages.

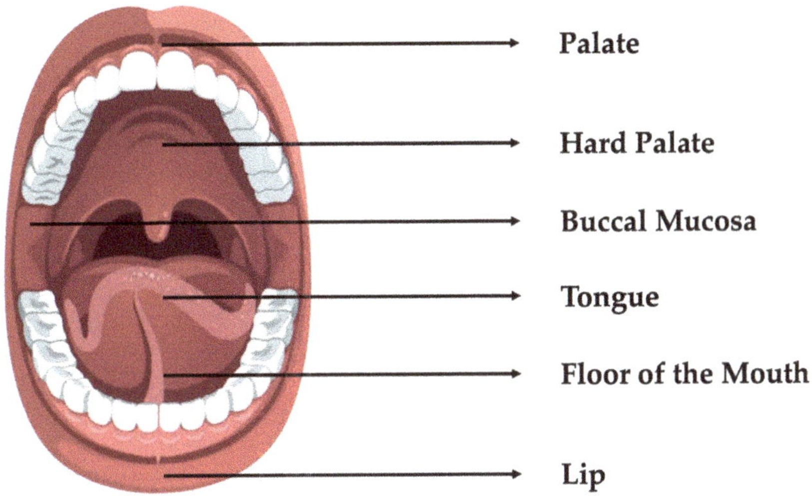


**Figure 1.** The most common sites of oral cancer.

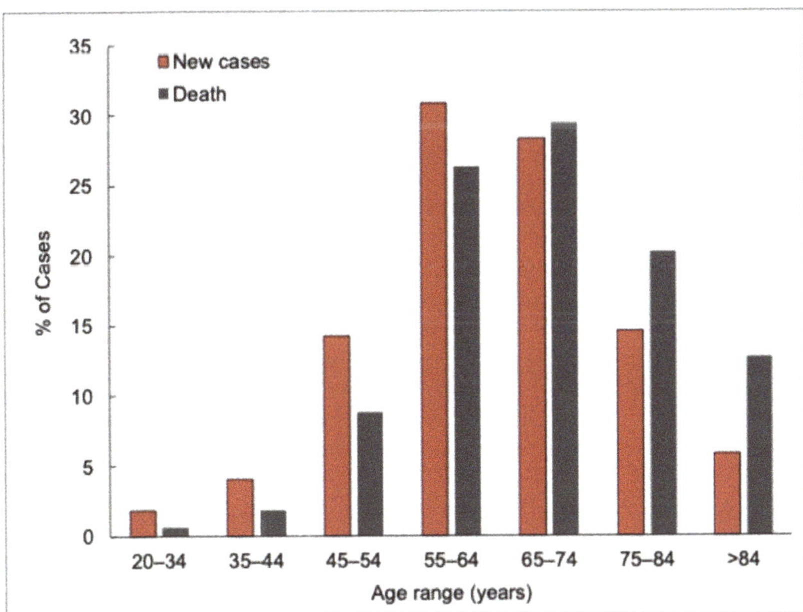


**Figure 2.** Reported new cases and deaths due to oral cancer across different age groups. (Available online https://seer.cancer.gov/statistics-network/explorer/. Data source(s): SEER Incidence Data, accessed on 20 December 2024) [7].

Studies have reported that oral visual screening can reduce mortality in individuals with significant risk factors (especially tobacco and alcohol users). Visual screening involves thoroughly examining the head, neck, and oral regions to identify premalignant cells [8].

Following this, histopathology also plays an essential role in analyzing the tissue sections by examining the structural and cellular characteristics to determine whether lesions are cancerous or pre-cancerous [9]. Beyond these conventional methods, several adjunctive techniques are also present to enhance oral cancer detection. For example, vital staining is a valuable tool where specific dyes highlight abnormal cells. Similarly, the brush biopsy offers a minimally invasive approach by collecting tissue samples from suspicious areas for detailed analysis. Chemiluminescence or auto-fluorescence is another diagnostic tool that employs a light-sensitive emitting tool to identify malignant tissues for early-stage determination. However, the cornerstone of detection has evolved towards advanced imaging technologies, which play pivotal roles in the precise assessment of malignancies. The magnetic resonance imaging (MRI) technique uses strong magnetic fields that provide extensive and detailed images of the mouth and surrounding tissues to determine the tumor size, location, and extent of invasion [10,11]. Alongside MRI, ultrasound assesses tumor growth in the throat and neck lymph nodes, and computed tomography (CT) provides comprehensive images of oral cross-sectional areas. Other methods, including optical coherence tomography and positron emission tomography, use light or radioactive material to identify cancerous growths throughout the body. Recent advances have introduced several sophisticated approaches, including photoacoustic imaging, which combines ultrasound with optical imaging properties for the staging of cancerous lesions [12,13]. However, these expensive and power-intensive methods have not penetrated the market to become readily available for people with low-income backgrounds. For example, X-rays struggle in soft tissue visualization, while CT scans involve radiation exposure among patients [14,15]. Meanwhile, MRI may not be able to distinguish between benign and malignant lesions consistently, and ultrasound imaging is limited by its penetration depth into the tissue, which compromises its utility in deeper and metastatic lymph nodes [16]. These issues collectively impact the feasibility of and equitable access to diagnostic methods, potentially hindering the cost-effective delivery of dental healthcare services. Raman et al. reviewed several studies on oral and salivary gland cancers, highlighting the effectiveness of oral cancer screening programs. According to the review, the total societal costs depend on several parameters, such as the screening strategies, treatment provided, location, etc. In a community-based program in the United States (US), the cost per screening was estimated at USD 50 to USD 100 per individual, depending on the resources obtained and utilized and the population [17]. Some studies have also reported the quality-adjusted life years (QALY), which vary from USD 20,000 to USD 50,000 in the US [17]. QALY is a measure of the quantity and quality of life used to assess health outcomes, and one QALY is equal to one year of a healthy life. This is usually considered cost-effective in many healthcare systems. However, a limitation of QALY is the availability and quality of the data analyzed. The findings of QALY studies cannot be generalized across different healthcare systems, countries, and populations. Ribeiro-Rotta et al. conducted a comprehensive assessment of the economic burden of oral cancer, with direct costs like medical and non-medical costs and indirect costs like early deaths. Several parameters were identified in the study, including the cost per patient, total cost in a period, and cost per treatment [18]. It also included the inpatient and outpatient costs and the stages of the cancer. It was found that the inpatient costs were 968% higher than the outpatient costs.

Thus, these limitations highlight the necessity of advanced diagnostic technologies and additional methods for improved early screening to enhance the clinical outcomes for patients. From the perspective of the patient, early detection offers profound psychological and physiological benefits, such as identifying malignant lesions at the early stages, reducing the psychological trauma of costly cancer treatment, and increasing the likelihood of complete recovery. Consequently, there is an ongoing demand for more advanced and comprehensive screening strategies. Biomarker-based detection techniques have emerged as a vital early oral cancer diagnosis approach. These molecular methods detect specific indicators of cancer development, potentially offering more precise and comprehensive screening strategies.

The Preferred Reporting Items for Systematic Reviews and Meta-Analyses (PRISMA) flow diagram was used to report the studies involved in this non-meta literature review (Figure 3). A comprehensive search was conducted using keywords on PUBMED, Web of Science, and individual websites using the Google search engine. The keywords used were ("oral cancer" OR "oral squamous cell carcinoma") AND ("biomarker") AND ("early detection" OR "early diagnosis"). The studies were searched from 2014 to 2024 and we excluded citations, conference proceedings, abstracts, and any systematic reviews. An additional search was conducted to analyze any commercially available kits for the early detection of oral cancer. The search also focused on studies specifically targeting the early detection or diagnosis of oral cancer biomarkers, with an additional searching criterion of electrochemical biosensors for early biomarker detection. This included the three main electrochemical methods reported in this review (amperometric, voltammetric, and impedimetric). The original research articles included were in the English language, and research papers published before 2014, book chapters, and those in other languages were excluded from this review. To illustrate the usability of this framework, the biomarkers specific to oral cancer were identified, their current molecular detection strategies were studied, and the need for electrochemical biosensors was analyzed. An overview of such biosensors was created, with their fabrication, the strategies involved, and the need for early detection biomarkers associated with oral cancer, as well as their limits of detection (LODs).

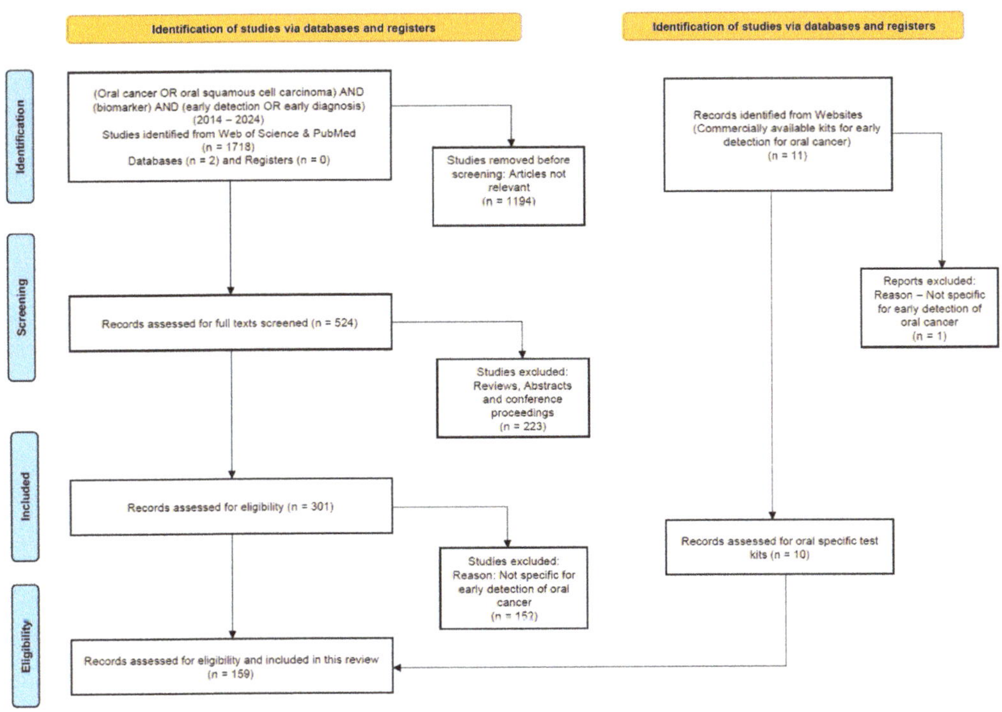


**Figure 3.** PRISMA flow chart of the study selection process.

## 2. Biomarkers: Significance, Types, and Molecular Detection Strategies

Biomarkers are essential molecular indicators of the normal or malignant functioning of cells and help to determine the diagnosis and prognosis of a particular disease. Biomarkers are valuable in screening individuals without clinical or histological signs of oral cancer, as well as healthy individuals. Their use in early detection significantly reduces patient morbidity and mortality, underscoring the importance of developing clinically validated biomarkers for effective oral cancer screening. Such biomarkers include deoxyribonucleic acid (DNA), microRNAs (miRNAs), long non-coding ribonucleic acids (lncRNAs), messenger RNAs (mRNAs), and proteins. DNA biomarkers involve alterations in the DNA sequence, such as mutations or methylation patterns, associated with oral cancer progression [19], whereas RNA biomarkers play crucial roles in cancer by modulating gene expression [20]. In addition, proteins are associated with various metabolic, structural, and regulatory functions in disease progression [21].

### 2.1. DNA Biomarkers

DNA biomarkers are complex molecules found in every cell of the body. They contain all of the necessary genetic information for the development and functioning of an organism. Gene mutations cause most malignancies, and molecular-based techniques are used to develop and improve diagnostic procedures [22]. Several studies have delved into the characterization of DNA biomarkers related to oral cancer. For instance, the overexpression of NCBP2 and TFRC has been identified in tumor cells, demonstrating significantly higher expression levels than most normal human tissues [23]. On the other hand, emerging DNA methylation markers, such as HOXA1 3′UTR methylation, have been recognized for their potential as predictive biomarkers for OSCC [24]. These findings underscore the critical role of DNA biomarkers in understanding the molecular landscape of oral cancer and their potential applications in diagnostic and predictive settings. The investigation of molecular markers for early-stage OSCC has also been highlighted, emphasizing the need for the comprehensive analysis of such biomarkers to enable optimal clinical treatments and improve patients' survival rates [25]. These studies collectively advance our understanding of the DNA biomarkers associated with oral cancer, shedding light on their regular and overexpressed levels and their potential significance in clinical practice. However, the performance is not always satisfactory due to DNA denaturation [26]. Although DNA biomarkers provide valuable insights into genetic predispositions and mutations, recently, miRNAs have offered a more nuanced understanding of post-transcriptional gene regulation [20]. Given the dynamic nature of biomarker research, the exploration of miRNAs opens up new avenues for precision medicine and personalized diagnostics, complementing the insights gained from DNA-based analyses. As research has advanced, the focus on biomarker discovery has expanded beyond DNA to explore the potential of miRNAs as novel indicators of biological processes and disease states.

### 2.2. RNA Biomarkers

RNA biomarkers are molecular indicators derived from RNA molecules that provide valuable insights into physiological and pathological conditions within the body. They provide essential information about specific biological conditions. RNA biomarkers are molecular signatures derived from RNA molecules, including mRNA, miRNA, and other non-coding RNAs [27]. RNA molecules play a crucial role in gene expression and regulatory mechanisms. They are used to diagnose diseases and monitor the progression of tumors, as well as the patient's response to therapy. mRNA biomarkers identify key oncogenes or tumor suppressor genes that are dysregulated in oral cancer [28]. For instance, epidermal growth factor receptor (EGFR) and cyclin D1 can be detected through mRNA analysis and can be correlated with tumor progression or a poor prognosis [29,30]. Similarly, lncRNAs are a group of non-coding RNAs longer than 200 nucleotides that modulate chromatin organization, transcription control, and overall stability. Any aberrant expression in lncRNAs, including HOTAIR and MALAT1, can be considered an early detection biomarker [31,32].

In recent years, miRNAs have emerged as pivotal elements in the pathogenesis of oral cancer, offering potential diagnostic utility and therapeutic value. miRNAs are small non-coding RNAs that govern several physiological processes and have been identified as critical elements in the pathological mechanisms of diverse diseases, including oral cancer [33,34]. These miRNAs are widely distributed in body fluids and demonstrate disease-specific expression patterns, making them potential biomarkers for early detection. Moreover, the inherent stability of miRNAs in bodily fluids, such as serum and saliva, underscores their possible applicability in early cancer diagnosis, highlighting their value as diagnostic biomarkers [33].

Circulating miRNAs, a subset of miRNAs, hold immense promise as diagnostic biomarkers for various types of cancer [35]. Their utility stems from their potential as molecular labels of tumor cells throughout tumorigenesis and cancer progression, with their expression patterns changing as the disease advances. Specific circulating miRNAs, such as miRNA-21, have demonstrated diagnostic potential for certain cancers, offering moderate sensitivity [36]. Notably, a profiling study revealed that miRNAs are differentially expressed in OSCC patients compared to healthy individuals. Therefore, specific miRNAs may serve as promising diagnostic biomarkers, as evidenced by the identification of various miRNAs that show potential for discrimination between oral cancer patients and healthy individuals [33]. Furthermore, specific miRNAs have been associated with clinical staging, metastasis, and overall survival in oral cancer patients [33]. He et al. identified miRNA-24-3p as a potential novel diagnostic salivary biomarker directly associated with the proliferation of cancer cells [37].

Continued research efforts and technological innovations in miRNA-based diagnostic tools are essential in advancing the detection and management of oral cancer, thereby improving patient outcomes and prognoses. miRNAs have exhibited strong potential diagnostic and prognostic value in the context of oral cancer, especially with their distinct expression profiles between healthy individuals and cancer patients and their association with clinical parameters and patient outcomes, underlining their significance as promising biomarkers for early diagnosis and prognostic assessment in oral cancer. As research progresses, it is becoming increasingly evident that miRNAs hold substantial promise, offering novel diagnostic and therapeutic avenues, thus signifying their potential as innovative tools to address the challenges in diagnosing and managing oral cancer. While miRNAs offer valuable information due to their stability and association with disease states [38], protein biomarkers also present a complementary avenue, leveraging proteins' diverse functions and interactions within biological systems.

### 2.3. Protein Biomarkers

Protein biomarkers play a crucial role in understanding the molecular landscape of oral cancer and its clinical implications. The analysis of the protein expression levels in oral cancer samples reveals the complex divergence between the normal and overexpressed states of specific proteins, providing insights for diagnostic and prognostic evaluations. Protein biomarkers can be detected in various biofluids and are used extensively for disease diagnosis, prognosis, and monitoring. As a promising alternative, oral fluids, such as saliva, human serum, and urine, have gained attention as potential bio-media for oral cancer diagnostics [39].

A large group of protein biomarkers, such as interleukin-6 (IL-6), -8, -1$\alpha$, and -1$\beta$, are commonly used indicators for OSCC diagnosis due to their association with lesion transformation in oral cancer [40]. The use of cytokeratin 19 fragment (CYFRA) 21-1 as a diagnostic biomarker for various cancers, including OSCC, head and neck cancer, bladder cancer, and intrahepatic cholangiocarcinoma, has been explored in several studies. Alali et al. and Liu et al. found that CYFRA 21-1 had high specificity in detecting OSCC and head and neck cancer, respectively, but its sensitivity was low [41,42]. These findings underscore the critical role of protein biomarkers in elucidating oral cancer's molecular intricacies and their potential as diagnostic and prognostic indicators. By illuminating specific proteins' normal and overexpressed levels, researchers are paving the way for the development of targeted therapeutic interventions and advancements in personalized medicine in oral cancer management. Huang et al. and Guowei et al. reported the high specificity of CYFRA 21-1 in diagnosing bladder cancer and intrahepatic cholangiocarcinoma, respectively [43,44]. However, the sensitivity of CYFRA 21-1 in these cancers was low. These findings indicate that, while CYFRA 21-1 holds potential as a valuable biomarker in detecting these cancers, its sensitivity and specificity show variability. Therefore, achieving clinical sensitivity and specificity is crucial in detecting protein biomarkers, particularly due to the challenges posed by low sample concentrations and non-specific interactions [45].

Specific proteins, including DEP Domain Containing 1B (DEPDC1B), demonstrate enhanced expression levels in oral cancer samples compared to normal adjacent tissues. This overexpression signifies the potential utility of DEPDC1B as a biomarker for oral cancer [19]. Moreover, proteomic studies have highlighted the differential expression patterns of several proteins in oral cancer. The overexpression of TYPH, an angiogenesis-associated protein, has been identified in many solid tumors, indicating its induction by multiple cytokines and significant contribution to angiogenesis [46]. Conversely, specific proteins in oral cancer samples have shown lower protein expression levels. For instance, the secreted protein acidic and rich in cysteine has shown decreased expression in OSCC tissues, indicating potential alterations in its protective role in the tumor microenvironment [47].

In disease monitoring and detection, biomarkers such as DNA, miRNA, and proteins are essential. DNA biomarkers provide information about the onset and course of the disease. In contrast, miRNA biomarkers control gene expression, and the stability of miRNAs in biofluids is noteworthy, indicating their potential for non-invasive diagnostic applications [35]. On the other hand, specific proteins linked to oral disease can be found in biofluids like saliva, offering important data for prognosis and early detection.

### 2.4. Presence of Biomarkers in Saliva, Serum, and Urine

Oral fluids, including saliva, serum, and urine, are valuable biofluids for the detection of oral cancer due to the presence of specific biomarkers. These biomarkers, including DNA molecules, proteins, and RNA molecules, can indicate the presence of oral cancer and aid in its early detection and monitoring. Salivary biomarkers, for instance, play a crucial role in facilitating the detection and monitoring of oral cancer. Similarly, biomarkers in urine and serum also contribute to effectively detecting oral cancer. Saliva, a biofluid that is abundant in proteins and enzymes, provides a non-invasive means of detecting oral cancer biomarkers. This valuable characteristic makes saliva an essential medium for the early detection and monitoring of oral cancer. Similarly, serum, the precise component of blood without clotting factors, contains specific proteins and molecules that indicate oral cancer, enhancing its utility as a biofluid for diagnostic purposes. Furthermore, despite not originating directly from the oral cavity, urine can offer insights into systemic changes associated with oral cancer, as it harbors metabolites and proteins linked to the disease [48]. These distinct biofluids collectively offer non-invasive and potentially early detection methods, significantly contributing to the management and treatment of oral cancer.

Saliva is a watery extracellular liquid usually secreted by the salivary glands inside the mouth. Whole saliva contains various cellular components, bacteria, viruses, and fungal information, which is critical for any disease analysis. The direct contact of saliva with oral lesions makes it a specific screening tool for the identification of hundreds of cancer biomarkers [49]. The development of screening technologies has shown a promising future for salivary biomarkers in the early detection of oral cancer. Saliva has been extensively studied to discover oral cancer biomarkers [49]. It contains various biomolecules that can indicate the presence of OSCC, offering non-invasive and convenient strategies for diagnosis [40]. Researchers have identified salivary metabolomic biomarkers correlating with oral cancer, aiding early detection [50].

Salivary protein biomarkers for the early diagnosis and prevention of oral cancer include IL-8, IL-6, and tumor necrosis factor $\alpha$ (TNF-$\alpha$) [51]. Among several groups of proteomic biomarkers in OSCC, interleukins play a vital role in the immune response and the transformation to malignancies [51]. In addition to interleukins, Ricardi et al. identified several critical sources of such biomarkers, including cytokines, growth factors, matrix metalloproteinases (MMPs), acute-phase proteins, and proline-rich proteins [52]. In recent years, saliva-based samples have become more accessible, reliable, and non-invasive. However, the utility of salivary biomarkers in the early diagnosis of OSCC is still under debate, with some studies showing the potential for discrimination between healthy and cancer patients [53].

Serum is another biofluid explored for its potential in oral cancer detection. Proteins, enzymes, hormones, and antibodies are a few of the many biomolecules found in serum, the liquid part of the blood left over after blood clotting [39]. These biomolecules can serve as valuable indicators of various physiological and pathological processes, including the presence and progression of oral cancer. These biomarkers can include tumor-associated proteins, circulating tumor DNA (ctDNA), miRNAs, and other metabolites [39]. Although serum biomarkers hold great promise, further research is needed to establish their clinical utility and reliability for oral cancer detection. While not directly from the oral cavity, serum contains proteins and molecules indicative of oral cancer. Serum contains useful biomarkers that are usually detected by biochemical analysis, providing essential insights into the presence of oral cancer [54].

Urine is another biofluid explored for its potential in oral cancer detection. Although oral cancer primarily affects the oral cavity, specific biomarkers associated with the disease can be detected in urine samples. Urine samples can be collected repeatedly, allowing for longitudinal monitoring and the tracking of changes in biomarker levels [55]. While less commonly studied for oral cancer detection, urine can provide systemic insights into the disease. The metabolites and proteins present in urine can reflect systemic changes associated with oral cancer, offering a potential avenue for diagnostic research [39]. Table 1 highlights the different biomarkers used in the detection of oral cancer.

Among the various biomarkers identified for OSCC, the most prominent ones are determined based on their specificity and utility at different stages of the cancer. Early-stage biomarkers are useful in detecting pre-cancerous lesions. For example, tumor protein 53 (TP53), miRNA-21, EGFR, cyclin D1, and MMPs are observed in oral tissues and are indicative of abnormal cell cycle regulation and extra cellular matrix degradation. The presence of TP53 is one of the earliest molecular events in OSCC progression and plays a major role in regulating cell cycle checkpoints, apoptosis, and DNA repair. Therefore, any TP53 mutation or loss of functionality often indicates pre-cancerous tissue. Similarly, the overexpression of miRNA-21 and cyclin D1 is a hallmark of tumor initiation. Some intermediate and late-stage biomarkers include vascular endothelial growth factor (VEGF), miRNA-10b, IL-6, cancer antigen 125, etc. These late-stage biomarkers are often related to tumor progression, invasion, and metastasis. Their elevated levels in the saliva or serum are linked to increased angiogenesis, disease progression, and localized spread to other body areas. By identifying and understanding the molecular mechanisms of these biomarkers, a critical foundation for early diagnosis and disease monitoring can be leveraged. Building on these insights, the application of advanced molecular techniques enables the precise identification of OSCC biomarkers to determine the required clinical actions.

**Table 1.** Biomarkers used for the detection of oral cancer.

| Biomarker | Candidate Biomarker | Sample Type | Type of Analysis | Sensitivity | References |
|---|---|---|---|---|---|
| | NOTCH1 | Tissue | qRT-PCR | - | [56] |
| | CDKN2A | Tissue | NGS | - | [57] |
| | CASP8 | Tissue | Whole-exome sequencing | - | [58] |
| | MMP9 (mRNA) | Blood/Serum | ELISA | NA | [59] |
| | VEGF (mRNA) | Serum | qRTPCR | NA | [60] |
| | MALAT1 (lncRNA) | Tissue | qRT-PCR | NA | [61] |
| | miRNA-21 | Saliva | qRT-PCR | >65% | [62] |
| | miRNA-31 | Saliva, Plasma, Serum | qRT-PCR | >70% | [63] |
| | miRNA-145 | Saliva | qRT-PCR | >60% | [62] |
| | miRNA-196a and miRNA-196b | Tissue | qRT-PCR | >95% | [64] |
| Protein-Based Biomarkers | p53 | Saliva Serum | ELISA | >87% | [65] |
| | Ki-67 | Tissue | IHC | - | [66,67] |
| | EGFR | Saliva, Buccal cells | ELISA, RT-PCR | - | [68] |
| | EGFR | Tissue | Fluorescence molecular imaging | 100% | [69] |
| | VEGF | Serum | ELISA | 64% | [70] |
| | IL-6 | Saliva, Tissue | ELISA, Western blotting, IHC | - | [71–73] |
| | IL-8 | Saliva | ELISA, qRT-PCR | - | [74,75] |
| | TNF-α | Saliva | ELISA | | [76,77] |
| | MMP9 | Saliva | ELISA | - | [78] |
| | MMP1 | Saliva | Immuno-MALDI | - | [79] |
| | CYFRA 21-1 | Saliva, Serum | ELISA, qRT-PCR | 93.8%, 88% | [80] |

Abbreviations: qRT-PCR—quantitative reverse transcription polymerase chain reaction; IHC—immunohistochemical detection; ELISA—enzyme-linked immunosorbent assay; NGS—next-generation sequencing.

Molecular and genomic-based approaches have been widely used for oral cancer diagnostics as they are precise, thereby offering the potential for targeted therapies. Similarly, liquid biopsy detects oral squamous cell molecular biomarkers such as ctDNA, miRNAs, lncRNAs, and proteins in bodily fluids like saliva, serum, and blood for the systemic monitoring of the progression of oral lesions. Complementing these advanced techniques, immunoassays provide useful diagnostic information, although they show moderate sensitivity for early-stage detection [12]. Enzymatic and biochemical tests offer a more cost-effective approach to biomarker detection, although they may lack the specificity of newer molecular methods. When combined with imaging and cytological methods, these molecular approaches create a comprehensive diagnostic framework that enhances the accuracy and reliability of oral cancer detection and monitoring.

Among several molecular analytical techniques, nucleic acid-based approaches such as quantitative PCR (qPCR) enable the real-time monitoring and quantification of specific DNA sequences, providing valuable information related to gene mutations, deletions, and amplifications [81]. Such gene alterations are associated with oral cancer; therefore, qPCR offers high sensitivity and specificity. Additionally, quantitative reverse transcriptase polymerase chain reaction (qRT-PCR) is also used for the early detection of gene alterations. It involves multiple steps, such as RNA isolation and reverse transcription [82]. DNA microarrays and next-generation sequencing allow the analysis of genetic mutations by binding DNA or RNA fragments to a solid surface, and the microarray is scanned to measure the expression of each gene printed on the experimental slide [83–85]. If a particular gene is different from the reference one, fluorescence is emitted, and the data analyzed provide information that is critical in understanding the presence of oral lesions. However, these techniques are extremely intrusive and cumbersome, require trained personnel, and are time-consuming.

Several user-friendly and non-invasive oral cancer detection kits have been developed recently, providing various screening approaches. Table 2 showcases the features of these commercially available detection kits. Although commercially available kits have been widely used, they are limited by certain characteristics. For example, the commercial kit ViziLite Plus offers insights specifically into HPV-related oral cancer, but non-HPV-related cancer information can be compromised [86]. Similarly, there is a risk of not identifying abnormal cells in mouth areas that are not inspected, and this could lead to lower sensitivity. The precision of such kits also depends on the quality, quantity, and type of the sample collected, which might lead to an inaccurate diagnosis. Some commercially available kits rely on visual inspection methods, either by light-based or dye-based examination. For example, kits such as OralID, Oral Scan Pro, and the Bio/Screen Oral Cancer Screening Kit use fluorescent imaging technologies. This introduces subjectivity, as the interpretation could vary between different examiners with differences in experience and expertise. In addition, a general drawback is false positives due to tissue inflammation or benign lesions rather than actual malignant ones. This can lead to unnecessary follow-up procedures, resulting in greater time consumption and higher expenses. Therefore, additional research is needed to ensure the utmost accuracy and reliability in minimizing the global impact of oral cancer. In this context, biosensors have emerged as novel diagnostic tools for the detection of oral carcinoma, seeking to investigate the biomarkers responsible for oral cancer malignancies and to diagnose it in the early stages.

**Table 2.** Commercially available kits for early detection of oral cancer detection.

| Test Kit Name and Location | Techniques Used | Sample Type | Dose/ Volume | Sensitivity | Specificity | Characteristics | References |
|---|---|---|---|---|---|---|---|
| OralCDx Brush Test Suffern, NY, USA | Brush biopsy combined with traditional cytopathology | Cells | No specific dose | Less than 4% | Less than 3% | Is non-painful, captures samples from all three layers of the epithelium, and shows accuracy in detecting pre-cancerous lesions | [87] |
| ViziLite Plus Phoenix, AZ, USA | Chemiluminescence using Toluidine Blue O (TBO) dye | Exfoliative cytology | No specific dose | 100% | - | Helps dental professionals to detect oral cancer at early stages; it is effective for patients at increased risk of oral cancer, especially those with HPV infection or those over 40 years old | [88] |
| OralID Houston, TX, USA | Fluorescence visualization technique | Blood or tissue biopsies | No specific dose | 96% | - | Helps to identify oral cancer and pre-cancerous lesions using fluorescence technology, highlighting abnormalities in the mouth | [89] |
| Oral Scan Pro Thiruvananthapuram, Kerala, India | Fluorescence imaging technology | Visual inspection | No sample required | 83% | 72% | Requires no direct biological samples like saliva or blood, enables early detection via a handheld multi-modal imaging system that captures real-time images of the oral tissue | [90] |

**Table 2.** *Cont.*

| Test Kit Name and Location | Techniques Used | Sample Type | Dose/ Volume | Sensitivity | Specificity | Characteristics | References |
|---|---|---|---|---|---|---|---|
| Bio/Screen Oral Cancer Screening Kit Danbury, CT, USA | Fluorescence visualization technique | Visual inspection | No specific dose—oral mucosa | - | - | Functions as an adjunct tool for detection of potential abnormalities, highly portable | [91] |
| OraRisk HPV Test Brentwood, TN, USA | PCR | Saliva | No specific dose—oral mucosa | - | - | Specifically identifies HPV types and is recommended for patients with risk factors such as a history of oral cancer, sexually active individuals, and those with suspicious oral lesions | [86] |
| OncAlert Miami, FL, USA | ELISA | Oral mucosa | No specific dose | Under development | Under development | Identifies high-risk individuals | [92] |
| VIOME CancerDetect Bellevue, WA, USA | Meta-transcriptomics sequencing | Saliva | A few mL | 90% | 95% | It is non-invasive and employs RNA sequencing technology combined with AI for high accuracy | [93] |
| Galleri—Multicancer early detection test Menlo Park, CA, USA | Next-generation sequencing combined with methylation analysis | Blood | 20 mL | Less than 1% | 51.5% | Recommended for individuals at higher cancer risk, especially aged 50 and older | [94] |
| Quick Blue Oral Care Pvt Ltd. KIIT, India | Microscopic evaluation using Toluidine Blue O | Oral mucosa | No specific dose | 93% | 100% | Detects oral cancer cells in minutes and reduces the need for multiple biopsies. Particularly useful for cases of non-healing oral ulcers | [95] |

## 3. Advancing Biomarker Detection with Electrochemical Biosensors

Biosensors are diagnostic tools that offer compelling advantages over conventional practices. By harnessing biological interactions, biosensors provide non-invasive, rapid, and cost-effective diagnostic solutions to improve oral healthcare standards. They are suitable due to their fast responses, reliability, and direct sample reading, without needing to rely on a laboratory for analysis. Their advantages also include portability, simplicity, sensitivity, a small size, low costs, rapidity, ease of use, and wide range of concentrations [96]. They comprise a biorecognition probe for the specific target analyte of interest, a transducer, and a signal processing system integrated into a device or assembly. Within biosensors, the biorecognition probe encompasses various components, such as proteins (natural or engineered antibodies, enzymes, plant proteins), whole cells (bacteria), and nucleic acids (DNA, RNA, aptamers), which are immobilized on a transducer platform [97]. The biological signal obtained from the interaction between the analyte and biorecognition molecule is transformed and measured in electrical, fluorometric, luminometric, or colorimetric signals, confirming the presence or absence of the bio-analytes of interest [98]. These bio-analytes are commonly found in biological fluids, including the blood, serum, urine, saliva, and other bodily fluids, and they serve as essential targets for biosensor detection [99]. Utilizing these bio-analytes as targets (analytes) for biosensing applications provides a rapid and selective analysis, which is crucial for medical diagnostics and pharmaceutical development.

Within the realm of biosensors, various biosensing platforms have been developed, including lateral flow assays (LFAs) and colorimetric, optical, and electrochemical biosensors. LFAs are portable devices that exhibit a color change response upon detecting the target analyte, characterized by their simplicity, cost-effectiveness, and rapid testing capabilities in identifying biomarkers in bodily fluids. They require minimal sample preparation and usually provide results within minutes. This makes LFAs valuable for early screening, without the need for complex instruments. However, they can pose a challenge in low concentrations. Recent studies have highlighted the application of LFAs using microfluidic-based colorimetric approaches with point-of-care devices for oral cancer [100]. Similarly, colorimetric biosensors are based on the principle of a visible color change when a reaction between the bioreceptor and the analyte occurs. An immediate color change indicates the presence of a biomarker; however, the color change could also be affected by complicated samples or environmental changes [22,101]. Likewise, optical biosensors detect changes in light properties during a biological interaction, thereby being useful for the real-time monitoring of biological reactive systems [102]. These techniques may include fluorescence and surface plasmon resonance, which can function in label-free modes; however, they require sophisticated equipment, such as lasers and photodetectors, which may not be appropriate in terms of portability.

On the other hand, electrochemical biosensors record and monitor changes in electrical signals due to electrochemical reactions between the target molecules and the biorecognition molecules (enzymes, antibodies, and synthetic molecules like aptamers, DNA fragments, peptides, etc.) [103]. Electrochemical biosensors can efficiently detect biomarkers in body fluids, such as sweat, blood, or feces, making them vital for health monitoring, disease diagnosis, and environmental analysis [104–106]. Understanding the principles and capabilities of such biosensors sets the stage for the exploration of their specific use in oral cancer detection.

In electrochemical biosensors, an electrode, known as a working electrode, serves as a platform for the immobilization of biomolecules and electron mobility, in addition to a counter and reference electrode. This working electrode is the primary electrode where the target interacts with the biorecognition element and is typically modified with nanomaterials and crosslinkers to increase the surface area and conductivity and ensure efficient immobilization [107]. The counter electrode completes the necessary circuit to sustain the current generated at the working electrode, whereas the reference electrode maintains a known and constant potential at the working electrode. Usually, a glassy carbon electrode, a gold electrode, a screen-printed electrode (SPE), or indium tin oxide (ITO) glass is used as the working electrode in the electrochemical setup. Nanomaterials such as graphene, gold nanoparticles, carbon nanotubes, etc., are deposited onto the electrode to increase electron transfer and amplify signals [108]. The immobilization of biomolecules on the electrode is achieved using molecular crosslinkers like carbodiimide 1-ethyl-3-(3-dimethylaminopropyl) carbodiimide (EDC)/N-hydroxysuccinimide (NHS), glutaraldehyde, 1-pyrenebutanoic acid, succinimidyl ester, (3-aminopropyl)triethoxysilane (APTES), etc. [109]. These crosslinkers secure the molecules onto the platform and maintain biological activity for target detection, which is related to the biosensor's sensitivity and LOD. In terms of electrochemical strategies, several techniques are typically chosen for biosensor development, including cyclic voltammetry (CV), differential pulse voltammetry (DPV), square wave voltammetry (SWV), amperometry, and electrochemical impedance spectroscopy (EIS) [110].

### 3.1. Amperometric Biosensors

Amperometric biosensors function by primarily measuring the current generated by a redox process at a fixed voltage. Such biosensors include an electrode surface that facilitates electron transfer reactions. Upon binding with the target analyte, the recognition element triggers an electrochemical reaction, resulting in a measurable current. The concentration of the desired analyte within the sample is directly correlated with the amplitude of the current. A typical electrochemical technique performed is chronoamperometry, where a constant potential is applied, and the resulting current is measured as a function of time. This change in current is proportional to the amount of redox species that are oxidized and reduced. The capability of amperometric biosensors to enable the reliable and non-invasive detection of biomarkers presents a transformative opportunity for early intervention and enhanced patient outcomes [111].

Amperometric biosensors provide a fast and cost-effective solution for the detection of proteins and DNA species associated with oral cancer [111]. Two gene-specific strategies (one DNA sensor and an immunosensor) were employed by Povedano et al. for the development of electrochemical biosensors for the quick detection of DNA methylation. They used functionalized MBs, attaching 5-methylcytosine (mC) as a bio-receptor and achieving amperometric detection by using a hydrogen peroxide/hydroquinone ($H_2O_2$/HQ) system (Figure 4(i)) [112]. The aberrant methylation of DNA is usually observed in tumor cells, which helps in the promotor regions of tumor suppressor genes. Therefore, the inactivation of tumor suppressor genes has been shown to be an important mechanism in the development of human oral cancer.

Hypoxia-inducible factor-1 alpha (HIF-1$\alpha$) is a transcription factor that plays a role in oral tumor growth and metastasis by regulating genes associated with the cellular response to hypoxia. Martin et al. developed an immunosensing amperometric detection tool to analyze HIF-1$\alpha$ on disposable screen-printed electrodes (Figure 4(ii)) [113]. Magnetic immunoconjugates of carboxylic acid-modified magnetic particles (HOOC-MBs) were utilized to selectively capture the target protein with an EDC/NHS crosslinking mechanism. The biosensor showed great selectivity and a low LOD of 76 pg/mL. This was followed by the determination of HIF-1$\alpha$ in raw saliva samples. Further research in more sensitive electrochemical sensors has paved the way for spatially resolved biosensing and imaging applications, indicating the potential of these biosensors for point-of-care testing in the diagnosis of various diseases. In another study, Torrente-Rodriquez et al. reported the simultaneous detection of IL-8 mRNA and IL-8 protein oral cancer biomarkers in undiluted human saliva using functionalized MBs (Figure 4(iii)) [114]. The use of MBs proves useful for improving the sensitivity and minimizing matrix effects, which makes them a relevant tool in the design of biosensors for complex samples.

Recent advances in amperometric biosensors for medical diagnostics have significantly expanded their applicability in disease detection. The fabrication of amperometric biosensors for the detection of MMP-7 has seen significant advancements, particularly with the use of palladium (Pd)-functionalized carbon nanocomposites as another biomarker. Palomar et al. studied the detection of MMP-7 by integrating peptide-decorated gold nanoparticles and carbon nanotubes, achieving an LOD of 6 pg/mL [115]. Wei et al. further improved the detection limit using a Pd-functionalized carbon nanocomposite, as shown in Figure 4(iv) [116]. A glassy carbon electrode was used to immobilize peptides on gold-modified reduced graphene oxide (rGO), which improved the biosensor's conductivity and provided more reaction sites for immobilization. Catalytic precipitation with Pd-functionalized carbon nanospheres amplified the signal obtained, resulting in an LOD of 17.38 fg/mL [116]. Table 3 highlights the recent developments in amperometric-based biosensors for the detection of oral cancer biomarkers.

Despite demonstrating significant advancements in the detection of oral cancer, amperometric biosensors possess certain limitations that need to be considered. Signal reduction may occur due to fouling agents and other interfering substances, which poses challenges in maintaining consistent and reliable detection [117]. Furthermore, some amperometric biosensors may have poor specificity, limiting their effectiveness in specific applications [118]. Another aspect of their use that must be considered is the possibility of cross-sensitivity to other substances. These issues emphasize the importance of calibrating the instruments precisely and considering possible interferences while performing detection. A bottleneck in the translation of such biosensors from the lab to industry lies in the stability and activity of the species involved in the electrochemical reaction. Additional efforts are required to address the effects of the temperature, buffer, and pH and the appropriate potential to maintain the stability of the system. Overall, careful calibration that considers the real environmental conditions and potential interferences is essential in accurately applying amperometric biosensors, especially in complex samples such as those for oral cancer detection.

**Table 3.** Amperometric biosensors and their experimental conditions for oral cancer.

| Electrode/Platform | Biomarker | Analyte | Method | Detection Limit | Normal and Cancer Patient Levels | Reference |
|---|---|---|---|---|---|---|
| SPCE/Strep–MBs | 5–mC | Human saliva, serum, and urine | Amperometry | 0.3–1 fg/mL | - | [112] |
| SPCE/HOOC–MBs | HIF–1α | Human saliva | Amperometry | 76 pg/mL | - | [113] |
| SPE/HOOC–MBs | IL–8 protein | Human saliva | Amperometry | 72.4 pg/mL | IL–8 protein—patient: 720 pg/mL, normal: 250 pg/mL | [114] |
| GCE/Au–rGO | MMP–7 | Human serum | Amperometry | 1 ng/mL | - | [116] |

Abbreviations: SPE—screen-printed electrode, HOOC—carboxylic acid, MBs—magnetic beads, SPCE—screen-printed carbon electrode, Strep-MBs—streptococcal magnetic beads, GCE—glassy carbon electrode, Au—gold, rGO—reduced graphene oxide, IL-8—interleukin-8, HIF-1α—hypoxia-inducible factor-1 alpha, 5-mC—5-methylated cytokine, MMP-7—matrix metalloproteinase-7.

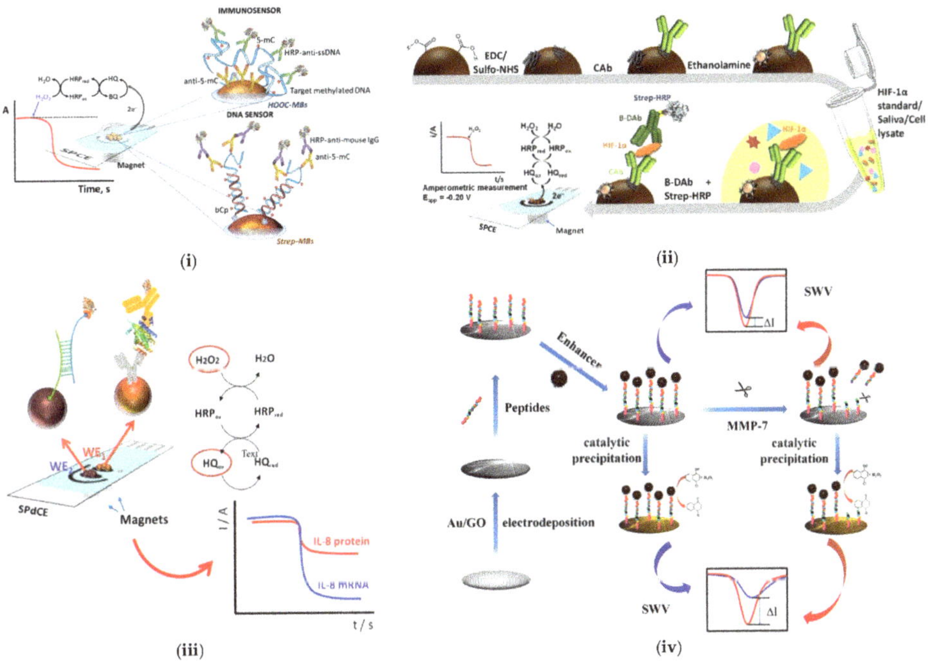


**Figure 4.** Amperometric biosensors using different immobilization strategies for detection of various oral cancer biomarkers. Schematic representation of the fabrication procedure for (**i**) 5–mC Reprinted with permission from [112]; (**ii**) HIF–1α Reprinted with permission from [113]; (**iii**) IL–8 protein and IL–8 mRNA Reprinted with permission from [114]; (**iv**) MMP–7 Reprinted with permission from [116].

## 3.2. Voltammetric Biosensors

Voltammetric biosensors are a class of electrochemical sensors that have gained significant attention in medical diagnostics. Voltammetric biosensors measure the current when the potential is applied at the working electrode, enabling the detection and quantification of target analytes. This binding event between the biorecognition element and the target analyte triggers an oxidation/reduction reaction that forms the basis of electrochemical detection. Several electrochemical techniques are employed, such as CV, DPV, and SWV, to analyze this redox behavior. In CV, the potential of the working electrode is swept linearly and reversed, which generates a current vs. voltage graph, providing the relevant information about the reaction that occurs at the electrode surface. In the case of DPV, small potential pulses are applied to enhance the signal-to-noise ratio, which improves the detection limits, while small steps are applied in SWV. The current generated in such biosensors is directly related to the concentration of the analyte. For example, Tiwari et al. developed a biosensor based on the CV and DPV methods to detect CYFRA-21-1 by immobilizing anti-CYFRA-21-1 antibodies on an ITO platform, as shown in Figure 5(i) [119]. Lanthanum hydroxide ($La(OH)_3$) has several electrochemical properties, including electron transfer mobility, which provides more free binding and adsorption sites for the attachment of molecules like L-cysteine. This increases the loading of the anti-CYFRA-21-1 antibodies, which enhances the stability of the electrode. The immunosensor showed a low detection limit of 0.001 ng/mL, with a response time of 5 min. This immunosensor has high sensitivity compared to other commercially available ELISA kits. In a similar context, Jafari et al. immobilized anti-CYFRA-21-1 antibodies on a gold electrode using a cysteamine and glutaraldehyde self-assembly composite [120]. This composite biosensor resulted in a low limit of quantitation of 2.5 ng/mL when using CV and SWV. Because of its quick, reliable, and user-friendly nature, the fabricated layer, with a high current intensity, can be used to monitor oral abnormalities. However, the detection limit is 100 times lower than that of the $La(OH_3)$-based nanomaterial reported, and it demonstrates a larger surface area, better redox reactivity, and excellent electrochemical behavior. However, given the need for a low-cost biosensor, $La(OH)_3$ could prove to be too expensive, since lanthanum is a rare earth metal. In this case, the biosensor fabricated by Jafari et al. could be useful.

Kumar et al. developed a biocompatible serine-functionalized nanostructured zirconia platform that showed a low detection limit for the oral cancer biomarker CYFRA-21-1 [121]. This biosensing platform was fabricated with nanostructured zirconia with rGO, which exhibited improved electron transfer kinetics and increased sensitivity [121]. In 2015, the same group utilized silanized nanostructured zirconia for the covalent immobilization of monoclonal anti-CYFRA-21-1 antibodies, resulting in a valuable and stable biosensing platform with a lower detection limit of 0.001 ng/mL [122]. More recently, Kumar et al. synthesized nanodot zirconia and developed an efficient biosensing platform with a broad linear detection range and excellent sensitivity for CYFRA-21-1 [123]. However, the LOD was lower as compared to that of the silanized nanostructured zirconia. This indicates that the zirconia surface, when functionalized with silane coupling agents, provides stronger adhesion to the biomolecules, thereby achieving higher surface reactivity and, hence, a lower limit of detection.

The label-free electrochemical detection of oral cancer-related biomarkers has been a recent research focus. Verma et al. demonstrated the fabrication of an electrochemical immunosensor to detect the salivary oral cancer biomarker IL-8 using a gold nanoparticle–rGO composite material (Figure 5(ii)) [124]. The synergy between the gold nanoparticles and rGO led to a faster response and higher sensitivity due to the enhanced electron transfer behavior of the nanocomposite, resulting in a detection limit of 72 pg/mL, as well as excellent specificity in human saliva samples. Liu et al. further advanced this work by creating an MMP-1 immunosensor based on a gold nanoparticle/polyethyleneimine/rGO nanocomposite [125]. Braiek et al. expanded the application of these nanocomposites to the detection of IL-8 using a boron-doped diamond electrode modified with magnetic nanocomposites [126]. These studies collectively demonstrate the potential of several nanocomposites for the sensitive and specific detection of salivary oral cancer biomarkers.

Another mechanism involved in electrochemical biosensors for oral cancer-based DNA detection is based on nicking endonuclease signal amplification (NESA). This is a technique used to detect specific DNA sequences by utilizing an enzyme that only cuts one strand of the DNA to generate a signal that can detected. An NESA-based approach on an ITO electrode was reported by Tan et al. for the detection of DNA to target ORAOV1 in saliva [127]. The DPV responses from the system were detected to evaluate the sensitivity and dynamic range of the biosensor, resulting in an LOD of 0.35 pM. Regarding similar NESA approaches, a sensitive biosensor was successfully constructed by Hu et al. with a novel host–guest recognition system consisting of tryptophan-linked DNA, methyl viologen (MV2+), and cucurbit uril for the sensitive detection of oral cancer genes, as shown in Figure 5(iii) [128]. Platinum (Pt)- and Pd-based nanomaterials were used to detect ORAOV1 using hairpin DNA labeled with tryptophan. The catalysis of the Pt and Pd–molybdenum disulfide composite was not only used to promote electron transfer but also for signal amplification, and the biosensor was reported as a promising tool for ORAOV1 detection on a glassy carbon electrode. Table 4 highlights the recent developments in voltammetric-based biosensors for the detection of oral cancer biomarkers.

One of the key advantages of voltammetric-based electrochemical biosensors is their high sensitivity, which enables the detection of oral cancer biomarkers at low concentrations [129]. Furthermore, these biosensors are known for their rapid detection capabilities, cost-effectiveness, and low production costs, making them accessible for widespread use [130]. However, voltammetric biosensors face limitations in complex oral environments, as well as having a restricted dynamic range and reproducibility [22]. These biosensors utilize electrochemical amplification, surface modifications, nanomaterial integration, specific recognition elements, and sophisticated signal processing to achieve exceptional sensitivity, even at trace biomarker levels. However, the complex oral environment presents challenges to the stability and robustness of voltammetric biosensors, necessitating the use of protective coatings, surface modifications, advanced materials, fluid handling systems, calibration and referencing, optimized recognition elements, environmental controls, signal processing techniques, robust design, and routine maintenance. By implementing these strategies, the reliability and performance of voltammetric biosensors can be enhanced in detecting oral cancer biomarkers, thereby supporting their potential application in early diagnosis and effective treatment management.

**Table 4.** Voltammetric biosensors and their experimental conditions for oral cancer.

| Electrode/Platform | Biomarker | Analyte | Method | Detection Limit | Normal and Cancer Patient Levels | Reference |
|---|---|---|---|---|---|---|
| ITO/APTES/nd$ZrO_2$ | CYFRA-21-1 | Human saliva | CV/DPV | 0.5 ng/mL | Normal—3.3 ng/mL CYFRA-21-1— 17.46 ± 1.46 ng/mL | [123] |
| ITO/L-Cys-La $(OH)_3$ | CYFRA21-1-1 | Human saliva | CV/DPV | 0.001 ng/mL | Normal—3.3 ng/mL CYFRA-21-1— 17.46 ± 1.46 ng/mL | [119] |
| ITO/Serine/n$ZrO_2$ | CYFRA-21-1 | Human saliva | CV/DPV | 0.01 ng/mL | Normal 3.3 ng/mL CYFRA-21-1— 17.46 ± 1.46 ng/mL | [121] |
| ITO/APTES/$ZrO_2$ rGO | CYFRA-21-1 | Human saliva | DPV | 0.122 ng/mL | Normal—3.3 ng/mL CYFRA-21-1— 17.46 ± 1.46 ng/mL | [131] |
| GE/Cys A | CYFRA-21-1 | Human saliva | CV/DPV/SWV | 2.5 ng/mL | Normal—3.3 ng/mL CYFRA-21-1— 17.46 ± 1.46 ng/mL | [120] |
| ITO/ZnO-rGO | IL-8 protein | Human saliva | CV/DPV | 51.53 ± 0.43 pg/mL | Normal—250 pg/mL IL-8 protein, patient—720 pg/mL | [132] |
| ITO/AuNPs-rGO | IL-8 | Human saliva | CV/DPV | 72.73 ± 0.18 pg/mL | Normal—250 pg/mL IL-8 protein, patient—720 pg/mL | [124] |
| GCE/Au/Pt and Pd-$MoS_2$ | ORAOV1 | Artificial saliva | CV/DPV | 3 fg/mL | - | [128] |
| ITO/eMB/Nt.BstNBI | ORAOV1 | Human saliva | DPV | 0.35 pg/mL | - | [127] |

Abbreviations: ITO—indium tin oxide, ZnO—zinc oxide, rGO—reduced graphene oxide, GE—gold electrode, Cys-A—cysteine-A, AuNPs—gold nanoparticles, APTES—(3-aminopropyl) triethoxysilane, $ZrO_2$—zirconium oxide, Cys—cysteine, La $(OH)_3$—lanthanum hydroxide, GO—graphene oxide, GCE—glassy carbon electrode, n$ZrO_2$—zirconium oxide, Au—gold, Pt and Pd-$MoS_2$—platinum and palladium molybdenum disulfide, eMB—methylene blue, Nt.BstNBI—nicking enzyme, L-Cys—L-cysteine, nd$ZrO_2$—nanodiamonds zirconium oxide, IL-8—interleukin-8, CYFRA-21-1—cytokeratin 19 fragment, ORAOV1—oral cancer overexpressed 1, CV—cyclic voltammetry, DPV—differential pulse voltammetry, SWV—square wave voltammetry.

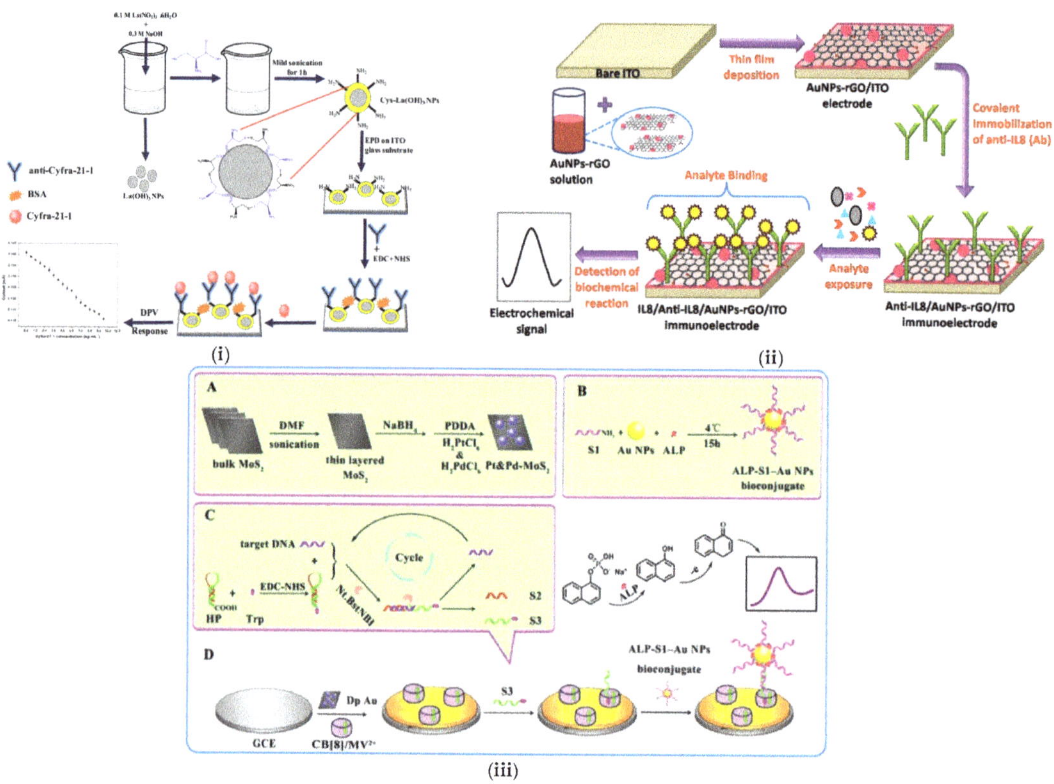


**Figure 5.** Voltammetric biosensors using different immobilization strategies for detection of various oral cancer biomarkers. Schematic representation of the fabrication procedure for (**i**) CYFRA–21–1 Reprinted with permission from [119]; (**ii**) IL–8 Reprinted with permission from [124]; (**iii**) ORAOV1: (**A–C**) represent the preparation steps for composite nanomaterial/bioconjugates and target amplification, while (**D**) shows the preparation process of the biosensor. Reprinted with permission from [128].

### *3.3. Impedimetric Biosensors*

Impedimetric biosensors are advantageous for point-of-care diagnostics because they can identify biorecognition events by measuring the changes in the electrical properties at the electrode/electrolyte interface over a range of frequencies. In impedimetric biosensors, the activity of the target analyte is proportional to the output of an electrical impedance, and a common reporting method is EIS. It can easily help to determine the properties of the processes in the bulk, as well as at the electrode interface, which are represented in the form of a Nyquist plot or Bode plot as a function of the frequency. A Nyquist plot generally consists of a semi-circular region and a straight line that represent the charge transfer process and the diffusion process, respectively, whereas a Bode plot consists of a logarithm of the absolute values of the impedance and phase plotted against the logarithm of the frequency [133]. These plots are used to understand the electrochemical processes involved in the biosensing mechanism, such as charge transfer, electroactive species adsorption, mass transfer, and electrolyte resistance, each represented by an equivalent electrical circuit. In a Nyquist plot, the change in impedance obtained indicates when the receptor binds to the target analyte, and it is proportional to the diameter of the semi-circle. This measurement does not depend on the redox species present in the electrolyte. In order to maintain stability and repeatability, nanomaterials functionalized with different crosslinkers are critical in attaching the relevant receptors. For example, Choudhary et al. developed a highly sensitive and selective label-free impedimetric biosensor on a gold electrode by detecting CD59 as an oral cancer biomarker (Figure 6(i)) [134]. CD59 is a serine protease and a complement restriction factor that can be used as a relevant early-stage biomarker for oral cancer diagnosis. The electrode was functionalized with an L-cysteine self-assembled monolayer to provide carboxyl functional groups for the attachment of CD59 antibodies using EDC-NHS functionalization. EDC-NHS was used as a coupling agent to immobilize the anti-CD59 antibodies by forming covalent bonds between the cysteine-functionalized -COOH groups and the -NH2 groups on the antibodies. This resulted in an LOD of 0.38 fg/mL and 1.46 fg/mL in standard buffer and untreated human saliva samples, respectively [134].

On the other hand, Kumar et al. explored and reported the detection of the CYFRA-21-1 biomarker, as shown in Figure 6(ii) [135]. Herein, the researchers used yttrium oxide ($nY_2O_3$) nanoparticles functionalized using APTES to immobilize anti-CYFRA-21-1 antibodies, followed by a blocking agent [135]. $nY_2O_3$ is known to have a high quantum yield, which creates a highly conductive thin film, thereby making it a potential candidate for the development of biosensors. It resulted in an LOD of 0.33 ng/mL. Similarly, the studies by Pachauri et al. presented innovative biosensing platforms for CYFRA-21-1, using a nanocomposite of $ncCeO_2$-rGO (cerium oxide nanocubes–rGO) [136]. The use of $ncCeO_2$-rGO provided a larger surface area and faster electron transfer due to the conducting nature of rGO. This also resulted in a better LOD of 0.625 pg/mL as compared to Kumar et al.'s work. However, such studies require a more critical analysis of the potential limitations and challenges in translating these platforms to clinical use. Additionally, the specificity of the biosensors to CYFRA-21-1 in the presence of other biomolecules, such as those found in saliva, needs to be thoroughly addressed.

In addition to CD59 and CYFRA-21-1, several other biomarkers have been used for the electrochemical detection of OSCC. For example, Cancerous Inhibitor of Protein Phosphatase 2A (CIP2A) is highly expressed in OSCC cell lines, in addition to lung and breast cancers. Figure 6(iii) highlights the fabrication steps for such a biosensor, where vertically aligned carbon nanotube (VACNT) array electrodes were used for the electrochemical detection of CIP2A [137]. They had a large surface area that favored electrical conductivity and reactivity, leading to higher biosensor sensitivity compared to the ELISA test. The target CIP2A was detected using CIP2A antibodies (anti-CIP2A) attached to interdigitated electrodes, which acted as electrochemical transducers. Although this biosensor achieved detection in the range of pg/mL, scaling up and maintaining the perfect alignment of VACNT arrays can be expensive and challenging. This is especially important when less expensive nanomaterials, such as graphene or cysteine, are available and offer a similar detection range.

The development of impedimetric biosensors also leverages advanced functional polymers, which play a critical role in the sensitivity and biocompatibility of oral cancer biomarkers. The studies by Aydin et al. present the development of an immunosensor for the detection of IL-8 using a variety of conjugated polymers containing epoxy side groups; see Figure 6(iv) [138]. This study demonstrates the successful immobilization of anti-IL-8 receptors on modified disposable ITO electrodes bound to the epoxy groups of star polymers via a covalent bond. The star polymers provide a convenient interface due to the presence of linear polymer chains in a branched fashion, and the system exhibits remarkable features like a low cost and good feasibility. It resulted in a detection limit of 3.3 fg/mL. Interestingly, Ma et al. developed a ratiometric electrochemical DNA biosensor for ORAOV1, with a detection limit of 12 fM in artificial saliva samples, as shown in Figure 6(v) [139]. The DNA is hybridized with a ferrocene-labeled hairpin probe and separated by exonuclease III, which then releases the target and triggers an amplification cycle, whereas the remaining probe is used for biosensing purposes. The larger the amount of the ORAOV1 DNA biomarker present in the saliva sample, the higher the response detected and, hence, the larger the impedance on the Nyquist plot. These biosensors exhibited a linear calibration curve over a wide concentration range; however, the authors did not critically analyze the potential for non-specific binding or interference from other biomarkers. Additionally, the comparison with commercial ELISA kits was limited to the detection limits and analysis costs, without considering other important factors such as the accuracy and precision. Further research is needed to address these limitations and confirm the clinical utility of these biosensors.

The capacity of impedimetric biosensors to detect the presence of cancer biomarkers makes them precise and sensitive instruments that are essential in the early identification of oral cancer. These biosensors can evaluate biomarkers without labeling the target molecules, simplifying sample preparation and preserving the integrity of biological samples while enhancing their non-invasiveness. In clinical settings, real-time analysis capabilities enable prompt diagnostic feedback, essential for rapid decision-making. Additionally, the small sample size required for impedimetric biosensor testing minimizes the risk of side effects during blood draws and improves patients' comfort. Table 5 highlights the recent developments in impedimetric-based biosensors for the detection of oral cancer biomarkers.

**Table 5.** Impedimetric biosensors and their experimental conditions for oral cancer.

| Electrode/Platform | Biomarker | Analyte | Method | Detection Limit | Normal and Cancer Patient Levels | Reference |
|---|---|---|---|---|---|---|
| GE/Cys | CD59 | Human saliva | CV/EIS | Treated saliva: 0.84 ± 0.04 fg/mL; raw saliva: 1.46 ± 0.05 fg/mL | Normal—7.8 ng/mL CD59 patients—27.3 ng/mL | [134] |
| Silicon oxide wafer/$Al_2O_3$ | CIP2A | Human saliva | CV/EIS | 0.24 pg/ml | - | [137] |
| ITO/nc$CeO_2$–rGO | CYFRA-21-1 | Human saliva | DPV/EIS | 0.625 pg/mL | Normal—3.3 ng/mL CYFRA-21-1—17.46 ± 1.46 ng/mL | [136] |
| ITO/APTES/n$Y_2O_3$ | CYFRA-21-1 | Human saliva | EIS | 0.01 ng/mL | Normal—3.3 ng/mL CYFRA-21-1—17.46 ± 1.46 ng/mL | [135] |
| ITO–PET | IL-8 | Human serum and saliva | CV/EIS | 3.3 fg/mL | IL-8 protein, patient: 720 pg/mL; normal: 250 pg/mL | [138] |
| GE/MB–PP1–MCH | ORAOV1 | Artificial saliva | EIS | 12.8 fg/mL | - | [139] |

Abbreviations: ITO—indium tin oxide, PET—polyethylene terephthalate, nc$CeO_2$—nanocubes cerium oxide, rGO—reduced graphene oxide, $Al_2O_3$—aluminum oxide, GE—gold electrode, MB-PP1—methylene blue-labeled hairpin probe, MCH—6-mercapto-1-hexanol, Cys—cysteine, APTES—(3-aminopropyl) triethoxysilane, n$Y_2O_3$—yttrium oxide, IL-8—interleukin-8, CYFRA-21-1—cytokeratin 19 fragment, CIP2A—Cancerous Inhibitor of Protein Phosphatase 2A, ORAOV1—oral cancer overexpressed 1, CD59—cluster of differentiation 59, CV—cyclic voltammetry, EIS—electrochemical impedance spectroscopy, DPV—differential pulse voltammetry.

Nonetheless, electrochemical biosensors' practical applications could be improved by addressing the issue of the non-specific binding of non-target compounds, resulting in reduced sensitivity and selectivity. The limitations also include linearity, stability, and the precision of the instruments utilized. Biological samples, like saliva, often contain a diverse range of substances that could interfere with measurements, potentially leading to inaccuracies. Additionally, the precision of these biosensors is highly dependent on stable environmental conditions, such as consistent temperature and pH levels, meaning that they might require a controlled testing environment. Another noteworthy concern is electrode fouling, where substances from the samples may adhere to the sensor's electrodes, impairing its performance and necessitating more maintenance, like frequent calibration. There are also practical hindrances, such as the sophistication and precision of the instrumentation necessary for accurate readings. This requirement can increase the costs and restrict the distribution and use of such biosensors, especially in low-resource settings.

Ongoing advancements in sensor technology, materials science, and signal processing are crucial to overcome these challenges, ultimately facilitating wider adoption and more effective point-of-care testing in various medical environments. Table 6 briefly describes the advantages and disadvantages of electrochemical biosensors.

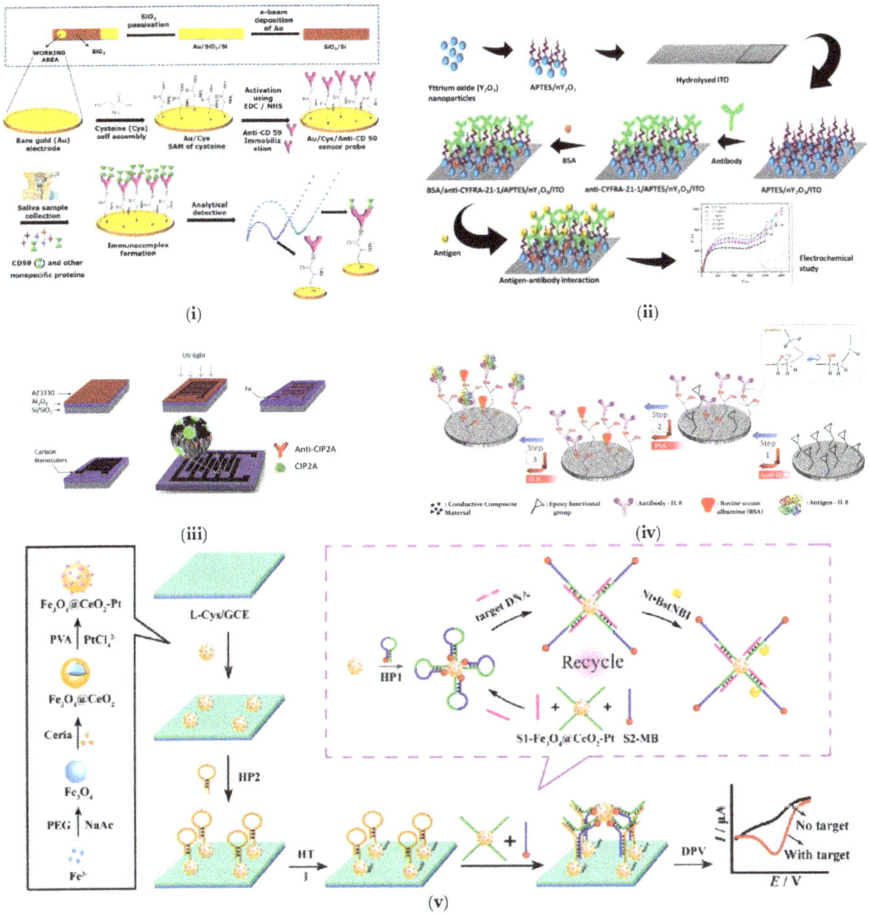


**Figure 6.** Impedimetric biosensors using different immobilization strategies for detection of various oral cancer biomarkers. Schematic representation of the fabrication procedure for (**i**) CD59 Reprinted with permission from [134]. (**ii**) CYFRA–21–1 Reprinted with permission from [135]; (**iii**) CIP2A Reprinted with permission from [137]. (**iv**) IL–8 Reprinted with permission from [138]; (**v**) ORAOV1 (DNA) Reprinted with permission from [139].

**Table 6.** The advantages and disadvantages of electrochemical biosensors for the detection of oral cancer biomarkers.

| Advantages | Disadvantages |
|---|---|
| High sensitivity with low levels of concentrations of cancer biomarkers | Interference from electroactive species in complex samples |
| Rapid response time | Batch-to-batch variability in fabricated sensors |
| Potential for miniaturization for point-of-care devices | Difficulty in integration into clinical practice |
| Real-time monitoring capabilities | Interfacial area degradation over time |
| No need for complex instrumentation | Accuracy can be affected by temperature and pH |
| Minimal sample preparation and consumption | Potential cross-reactivity and complex data interpretation |

## 4. Challenges and Future Perspectives

The potential of non-invasive electrochemical biosensors for the detection of oral cancer is vast. This review paper explores the cutting-edge developments in electrochemical biosensors for early and accurate oral cancer diagnosis. While significant progress has been made, several challenges persist in this field. One of the main challenges is identifying specific and sensitive biomarkers that can reliably indicate the presence of oral cancer. Translating laboratory research into practical, user-friendly biosensing devices presents technical and engineering obstacles. The standardization of protocols and the validation of biosensors' performance across different patient populations and clinical settings are crucial in establishing their clinical utility. Moreover, integrating electrochemical biosensors into the existing healthcare infrastructure requires collaborative efforts from researchers, clinicians, and regulatory authorities. However, the outlook remains promising, as advances in nanotechnology, biomarker discovery, and machine learning algorithms hold the potential to overcome these challenges. With continued research and innovation, electrochemical biosensors are poised to revolutionize oral cancer detection, enabling early intervention and personalized treatments and ultimately contributing to the improved management and prognosis of oral cancer patients. Amperometric, voltammetric, and impedimetric biosensors offer specific advantages and limitations in the context of oral cancer detection.

Amperometric biosensors offer the advantage of simplicity, easy on-chip integration, and potentially low costs and fast response times [140,141]. Their sensitivity and versatility enable the accurate and early detection of cancer biomarkers, contributing to improved diagnostic processes and treatment management, making them a promising tool for oral cancer and other malignancies [26]. However, challenges such as regular calibration and potential interference in the detection process should be carefully addressed for their effective use in complex samples, as in oral cancer detection [142]. Voltammetric biosensors have the advantages of simple on-chip integration and potentially low costs, serving as beneficial point-of-care diagnostic tools [140]. They can also provide comprehensive insights for the targeting and detection of cancer biomarkers. However, they require careful electrode design and corresponding materials to ensure the highly selective and sensitive detection of cancer biomarkers [143]. Impedimetric biosensors, on the other hand, possess the potential for the early detection and diagnosis of oral cancer, attributed to their high sensitivity, non-invasive nature, label-free detection methods, and capabilities for real-time monitoring [144]. On the other hand, they could be more valuable in other biomarker categories, where difficulties restricting their use are not present. Complex sample matrices and ambient conditions can also impact them, and they need precise devices to measure exact values. To summarize, impedimetric biosensors provide non-invasive and real-time monitoring capabilities, amperometric biosensors offer sensitivity and simplicity for oral cancer diagnosis, and voltammetric biosensors enable point-of-care diagnostics. Selecting a biomarker type for a given biomarker class is contingent upon the requirements for precision, user-friendliness, and environmental factors.

Furthermore, a range of studies have explored the potential of using machine learning and electronic tongues to discriminate between saliva samples from oral cancer patients and healthy individuals. Braz et al. demonstrated that supervised machine learning algorithms, especially support vector machines and random forest, achieved high accuracy in this discrimination [145]. Zlotogorski et al. similarly discovered that the Fourier transform-based spectrum of salivary exosomes and computer-aided discrimination analysis could be used to accurately distinguish between these two groups [146]. Kouznetsova et al. further expanded on this work by using saliva metabolites and machine learning to distinguish between oral cancer and periodontitis, achieving accuracy of 79.54% [147]. Lastly, Ishikawa et al. identified specific salivary metabolite biomarkers for oral cancer screening, which could be integrated into a non-invasive diagnostic method [50]. These studies collectively highlight the potential of machine learning and electronic tongues in improving the accuracy and efficiency of oral cancer diagnosis. In addition, future studies should also focus on recurring predictor biomarker candidates to reduce the mortality rates.

Modern AI systems implemented in image analysis and pattern recognition are attracting attention. They use deep learning algorithms to analyze clinical and microscopic images so as to identify subtle changes within malignant tissue. These algorithms are trained on vast datasets, potentially distinguishing between benign and malignant tissues. This will also be a valuable resource during clinical examinations so that healthcare providers can give a second opinion and monitor the patient's health for better prospects. Smart devices and mobile health applications also show promising results for preliminary oral cancer screening [148,149]. These systems include either mouth safeguards or intraoral devices that are integrated with biomarkers, highly specialized cameras, and AI-powered technologies and can help to regularly monitor any potential risk [150]. Popovic et al. presented a platform that primarily addresses xerostomia via a personalized approach that involves a microfluidic chip embedded with a tooth model for targeted therapeutic delivery based on real-time oral cavity measurements [151]. Similarly, Yaduvanshi et al. proposed a machine learning-based approach that utilizes a modified local binary pattern (MLBP) to analyze oral lesions using histopathological images. The MLBP technique is employed for texture feature extraction, using deep convolutional neural networks to detect changes in the oral cavity due to cancer [152]. There is potential to extend this work by using other frameworks to identify diseased regions in oral cancer images. Deep learning (DL) is a subset of machine learning that simplifies complex data algorithms and helps to identify similar patterns; it could be used by healthcare professionals to monitor the conditions of patients. A notable increase has been observed in DL-based techniques for oral cancer diagnosis and prognostic prediction. Warin et al. discussed and reviewed the application of DL techniques by analyzing various DL models, focusing on the accuracy, sensitivity, and specificity across different imaging modalities [153]. The authors recommended the development of specific reporting protocols in a standardized fashion to improve the transparency and address issues like heterogeneity for improved diagnostic accuracy in DL applications. Hence, there is a need for future studies to implement standard methodologies to verify clinical images consisting of both cancerous and non-cancerous stages, so as to ensure the accuracy of the data while establishing the need for its usage in medical AI studies. This will provide a promising approach to improved DL design so as to potentially increase oral cancer patients' survival rates. Interestingly, neural networks and particularly convolutional neural networks (CNNs) can help to enhance the accuracy of early detection by identifying subtle patterns indicative of early-stage lesions. By integrating genetic biomarkers, histopathological images, and clinical parameters with DL architectures, the development of predictive models for treatment response monitoring and prognosis is facilitated. Huang et al. utilized an oral cancer (lips and tongue) dataset consisting of 87 sets of cancerous and 44 sets of non-cancerous oral images [154]. They used a CNN-optimized combined algorithm to extract features from the images, removing noise while enhancing the visibility and diversity of the data. However, several challenges, such as limited and unequal sample sizes, can hinder such a model's ability to create a generalized algorithm for future use. Building upon similar models, Lin et al. presented an effective smartphone-based oral cancer imaging diagnosis technique, using 688 lesion images and 760 normal mucosa images. A high-resolution network (HRNet-W18), an advanced DL model, was used to classify positive and negative cases of oral cancer, potentially creating a viable option for healthcare workers and patients in remote areas [155]. Thakuria et al. followed a similar approach to the application of DL techniques in the diagnosis of oral cancer with the use of smartphone and DSLR image analysis. The authors conducted a systematic review of 25 papers, highlighting several models for multiclass classification and object detection and focusing on parameters like precision, specificity, and accuracy [156]. However, a similar limitation could be observed regarding the need for larger and more diverse datasets from multiple healthcare facilities that can be integrated with lesion characteristics for improved diagnostic capabilities. As the research continues to evolve, there is immense potential for breakthroughs in oral cancer treatment via artificial intelligence-driven diagnostics and more advanced imaging techniques, leading to quicker and less invasive procedures. Moreover, the discovery of appropriate biomarkers could be helpful for the early diagnosis of oral cancer. This could eventually result in higher survival rates for patients.

## 5. Conclusions

This review delves into the various applications of biosensors with non-invasive diagnostics for oral cancer. Integrating biomarkers and biofluids into the development and application of biosensors enhances the potential for accurate, reliable, and non-invasive oral cancer detection methods. While each type of biosensor offers distinct advantages, further research and development are essential to address challenges such as optimization for clinical settings, the validation of real-world applications, and integration into established diagnostic pathways. Understanding these biosensors' unique capabilities and limitations is crucial in harnessing their full potential in oral cancer detection, ultimately contributing to improved patient care and outcomes. As the field of electrochemical biosensing continues to advance, the strategic integration of these biosensors into clinical practice holds promise to transform the landscape of oral cancer detection and management.

**Author Contributions:** This manuscript was written with the contributions of all authors. All authors have read and agreed to the published version of the manuscript.

**Funding:** This research received no external funding.

**Conflicts of Interest:** The authors declare no conflicts of interest.

## References

1. Badwelan, M.; Muaddi, H.; Ahmed, A.; Lee, K.T.; Tran, S.D. Oral Squamous Cell Carcinoma and Concomitant Primary Tumors, What Do We Know? A Review of the Literature. *Curr. Oncol.* **2023**, *30*, 3721–3734. [CrossRef] [PubMed]
2. Omura, K. Current status of oral cancer treatment strategies: Surgical treatments for oral squamous cell carcinoma. *Int. J. Clin. Oncol.* **2014**, *19*, 423–430. [CrossRef] [PubMed]
3. Hashibe, M. *Risk Factors for Cancer of the Mouth: Tobacco, Betel Quid, and Alcohol*; Springer International Publishing: London, UK, 2020; pp. 23–30.
4. National Institute of Dental and Craniofacial Research. Oral Cancer. Available online: https://www.nidcr.nih.gov/health-info/oral-cancer (accessed on 28 February 2024).
5. Khani Jeihooni, A.; Jafari, F. *Oral Cancer: Epidemiology, Prevention, Early Detection, and Treatment*; IntechOpen: Oxford, UK, 2022.
6. Healthline. Oral Cancers. Available online: https://www.healthline.com/health/oral-cancer#by-geographic-region (accessed on 28 February 2024).
7. National Cancer Institute. Cancer Stat Facts: Oral Cavity and Pharynx Cancer. Available online: https://seer.cancer.gov/statfacts/html/oralcav.html (accessed on 28 February 2024).
8. The Oral Cancer Foundation. Early Detection, Diagnosis and Staging. Available online: https://oralcancerfoundation.org/cdc/early-detection-diagnosis-staging/ (accessed on 28 February 2024).
9. Rivera, C.; Venegas, B. Histological and molecular aspects of oral squamous cell carcinoma (Review). *Oncol. Lett.* **2014**, *8*, 7–11. [CrossRef] [PubMed]
10. Singh, A.; Thukral, C.L.; Gupta, K.; Sood, A.S.; Singla, H.; Singh, K. Role of MRI in Evaluation of Malignant Lesions of Tongue and Oral Cavity. *Pol. J. Radiol.* **2017**, *82*, 92–99. [CrossRef]

11. Kaltoft, M.; Hahn, C.H.; Wessman, M.; Hansen, M.L.; Agander, T.K.; Makouei, F.; Wessel, I.; Todsen, T. Intraoral Ultrasound versus MRI for Depth of Invasion Measurement in Oral Tongue Squamous Cell Carcinoma: A Prospective Diagnostic Accuracy Study. *Cancers* **2024**, *16*, 637. [CrossRef]
12. Wang, S.; Yang, M.; Li, R.; Bai, J. Current advances in noninvasive methods for the diagnosis of oral squamous cell carcinoma: A review. *Eur. J. Med. Res.* **2023**, *28*, 53. [CrossRef]
13. Warnakulasuriya, S.; Kerr, A.R. Oral Cancer Screening: Past, Present, and Future. *J. Dent. Res.* **2021**, *100*, 1313–1320. [CrossRef]
14. De Felice, F.; Di Carlo, G.; Saccucci, M.; Tombolini, V.; Polimeni, A. Dental Cone Beam Computed Tomography in Children: Clinical Effectiveness and Cancer Risk due to Radiation Exposure. *Oncology* **2019**, *96*, 173–178. [CrossRef]
15. Tripathy, S.; Mathur, A.; Mehta, V. Revolutionizing Oral Cancer Screening: New Approaches and Emerging Technologies. *Asian Pac. J. Cancer Prev.* **2023**, *24*, 4007–4008. [CrossRef]
16. Ahlawat, S.; Fritz, J.; Morris, C.D.; Fayad, L.M. Magnetic resonance imaging biomarkers in musculoskeletal soft tissue tumors: Review of conventional features and focus on nonmorphologic imaging. *J. Magn. Reson. Imaging* **2019**, *50*, 11–27. [CrossRef]
17. Raman, S.; Shafie, A.A.; Tan, B.Y.; Abraham, M.T.; Chen Kiong, S.; Cheong, S.C. Economic Evaluation of Oral Cancer Screening Programs: Review of Outcomes and Study Designs. *Healthcare* **2023**, *11*, 1198. [CrossRef] [PubMed]
18. Ribeiro-Rotta, R.F.; Rosa, E.A.; Milani, V.; Dias, N.R.; Masterson, D.; da Silva, E.N.; Zara, A. The cost of oral cancer: A systematic review. *PLoS ONE* **2022**, *17*, e0266346. [CrossRef] [PubMed]
19. Cervino, G.; Fiorillo, L.; Herford, A.S.; Romeo, U.; Bianchi, A.; Crimi, S.; D'Amico, C.; De Stefano, R.; Troiano, G.; Santoro, R.; et al. Molecular Biomarkers Related to Oral Carcinoma: Clinical Trial Outcome Evaluation in a Literature Review. *Dis. Markers* **2019**, *2019*, 8040361. [CrossRef]
20. Condrat, C.E.; Thompson, D.C.; Barbu, M.G.; Bugnar, O.L.; Boboc, A.; Cretoiu, D.; Suciu, N.; Cretoiu, S.M.; Voinea, S.C. miRNAs as Biomarkers in Disease: Latest Findings Regarding Their Role in Diagnosis and Prognosis. *Cells* **2020**, *9*, 276. [CrossRef]
21. Vanova, V.; Mitrevska, K.; Milosavljevic, V.; Hynek, D.; Richtera, L.; Adam, V. Peptide-based electrochemical biosensors utilized for protein detection. *Biosens. Bioelectron.* **2021**, *180*, 113087. [CrossRef]
22. Umapathy, V.R.; Natarajan, P.M.; Swamikannu, B.; Moses, J.; Jones, S.; Chandran, M.P.; Anbumozhi, M.K. Emerging Biosensors for Oral Cancer Detection and Diagnosis-A Review Unravelling Their Role in Past and Present Advancements in the Field of Early Diagnosis. *Biosensors* **2022**, *12*, 498. [CrossRef]
23. Arora, R.; Haynes, L.; Kumar, M.; McNeil, R.; Ashkani, J.; Nakoneshny, S.C.; Matthews, T.W.; Chandarana, S.; Hart, R.D.; Jones, S.J.M.; et al. NCBP2 and TFRC are novel prognostic biomarkers in oral squamous cell carcinoma. *Cancer Gene Ther.* **2023**, *30*, 752–765. [CrossRef]
24. Sorroche, B.P.; Miranda, K.C.; Beltrami, C.M.; Arantes, L.; Kowalski, L.P.; Marchi, F.A.; Rogatto, S.R.; Almeida, J.D. HOXA1 3′UTR Methylation Is a Potential Prognostic Biomarker in Oral Squamous cell Carcinoma. *Cancers* **2024**, *16*, 874. [CrossRef]
25. Alafaria, H.A.A.; Jalal, A.S. Novel DNA methylation biomarkers for early diagnosis of oral tongue squamous cell carcinoma (OTSCC). *J. Appl. Genet.* **2024**, *65*, 541–548. [CrossRef]
26. Lin, Y.T.; Darvishi, S.; Preet, A.; Huang, T.Y.; Lin, S.H.; Girault, H.H.; Wang, L.G.; Lin, T.E. A Review: Electrochemical Biosensors for Oral Cancer. *Chemosensors* **2020**, *8*, 54. [CrossRef]
27. Vageli, D.; Doukas, P.; Judson, B. Discovering Novel and Unique Saliva and Serum miRNA and mRNA Signatures for Oral Cancer Detection Using Whole Transcriptome and Small Non-Coding RNA Sequencing: Prediction of Their Association With the PI3K/AKT Pathway. *JCO Glob. Oncol.* **2024**, *10* (Suppl. 1), 52. [CrossRef]
28. Wu, J.; Zhang, C.; Li, H.; Zhang, S.; Chen, J.; Qin, L. Competing endogenous RNAs network dysregulation in oral cancer: A multifaceted perspective on crosstalk and competition. *Cancer Cell Int.* **2024**, *24*, 431. [CrossRef] [PubMed]
29. Oh, S.Y.; Kang, S.M.; Kang, S.H.; Lee, H.J.; Kwon, T.G.; Kim, J.W.; Lee, S.T.; Choi, S.Y.; Hong, S.H. Potential Salivary mRNA Biomarkers for Early Detection of Oral Cancer. *J. Clin. Med.* **2020**, *9*, 243. [CrossRef] [PubMed]
30. Khan, H.; Gupta, S.; Husain, N.; Misra, S.; Mps, N.; Jamal, N.; Ghatak, A. Correlation between expressions of Cyclin-D1, EGFR and p53 with chemoradiation response in patients of locally advanced oral squamous cell carcinoma. *BBA Clin.* **2015**, *3*, 11–17. [CrossRef]
31. Zhou, X.; Liu, S.; Cai, G.; Kong, L.; Zhang, T.; Ren, Y.; Wu, Y.; Mei, M.; Zhang, L.; Wang, X. Long Non Coding RNA MALAT1 Promotes Tumor Growth and Metastasis by inducing Epithelial-Mesenchymal Transition in Oral Squamous Cell Carcinoma. *Sci. Rep.* **2015**, *5*, 15972. [CrossRef]
32. Zhu, C.; Wang, X.; Wang, Y.; Wang, K. Functions and underlying mechanisms of lncRNA HOTAIR in cancer chemotherapy resistance. *Cell Death Discov.* **2022**, *8*, 383. [CrossRef]
33. Wang, J.; Lv, N.; Lu, X.; Yuan, R.; Chen, Z.; Yu, J. Diagnostic and therapeutic role of microRNAs in oral cancer (Review). *Oncol. Rep.* **2021**, *45*, 58–64. [CrossRef]
34. Eslami, M.; Khazeni, S.; Khanaghah, X.M.; Asadi, M.H.; Ansari, M.A.; Garjan, J.H.; Lotfalizadeh, M.H.; Bayat, M.; Taghizadieh, M.; Taghavi, S.P.; et al. MiRNA-related metastasis in oral cancer: Moving and shaking. *Cancer Cell Int.* **2023**, *23*, 182. [CrossRef]

35. Smolarz, B.; Durczynski, A.; Romanowicz, H.; Szyllo, K.; Hogendorf, P. miRNAs in Cancer (Review of Literature). *Int. J. Mol. Sci.* **2022**, *23*, 2805. [CrossRef]
36. Bautista-Sanchez, D.; Arriaga-Canon, C.; Pedroza-Torres, A.; De La Rosa-Velazquez, I.A.; Gonzalez-Barrios, R.; Contreras-Espinosa, L.; Montiel-Manriquez, R.; Castro-Hernandez, C.; Fragoso-Ontiveros, V.; Alvarez-Gomez, R.M.; et al. The Promising Role of miR-21 as a Cancer Biomarker and Its Importance in RNA-Based Therapeutics. *Mol. Ther. Nucleic Acids* **2020**, *20*, 409–420. [CrossRef]
37. He, L.; Ping, F.; Fan, Z.N.; Zhang, C.; Deng, M.; Cheng, B.; Xia, J. Salivary exosomal miR-24-3p serves as a potential detective biomarker for oral squamous cell carcinoma screening. *Biomed. Pharmacother.* **2020**, *121*, 109553. [CrossRef] [PubMed]
38. Saliminejad, K.; Khorshid, H.R.K.; Ghaffari, S.H. Why have microRNA biomarkers not been translated from bench to clinic? *Future Oncol.* **2019**, *15*, 801–804. [CrossRef]
39. Wang, A.; Wang, C.P.; Tu, M.; Wong, D.T. Oral Biofluid Biomarker Research: Current Status and Emerging Frontiers. *Diagnostics* **2016**, *6*, 45. [CrossRef]
40. Goldoni, R.; Scolaro, A.; Boccalari, E.; Dolci, C.; Scarano, A.; Inchingolo, F.; Ravazzani, P.; Muti, P.; Tartaglia, G. Malignancies and Biosensors: A Focus on Oral Cancer Detection through Salivary Biomarkers. *Biosensors* **2021**, *11*, 396. [CrossRef]
41. AlAli, A.M.; Walsh, T.; Maranzano, M. CYFRA 21-1 and MMP-9 as salivary biomarkers for the detection of oral squamous cell carcinoma: A systematic review of diagnostic test accuracy. *Int. J. Oral. Max Surg.* **2020**, *49*, 973–983. [CrossRef] [PubMed]
42. Liu, L.; Xie, W.; Xue, P.; Wei, Z.; Liang, X.; Chen, N. Diagnostic accuracy and prognostic applications of CYFRA 21-1 in head and neck cancer: A systematic review and meta-analysis. *PLoS ONE* **2019**, *14*, e0216561. [CrossRef]
43. Huang, Y.L.; Chen, J.; Yan, W.; Zang, D.; Qin, Q.; Deng, A.M. Diagnostic accuracy of cytokeratin-19 fragment (CYFRA 21-1) for bladder cancer: A systematic review and meta-analysis. *Tumour Biol.* **2015**, *36*, 3137–3145. [CrossRef]
44. Guowei, H.; Yuan, L.; Ma, L.; Zhongyang, L.; Zhixing, S.; Lin, L.; Minqi, L. The diagnostic efficacy of CYFRA21-1 on intrahepatic cholangiocarcinoma: A meta-analysis. *Clin. Res. Hepatol. Gastroenterol.* **2019**, *43*, 266–272. [CrossRef]
45. Lichtenberg, J.Y.; Ling, Y.; Kim, S. Non-Specific Adsorption Reduction Methods in Biosensing. *Sensors* **2019**, *19*, 2488. [CrossRef]
46. Pillai, J.; Chincholkar, T.; Dixit, R.; Pandey, M. A systematic review of proteomic biomarkers in oral squamous cell cancer. *World J. Surg. Oncol.* **2021**, *19*, 315. [CrossRef]
47. Mumtaz, M.; Bijnsdorp, I.V.; Bottger, F.; Piersma, S.R.; Pham, T.V.; Mumtaz, S.; Brakenhoff, R.H.; Akhtar, M.W.; Jimenez, C.R. Secreted protein markers in oral squamous cell carcinoma (OSCC). *Clin. Proteom.* **2022**, *19*, 4. [CrossRef] [PubMed]
48. Ferrari, E.; Wittig, A.; Basilico, F.; Rossi, R.; De Palma, A.; Di Silvestre, D.; Sauerwein, W.A.G.; Mauri, P.L. Urinary Proteomics Profiles Are Useful for Detection of Cancer Biomarkers and Changes Induced by Therapeutic Procedures. *Molecules* **2019**, *24*, 794. [CrossRef] [PubMed]
49. Khurshid, Z.; Zafar, M.S.; Khan, R.S.; Najeeb, S.; Slowey, P.D.; Rehman, I.U. Role of Salivary Biomarkers in Oral Cancer Detection. *Adv. Clin. Chem.* **2018**, *86*, 23–70. [CrossRef] [PubMed]
50. Ishikawa, S.; Sugimoto, M.; Kitabatake, K.; Sugano, A.; Nakamura, M.; Kaneko, M.; Ota, S.; Hiwatari, K.; Enomoto, A.; Soga, T.; et al. Identification of salivary metabolomic biomarkers for oral cancer screening. *Sci. Rep.* **2016**, *6*, 31520. [CrossRef]
51. Sahibzada, H.A.; Khurshid, Z.; Khan, R.S.; Naseem, M.; Siddique, K.M.; Mali, M.; Zafar, M.S. Salivary IL-8, IL-6 and TNF-alpha as Potential Diagnostic Biomarkers for Oral Cancer. *Diagnostics* **2017**, *7*, 21. [CrossRef]
52. Riccardi, G.; Bellizzi, M.G.; Fatuzzo, I.; Zoccali, F.; Cavalcanti, L.; Greco, A.; Vincentiis, M.; Ralli, M.; Fiore, M.; Petrella, C.; et al. Salivary Biomarkers in Oral Squamous Cell Carcinoma: A Proteomic Overview. *Proteomes* **2022**, *10*, 37. [CrossRef]
53. Gualtero, D.F.; Castillo, A.S. Biomarkers in saliva for the detection of oral squamous cell carcinoma and their potential use for early diagnosis: A systematic review. *Acta Odontol. Scand.* **2016**, *74*, 170–177. [CrossRef]
54. Khurshid, Z.; Warsi, I.; Moin, S.F.; Slowey, P.D.; Latif, M.; Zohaib, S.; Zafar, M.S. Biochemical analysis of oral fluids for disease detection. *Adv. Clin. Chem.* **2021**, *100*, 205–253. [CrossRef]
55. Prasad, S.; Tyagi, A.K.; Aggarwal, B.B. Detection of inflammatory biomarkers in saliva and urine: Potential in diagnosis, prevention, and treatment for chronic diseases. *Exp. Biol. Med.* **2016**, *241*, 783–799. [CrossRef]
56. Sadeghi, E.S.; Nematpour, F.S.; Mohtasham, N.; Mohajertehran, F. The oncogenic role of NOTCH1 as biomarker in oral squamous cell carcinoma and oral lichen planus. *Dent. Res. J.* **2023**, *20*, 102. [CrossRef]
57. Xue, L.; Tang, W.; Zhou, J.; Xue, J.; Li, Q.; Ge, X.; Lin, F.; Zhao, W.; Guo, Y. Next-generation sequencing identifies CDKN2A alterations as prognostic biomarkers in recurrent or metastatic head and neck squamous cell carcinoma predominantly receiving immune checkpoint inhibitors. *Front. Oncol.* **2023**, *13*, 1276009. [CrossRef] [PubMed]
58. Lu, H.-J.; Wu, M.-F.; Yang, S.-F. Prognostic impact of *caspase-8* mutation in oral cavity squamous cell carcinoma: A real-world cohort study. *J. Clin. Oncol.* **2024**, *42* (Suppl. 16), e18029. [CrossRef]
59. Saini, J.; Bakshi, J.; Panda, N.K.; Sharma, M.; Yadav, A.K.; Kamboj, K.; Goyal, A.K. Serum Concentration of MMP-9 as a Predictive Biomarker for the Progression of Oral Cancer. *J. Maxillofac. Oral Surg.* **2024**, *23*, 1079–1088. [CrossRef]
60. He, K.; Zhu, Z.B.; Shu, R.; Hong, A. LncRNA NEAT1 mediates progression of oral squamous cell carcinoma via VEGF-A and Notch signaling pathway. *World J. Surg. Oncol.* **2020**, *18*, 261. [CrossRef]

61. Hu, Y.; Zheng, L.; Zhang, J.; Shen, Y.; Zhang, X.; Lin, L. LncRNA-MALAT1 is a promising biomarker for prognostic evaluation of tongue squamous cell carcinoma. *Eur. Arch. Oto-Rhino-Laryngol.* **2020**, *277*, 3155–3160. [CrossRef] [PubMed]
62. Zahran, F.; Ghalwash, D.; Shaker, O.; Al-Johani, K.; Scully, C. Salivary microRNAs in oral cancer. *Oral Dis.* **2015**, *21*, 739–747. [CrossRef]
63. Mazumder, S.; Basu, B.; Ray, J.G.; Chatterjee, R. MiRNAs as non-invasive biomarkers in the serum of Oral Squamous Cell Carcinoma (OSCC) and Oral Potentially Malignant Disorder (OPMD) patients. *Arch. Oral Biol.* **2023**, *147*, 105627. [CrossRef]
64. Singh, K.; Urs, A.B.; Koner, B.C.; Augustine, J.; Shrivastava, R.; Narayan, B. Upregulation of miRNA-196a and miRNA-196b correlates with Bryne's prognostic score in oral squamous cell carcinoma. *Pathol. Res. Pract.* **2024**, *253*, 154954. [CrossRef]
65. Rebaudi, F.; De Rosa, A.; Greppi, M.; Pistilli, R.; Pucci, R.; Govoni, F.A.; Iacoviello, P.; Broccolo, F.; Tomasello, G.; Pesce, S.; et al. A new method for oral cancer biomarkers detection with a non-invasive cyto-salivary sampling and rapid-highly sensitive ELISA immunoassay: A pilot study in humans. *Front. Immunol.* **2023**, *14*, 1216107. [CrossRef]
66. Takkem, A.; Barakat, C.; Zakaraia, S.; Zaid, K.; Najmeh, J.; Ayoub, M.; Seirawan, M.Y. Ki-67 Prognostic Value in Different Histological Grades of Oral Epithelial Dysplasia and Oral Squamous Cell Carcinoma. *Asian Pac. J. Cancer Prev.* **2018**, *19*, 3279–3286. [CrossRef]
67. Mohamed, A.A.; Abbas, M.Y.; Alharbi, H.; Babiker, A.Y. Assessment of Expression of Ki-67 in Benign and Malignant Prostatic Lesions among Sudanese Patients. *Open Access Maced. J. Med. Sci.* **2018**, *6*, 1809–1812. [CrossRef] [PubMed]
68. Moorthy, A.; Venugopal, D.C.; Shyamsundar, V.; Madhavan, Y.; Ravindran, S.; Kuppuloganathan, M.; Krishnamurthy, A.; Sankarapandian, S.; Ganapathy, V.; Ramshankar, V. Identification of EGFR as a Biomarker in Saliva and Buccal Cells from Oral Submucous Fibrosis Patients-A Baseline Study. *Diagnostics* **2022**, *12*, 1935. [CrossRef] [PubMed]
69. de Wit, J.G.; Vonk, J.; Voskuil, F.J.; de Visscher, S.A.H.J.; Schepman, K.-P.; Hooghiemstra, W.T.R.; Linssen, M.D.; Elias, S.G.; Halmos, G.B.; Plaat, B.E.C.; et al. EGFR-targeted fluorescence molecular imaging for intraoperative margin assessment in oral cancer patients: A phase II trial. *Nat. Commun.* **2023**, *14*, 4952. [CrossRef] [PubMed]
70. Payehghadr, S.; Bahrami, N.; Naji, T.; Mohamadnia, A.; Mohammadi, F. Detection of miR-21, MUC1mRNA and VEGF Protein Biomarkers Expression Changes in Oral Squamous Cell Carcinomas (OSCC) in Peripheral Blood. *Asian Pac. J. Cancer Biol.* **2018**, *3*, 59–64. [CrossRef]
71. Rani, N.A.J.; Vardhan, B.G.H.; Srinivasan, S.; Gopal, S.K. Evaluation of Salivary Interleukin-6 in Patients with Oral Squamous Cell Carcinoma, Oral Potentially Malignant Disorders, Chronic Periodontitis and in Healthy Controls—A Cross-Sectional Comparative Study. *Ann. Maxillofac. Surg.* **2023**, *13*, 70–75. [CrossRef]
72. Xiao, L.; Li, X.; Cao, P.; Fei, W.; Zhou, H.; Tang, N.; Liu, Y. Interleukin-6 mediated inflammasome activation promotes oral squamous cell carcinoma progression via JAK2/STAT3/Sox4/NLRP3 signaling pathway. *J. Exp. Clin. Cancer Res.* **2022**, *41*, 166. [CrossRef]
73. Chundru, V.N.S.; Madhavan, R.N.; Chintala, L.; Boyapati, R.; Srikar, M. Evaluation of salivary biomarker interleukin-6 in oral squamous cell carcinoma and oral potentially malignant disorders—A comparative cross-sectional South Indian study. *J. Oral. Maxillofac. Pathol.* **2024**, *28*, 387–392. [CrossRef]
74. Piyarathne, N.S.; Weerasekera, M.M.; Fonseka, P.F.D.; Karunatilleke, A.; Liyanage, R.; Jayasinghe, R.D.; De Silva, K.; Yasawardene, S.; Gupta, E.; Jayasinghe, J.A.P.; et al. Salivary Interleukin Levels in Oral Squamous Cell Carcinoma and Oral Epithelial Dysplasia: Findings from a Sri Lankan Study. *Cancers* **2023**, *15*, 1510. [CrossRef]
75. Principe, S.; Zapater-Latorre, E.; Arribas, L.; Garcia-Miragall, E.; Bagan, J. Salivary IL-8 as a putative predictive biomarker of radiotherapy response in head and neck cancer patients. *Clin. Oral Investig.* **2022**, *26*, 437–448. [CrossRef]
76. Tarek, H.E.; Shalash, H.N.; Morsy, R.A.A.; Mostafa, B.; Hassan, M.; Abbas, R.M.; Ellithy, M.M. Salivary and serum expression of TNF-$\alpha$ and Ki-67 in oral potentially malignant lesions. *Bull. Natl. Res. Cent.* **2022**, *46*, 241. [CrossRef]
77. Abdul Aziz Shaikh, S.; Denny, E.C.; Kumarchandra, R.; Natarajan, S.; Sunny, J.; Shenoy, N.; Nandita, K.P. Evaluation of salivary tumor necrosis factor alpha as a diagnostic biomarker in oral submucosal fibrosis and squamous cell carcinoma of the oral cavity and oropharynx: A cross sectional observational study. *Front. Oral Health* **2024**, *5*, 1375162. [CrossRef] [PubMed]
78. Honarmand, M.H.; Farhad-Mollashahi, L.; Nakhaee, A.; Nehi, M. Salivary Levels of ErbB2 and CEA in Oral Squamous Cell Carcinoma Patients. *Asian Pac. J. Cancer Prev.* **2016**, *17*, 77–80. [CrossRef] [PubMed]
79. Hsiao, Y.C.; Lin, S.Y.; Chien, K.Y.; Chen, S.F.; Wu, C.C.; Chang, Y.T.; Chi, L.M.; Chu, L.J.; Chiang, W.F.; Chien, C.Y.; et al. An immuno-MALDI mass spectrometry assay for the oral cancer biomarker, matrix metalloproteinase-1, in dried saliva spot samples. *Anal. Chim. Acta* **2020**, *1100*, 118–130. [CrossRef] [PubMed]
80. Malhotra, R.; Urs, A.B.; Chakravarti, A.; Kumar, S.; Gupta, V.K.; Mahajan, B. Correlation of Cyfra 21-1 levels in saliva and serum with CK19 mRNA expression in oral squamous cell carcinoma. *Tumour Biol.* **2016**, *37*, 9263–9271. [CrossRef]
81. Abdul, N.S. Role of Advanced Diagnostic Aids in the Detection of Potentially Malignant Disorders and Oral Cancer at an Early Stage. *Cureus* **2023**, *15*, e34113. [CrossRef]
82. İlhan, B.; Vural, C.; Gürhan, C.; Vural, C.; Veral, A.; Wilder-Smith, P.; Özdemir, G.; Güneri, P. Real-Time PCR Detection of Candida Species in Biopsy Samples from Non-Smokers with Oral Dysplasia and Oral Squamous Cell Cancer: A Retrospective Archive Study. *Cancers* **2023**, *15*, 5251. [CrossRef]

83. Jessri, M.; Farah, C.S. Next generation sequencing and its application in deciphering head and neck cancer. *Oral Oncol.* **2014**, *50*, 247–253. [CrossRef]
84. Li, G.; Li, X.; Yang, M.; Xu, L.; Deng, S.; Ran, L. Prediction of biomarkers of oral squamous cell carcinoma using microarray technology. *Sci. Rep.* **2017**, *7*, 42105. [CrossRef]
85. Ramaswamyreddy, S.H.; Smitha, T. Microarray-based gene expression profiling for early detection of oral squamous cell carcinoma. *J. Oral. Maxillofac. Pathol.* **2018**, *22*, 293–295. [CrossRef]
86. OralDNA Labs. OralDNA Labs—Innovations in Salivary Diagnostics. Available online: https://www.oraldna.com (accessed on 15 December 2024).
87. CDx Diagnostics. OralCDx Brush Biopsy to Rule out Oral Precancer for Unhealthy Cell Test. Available online: https://www.cdxdiagnostics.com/oralcdx (accessed on 2 December 2024).
88. DenMat. ViziLite Plus Exam 10 Pack. Available online: https://www.denmat.com/product/vizilite-plus-exam-10-pack/01t800000036oGpAAI (accessed on 4 December 2024).
89. OralID. Forward Science. Available online: https://forwardscience.com/oralid (accessed on 10 December 2024).
90. Sascan Saving Lives Meditech. Sascan Saving Lives—Early Detection Saves Lives. Available online: https://sascan.in/ (accessed on 10 December 2024).
91. Incorporated Addent. BioSCREEN—Oral Cancer Examination. Available online: https://addent.com/bioscreen/ (accessed on 20 December 2024).
92. Vigilant Biosciences. Oral Cancer Screening • Early Detection. Available online: https://vigilantbiosciences.com/ (accessed on 2 December 2024).
93. CancerDetect. Oral Cancer Test Kit & Throat Cancer Test at Home. Available online: https://cancerdetect.viome.com/ (accessed on 5 December 2024).
94. Galleri. Galleri Test. Available online: https://www.galleri.com/ (accessed on 20 December 2024).
95. Rout, S.; Panda, A. Synthesis of Toluidine Blue O and A Kit for Oral Cancer Screening. Patent No. 202131053258, 21 December 2023. Available online: https://tto.kiitincubator.in/wp-content/uploads/2023/05/QuickBlue-Tech-Brief-by-Dhruti-sir-2.pdf (accessed on 16 December 2024).
96. Bhalla, N.; Jolly, P.; Formisano, N.; Estrela, P. Introduction to biosensors. *Essays Biochem.* **2016**, *60*, 1–8. [CrossRef]
97. Singhal, J.; Verma, S.; Kumar, S.; Mehrotra, D. Recent Advances in Nano-Bio-Sensing Fabrication Technology for the Detection of Oral Cancer. *Mol. Biotechnol.* **2021**, *63*, 339–362. [CrossRef] [PubMed]
98. Vigneshvar, S.; Sudhakumari, C.C.; Senthilkumaran, B.; Prakash, H. Recent Advances in Biosensor Technology for Potential Applications—An Overview. *Front. Bioeng. Biotechnol.* **2016**, *4*, 11. [CrossRef] [PubMed]
99. Asal, M.; Özen, Ö.; Sahinler, M.; Baysal, H.T.; Polatoglu, I. An overview of biomolecules, immobilization methods and support materials of biosensors. *Sens. Rev.* **2019**, *39*, 377–386. [CrossRef]
100. Balapure, A.; Dubey, S.K.; Javed, A.; Chattopadhyay, S.; Goel, S. A review: Early detection of oral cancer biomarkers using microfluidic colorimetric point-of-care devices. *Anal. Methods* **2024**, *16*, 6098–6118. [CrossRef]
101. Carneiro, M.; Rodrigues, L.R.; Moreira, F.T.C.; Sales, M.G.F. Colorimetric Paper-Based Sensors against Cancer Biomarkers. *Sensors* **2022**, *22*, 3221. [CrossRef]
102. Moorthy, D.N.; Dhinasekaran, D.; Rebecca, P.N.B.; Rajendran, A.R. Optical Biosensors for Detection of Cancer Biomarkers: Current and Future Perspectives. *J. Biophotonics* **2024**, *17*, e202400243. [CrossRef]
103. Khanmohammadi, A.; Aghaie, A.; Vahedi, E.; Qazvini, A.; Ghanei, M.; Afkhami, A.; Hajian, A.; Bagheri, H. Electrochemical biosensors for the detection of lung cancer biomarkers: A review. *Talanta* **2020**, *206*, 120251. [CrossRef]
104. Hansen, L.; Nagdeve, S.N.; Suganthan, B.; Ramasamy, R.P. An Electrochemical Nucleic Acid Biosensor for Triple-Negative Breast Cancer Biomarker Detection. *Sensors* **2024**, *24*, 5747. [CrossRef]
105. Kaur, D.; Esseili, M.A.; Ramasamy, R.P. A Cell-Based Electrochemical Biosensor for the Detection of Infectious Hepatitis A Virus. *Biosensors* **2024**, *14*, 576. [CrossRef]
106. Zolti, O.; Suganthan, B.; Ramasamy, R.P. Lab-on-a-Chip Electrochemical Biosensors for Foodborne Pathogen Detection: A Review of Common Standards and Recent Progress. *Biosensors* **2023**, *13*, 215. [CrossRef]
107. Kim, J.; Jeong, J.; Ko, S.H. Electrochemical biosensors for point-of-care testing. *Bio-Des. Manuf.* **2024**, *7*, 548–565. [CrossRef]
108. Cho, I.-H.; Kim, D.H.; Park, S. Electrochemical biosensors: Perspective on functional nanomaterials for on-site analysis. *Biomater. Res.* **2020**, *24*, 6. [CrossRef] [PubMed]
109. Smutok, O.; Katz, E. Biosensors: Electrochemical Devices-General Concepts and Performance. *Biosensors* **2022**, *13*, 44. [CrossRef] [PubMed]
110. Ashraf, G.; Chen, W.; Asif, M.; Aziz, A.; Zhong, Z.T.; Iftikhar, T.; Zhao, Y.D. Topical advancements in electrochemical and optical signal amplification for biomolecules detection: A comparison. *Mater. Today Chem.* **2022**, *26*, 101119. [CrossRef]
111. Yadav, A.K.; Dhiman, T.K.; Lakshmi, G.B.V.S.; Berlina, A.N.; Solanki, P.R. A highly sensitive label-free amperometric biosensor for norfloxacin detection based on chitosan-yttria nanocomposite. *Int. J. Biol. Macromol.* **2020**, *151*, 566–575. [CrossRef]

112. Povedano, E.; Vargas, E.; Ruiz-Valdepeñas, V.; Torrente-Rodríguez, R.M.; Pedrero, M.; Barderas, R.; San Segundo-Acosta, P.; Peláez-García, A.; Mendiola, M.; Hardisson, D.; et al. Electrochemical affinity biosensors for fast detection of gene-specific methylations with no need for bisulfite and amplification treatments. *Sci. Rep.* **2018**, *8*, 6418. [CrossRef]
113. Muñoz-San Martín, C.; Gamella, M.; Pedrero, M.; Montero-Calle, A.; Barderas, R.; Campuzano, S.; Pingarrón, J.M. Magnetic beads-based electrochemical immunosensing of HIF-1$\alpha$, a biomarker of tumoral hypoxia. *Sens. Actuat B Chem.* **2020**, *307*, 127623. [CrossRef]
114. Torrente-Rodríguez, R.M.; Campuzano, S.; Montiel, V.R.V.; Gamella, M.; Pingarrón, J.M. Electrochemical bioplatforms for the simultaneous determination of interleukin (IL)-8 mRNA and IL-8 protein oral cancer biomarkers in raw saliva. *Biosens. Bioelectron.* **2016**, *77*, 543–548. [CrossRef]
115. Palomar, Q.; Xu, X.X.; Selegård, R.; Aili, D.; Zhang, Z. Peptide decorated gold nanoparticle/carbon nanotube electrochemical sensor for ultrasensitive detection of matrix metalloproteinase-7. *Sens. Actuat B Chem.* **2020**, *325*, 128769. [CrossRef]
116. Wei, Z.; Wang, H.; Ma, Z.; Han, H. Amperometric Biosensor of Matrix Metalloproteinase-7 Enhanced by Pd-Functionalized Carbon Nanocomposites. *Nanoscale Res. Lett.* **2018**, *13*, 375. [CrossRef]
117. Rocchitta, G.; Spanu, A.; Babudieri, S.; Latte, G.; Madeddu, G.; Galleri, G.; Nuvoli, S.; Bagella, P.; Demartis, M.I.; Fiore, V.; et al. Enzyme Biosensors for Biomedical Applications: Strategies for Safeguarding Analytical Performances in Biological Fluids. *Sensors* **2016**, *16*, 780. [CrossRef]
118. Andriukonis, E.; Celiesiute-Germaniene, R.; Ramanavicius, S.; Viter, R.; Ramanavicius, A. From Microorganism-Based Amperometric Biosensors towards Microbial Fuel Cells. *Sensors* **2021**, *21*, 2442. [CrossRef] [PubMed]
119. Tiwari, S.; Gupta, P.K.; Bagbi, Y.; Sarkar, T.; Solanki, P.R. L-cysteine capped lanthanum hydroxide nanostructures for non-invasive detection of oral cancer biomarker. *Biosens. Bioelectron.* **2017**, *89*, 1042–1052. [CrossRef] [PubMed]
120. Jafari, M.; Hasanzadeh, M. Non-invasive bioassay of Cytokeratin Fragment 21.1 (Cyfra 21.1) protein in human saliva samples using immunoreaction method: An efficient platform for early-stage diagnosis of oral cancer based on biomedicine. *Biomed. Pharmacother.* **2020**, *131*, 110671. [CrossRef]
121. Kumar, S.; Sharma, J.G.; Maji, S.; Malhotra, B.D. A biocompatible serine functionalized nanostructured zirconia based biosensing platform for non-invasive oral cancer detection. *Rsc Adv.* **2016**, *6*, 77037–77046. [CrossRef]
122. Kumar, S.; Kumar, S.; Tiwari, S.; Srivastava, S.; Srivastava, M.; Yadav, B.K.; Kumar, S.; Tran, T.T.; Dewan, A.K.; Mulchandani, A.; et al. Biofunctionalized Nanostructured Zirconia for Biomedical Application: A Smart Approach for Oral Cancer Detection. *Adv. Sci.* **2015**, *2*, 1500048. [CrossRef]
123. Kumar, S.; Chauhan, D.; Renugopalakrishnan, V.; Malhotra, B.D. Biofunctionalized nanodot zirconia-based efficient biosensing platform for noninvasive oral cancer detection. *Mrs Commun.* **2020**, *10*, 652–659. [CrossRef]
124. Verma, S.; Singh, A.; Shukla, A.; Kaswan, J.; Arora, K.; Ramirez-Vick, J.; Singh, P.; Singh, S.P. Anti-IL8/AuNPs-rGO/ITO as an Immunosensing Platform for Noninvasive Electrochemical Detection of Oral Cancer. *ACS Appl. Mater. Inter.* **2017**, *9*, 27462–27474. [CrossRef]
125. Liu, X.K.; Lin, L.Y.; Tseng, F.Y.; Tan, Y.C.; Li, J.; Feng, L.; Song, L.J.; Lai, C.F.; Li, X.H.; He, J.H.; et al. Label-free electrochemical immunosensor based on gold nanoparticle/polyethyleneimine/reduced graphene oxide nanocomposites for the ultrasensitive detection of cancer biomarker matrix metalloproteinase-1. *Analyst* **2021**, *146*, 4066–4079. [CrossRef]
126. Braiek, M.; Yang, Y.; Farre, C.; Chaix, C.; Bessueille, F.; Baraket, A.; Errachid, A.; Zhang, A.D.; Jaffrezic-Renault, N. Boron-doped Diamond Electrodes Modified with Fe.O. @Au Magnetic Nanocomposites as Sensitive Platform for Detection of a Cancer Biomarker, Interleukin-8. *Electroanalysis* **2016**, *28*, 1810–1816. [CrossRef]
127. Tan, Y.; Wei, X.F.; Zhao, M.M.; Qiu, B.; Guo, L.H.; Lin, Z.Y.; Yang, H.H. Ultraselective Homogeneous Electrochemical Biosensor for DNA Species Related to Oral Cancer Based on Nicking Endonuclease Assisted Target Recycling Amplification. *Anal. Chem.* **2015**, *87*, 9204–9208. [CrossRef] [PubMed]
128. Hu, Y.R.; Chang, Y.Y.; Chai, Y.Q.; Yuan, R. An Electrochemical Biosensor for Detection of DNA Species Related to Oral Cancer Based on a Particular Host-Guest Recognition-Assisted Strategy for Signal Tag In Situ. *J. Electrochem. Soc.* **2018**, *165*, B289–B295. [CrossRef]
129. Cui, F.Y.; Zhou, Z.R.; Zhou, H.S. Review-Measurement and Analysis of Cancer Biomarkers Based on Electrochemical Biosensors. *J. Electrochem. Soc.* **2019**, *167*, 037525. [CrossRef]
130. Kim, J.H.; Suh, Y.J.; Park, D.; Yim, H.; Kim, H.; Kim, H.J.; Yoon, D.S.; Hwang, K.S. Technological advances in electrochemical biosensors for the detection of disease biomarkers. *Biomed. Eng. Lett.* **2021**, *11*, 309–334. [CrossRef]
131. Kumar, S.; Sharma, J.G.; Maji, S.; Malhotra, B.D. Nanostructured zirconia decorated reduced graphene oxide based efficient biosensing platform for non-invasive oral cancer detection. *Biosens. Bioelectron.* **2016**, *78*, 497–504. [CrossRef]
132. Verma, S.; Singh, S.P. Non-invasive oral cancer detection from saliva using zinc oxide-reduced graphene oxide nanocomposite based bioelectrode. *Mrs Commun.* **2019**, *9*, 1227–1234. [CrossRef]
133. Wu, J.; Liu, H.; Chen, W.; Ma, B.; Ju, H. Device integration of electrochemical biosensors. *Nat. Rev. Bioeng.* **2023**, *1*, 346–360. [CrossRef]

134. Choudhary, M.; Yadav, P.; Singh, A.; Kaur, S.; Ramirez-Vick, J.; Chandra, P.; Arora, K.; Singh, S.P. CD 59 Targeted Ultrasensitive Electrochemical Immunosensor for Fast and Noninvasive Diagnosis of Oral Cancer. *Electroanalysis* **2016**, *28*, 2565–2574. [CrossRef]
135. Kumar, S.; Panwar, S.; Kumar, S.; Augustine, S.; Malhotra, B.D. Biofunctionalized Nanostructured Yttria Modified Non-Invasive Impedometric Biosensor for Efficient Detection of Oral Cancer. *Nanomaterials* **2019**, *9*, 1190. [CrossRef]
136. Pachauri, N.; Dave, K.; Dinda, A.; Solanki, P.R. Cubic CeO(2) implanted reduced graphene oxide-based highly sensitive biosensor for non-invasive oral cancer biomarker detection. *J. Mater. Chem. B* **2018**, *6*, 3000–3012. [CrossRef]
137. Ding, S.; Das, S.R.; Brownlee, B.J.; Parate, K.; Davis, T.M.; Stromberg, L.R.; Chan, E.K.L.; Katz, J.; Iverson, B.D.; Claussen, J.C. CIP2A immunosensor comprised of vertically-aligned carbon nanotube interdigitated electrodes towards point-of-care oral cancer screening. *Biosens. Bioelectron.* **2018**, *117*, 68–74. [CrossRef] [PubMed]
138. Aydin, M.; Aydin, E.B.; Sezginturk, M.K. A highly selective electrochemical immunosensor based on conductive carbon black and star PGMA polymer composite material for IL-8 biomarker detection in human serum and saliva. *Biosens. Bioelectron.* **2018**, *117*, 720–728. [CrossRef] [PubMed]
139. Li, Y.; Chang, Y.; Yuan, R.; Chai, Y. Highly Efficient Target Recycling-Based Netlike Y-DNA for Regulation of Electrocatalysis toward Methylene Blue for Sensitive DNA Detection. *ACS Appl. Mater. Interfaces* **2018**, *10*, 25213–25218. [CrossRef]
140. Dong, T.; Pires, N.M.M.; Yang, Z.C.; Jiang, Z.D. Advances in Electrochemical Biosensors Based on Nanomaterials for Protein Biomarker Detection in Saliva. *Adv. Sci.* **2023**, *10*, e2205429. [CrossRef]
141. Bhardwaj, H.; Archana; Noumani, A.; Himanshu, J.K.; Chakravorty, S.; Solanki, P.R. Recent advancement in the detection of potential cancer biomarkers using the nanomaterial integrated electrochemical sensing technique: A detailed review. *Mater. Adv.* **2024**, *5*, 475–503. [CrossRef]
142. Lin, X.; Shi, H.; Zheng, F.; Zeng, J. Simple and sensitive sandwich-like voltammetric immunosensing of procalcitonin. *Anal. Sci.* **2024**, *40*, 541–547. [CrossRef]
143. Sanko, V.; Kuralay, F. Label-Free Electrochemical Biosensor Platforms for Cancer Diagnosis: Recent Achievements and Challenges. *Biosensors* **2023**, *13*, 333. [CrossRef]
144. Topkaya, S.N.; Azimzadeh, M.; Ozsoz, M. Electrochemical Biosensors for Cancer Biomarkers Detection: Recent Advances and Challenges. *Electroanalysis* **2016**, *28*, 1402–1419. [CrossRef]
145. Braz, D.C.; Neto, M.P.; Shimizu, F.M.; Sá, A.C.; Lima, R.S.; Gobbi, A.L.; Melendez, M.E.; Arantes, L.M.R.B.; Carvalho, A.L.; Paulovich, F.V.; et al. Using machine learning and an electronic tongue for discriminating saliva samples from oral cavity cancer patients and healthy individuals. *Talanta* **2022**, *243*, 123327. [CrossRef]
146. Zlotogorski-Hurvitz, A.; Dekel, B.; Malonek, D.; Yahalom, R.; Vered, M. FTIR-based spectrum of salivary exosomes coupled with computational-aided discriminating analysis in the diagnosis of oral cancer. *J. Cancer Res. Clin.* **2019**, *145*, 685–694. [CrossRef]
147. Kouznetsova, V.L.; Li, J.; Romm, E.; Tsigelny, I.F. Finding distinctions between oral cancer and periodontitis using saliva metabolites and machine learning. *Oral Dis.* **2021**, *27*, 484–493. [CrossRef]
148. Li, Y.; Tang, H.; Liu, Y.; Qiao, Y.; Xia, H.; Zhou, J. Oral wearable sensors: Health management based on the oral cavity. *Biosens. Bioelectron. X* **2022**, *10*, 100135. [CrossRef]
149. Dhane, A.S. The promise of artificial intelligence and internet of things in oral cancer detection. *J. Med. Surg. Public Health* **2024**, *3*, 100130. [CrossRef]
150. Pappot, H.; Steen-Olsen, E.B.; Holländer-Mieritz, C. Experiences with Wearable Sensors in Oncology during Treatment: Lessons Learned from Feasibility Research Projects in Denmark. *Diagnostics* **2024**, *14*, 405. [CrossRef] [PubMed]
151. Popovic, Z.V.; Thiha, A.; Ibrahim, F.; Petrovic, B.B.; Dahlan, N.A.; Milic, L.; Kojic, S.; Stojanovic, G.M. Oral micro-electronic platform for temperature and humidity monitoring. *Sci. Rep.* **2023**, *13*, 21277. [CrossRef]
152. Yaduvanshi, V.; Murugan, R.; Goel, T. Automatic oral cancer detection and classification using modified local texture descriptor and machine learning algorithms. *Multimed. Tools Appl.* **2024**, *84*, 1031–1055. [CrossRef]
153. Warin, K.; Suebnukarn, S. Deep learning in oral cancer- a systematic review. *BMC Oral Health* **2024**, *24*, 212. [CrossRef]
154. Huang, Q.; Ding, H.; Razmjooy, N. Oral cancer detection using convolutional neural network optimized by combined seagull optimization algorithm. *Biomed. Signal Process. Control* **2024**, *87*, 105546. [CrossRef]
155. Lin, H.; Chen, H.; Weng, L.; Shao, J.; Lin, J. Automatic detection of oral cancer in smartphone-based images using deep learning for early diagnosis. *J. Biomed. Opt.* **2021**, *26*, 086007. [CrossRef]
156. Thakuria, T.; Rahman, T.; Mahanta, D.R.; Khataniar, S.K.; Goswami, R.D.; Rahman, T.; Mahanta, L.B. Deep learning for early diagnosis of oral cancer via smartphone and DSLR image analysis: A systematic review. *Expert. Rev. Med. Devices* **2024**, *21*, 1189–1204. [CrossRef]

*sensors*

MDPI

*Review*

# Plasma-Treated Nanostructured Resistive Gas Sensors: A Review

**Mahmoud Torkamani Cheriani and Ali Mirzaei ***

Department of Materials Science and Engineering, Shiraz University of Technology, Shiraz 71557-13876, Iran; m.torkamani@sutech.ac.ir

* Correspondence: mirzaei@sutech.ac.ir

**Abstract:** Resistive gas sensors are among the most widely used sensors for the detection of various gases. In this type of gas sensor, the gas sensing capability is linked to the surface properties of the sensing layer, and accordingly, modification of the sensing surface is of importance to improve the sensing output. Plasma treatment is a promising way to modify the surface properties of gas sensors, mainly by changing the amounts of oxygen ions, which have a central role in gas sensing reactions. In this review paper, we focus on the role of plasma treatment in the gas sensing features of resistive gas sensors. After an introduction to air pollution, toxic gases, and resistive gas sensors, the main concepts regarding plasma are presented. Then, the impact of plasma treatment on the sensing characteristics of various sensing materials is discussed. As the gas sensing field is an interdisciplinary field, we believe that the present review paper will be of significant interest to researchers with various backgrounds who are working on gas sensors.

**Keywords:** plasma treatment; toxic gas; gas sensor; sensing mechanism

## 1. Introduction to Toxic Gases

Air pollution is due to the existence of unwanted substances in the air, affecting its cleanness and quality. It is a serious issue in most countries and led to 4.14 million premature deaths worldwide in 2019 [1]. In addition to particulate matter, toxic gases are among the main components of polluted air. $NO_2$, $SO_x$, $O_3$, and CO gases are among the most dangerous gases often found in polluted air. Natural air pollution sources include volcano eruptions and wind-blown dust, and anthropogenic sources include the burning of fossil fuels, agricultural activities, waste management, and so on (Figure 1) [2].

Air pollution has many negative effects on animals [3] as well as on the environment [4]. For example, $NO_x$ and $SO_2$ gases in polluted air can directly affect photosynthesis and bring about premature leaf senescence, eventually decreasing crop yields [5,6]. In addition, it has detrimental effects on human health. Air pollutants may influence influenza transmission [7], intensify COVID-19 mortality [8], induce respiratory diseases [9], such as bronchoconstriction [10], asthma [11], and lung cancer [12], affect pregnancies and related things such as low birth weight, preterm birth, and fetal hyperinsulinism [13], cause cardiovascular diseases [14,15], and impact the immune system [16]. Since about 90% of people live in places with polluted air [17], it is necessary to monitor the air quality precisely. In this regard, the development of reliable gas sensors is vital.



https://doi.org/10.3390/s25072307

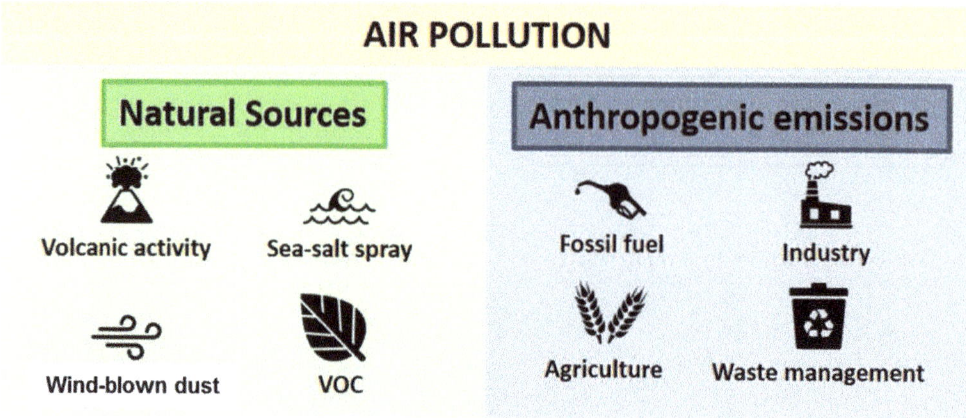


**Figure 1.** Sources of air pollution [2]. With permission from MDPI.

## 2. Resistive Gas Sensors: An Introduction

There are various types of gas sensors that can be used to detect toxic gases. The most common gas sensors are optical [18], electrochemical [19], surface acoustic wave [20], and resistive [21]. Resistive sensors have high response, high stability, swift dynamics, ease of design and fabrication, compact size, and low price. They are realized from semiconducting materials, and currently, semiconducting metal oxides are widely used for this purpose. However, metal oxide gas sensors have some drawbacks, such as poor selectivity, high sensing temperature, and humidity interference [22]. Thus, recently, other semiconductors, such as carbon-based materials, including carbon nanotubes [23], graphene [24], reduced oxide graphene [25], conducting polymers (CPs) [26], transition metal dichalcogenides (TMDs) [27], and MXenes, have been employed for the fabrication of resistive gas sensors in order to reduce sensing temperature and increase the selectivity to a specific gas.

In resistive gas sensors, the sensing layer is applied on the surface of an insulating substrate, such as alumina, which is equipped with electrodes. Also, sometimes a micro-heater is attached to a substrate to provide sufficient heat to increase the sensing device to the desired temperature [28]. The principle of the gas sensing mechanisms of resistive sensors is based on a variation in resistance in the presence of target gas. In general, there are two types of semiconducting materials based on the majority of charge carriers. In n- and p-type gas sensors, electrons and holes are the main charge carriers, respectively. When a resistive gas sensor is exposed to air, oxygen gas will be adsorbed on it, and thanks to its high electrophilic nature, it takes electrons from the sensor surface. Accordingly, on n-type gas sensors, a so-called electron depletion layer (EDL) will appear in which the concentration of electrons is lower relative to the inner part. Hence, the resistance of n-type gas sensors increases in air relative to vacuum conditions. Furthermore, a hole accumulation layer (HAL) will appear on the surface of p-type gas sensors in which the concentrations of holes are higher than those in the core part of the sensor. When an n-type sensor is put in a reducing gas atmosphere, the gas reacts with adsorbed oxygen on the sensor surface, releasing electrons back to the sensor. Hence, the thickness of the EDL decreases, bringing about a decrease in sensor resistance. Upon exposure to oxidizing gas, the gas takes further electrons from the sensor surface, leading to the expansion of the EDL and an increase in sensor resistance (Figure 2a). For a p-type sensor, the release of electrons in the presence of an n-type gas leads to the narrowing of the HAL and an increase in sensor resistance, whereas in an oxidizing gas atmosphere, the further abstraction of electrons leads to an expansion of the HAL and a decrease in resistance [29–31] (Figure 2b).

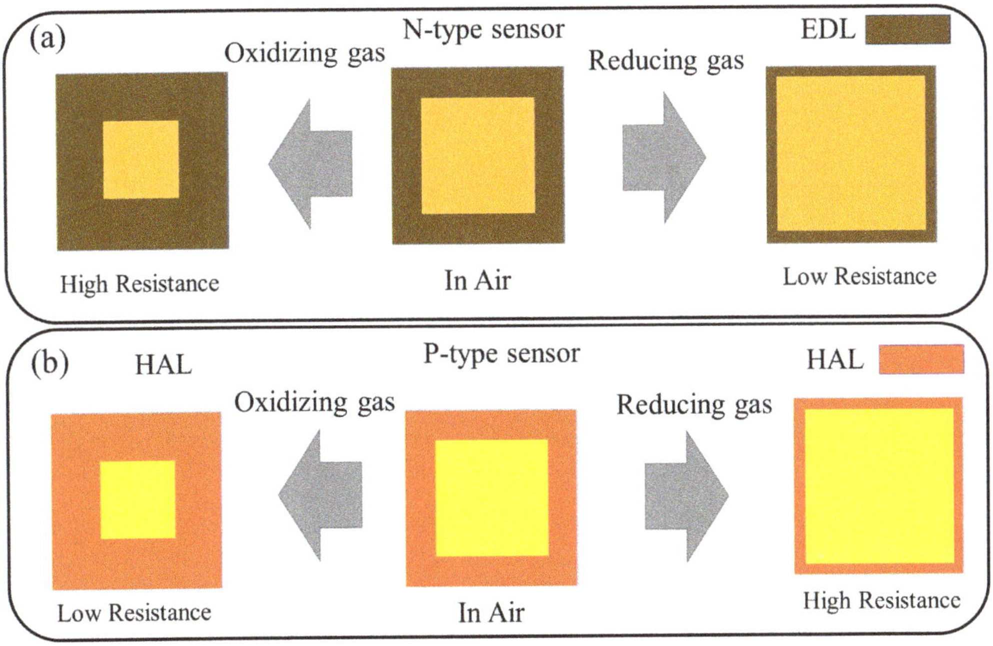


**Figure 2.** Basic gas sensing mechanism of resistive gas sensors: (**a**) n- and (**b**) p-type sensors.

Currently, the main challenges of nanostructured resistive gas sensors are as follows: (i) the development of highly sensitive gas sensors with the capability of gas sensing down to ppt level; (ii) the development of highly selective gas sensors; (iii) the development of humidity-resistant gas sensors; and (iv) the development of gas sensors with low power consumption. To address the above-mentioned challenges, the synthesis of sensing materials with high surface areas, the combination of various sensing materials, functionalization with appropriate noble metals, functionalization using plasma treatment, UV illumination, and operation of the sensors in self-heating mode have been proposed.

In more detail, there are various techniques to boost the sensing performance of resistive sensors. Heterojunction formation [32], doping [33], noble metal decoration [34], morphology engineering [35], UV illumination [36], high-energy irradiation [37], and plasma exposure are among the most widely used techniques. In particular, some problems of resistive gas sensors, namely selectivity and high operating temperature, can be at least partially addressed by plasma treatment. Plasma exposure causes a change in the amount of oxygen vacancies, thereby facilitating the adsorption of oxygen gas on the surface of the plasma-treated sensor. Since oxygen species engage in sensing reactions, better gas sensing properties can be obtained. Furthermore, by the right selection of plasma type, other species such as –F, -C can be added on the surface of the sensor, acting as favorable adsorption sites for target gas molecules. Hence, both selectivity and working temperature can be improved.

So far, there have been no published review papers dealing with the effect of plasma treatment on the gas sensing properties of resistive sensors. Hence, in this review paper, we will review the effect of plasma treatment on the sensing features of resistance sensors.

## 3. The Plasma Concept

When energy is supplied to a gas, its temperature gradually increases, and by the further provision of energy, the kinetic energy of gas molecules significantly increases, leading to the collision of more gas molecules. Hence, electrons and ions are formed in the gas, leading to the existence of an electrical charge in the gas. This state of matter is known as plasma, originating from Greek, which means 'something molded', indicating a glowing gas which alters its shape based on the container. Gases are often electrical insulators, while plasmas have an equal amount of positive and negative charge carriers along with neutral particles, giving them electrical conductivity [38].

In equilibrium plasma, local thermodynamic equilibrium exists among the plasma species and collision processes, where heavy particles and electrons will be at almost identical temperatures. In contrast, non-equilibrium plasma or cold plasma involves a thermodynamic imbalance among the electrons and heavy particles, and the temperature of the heavy particles is much lower than that of electrons. During cold plasma production, heating the entire gas stream (air or individual gases like Ar and He) is undesirable; thus, energy is directed to the electrons via the electrical discharge in the gas [39]. Corona discharge, dielectric barrier discharge (DBD), and cold plasma jet are among the most common ways to generate cold plasma. To generate cold plasma at atmospheric pressure, a high voltage is applied for the generation of a gas discharge, and the discharge easily proceeds to arc discharge. Furthermore, energy should be selectively transferred to electrons using effective methods without raising the temperature of the gas [38].

Cold plasma treatment is an environmentally friendly technique without the production of toxic waste, and thanks to its operation under atmospheric pressure, it is an appropriate technique for the treatment of low-melting-point or heat-sensitive materials and substrates [40]. In particular, flexible polymeric substrates have low surface energy and poor wettability. Therefore, the adhesion between electrodes and a polymeric substrate is weak. Accordingly, plasma surface modification of polymeric substances can overcome this shortage [41].

In both corona discharge and DBD, the sample to be plasma treated is put between electrodes in a fixed space under atmospheric pressure. In corona discharge, by applying a DC electrical source in a pulsed mode, a lighting crown is built out of many streamers, while in DBD, a high-frequency source is employed for this purpose. During corona discharge, the cathode is a conductive wire, and the anode is the sample. A DBD reactor usually has two parallel metal electrodes at a fixed distance covered with a dielectric material, and the sample is placed between them. The formed plasma has many homogeneously distributed micro-streamers across the electrodes [42].

Plasma treatment is a flexible, fast, green, and non-contaminating method of changing surface morphology and composition. Compared to conventional routes, this method is faster and needs fewer reagents. Also, only the surface area is affected by the plasma treatment, without affecting the bulk region. By optimizing the plasma parameters, including plasma power, exposure time, and the type of gas, various functional groups with different amounts can be added on the surface of the host material [43]. Furthermore, plasma treatment can be performed at atmospheric pressure, facilitating large-scale treatment for mass production [44]. Moreover, plasma can be employed to deposit thin layers over the substrate in a process called plasma spray [45,46]. In the following sections, we will discuss the impact of plasma treatment on the gas sensing properties of resistive gas sensors.

## 4. Plasma-Treated Gas Sensors

### 4.1. Plasma-Treated Carbon Nanotube Gas Sensors

Carbon nanotubes (CNTs) are one-dimensional materials with high conductivity, a large surface area, and the possibility of functionality [47,48]. Nevertheless, homogeneous dispersion of CNTs is a challenge owing to the presence of attractive Van der Waals forces among CNTs, leading to agglomeration and weak solubility in most solvents. Hence, it is required to change the surface properties of CNTs via surface treatment or chemical functionalization [49]. Compared to the surface treatment of CNTs using strong acids such as $HNO_3$ and $H_2SO_4$, which is time-consuming and dangerous, plasma treatment using oxygen is a simple, clean, and effective way to functionalize CNTs. Acid treatment leads to the presence of carboxylic acids, ethers, and so on on the surface of CNTs, while oxygen plasma treatment increases the number of oxygen-bearing defects on the entire surface of the CNTs. Also, the hydrophilic nature of CNTs can be boosted thanks to the presence of oxygen-containing species on the surface of CNTs. Furthermore, the bulk features of CNTs remains untouched during plasm treatment, without any structural destruction [50]. As a result of oxygen plasma treatment, oxygenated vacancies and functional groups will be present on the surface of CNTs, which are very reactive species and act as favorable adsorption sites for gas molecules [51]. Therefore, plasma treatment has been extensively applied on CNTs to increase their gas sensing performance [52–56].

In this regard, Bannov et al. [57] functionalized the surface of multi-walled CNTs (MWCNTs) using oxygen plasma exposure followed by $C_2H_2(CO)_2O$ plasma treatment. The MWCNTs were comprised of intertwined MWCNTs with a mean diameter of 20–50 nm. After plasma treatment, most of the MWCNTs were strongly etched by oxygen plasma, and only a few MWCNT bundles remained. Based on an XPS study, the plasma treatment led to the presence of a high amount of oxygen-containing surface groups on the MWCNTs. The sensor resistance was increased after the plasma treatment due to the oxidation of the MWCNTs and the loss of the connections among MWCNT networks. At room temperature (RT), the response of the fabricated sensor to 500 ppm $NH_3$ was only 11.7%, while after plasma treatment it increased to 31.4%, demonstrating the promising role of plasma treatment. The increase in sensor response was related to the enhanced $NH_3$ adsorption by the oxygen rich surfaces as a result of the plasma treatment. Due to the reducing nature of $NH_3$ gas, it should react with adsorbed oxygen to release the electrons on the sensor surface. Hence, the higher amounts of oxygen species on the sensor surface as a result of plasma exposure led to a higher probability of the reaction with $NH_3$, resulting in a higher response. The same group [58] investigated the effect of oxygen plasma exposure time (3, 5, and 7 min) on the $NH_3$ gas sensing properties of MWCNTs. The sensor exposed to oxygen plasma for 5 min exhibited the highest response to $NH_3$ gas, which was related to the presence of the highest amount of adsorbed oxygen species on the surface of CNTs. In another study, Dong et al. [59] studied the effect of various plasma types using Ar, $O_2$, $CF_4$, and $SF_6$ gases on the gas sensing properties of single-walled CNTs (SWCNTs). Thanks to reactive ion etching, defects were generated on the SWCNTs. Based on a Raman analysis, the intensity of the D-band to G-band ($I_D/I_G$) ratio of the pristine sample was only 0.14, while after plasma treatment by the above-mentioned gases, it was changed to 0.23, 0.36, 0.33, and 0.5, respectively. Therefore, more defects were generated on the surface of the plasma-treated samples. The pristine sensor not only showed a very low response to both $NO_2$ and $NH_3$ gases, but also the recovery time was very long (more than 20 min). Also, the resistance did not completely return to its initial value. In contrast, the plasma-treated sensors showed better sensing performance. The sensor treated with $O_2$ plasma revealed a higher response to $NO_2$ gas thanks to the presence of defect sites and adsorbed oxygen species groups, which led to the better adsorption and reaction of $NO_2$ gas on the sensor surface. Also, the responses of the sensors treated with $CF_4$ and $SF_6$ were higher than $NH_3$ gas relative to other gases, which was attributed to the sufficient adsorption energies and easy charge transfer between the $NH_3$ and C–F bonds ($CF_4$ and $SF_6$) of the plasma-treated MWCNTs.

Santosh et al. [60] used Ar and oxygen plasma treatment using a fixed 100 sccm of gas for 3 min on MWCNTs for improvement of the gas sensing capacity. The sensor treated with

Ar plasma revealed a higher response to other gas sensors, which was related to the greater extent of the surface modifications by the Ar plasma. At 65 °C, the maximum response to ethanol gas was observed with a response $[(R_a - R_g)/R_a]$ of 1.7 to 100 ppm ethanol. Argon is much heavier than He, and hence, more defects were generated on the MWCNTs after Ar treatment. The diameter of the MWCNTs decreased after plasma exposure due to the etching effect of plasma (Figure 3a–c). In addition, based on Raman analysis, the amorphous wrinkled layer on the pristine sensor was removed after plasma treatment, which eventually improved the crystalline behavior of the MWCNTs. Furthermore, thanks to the higher crystallinity and high amount of carbon defects, the conductivity increased after plasma exposure and enhanced the interaction of the MWCNTs with ethanol. Finally, the plasma treatment led to the formation of dangling bonds, which acted as favorable sites for ethanol gas adsorption.

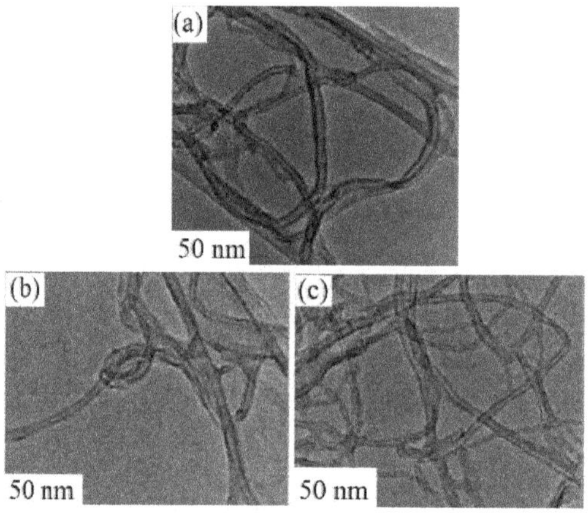


**Figure 3.** TEM views of MWCNTs: (**a**) pristine, (**b**) He-, and (**c**) Ar-treated MWCNTs [60]. With permission from Elsevier. Copyright (2020).

Ham et al. [61] modified MWCNTs by oxygen plasma for 10–50 s. The morphology of the MWCNTs did not change in up to 20 s of plasma exposure. However, by increasing the treatment time to more than 30 s, the surface became highly rough, and the MWCNTs were partially etched. The sensor treated with plasma for 20 s showed a higher response to $NH_3$ gas relative to other gas sensors, indicating that the sensitivity of the MWCNT gas sensors can be enhanced through defect generation and the adding of oxygenated functional groups. The enhanced sensitivity was ascribed to the generation of hydrogen bonds between $NH_3$ gas and surface oxygen groups on the MWCNTs.

The sensing performance of nitrogen-plasma-treated SWCNT gas sensors has rarely been investigated. In this regard, Zamansky et al. [62] synthesized SWCNTs via a CVD method and then used nitrogen plasma to modify their surfaces. Based on characterization results, the defects were introduced on SWCNTs after plasma treatment. In particular, with longer plasma exposure, the amount of substitutional N defects relative to -$NH_2$ surface groups increased, indicating the incorporation of N into the SWCNT lattice. Also, due to the exposure of the MWCNTs to air after plasma treatment, many oxygen-containing defects were detected. The pristine sensor was almost insensitive to the gases. In contrast, the sensor treated for 19 min exhibited a response of 121% to 50 ppm $NO_2$, and its response to 50 ppm $NH_3$ was 36% at RT. The enhanced performance was related to the existence of oxygen- and nitrogen-related defects, which served as desirable adsorption sites for gas molecules. Also, the amount of metallic SWCNTs with poor gas sensing properties decreased after etching. Based on DFT calculations, the gas adsorption on defect sites was more favorable compared to basal plane sites. In addition, while $NH_3$ sensing was accelerated by hydrogen bond formation with surface groups such as COOH, the adsorption of $NO_2$ was mainly caused by the oxidation of carbon defect regions and physisorption.

Ham et al. [63] investigated the impact of plasma treatment on the $NH_3$ sensing properties of SWCNTs with different amounts of semiconducting SWCNTs (66 and 90 wt.%). After oxygen plasma treatment, the sensor with 66 wt.% semiconducting SWCNTs exhibited a 5.5-times increase in sensitivity relative to the pristine sensor, while the sensor with 90 wt.% semiconducting SWCNTs revealed a 13-times increase in sensitivity compared to the pristine sensor. The pristine SWCNT sensor revealed a very long response time and incomplete recovery, while the plasma-treated sensors showed much quicker dynamics, with complete recovery of baseline resistance after treatment. Based on an XPS study, the amount of oxygen functional groups was significantly improved after plasma treatment. The $NH_3$ molecules formed strong hydrogen bonds with oxygen ions on the oxidized SWCNTs. Therefore, a significant response improvement was observed. Also, in the sample with 90 wt.% semiconducting SWCNTs, the $sp^3/sp^2$ ratio increased from 0.256 to 0.611 after the plasma treatment, indicating that the higher semiconducting nature of plasma-treated SWCNTs and the existence of $sp^3$ defects provided favorable adsorption sites for $NH_3$ gas.

CPs with high conductance, flexibility, simple synthesis procedures, and low cost are among the most promising materials for RT gas sensing applications [64,65]. Hence, composite formation between CNTs and CPs is a favorable strategy for RT gas sensing, while plasma exposure can further increase their performance [66]. In this regard, Yoo et al. [67] studied the effect of oxygen plasma treatment (10, 30, 60, and 90 s) on the $NH_3$ sensing capability of an MWCNT–polyaniline (PANI) composite. Based on a Raman analysis, the number of defects on the MWCNTs increased with the increase in plasma time (Figure 4a). During plasma exposure, oxygen ions destroyed the structure of the MWCNTs by turning them into carbon particles and amorphous carbon, along with the creation of more defects relative to the pristine MWCNTs. Based on an XPS study, the concentration of surface oxygen increased with the increase in the plasma exposure time up to 60 s and then decreased by a longer treatment of 90 s, which was attributed to chemical etching of the MWCNTs (Figure 4b). This resulted in the thinning or bending of the MWCNTs. At RT, the response of the plasma-treated MWCNTs was three times that of the pristine sensor thanks to the formation of hydrogen bonds between polar $NH_3$ and oxygen-containing defects on the MWCNTs. Also, the plasma-treated MWCNT-PANI composite sensor revealed a higher response to the plasma-treated MWCNTs thanks to the presence of PANI with a high intrinsic response to $NH_3$ gas. The $NH_3$ molecules abstracted protons from the PANI, forming energetically more favorable $NH_4^+$ ions, while the PANI changed into its base form with a different conductivity, resulting in the generation of a high sensing signal.

During the synthesis of CNTs, some impurities and contaminants are introduced into the CNTs. Even though purification procedures can be used, sometimes they are detrimental to the gas sensing properties of CNTs. In this regard, Kim et al. [68] investigated the impact of thermal annealing ($T > 300$ °C) and plasma treatment with oxygen on the characteristics and $NH_3$ gas sensing properties of CNTs. The pristine SWCNTs had a hydrophobic nature with a water contact angle (WCA) of 84.91°. The thermally treated SWCNTs again showed a hydrophobic nature with a WCA of 79.07°, while the plasma-treated SWCNTs showed a WCA of only 5.15°, indicating an increase in the hydrophobic nature after plasma treatment due to the adding of oxygen surface groups on the SWCNTs. The plasma-treated SWCNTs showed a decrease in $sp^2$ bonding with an increase in $sp^3$ bonding, indicating a change in electrical conductivity. Among the three sensors, the plasma-treated SWNT sensor exhibited the highest response and the fastest response time to $NH_3$ gas. In addition, both the pristine and thermally treated SWNT sensors exhibited incomplete recovery of their resistance. While thermal cleaning of the SWCNTs removed impurities from the surface of the SWCNTs, the plasma treatment included cleaning and functionalization of the SWNTs at the same time to a greater extent, resulting in better sensing capability after plasma treatment.

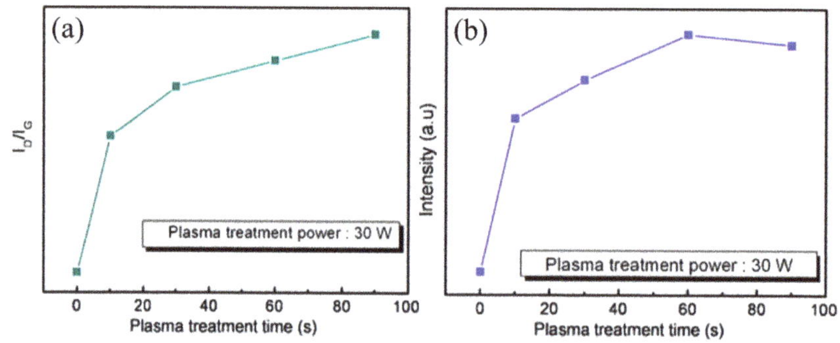


**Figure 4.** (**a**) Changes in the $I_D/I_G$ and (**b**) oxygen vacancy of SWCNTs as a function of plasma treatment time [67]. With permission from Elsevier. Copyright (2009).

Zhao et al. [69] applied plasma on CNTs for CO gas sensing. While the pristine CNTs showed no response to this gas, the plasma-treated CNTs were able to detect down to 5 ppm CO at RT. The improved sensing response was related to the conversion of metallic CNTs to semiconducting CNTs after plasma treatment, along with the promising effect of surface oxygen addition for sensing reactions with CO gas.

### *4.2. Plasma-Treated Graphene and Graphene Oxide Gas Sensors*

Pristine graphene (G) has a high surface area and high conductivity, demonstrating its potential for sensing applications [70]. However, it generally has poor selectivity since the gas adsorption on G is based on Van der Waals interactions with gases, which limit its selectivity [24]. To address this issue, plasma treatment can be used [71]. In this regard, Masterapa et al. [72] applied plasma treatment on CVD-grown G for 5, 10, 20, and 30 s. The sample treated with plasma for 30 s revealed higher amounts of defects, as demonstrated by a Raman analysis, and hence it showed a higher response to $NO_2$ and $NH_3$ gases relative to other sensors. However, the response time of all the sensors was very long (10 min), and the recovery curves were not shown possibly due to very long recovery times.

Fluorination surface treatment could change the intrinsic properties of G. In this regard, Zhang et al. [73] synthesized monolayer fluorinated graphene (FG) by a $SF_6$ plasma treatment (5–90 s). The concentration of F in the samples increased with the increase in plasma treatment time up to 20 s, and then it decreased. During the plasma treatment, the fluorine atoms attached to carbon atoms to form C−F bonds on the surface of G. After a critical time, the F atoms broke down some previously formed C−F bonds, and hence, F atoms were released from the surface of G. The pristine G sensor exhibited slow dynamics, and even after 500 s of recovery, only ~66.7% of the initial resistance was recovered. The sensor treated with plasma for 20 s exhibited a response of 3.8% to 100 ppm $NH_3$ gas at RT, with complete recovery of baseline resistance in less than 200 s. Based on DFT results, the improved performance was ascribed to the opening up of the band gap after fluorination and the enhanced adsorption of $NH_3$ in the presence of surface functional groups.

Chung et al. [74] synthesized G films using the CVD route, and they were then treated with plasma ozone for 60, 70, 80, and 90 s. Among the fabricated sensors, the sensor with the ozone treatment time of 70 s showed a response of 19.7% to 10 ppm $NO_2$ gas at RT, which was two times higher than that of the pristine G sensor. Also, the sensor was able to detect as low as 200 ppb $NO_2$ gas. Further increasing the plasma treatment time resulted in a decrease in the sensing response due to extensive oxidation of G with high baseline resistance. The presence of sufficient amounts of oxygen groups on the surface of G resulted in an increase in adsorption sites and sensing reactions with the $NO_2$ gas. However, the sensors showed long dynamics, and all sensors showed a long recovery time of ~20 min or longer.

$CO_2$ is the main gas responsible for the greenhouse effect [75]. Hence, the development of sensitive $CO_2$ sensors is crucial for environmental and industrial applications. Casanova-Chafer et al. [76] synthesized a G-$CsPbBr_3$ nanocomposite and subsequently applied oxygen plasma treatment on it. The sensor exposed for 5 min to oxygen plasma exhibited a 3-fold improvement in gas sensing compared to the pristine sensor, with a limit of detection of 6.9 ppm. The improved sensing performance was attributed to the promising role of oxygen species, facilitating sensing reactions with $CO_2$ gas on the sensor surface.

Graphene oxide is the oxidized form of G with two key advantages. First, the synthesis route of GO is easy using graphite as raw material, and hence its large-scale production is feasible. Second, in contrast to G, GO exhibits good hydrophilicity, making it possible to prepare stable aqueous colloids [77,78]. Nonetheless, the main shortcoming of GO is its high resistance, making it unsuitable for sensing applications [79]. In this regard, plasma treatment is a promising technique, allowing the reduction of GO by removing oxygen atoms during plasma exposure, without disrupting the GO lattice. Hydrogen or hydrogen-containing plasma with mild treatment conditions is an efficient and alternative route to the complex procedures for GO reduction. The hydrogen plasma contains radicals and atoms, which provide energy for the dissociation of oxygen functional groups. It effectively removes the oxygen functional groups at the edge sites and both basal planes while restoring C=C bonds [80]. After the reduction of GO, it becomes converted to reduced graphene oxide (rGO) with high conductivity, high amounts of surface defects, and also some oxygen surface groups along with a high surface area, all making it a good choice for gas sensing applications.

Hamzaj et al. [81] used hydrogen plasma treatment for 10, 20, 40, 120, and 240 s on GO to reduce it for gas sensing applications. The surface morphology of pristine GO showed some mild wrinkles, and it was not changed after plasma treatment for 10 and 240 s (Figure 5a–c). It should be noted that generally, plasma treatment does not significantly change the surface morphology.

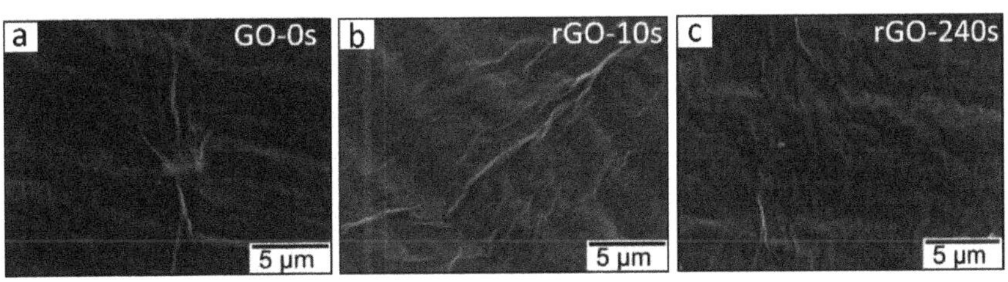


**Figure 5.** Surface morphology of (**a**) pristine GO and GO after plasma treatment for (**b**) 10 and (**c**) 240 s. With permission from Elsevier. Copyright (2024).

Based on resistance measurement studies, the resistance gradually decreased with the increase in hydrogen plasma treatment (Figure 6), which effectively removed the oxygen groups from GO and partially restored the $sp^2$-bounded carbon network, resulting in enhanced conductivity. In addition, during plasma exposure, the amorphous phases were etched, which contributed to the improved conductance. Based on various characterizations, the pristine GO had an oxygen content of 30 at.%, and after plasma treatment for 240 s, it decreased to 20 at%, confirming the reducing effect of plasma exposure on GO.

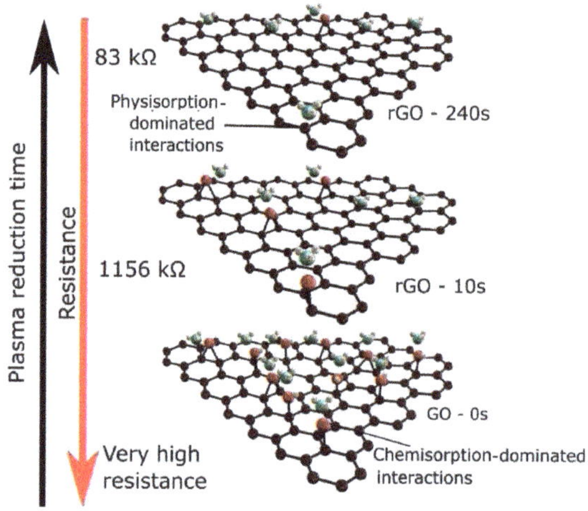


**Figure 6.** A possible physisorption/chemisorption-assisted sensing mechanism towards ammonia in plasma rGO sensors [81]. With permission from Elsevier. Copyright (2024).

Among the different gas sensors, the sensor exposed to plasma for 20 s revealed the highest response to $NH_3$ gas at RT. After 20 s of plasma exposure, the oxygen groups were not extensively removed, and hence, they provided sufficient channels for the physisorption of $NH_3$. Meanwhile, chemisorption of $NH_3$ was facilitated due to the presence of oxygen groups. This led to achieving the optimal sensing performance by providing a balance between both the chemisorption and physisorption phenomena.

### 4.3. Plasma-Treated ZnO Gas Sensors

The response of ZnO sensors also can be remarkably increased by plasma treatment [82–84]. In this context, Hou et al. [85] prepared ZnO thin films by sol–gel spin-coating deposition. Then, they were treated with $O_2$ plasma for 3, 5, 8, 11, and 15 min. During plasma treatment for 3 and 5 min, the crystallinity increased thanks to the decrease in oxygen vacancies, while a further increase in plasma exposure time led to a decrease in the crystallinity due to the formation of zinc vacancies. Also, the roughness of the pristine ZnO thin film was 5.5 nm, which decreased to 3.6 nm after plasma treatment and then increased to 4.3 and 5 nm with further increase in the treatment time to 8 and 11 min, respectively. The sensor that underwent 8 min plasma exposure revealed a higher response of 65% to 50 ppm $NH_3$ at RT compared to the other sensors. It manifested a higher baseline resistance relative to the pristine sensor, and hence, more reactions occurred between the adsorbed oxygen and $NH_3$ gas, contributing to a higher sensing response.

Gui et al. [86] produced ZnO nanorods (NRs) with average diameters of 300 nm on ceramic tubes by an in situ hydrothermal growth method at 140 °C. Then, they were exposed to oxygen plasma for 30, 60, and 90 s. Upon plasma exposure, not only did the surface become rough, but also the content of oxygen vacancies increased up to a plasma exposure time of 60 s. The sensor exposed to plasma for 60 s manifested a high response of 198 to 100 ppm N-methyl pyrrolidone (NMP) at 210 °C, which was three times higher than that of the pristine sensor. The improved performance originated from the presence of the highest amount of oxygen vacancies, which acted as highly active sites for oxygen adsorption, and the subsequent increase in reactions with target gas molecules. Based on DFT calculations, the adsorption energy of NMP on the oxygen-plasma-treated ZnO was higher than that of the pristine ZnO. Furthermore, the adsorption energy of NMP on ZnO was the largest (−1.06 eV) compared to other gases, leading to better selectivity to NMP gas (Figure 7).

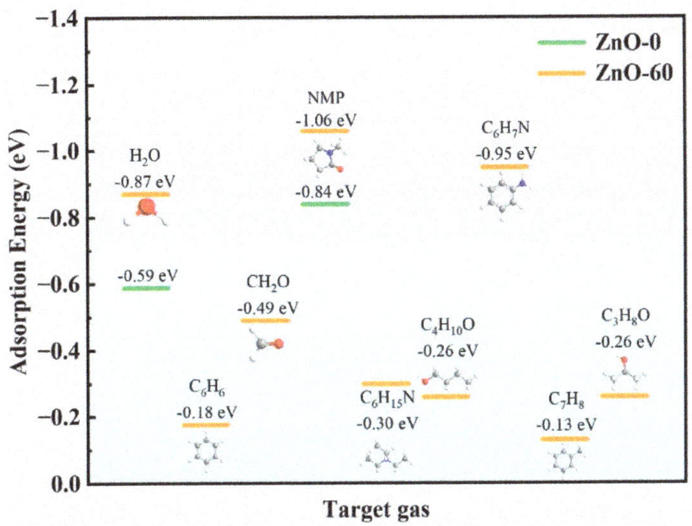


**Figure 7.** Adsorption energies of different gases on pristine and plasma-exposed (60 s) ZnO NRs [86]. With permission from Elsevier. Copyright (2024).

Although the chemical solution method is widely used for sensing film deposition, it still suffers from poor adhesion between the film and substrate along with poor reproducibility. In this regard, atomic layer deposition (ALD) is a highly reliable method of film deposition, allowing precise control of the thickness of the film by adjusting the number of ALD cycles. Furthermore, it can be used for the deposition of uniform sensing layers on a substrate with high reproducibility [87,88]. In this regard, Li et al. [89] deposited ultrathin ZnO films (20 nm) by the ALD technique followed by Ar plasma treatment for 1, 5, and 10 min. Among the different samples, the one exposed to plasma for 5 min exhibited the highest amount of oxygen vacancies, as confirmed by XPS and EPR analyses. It revealed a maximum response of 21.6 to 100 ppm TEA at 250 °C with a low limit of detection of 22 ppb. The high selectivity to TEA was ascribed to the presence of active C–N bonds in TEA and the high electron-denoting properties of TEA. Furthermore, oxygen vacancies acted as electron donors and decreased the band gap of ZnO, eventually facilitating the adsorption and activation of TEA.

### 4.4. Plasma-Treated $SnO_2$ Gas Sensors

$SnO_2$ is among the most widely used sensing materials, thanks to its high stability, high mobility of electrons, ease of the synthesis, low price, nontoxicity, and abundance [90,91]. Plasma treatment has been used on $SnO_2$ to modify its sensing properties [92,93]. Srivastava et al. [94] are among the leading researchers reporting the enhanced gas sensing properties of $SnO_2$ sensors under oxygen and hydrogen plasma exposure. The sensitivity of a sensor treated with oxygen plasma was found to be about 10 times more than that of the pristine sensor, while in the case of hydrogen plasma, the response of the plasma-treated (15 min) sensor was seven times higher than that of the pristine sensor. Also, the same group [95] reported the enhanced gas sensing response of elemental-doped $SnO_2$ gas sensors.

Acharyya et al. [96] synthesized $SnO_2$ nanosheets (NSs) through a hydrothermal route at 200 °C for 12 h. Then, the synthesized materials were exposed to Ar plasma for 2, 4, 7, and 10 min. At 270 °C, the sensor treated with Ar for 7 min revealed the highest response of 25 to 10 ppm HCHO gas. Also, the smaller molecule size and lowest activation energy of the HCHO compared to other gases accounted for the selective response to gas. The content of oxygen vacancies was highest in the sensor exposed to plasma for 7 min. This caused more oxygen and HCHO gas adsorption on the surface of the $SnO_2$ NSs, leading to a higher response relative to the pristine sensor (Figure 8a–d). Furthermore, as indicated in Figure 8e, the $SnO_2$-$SnO_2$ homojunctions were formed in air, and the height of barriers was lower relative to that of the plasma-exposed sensor. Hence, in the presence of HCHO gas, the significant reduction in homojunction height in the case of the plasma-treated sensor led to the generation of a higher sensing response relative to the pristine sensor.

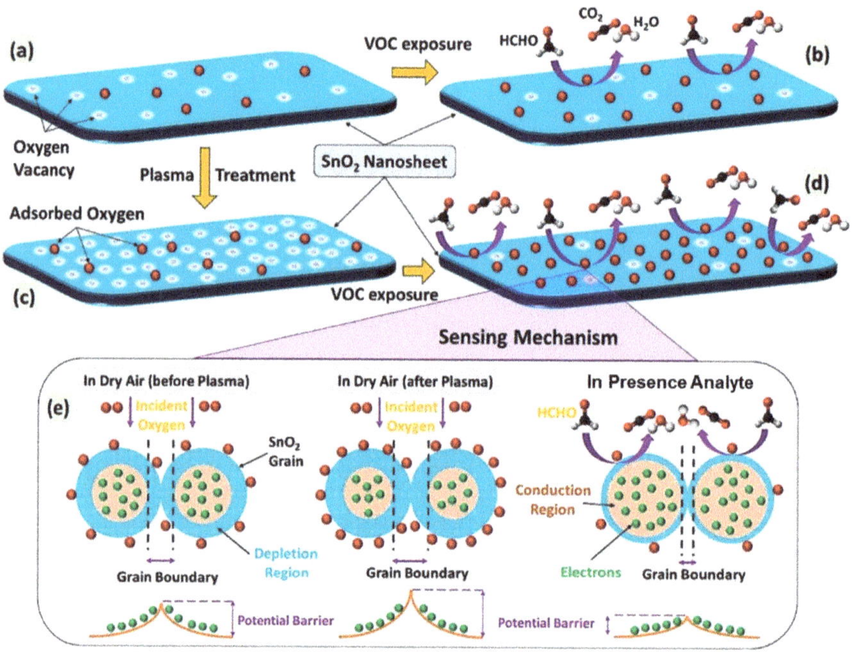


**Figure 8.** Schematic illustration of sensing mechanism of $SnO_2$ NSs to VOCs: (**a**,**b**) pristine $SnO_2$ NSs; (**c**,**d**) plasma-treated $SnO_2$ NSs; (**e**) modulation of double Schottky barrier in the presence of plasma and VOC [96]. With permission from Elsevier. Copyright (2024).

Pd is a good catalyst for $H_2$ gas dissociation, and therefore it is widely used as a decoration on the surface of resistive gas sensors [97,98]. Hu et al. [99] synthesized Pd-decorated $SnO_2$ nanofibers (NFs) via electrospinning of $SnO_2$ NFs followed by the decoration of Pd NPs using sputtering. Then, the samples were exposed to Ar plasma treatment for 5, 60, and 300 s. Based on an XPS analysis, the content of oxygen vacancies and adsorbed oxygen species increased after plasma treatment. The Sn-O bonds in the $SnO_2$ dissociated during the collision with Ar ions, resulting in the generation of oxygen vacancies. Furthermore, the dissociated oxygen was chemisorbed on the oxygen vacancy sites. The sensor exposed to plasma for 60 s revealed the highest response of 53 to 500 ppm $H_2$ gas at 130 °C. Figure 9a,b show the amounts of different enlacements as a function of plasma exposure time. The sensor exposed to plasma for 60 s exhibited the highest amount of oxygen vacancy and adsorbed oxygen species, both of which were highly beneficial for $H_2$ gas sensing. However, the extension of plasma exposure to 300 s led to the degradation of sensing performance due to the decrease in both oxygen vacancy and adsorbed oxygen species. Also, the catalytic effect of Pd towards $H_2$ dissociation was effective on the high sensing response towards $H_2$ gas.

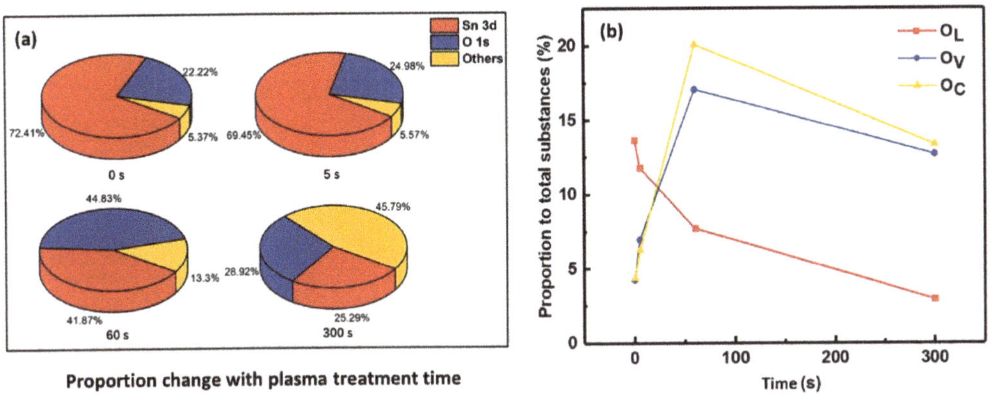


**Figure 9.** (**a**,**b**) The amounts of different species on Pd-$SnO_2$ NFs versus plasma exposure time [99]. With permission from Elsevier. Copyright (2020).

Chaturvedi et al. [100] used plasma treatment on Pd-doped $SnO_2$ gas sensors. The synthesized materials were exposed to $O_2$, $H_2$, $N_2$, and Ar plasma for 15 min. In all cases, the plasma-treated sensors revealed a higher response to $CCl_4$, CO, LPG, $C_3H_7OH$, $N_2O$, and $CH_4$ gases relative to the pristine sensor due to the release of a greater number of electrons upon interaction with the adsorbed gas molecules. The oxygen-treated sensor showed a higher response relative to other gas sensors; however, it showed weak selectivity. The non-stoichiometry was the highest in the case of the oxygen-plasma-treated sensor, where the sensitivity was maximum. At RT, the hydrogen-plasma-treated sensor was highly selective to CO gas, while the nitrogen treated sensor manifested a moderate response to all the gases, without selectivity. Also the argon-plasma-treated sensor did not show noticeable sensitivity to any gas.

Nanowires (NWs) are among the most popular morphologies for gas sensing applications thanks to their high surface area and numerous adsorption sites for gas adsorption. In this context, Pan et al. [101] synthesized $SnO_2$ NWs through a CVD method and then used $O_2$/Ar plasma to change the compositions to be more non-stoichiometric. The plasma power was varied between 10 and 80 W, while the plasma duration was fixed to 240 s. The samples exposed to plasma under 10, 20, and 40 W had some amounts of SnO, $Sn_2O_3$, and $Sn_3O_4$ phases due to the gradual reduction of tetragonal $SnO_2$ and removing of oxygen atoms from $SnO_2$. Also, the further increase in plasma power (80 W) resulted in extensive reduction of $SnO_2$ to metallic Sn, resulting in poor sensing performance. Among the different gas sensors, the sensor treated with a power of 40 W revealed an enhanced response to ethanol gas, thanks to the co-existence of $SnO_2$-$Sn_3O_4$ phases, in which potential barriers at interfaces acted as powerful sources of resistance modulation, in addition to the high surface area thanks to NW morphology and the presence of oxygen vacancies.

In another study, Huang et al. [102] synthesized $SnO_2$ thin films using plasma-enhanced CVD (PECVD) and then exposed it to oxygen plasma for 20 min. The pristine sensor manifested a low response of 3.9 to 1000 ppm CO at 330 °C. Also, the plasma-treated sensor showed the highest response of 31.7 to the same gas concentrations at 250 °C. Interestingly, $SnO_2$ nanorods (NRs) were grown on $SnO_2$ thin films after the plasma treatment by sputtering followed by a redeposition mechanism. In fact, the films were sputtered by the bombardment of heavy ions in the plasma, and then $SnO_2$ NRs were grown by the sputtering, redeposition, and rearrangement on the films. Hence, the surface area was significantly increased relative to the pristine $SnO_2$ thin film due to the presence of both the 1D NRs and 2D thin film. Accordingly, numerous adsorption sites were available for gas molecules, resulting in a boosted sensing response.

Huang et al. [103] synthesized $SnO_2$ nanocolumn arrays with aspect ratios of 20 using liquid immersion PECVD, and the impacts of thermal annealing (600 °C/2 h) and $O_2$ plasma treatment on the sensing response toward CO and $H_2$ gases were investigated. The response of the pristine sensor to 1000 ppm $H_2$ at 400 °C was 17, which was higher than the response to CO gas. Based on a compositional analysis, some residual carbon species remained on the pristine sensor, decreasing its sensing performance. After thermal annealing, the response was increased due to the removal of carbon impurities. Also, after plasma treatment for 40 min, the sensing response to both 1000 ppm CO and $H_2$ increased around seven times. The compositional analysis demonstrated that the amount of surface oxygen species significantly increased due to chemisorbed oxygen species on the surface during the plasma treatment, which reacted with target gases to release electrons on the sensor surface.

Hu et al. [104] synthesized ZnO-$SnO_2$ heterojunction NFs (200–500 nm) using electrospinning followed by Ar plasma exposure for 5, 20, and 60 min. Overall, all sensors treated with plasma exhibited higher responses than the pristine sensor. Also, at 300 °C, the sensor exposed to plasma for 20 min revealed a response of 18 to 100 ppm $H_2$ gas (Figure 10a,b).

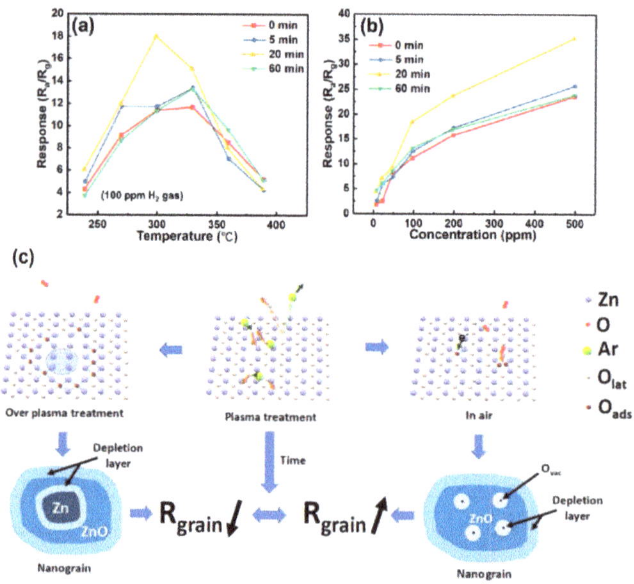


**Figure 10.** (**a**) Response to $H_2$ gas versus temperature and (**b**) calibration curves of plasma-treated gas sensors. (**c**) Mechanism of plasma treatment on a ZnO nanograin [104]. With permission from Elsevier. Copyright (2020).

Based on an XPS analysis, the amount of adsorbed oxygen species was highest in the optimal sensor. When the plasma was exposed to ZnO, some Zn-O bonds were broken, resulting in the formation of oxygen vacancies. Then, oxygen molecules from the air were adsorbed on the oxygen vacancy sites, and thanks to the highly electrophilic nature of oxygen, they abstracted the electrons from the conduction band of ZnO, leading to the expansion of EDL relative to the pristine ZnO and an increase in resistance. Excess plasma exposure led to the reduction of ZnO to Zn, reducing the overall resistance (Figure 10c). In the case of the optimal gas sensor, plasma exposure caused the formation of EDL with high thickness, and when the sensor was exposed to gas, the release of electrons significantly modulated the sensor resistance. Furthermore, heterojunctions were formed between ZnO and $SnO_2$, acting as resistance sources for the gas sensor.

In another study [105], $SnO_2/In_2O_3$ composite NFs were produced using electrospinning, and then they were exposed to oxygen plasma for 30 min. After plasma exposure, the morphology of $SnO_2$ changed to nanoneedles, while that of $In_2O_3$ changed to nanotapers. The surface area before the plasma treatment was 16.5 $m^2/g$, and after plasma exposure it increased to 31 $m^2/g$. This was due to the fact that the surface was rough and porous after plasma exposure. While the pristine sensor showed a response of 8 to 10 ppm formaldehyde at 375 °C, the response of the plasma-treated sensor was 14 to the same gas concentration at 290 °C. Furthermore, this selective response was related to the small bond dissociation energy of H-CHO, where it was easily broken and reacted with adsorbed oxygen species, releasing electrons on the sensor surface. Due to the plasma treatment, more oxygen species were adsorbed on the surface of sensor, leading to more sensing reactions with formaldehyde gas. In another similar study performed by the same group [106], $SnO_2$ NFs revealed an enhanced response to HCHO gas after oxygen plasma treatment. The response of pristine $SnO_2$ NFs was only 4.5 to 100 ppm HCHO at 300 °C, while after plasma treatment it was increased to 6.9 at 200 °C.

### 4.5. Plasma-Treated $In_2O_3$ Gas Sensors

One of main shortages of metal oxide gas sensors is humidity interference, which limits their applications in humid environments [107]. Hence, the development of anti-humidity gas sensors is vital. Du et al. [108] synthesized $In_2O_3$ by roasting $In_2SO_4$ at 550 °C, and then fluorocarbon (CF) was grafted onto it by the RF magnetron sputtering technique. The surface of the CF-$In_2O_3$ was evenly wrapped by CF layers with thicknesses of ~2 nm. The $In_2O_3$ film exhibited a low WCA of ∼16° and the CF-$In_2O_3$ film showed a hydrophobic nature with a large WCA of ∼137° thanks to the presence of low-energy CF on the surface of the $In_2O_3$. The $In_2O_3$ recorded a response of ∼18 to 1 ppm $NO_2$ gas at 200 °C. However, the CF-$In_2O_3$ sensor revealed a lower response of 13 at an optimal temperature of 100 °C due to the covering of CF on the surface of the $In_2O_3$ with lower sensing properties relative to $In_2O_3$. In the presence of 92% relative humidity, the response of the CF-$In_2O_3$ was not significantly decreased, demonstrating the anti-humidity properties of CF-$In_2O_3$. However, the response of the $In_2O_3$ dramatically decreased. Two reasons can account for the humidity-resistant nature of the optimal sensor: (i) the hydrophobic CF layer absorbed a sufficient amount of $H_2O$ molecules to increase the electron concentration, and hence more $NO_2$ molecules were adsorbed; (ii) the hydrophobic CF layer suppressed the reaction between $NO_2$ with $H_2O$ molecules, and therefore the concentration of the adsorbed and reacted $NO_2$ gas molecules on the surface of the sensor did not change.

In another study, Du et al. [109] synthesized $In_2O_3$ NFs and then exposed them to hydrogen and oxygen plasma for 30 min. The surface of the oxygen-plasma-treated $In_2O_3$ was rougher, and the diameters of the NFs were thicker than those of hydrogen plasma $In_2O_3$. However, the diameters of the nanograins on the surface of the NFs were smaller in the case of the oxygen-plasma-treated sample. Also, the surface areas of the pristine, oxygen-, and hydrogen-plasma-treated samples were 18, 32, and 29 $m^2/g$, respectively.

Thus, the surface area was increased thanks to the formation of many new small pores on the surface of the $In_2O_3$ NFs. Based on an XPS study, the oxygen content was greatly increased by the oxygen plasma, which is vital for sensing reactions with acetone gas. As expected, the sensor exposed to oxygen plasma revealed the largest response of 37 to 500 ppm acetone at 275 °C. The high surface area and the presence of a large amount of adsorbed oxygen species contributed to the enhanced sensing response to acetone.

### 4.6. Other Plasma-Treated Gas Sensors

$ZnGa_2O_4$ is a semiconducting material (5.1 eV) with features like the ease of fabrication, low cost, and high stability [110]. Chang et al. [111] synthesized a $ZnGa_2O_4$ epilayer (125 nm thick) on a sapphire substrate using a metal–organic CVD technique. Then, Ar plasma was applied for 5, 10, and 15 min on it. Spindle nanostructures changed to smaller sizes and near-spherical particles after Ar plasma treatment for 15 min due to heavy Ar plasma bombardment and the coalescence of nanostructures (Figure 11a–d). Furthermore, the Ar plasma treatment introduced Ar atoms, radicals, and ions on the epilayer surface, resulting in chemical changes after the plasma treatment.

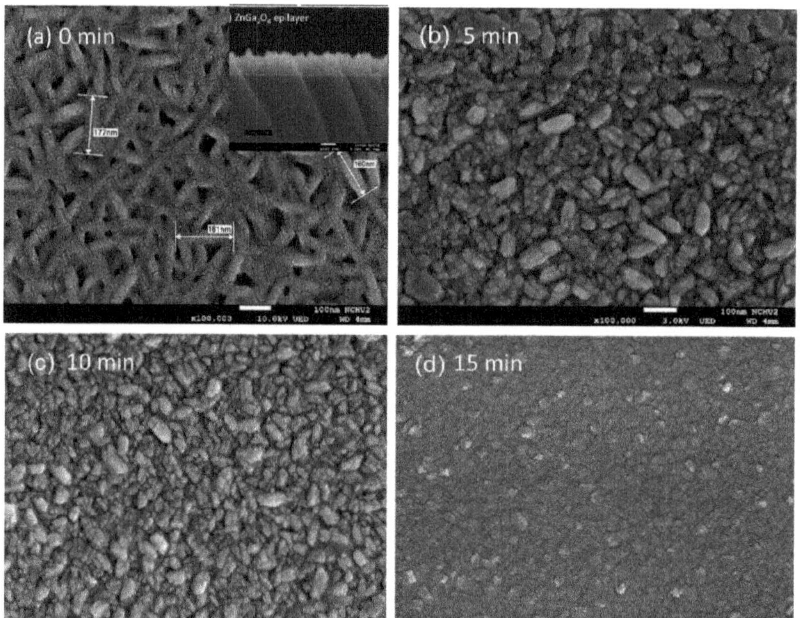


**Figure 11.** SEM micrographs of $ZnGa_2O_4$ epilayer: (**a**) before and after Ar plasma treatment and (**b**) 5, (**c**) 10, and (**d**) 15 min [111]. With permission from Elsevier. Copyright (2023).

Among e different sensors, the sensor treated with plasma for 10 min revealed an enhanced response at 300 °C with a response of 1300% to 5 ppm NO gas. The main reasons for sensing enhancement were related to the higher surface area and the presence of more oxygen dangling bonds, leading to an increase in the reactions with NO gas. Also, based on DFT calculations, $ZnGa_2O_4$ with surface oxygen groups had a greater tendency to adsorb NO molecules.

Polypyrrole (PPy) as a CP is a promising sensing material thanks to its high conductance, simple preparation methods, high sensitivity, and possibility of RT operation [112]. Similar to metal oxides, plasma treatment on CPs can increase their gas sensing performance [113]. Zhang et al. [114] applied hydrogen and oxygen plasma on PPy for 20 min and investigated the response to various gases. At 25 °C, the response of the hydrogen-treated sensor to 50 ppm $NO_2$ gas was 6, which was 1.6 and 1.2 times higher than that of the pristine and oxygen-treated sensors, respectively. Based on DFT calculations, the adsorption energy of $NO_2$ on the hydrogen-treated sensor was significantly higher (−1.72 eV) than that on the pristine (−0.58 eV) and oxygen-treated (−0.69 eV) sensors, respectively. This implied that the hydrogen plasma treatment was more efficient for $NO_2$ gas adsorption enhancement. In addition, the increase in surface area by the formation of pores after plasma exposure contributed to the sensing improvement. In another study related to oxygen-plasma-treated PANI, the response to hydrogen at RT was significantly improved relative to the pristine sensor [115].

MXenes are a new category of 2D materials with high conductivity, a large surface area, and tunable band gaps [116]. They have a general formula of $M_{n+1}X_nT_x$, in which A is a transition metal, X is C/or N, and $T_x$ shows the surface functional groups. They are synthesized from their parent MAX phases, which can be represented as $M_{n+1}A_nX$, where A is an element in group IIIA or group IVA [117,118]. In this context, Wang et al. [119] synthesized $Ti_3C_2T_x$ MXene via liquid exfoliation and subsequently exposed it to oxygen plasma treatment. The sensor exhibited a response of 13.8% to 10 ppm $NO_2$ gas at RT. The enhanced sensing performance was related to the presence of numerous oxygen surface functional groups as a result of the plasma treatment.

Transition dichalcogenides are 2D semiconductors with high conductivity and large surface areas. They have a general formula of $MX_2$, in which M is a transition metal and X is a chalcogenide such as S, Se, or Te [120,121]. Seo et al. [122] applied Ar plasma treatment on $MoS_2$ NSs for 2 s. As a result of plasma exposure, sulfur vacancies were created on the $MoS_2$. Then, it was exposed to a 3-mercaptopropionic acid (MPA) solution to form coordinate bonds between the HS groups in MPA and sulfur vacancies. Based on an XPS study, the pristine $MoS_2$ exhibited an ideal S/Mo ratio of 1.91, while after plasma exposure it was decreased to 1.51, indicating sulfur vacancy formation. Based on $NH_3$ gas sensing studies, the pristine sensor revealed a response of 1.25 to 130 ppm $NH_3$ gas at RT, and after treatment by plasma and MPA, the response increased to 4.45. The boosted sensing capability was related to the presence of oxygen and carboxyl groups (−COO) on the surface of the sensor.

Table 1 summarizes the gas sensing properties of plasma-treated gas sensors. Overall, plasma-treated gas sensors have been successfully used for the detection of various toxic gases.

**Table 1.** Gas sensing properties of plasma-treated gas sensors.

| Sensing Material | Synthesis Method | Plasma Condition | Gas | Conc. (ppm) | T (°C) | Response ($R_a/R_g$) or ($R_g/R_a$) or [($R_a$ − $R_g$)/$R_a$] × 100 | Ref. |
|---|---|---|---|---|---|---|---|
| MWCNTs | CVD | $O_2/C_2H_2(CO)_2O$ | $NH_3$ | 500 | RT | 31.4% | [57] |
| MWCNTs | CVD | Ar | Ethanol | 100 | 65 | 170% | [60] |
| SWCNTs | CVD | $N_2$ | $NO_2$ | 50 | RT | 121% | [62] |
| SWCNTs | CVD | $N_2$ | $NH_3$ | 50 | RT | 36% | [62] |
| Graphene | CVD | $CF_6$ for 20 s | $NH_3$ | 100 | RT | 3.8% | [73] |
| Graphene | CVD | $O_3$ for 70 s | $NO_2$ | 10 | RT | 19.7% | [74] |
| ZnO thin films | Sol–gel spin coating | $O_2$ for 8 min | $NH_3$ | 50 | RT | 65% | [85] |
| ZnO NRs | Hydrothermal | $O_2$ for 60 s | NMP | 100 | 210 | 198 | [86] |

**Table 1.** *Cont.*

| Sensing Material | Synthesis Method | Plasma Condition | Gas | Conc. (ppm) | T (°C) | Response ($R_a/R_g$) or ($R_g/R_a$) or [($R_a - R_g$)/$R_a$] × 100 | Ref. |
|---|---|---|---|---|---|---|---|
| ZnO film | ALD | Ar | TEA | 100 | 250 | 21.6 | [89] |
| $SnO_2$ NSs | Hydrothermal | Ar for 7 min | HCHO | 10 | 270 | 25 | [96] |
| Pd-$SnO_2$ NFs | Electrospinning | Ar for 60 s | $H_2$ | 500 | 130 | 53 | [99] |
| $SnO_2$ thin film | Plasma enhance CVD | $O_2$ for 20 min | CO | 1000 | 250 | 31.7 | [102] |
| $SnO_2$ nanocolumn arrays | Liquid immersion PECVD | $O_2$ | $H_2$ | 1000 | 400 | 17 | [103] |
| ZnO-$SnO_2$ heterojunction NFs | Electrospinning | Ar for 20 min | $H_2$ | 100 | 300 | 18 | [104] |
| $SnO_2/In_2O_3$ composite NFs | Electrospinning | $O_2$ for 30 min | HCHO | 100 | 290 | 14 | [105] |
| $SnO_2$ NFs | Electrospinning | $O_2$ | HCHO | 100 | 200 | 6.9 | [106] |
| $In_2O_3$ | Roasting of $In_2SO_4$ | Fluorocarbon CF | $NO_2$ | 1 | 100 | 13 | [108] |
| $In_2O_3$ NFs | Electrospinning | $O_2$ | $C_3H_6O$ | 500 | 275 | 37 | [109] |
| $ZnGa_2O_4$ | MOCVD technique | Ar for 10 min | $NO_2$ | 5 | 300 | 1300% | [111] |
| PPy | Polymerization | $O_2$ for 20 min | $NO_2$ | 50 | 25 | 6 | [114] |
| $Ti_3C_2T_x$ MXene | Liquid exfoliation | $O_2$ | $NO_2$ | 10 | 25 | 13.8% | [119] |
| $MoS_2$ | CVD | Ar for 2 s | $NH_3$ | 130 | 25 | 1.25 | [122] |

## 5. Conclusions and Outlooks

We reviewed the effect of plasma treatment on the gas sensing characteristics of gas sensors. In general, plasma exposure affects the amount of oxygen species on the sensor surface, and since the oxygen ions are highly required for gas sensing reactions, plasma treatment significantly affects the gas sensing characteristics of resistive sensors through modulation of the amount of oxygen ions. Generally, oxygen plasma causes the addition of surface oxygen functional groups on the sensor surface, and hence, the reactions between adsorbed gases with oxygen increase, leading to a higher sensing performance relative to pristine sensors. Also, exposure to other plasma atmospheres such as Ar or He causes the generation of oxygen defects, which act as favorable sites for oxygen adsorption and accordingly contribute to the enhanced sensing performance. Overall, plasma treatment can cause morphology changes when its power and treatment time are sufficiently high. Also, it causes changes in the amount of oxygen vacancies and adsorbed oxygen species. In some cases, it can add new functional groups on the sensor surface, which act as adsorption sites for gas molecules. Thus, when plasma treatment conditions such as plasma type, time, and power are optimized, it is expected that the sensing properties such as sensitivity, selectivity, and working temperature improve relative to pristine sensors.

Regardless of the type of plasma used, both plasma power and plasma exposure times should be optimized to achieve the highest gas sensing performance. However, in most cases, the focus is on the optimization of plasma time rather than plasma power. Thus, this aspect needs to be more explored in future studies. Different sensing materials such as metal oxides, TMDs, MXenes, CNTs, graphene, and CPs have been subjected to plasma treatment. In this regard, the combination of plasma exposure with other high irradiation techniques such as ion beams, electron beams, and gamma rays can lead to interesting results. Thus, future research directions on plasma-treated gas sensors can be summarized as follows: (i) the development of cheap plasma treatment devices with high availability across the world; (ii) the study of the optimal plasma power and time for various gas sensing materials; (iii) the study of various sensor parameters such as stability, response in humid environments, and reproducibility; and (iv) the reduction in sensing temperature on plasma-treated gas sensors by operation of the sensor in self-heating mode or under UV light illumination.

**Author Contributions:** M.T.C.: investigation, writing—original paper; A.M.: investigation, conceptualization, writing—review and editing. All authors have read and agreed to the published version of the manuscript.

**Funding:** This research received no external funding.

**Acknowledgments:** The authors are thankful to the Shiraz University of Technology.

**Conflicts of Interest:** The authors declare no conflicts of interest.

## References

1. Vos, T.; Lim, S.S.; Abbafati, C.; Abbas, K.M.; Abbasi, M.; Abbasifard, M.; Abbasi-Kangevari, M.; Abbastabar, H.; Abd-Allah, F.; Abdelalim, A.; et al. Global Burden of 369 Diseases and Injuries in 204 Countries and Territories, 1990–2019: A Systematic Analysis for the Global Burden of Disease Study 2019. *Lancet* **2020**, *396*, 1204–1222. [CrossRef] [PubMed]
2. Le Quilliec, E.; Fundere, A.; Al-U'datt, D.G.F.; Hiram, R. Pollutants, Including Organophosphorus and Organochloride Pesticides, May Increase the Risk of Cardiac Remodeling and Atrial Fibrillation: A Narrative Review. *Biomedicines* **2023**, *11*, 2427. [CrossRef] [PubMed]
3. Gal, O.; Eitan, O.; Rachum, A.; Weinberg, M.; Zigdon, D.; Assa, R.; Price, C.; Yovel, Y. Air Pollution Likely Reduces Hemoglobin Levels in Urban Fruit Bats. *iScience* **2025**, *28*, 111997. [CrossRef] [PubMed]
4. Manisalidis, I.; Stavropoulou, E.; Stavropoulos, A.; Bezirtzoglou, E. Environmental and Health Impacts of Air Pollution: A Review. *Front. Public Health* **2020**, *8*, 14. [CrossRef]
5. Hong, C.; Zhong, R.; Xu, M.; He, P.; Mo, H.; Qin, Y.; Shi, D.; Chen, X.; He, K.; Zhang, Q. Interactions Among Food Systems, Climate Change, and Air Pollution: A Review. *Engineering* **2025**, *44*, 215–233. [CrossRef]
6. Bell, J.N.B.; Honour, S.L.; Power, S.A. Effects of Vehicle Exhaust Emissions on Urban Wild Plant Species. *Environ. Pollut.* **2011**, *159*, 1984–1990. [CrossRef]
7. Gao, Q.; Jiang, B.; Tong, M.; Zuo, H.; Cheng, C.; Zhang, Y.; Song, S.; Lu, L.; Li, X. Effects and Interaction of Humidex and Air Pollution on Influenza: A National Analysis of 319 Cities in Mainland China. *J. Hazard. Mater.* **2025**, *490*, 137865. [CrossRef]
8. Semczuk-Kaczmarek, K.; Rys-Czaporowska, A.; Sierdzinski, J.; Kaczmarek, L.D.; Szymanski, F.M.; Platek, A.E. Association between Air Pollution and COVID-19 Mortality and Morbidity. *Intern. Emerg. Med.* **2022**, *17*, 467–473. [CrossRef]
9. Aithal, S.S.; Sachdeva, I.; Kurmi, O.P. Air Quality and Respiratory Health in Children. *Breathe* **2023**, *19*, 230040. [CrossRef]
10. Van Meerbeke, S.W.; McCarty, M.; Petrov, A.A.; Schonffeldt-Guerrero, P. The Impact of Climate, Aeroallergens, Pollution, and Altitude on Exercise-Induced Bronchoconstriction. *Immunol. Allergy Clin.* **2025**, *45*, 77–88. [CrossRef]
11. Tiotiu, A.I.; Novakova, P.; Nedeva, D.; Chong-Neto, H.J.; Novakova, S.; Steiropoulos, P.; Kowal, K. Impact of Air Pollution on Asthma Outcomes. *Int. J. Environ. Res. Public Health* **2020**, *17*, 6212. [CrossRef] [PubMed]
12. Chen, K.-C.; Tsai, S.-W.; Shie, R.-H.; Zeng, C.; Yang, H.-Y. Indoor Air Pollution Increases the Risk of Lung Cancer. *Int. J. Environ. Res. Public Health* **2022**, *19*, 1164. [CrossRef] [PubMed]
13. Syama, K.P.; Blais, E.; Kumarathasan, P. Maternal Mechanisms in Air Pollution Exposure-Related Adverse Pregnancy Outcomes: A Systematic Review. *Sci. Total Environ.* **2025**, *970*, 178999. [CrossRef] [PubMed]
14. Bayart, N.-E.; Pereira, G.; Reid, C.M.; Nyadanu, S.D.; Badamdorj, O.; Lkhagvasuren, B.; Rumchev, K. Effects of Outdoor Air Pollution on Hospital Admissions from Cardiovascular Diseases in the Capital City of Mongolia. *Atmos. Pollut. Res.* **2024**, *16*, 102338. [CrossRef]
15. Aryal, A.; Harmon, A.C.; Dugas, T.R. Particulate Matter Air Pollutants and Cardiovascular Disease: Strategies for Intervention. *Pharmacol. Ther.* **2021**, *223*, 107890. [CrossRef]
16. Glencross, D.A.; Ho, T.-R.; Camiña, N.; Hawrylowicz, C.M.; Pfeffer, P.E. Air Pollution and Its Effects on the Immune System. *Free Radic. Biol. Med.* **2020**, *151*, 56–68. [CrossRef]

17. Sicard, P.; Agathokleous, E.; Anenberg, S.C.; Marco, A.D.; Paoletti, E.; Calatayud, V. Trends in Urban Air Pollution over the Last Two Decades: A Global Perspective. *Sci. Total Environ.* **2023**, *858*, 160064. [CrossRef]
18. Li, P.; Li, J.; Song, S.; Chen, J.; Zhong, N.; Xie, Q.; Liu, Y.; Wan, B.; He, Y.; Karimi-Maleh, H. Recent Advances in Optical Gas Sensors for Carbon Dioxide Detection. *Measurement* **2025**, *239*, 115445. [CrossRef]
19. Gorbova, E.; Tzorbatzoglou, F.; Molochas, C.; Chloros, D.; Demin, A.; Tsiakaras, P. Fundamentals and Principles of Solid-State Electrochemical Sensors for High Temperature Gas Detection. *Catalysts* **2022**, *12*, 1. Available online: https://www.mdpi.com/2073-4344/12/1/1 (accessed on 4 January 2025).
20. Kumar, A.; Prajesh, R. The Potential of Acoustic Wave Devices for Gas Sensing Applications. *Sens. Actuators A Phys.* **2022**, *339*, 113498. [CrossRef]
21. El-Muraikhi, M.D.; Ayesh, A.I.; Mirzaei, A. Resistive Gas Sensors Based on Inorganic Nanotubes: A Review. *J. Alloys Compd.* **2025**, *1020*, 179585. [CrossRef]
22. Mirzaei, A.; Neri, G. Microwave-Assisted Synthesis of Metal Oxide Nanostructures for Gas Sensing Application: A Review. *Sens. Actuators B Chem.* **2016**, *237*, 749–775. [CrossRef]
23. Guo, S.-Y.; Hou, P.-X.; Zhang, F.; Liu, C.; Cheng, H.-M. Gas Sensors Based on Single-Wall Carbon Nanotubes. *Molecules* **2022**, *27*, 5381. [CrossRef]
24. Mirzaei, A.; Bharath, S.P.; Kim, J.-Y.; Pawar, K.K.; Kim, H.W.; Kim, S.S. N-Doped Graphene and Its Derivatives as Resistive Gas Sensors: An Overview. *Chemosensors* **2023**, *11*, 334. [CrossRef]
25. Kiranakumar, H.V.; Thejas, R.; Naveen, C.S.; Khan, M.I.; Prasanna, G.D.; Reddy, S.; Oreijah, M.; Guedri, K.; Bafakeeh, O.T.; Jameel, M. A Review on Electrical and Gas-Sensing Properties of Reduced Graphene Oxide-Metal Oxide Nanocomposites. *Biomass Convers. Biorefinery* **2024**, *14*, 12625–12635. [CrossRef]
26. Matindoust, S.; Farzi, G.; Nejad, M.B.; Shahrokhabadi, M.H. Polymer-Based Gas Sensors to Detect Meat Spoilage: A Review. *React. Funct. Polym.* **2021**, *165*, 104962. [CrossRef]
27. Hu, Y.; Zheng, W.; Fan, S.; Zhang, J.; Liu, X. Noble-Transition-Metal Dichalcogenides-Emerging Two-Dimensional Materials for Sensor Applications. *Appl. Phys. Rev.* **2023**, *10*, 031306. [CrossRef]
28. Amiri, V.; Roshan, H.; Mirzaei, A.; Neri, G.; Ayesh, A.I. Nanostructured Metal Oxide-Based Acetone Gas Sensors: A Review. *Sensors* **2020**, *20*, 3096. [CrossRef]
29. Bonyani, M.; Zebarjad, S.M.; Mirzaei, A.; Kim, T.-U.; Kim, H.W.; Kim, S.S. Electrospun ZnO Hollow Nanofibers Gas Sensors: An Overview. *J. Alloys Compd.* **2024**, *1001*, 175201. [CrossRef]
30. El-Muraikhi, M.D.; Ayesh, A.I.; Mirzaei, A. Review of Nanostructured $Bi_2O_3$, $Bi_2WO_6$, and $BiVO_4$ as Resistive Gas Sensors. *Surf. Interfaces* **2025**, *60*, 106003. [CrossRef]
31. El-Muraikhi, M.D.; Mirzaei, A.; Ayesh, A.I. Nanostructured $Nb_2O_5$ as Chemiresistive Gas Sensors. *Ceram. Int.* **2024**, *50*, 54897–54911. [CrossRef]
32. Zappa, D.; Galstyan, V.; Kaur, N.; Arachchige, H.M.M.M.; Sisman, O.; Comini, E. "Metal Oxide -Based Heterostructures for Gas Sensors"—A Review. *Anal. Chim. Acta* **2018**, *1039*, 1–23. [CrossRef] [PubMed]
33. Degler, D.; Weimar, U.; Barsan, N. Current Understanding of the Fundamental Mechanisms of Doped and Loaded Semiconducting Metal-Oxide-Based Gas Sensing Materials. *ACS Sens.* **2019**, *4*, 2228–2249. [CrossRef] [PubMed]
34. Zhu, L.-Y.; Ou, L.-X.; Mao, L.-W.; Wu, X.-Y.; Liu, Y.-P.; Lu, H.-L. Advances in Noble Metal-Decorated Metal Oxide Nanomaterials for Chemiresistive Gas Sensors: Overview. *Nano-Micro Lett.* **2023**, *15*, 89. [CrossRef]
35. Mirzaei, A.; Ansari, H.R.; Shahbaz, M.; Kim, J.-Y.; Kim, H.W.; Kim, S.S. Metal Oxide Semiconductor Nanostructure Gas Sensors with Different Morphologies. *Chemosensors* **2022**, *10*, 289. [CrossRef]
36. Liu, M.; Song, P.; Wang, Q.; Yan, M. CsPbBr3 Quantum Dot Modified $In_2O_3$ Nanofibers for Effective Detection of Ppb-Level HCHO at Room Temperature under UV Illumination. *ACS Sens.* **2024**, *9*, 6040–6050. [CrossRef]
37. Majhi, S.M.; Mirzaei, A.; Navale, S.; Kim, H.W.; Kim, S.S. Boosting the Sensing Properties of Resistive-Based Gas Sensors by Irradiation Techniques: A Review. *Nanoscale* **2021**, *13*, 4728–4757. [CrossRef]
38. Kim, J.H.; Lee, M.A.; Han, G.J.; Cho, B.H. Plasma in Dentistry: A Review of Basic Concepts and Applications in Dentistry. *Acta Odontol. Scand.* **2013**, *72*, 1–12. [CrossRef]
39. Bayati, M.; Lund, M.N.; Tiwari, B.K.; Poojary, M.M. Chemical and Physical Changes Induced by Cold Plasma Treatment of Foods: A Critical Review. *Compr. Rev. Food Sci. Food Saf.* **2024**, *23*, e13376. [CrossRef]
40. Bezerra, J.d.A.; Lamarão, C.V.; Sanches, E.A.; Rodrigues, S.; Fernandes, F.A.N.; Ramos, G.L.P.A.; Esmerino, E.A.; Cruz, A.G.; Campelo, P.H. Cold Plasma as a Pre-Treatment for Processing Improvement in Food: A Review. *Food Res. Int.* **2023**, *167*, 112663. [CrossRef]
41. Nimbekar, A.A.; Deshmukh, R.R. Plasma Surface Modification of Flexible Substrates to Improve Grafting for Various Gas Sensing Applications: A Review. *IEEE Trans. Plasma Sci.* **2022**, *50*, 1382–1394. [CrossRef]
42. Alemán, C.; Fabregat, G.; Armelin, E.; Buendía, J.J.; Llorca, J. Plasma Surface Modification of Polymers for Sensor Applications. *J. Mater. Chem. B* **2018**, *6*, 6515–6533. [CrossRef] [PubMed]

43. Saka, C. Overview on the Surface Functionalization Mechanism and Determination of Surface Functional Groups of Plasma Treated Carbon Nanotubes. *Crit. Rev. Anal. Chem.* **2017**, *48*, 1–14. [CrossRef] [PubMed]
44. Leghrib, R.; Felten, A.; Demoisson, F.; Reniers, F.; Pireaux, J.-J.; Llobet, E. Room-Temperature, Selective Detection of Benzene at Trace Levels Using Plasma-Treated Metal-Decorated Multiwalled Carbon Nanotubes. *Carbon* **2010**, *48*, 3477–3484. [CrossRef]
45. Ambardekar, V.; Sahoo, S.; Srivastava, D.K.; Majumder, S.B.; Bandyopadhyay, P.P. Plasma Sprayed CuO Coatings for Gas Sensing and Catalytic Conversion Applications. *Sens. Actuators B Chem.* **2021**, *331*, 129404. [CrossRef]
46. Ambardekar, V.; Bandyopadhyay, P.P.; Majumder, S.B. Plasma Sprayed Copper Oxide Sensor for Selective Sensing of Carbon Monoxide. *Mater. Chem. Phys.* **2021**, *258*, 123966. [CrossRef]
47. Choi, M.S.; Bang, J.H.; Mirzaei, A.; Na, H.G.; Kwon, Y.J.; Kang, S.Y.; Choi, S.-W.; Kim, S.S.; Kim, H.W. Dual Sensitization of MWCNTs by Co-Decoration with p- and n-Type Metal Oxide Nanoparticles. *Sens. Actuators B Chem.* **2018**, *264*, 150–163. [CrossRef]
48. Kwon, Y.J.; Mirzaei, A.; Kang, S.Y.; Choi, M.S.; Bang, J.H.; Kim, S.S.; Kim, H.W. Synthesis, Characterization and Gas Sensing Properties of ZnO-Decorated MWCNTs. *Appl. Surf. Sci.* **2017**, *413*, 242–252. [CrossRef]
49. Lavagna, L.; Nisticò, R.; Musso, S.; Pavese, M. Functionalization as a Way to Enhance Dispersion of Carbon Nanotubes in Matrices: A Review. *Mater. Today Chem.* **2021**, *20*, 100477. [CrossRef]
50. Jian, J.; Guo, X.; Lin, L.; Cai, Q.; Cheng, J.; Li, J. Gas-Sensing Characteristics of Dielectrophoretically Assembled Composite Film of Oxygen Plasma-Treated SWCNTs and PEDOT/PSS Polymer. *Sens. Actuators B Chem.* **2013**, *178*, 279–288. [CrossRef]
51. Leghrib, R.; Pavelko, R.; Felten, A.; Vasiliev, A.; Cané, C.; Gràcia, I.; Pireaux, J.-J.; Llobet, E. Gas Sensors Based on Multiwall Carbon Nanotubes Decorated with Tin Oxide Nanoclusters. *Sens. Actuators B Chem.* **2010**, *145*, 411–416. [CrossRef]
52. Min, S.; Kim, J.; Park, C.; Jin, J.-H.; Min, N.K. Long-Term Stability of Superhydrophilic Oxygen Plasma-Modified Single-Walled Carbon Nanotube Network Surfaces and the Influence on Ammonia Gas Detection. *Appl. Surf. Sci.* **2017**, *410*, 105–110. [CrossRef]
53. Yeo, S.; Choi, C.; Woong Jang, C.; Lee, S.; Min Jhon, Y. Sensitivity Enhancement of Carbon Nanotube Based Ammonium Ion Sensors through Surface Modification by Using Oxygen Plasma Treatment. *Appl. Phys. Lett.* **2013**, *102*, 073108. [CrossRef]
54. Zhang, X.; Wu, X.; Yang, B.; Xiao, H. Enhancement of Gas Sensing Characteristics of Multiwalled Carbon Nanotubes by CF4 Plasma Treatment for SF6 Decomposition Component Detection. *J. Nanomater.* **2015**, *2015*, 171545. [CrossRef]
55. Clément, P.; Ramos, A.; Lazaro, A.; Molina-Luna, L.; Bittencourt, C.; Girbau, D.; Llobet, E. Oxygen Plasma Treated Carbon Nanotubes for the Wireless Monitoring of Nitrogen Dioxide Levels. *Sens. Actuators B Chem.* **2015**, *208*, 444–449. [CrossRef]
56. Zhang, X.; Yang, B.; Wang, X.; Luo, C. Effect of Plasma Treatment on Multi-Walled Carbon Nanotubes for the Detection of $H_2S$ and $SO_2$. *Sensors* **2012**, *12*, 9375–9385. [CrossRef]
57. Bannov, A.G.; Jašek, O.; Manakhov, A.; Márik, M.; Nečas, D.; Zajíčková, L. High-Performance Ammonia Gas Sensors Based on Plasma Treated Carbon Nanostructures. *IEEE Sens. J.* **2017**, *17*, 1964–1970. [CrossRef]
58. Bannov, A.G.; Manakhov, A.M.; Shtansky, D.V. Plasma Functionalization of Multi-Walled Carbon Nanotubes for Ammonia Gas Sensors. *Materials* **2022**, *15*, 7262. [CrossRef]
59. Dong, K.-Y.; Ham, D.-J.; Kang, B.H.; Lee, K.; Choi, J.; Lee, J.-W.; Choi, H.H.; Ju, B.-K. Effect of Plasma Treatment on the Gas Sensor with Single-Walled Carbon Nanotube Paste. *Talanta* **2012**, *89*, 33–37. [CrossRef]
60. Santhosh, N.M.; Vasudevan, A.; Jurov, A.; Korent, A.; Slobodian, P.; Zavašnik, J.; Cvelbar, U. Improving Sensing Properties of Entangled Carbon Nanotube-Based Gas Sensors by Atmospheric Plasma Surface Treatment. *Microelectron. Eng.* **2020**, *232*, 111403. [CrossRef]
61. Ham, S.W.; Hong, H.P.; Kim, J.H.; Min, S.J.; Min, N.K. Effect of Oxygen Plasma Treatment on Carbon Nanotube-Based Sensors. *J. Nanosci. Nanotechnol.* **2014**, *14*, 8476–8481. [CrossRef] [PubMed]
62. Zamansky, K.K.; Osipova, A.A.; Fedorov, F.S.; Kopylova, D.S.; Shunaev, V.; Alekseeva, A.; Glukhova, O.E.; Nasibulin, A.G. Sensitivity Enhancement of SWCNT Gas Sensors by Nitrogen Plasma Treatment. *Appl. Surf. Sci.* **2023**, *640*, 158334. [CrossRef]
63. Ham, S.W.; Hong, H.P.; Kim, J.W.; Kim, J.H.; Kim, K.B.; Park, C.W.; Min, N.K. Comparison of Gas Sensors Based on Oxygen Plasma-Treated Carbon Nanotube Network Films with Different Semiconducting Contents. *J. Electron. Mater.* **2015**, *44*, 1344–1350. [CrossRef]
64. Farea, M.A.; Mohammed, H.Y.; Shirsat, S.M.; Sayyad, P.W.; Ingle, N.N.; Al-Gahouari, T.; Mahadik, M.M.; Bodkhe, G.A.; Shirsat, M.D. Hazardous Gases Sensors Based on Conducting Polymer Composites: Review. *Chem. Phys. Lett.* **2021**, *776*, 138703. [CrossRef]
65. Liu, X.; Zheng, W.; Kumar, R.; Kumar, M.; Zhang, J. Conducting Polymer-Based Nanostructures for Gas Sensors. *Coord. Chem. Rev.* **2022**, *462*, 214517. [CrossRef]
66. Guo, X.; Jian, J.; Lin, L.; Zhu, H.; Zhu, S. $O_2$ Plasma-Functionalized SWCNTs and PEDOT/PSS Composite Film Assembled by Dielectrophoresis for Ultrasensitive Trimethylamine Gas Sensor. *Analyst* **2013**, *138*, 5265–5273. [CrossRef]
67. Yoo, K.-P.; Kwon, K.-H.; Min, N.-K.; Lee, M.J.; Lee, C.J. Effects of $O_2$ Plasma Treatment on NH3 Sensing Characteristics of Multiwall Carbon Nanotube/Polyaniline Composite Films. *Sens. Actuators B Chem.* **2009**, *143*, 333–340. [CrossRef]

68. Kim, J.H.; Song, M.-J.; Kim, K.B.; Jin, J.-H.; Min, N.K. Evaluation of Surface Cleaning Procedures in Terms of Gas Sensing Properties of Spray-Deposited CNT Film: Thermal- and $O_2$ Plasma Treatments. *Sensors* **2017**, *17*, 73. [CrossRef]
69. Zhao, W.; Fam, D.W.H.; Yin, Z.; Sun, T.; Tan, H.T.; Liu, W.; Tok, A.I.Y.; Boey, Y.C.F.; Zhang, H.; Hng, H.H.; et al. A Carbon Monoxide Gas Sensor Using Oxygen Plasma Modified Carbon Nanotubes. *Nanotechnology* **2012**, *23*, 425502. [CrossRef]
70. Schedin, F.; Geim, A.K.; Morozov, S.V.; Hill, E.W.; Blake, P.; Katsnelson, M.I.; Novoselov, K.S. Detection of Individual Gas Molecules Adsorbed on Graphene. *Nat. Mater.* **2007**, *6*, 652–655. [CrossRef]
71. Wu, H.; Bu, X.; Deng, M.; Chen, G.; Zhang, G.; Li, X.; Wang, X.; Liu, W. A Gas Sensing Channel Composited with Pristine and Oxygen Plasma-Treated Graphene. *Sensors* **2019**, *19*, 625. [CrossRef] [PubMed]
72. Capote Mastrapa, G.; Freire, F.L., Jr. Plasma-Treated CVD Graphene Gas Sensor Performance in Environmental Condition: The Role of Defects on Sensitivity. *J. Sens.* **2019**, *2019*, 5492583. [CrossRef]
73. Zhang, H.; Fan, L.; Dong, H.; Zhang, P.; Nie, K.; Zhong, J.; Li, Y.; Guo, J.; Sun, X. Spectroscopic Investigation of Plasma-Fluorinated Monolayer Graphene and Application for Gas Sensing. *ACS Appl. Mater. Interfaces* **2016**, *8*, 8652–8661. [CrossRef] [PubMed]
74. Chung, M.G.; Kim, D.H.; Lee, H.M.; Kim, T.; Choi, J.H.; Seo, D.k.; Yoo, J.-B.; Hong, S.-H.; Kang, T.J.; Kim, Y.H. Highly Sensitive $NO_2$ Gas Sensor Based on Ozone Treated Graphene. *Sens. Actuators B Chem.* **2012**, *166–167*, 172–176. [CrossRef]
75. Zhang, C.; Xu, K.; Liu, K.; Xu, J.; Zheng, Z. Metal Oxide Resistive Sensors for Carbon Dioxide Detection. *Coord. Chem. Rev.* **2022**, *472*, 214758. [CrossRef]
76. Casanova-Chafer, J.; Garcia-Aboal, R.; Llobet, E.; Atienzar, P. Enhanced $CO_2$ Sensing by Oxygen Plasma-Treated Perovskite–Graphene Nanocomposites. *ACS Sens.* **2024**, *9*, 830–839. [CrossRef]
77. Chowdhury, I.; Duch, M.C.; Mansukhani, N.D.; Hersam, M.C.; Bouchard, D. Colloidal Properties and Stability of Graphene Oxide Nanomaterials in the Aquatic Environment. *Environ. Sci. Technol.* **2013**, *47*, 6288–6296. [CrossRef]
78. Dimiev, A.M.; Tour, J.M. Mechanism of Graphene Oxide Formation. *ACS Nano* **2014**, *8*, 3060–3068. [CrossRef]
79. Thangamani, G.J.; Deshmukh, K.; Kovářík, T.; Nambiraj, N.A.; Ponnamma, D.; Sadasivuni, K.K.; Khalil, H.P.S.A.; Pasha, S.K.K. Graphene Oxide Nanocomposites Based Room Temperature Gas Sensors: A Review. *Chemosphere* **2021**, *280*, 130641. [CrossRef]
80. Hafiz, S.M.; Ritikos, R.; Whitcher, T.J.; Razib, N.M.; Bien, D.C.S.; Chanlek, N.; Nakajima, H.; Saisopa, T.; Songsiriritthigul, P.; Huang, N.M.; et al. A Practical Carbon Dioxide Gas Sensor Using Room-Temperature Hydrogen Plasma Reduced Graphene Oxide. *Sens. Actuators B Chem.* **2014**, *193*, 692–700. [CrossRef]
81. Hamzaj, A.K.; Donà, E.; Santhosh, N.M.; Shvalya, V.; Košiček, M.; Cvelbar, U. Plasma-Modification of Graphene Oxide for Advanced Ammonia Sensing. *Appl. Surf. Sci.* **2024**, *660*, 160006. [CrossRef]
82. Gruber, D.; Kraus, F.; Müller, J. A Novel Gas Sensor Design Based on $CH_4/H_2/H_2O$ Plasma Etched ZnO Thin Films. *Sens. Actuators B Chem.* **2003**, *92*, 81–89. [CrossRef]
83. Law, J.B.K.; Thong, J.T.L. Improving the NH3 Gas Sensitivity of ZnO Nanowire Sensors by Reducing the Carrier Concentration. *Nanotechnology* **2008**, *19*, 205502. [CrossRef] [PubMed]
84. Wang, G.; Chen, T.; Guo, L.; Wang, W.; Wang, H.; Wang, Y.; Zeng, H.; Liu, X.; Wang, J.; Yang, Y. Highly Response Gas Sensor Based the Au-ZnO Films Processed by Combining Magnetron Sputtering and Ar Plasma Treatment. *Phys. Scr.* **2023**, *98*, 075609. [CrossRef]
85. Hou, Y.; Jayatissa, A.H. Enhancement of Gas Sensor Response of Nanocrystalline Zinc Oxide for Ammonia by Plasma Treatment. *Appl. Surf. Sci.* **2014**, *309*, 46–53. [CrossRef]
86. Gui, Y.; Zhao, S.; Tian, K.; Wu, J.; Guo, H.; Qin, X.; Qin, X.; Guo, D.; Zheng, G.; Guo, Y. Oxygen Plasma Treatment to Enhance the Gas-Sensing Performance of ZnO to N-Methyl Pyrrolidone: Experimental and Computational Study. *Ceram. Int.* **2024**, *50*, 47418–47427. [CrossRef]
87. George, S.M. Atomic Layer Deposition: An Overview. *Chem. Rev.* **2010**, *110*, 111–131. [CrossRef]
88. Xu, Y.; Zheng, W.; Liu, X.; Zhang, L.; Zheng, L.; Yang, C.; Pinna, N.; Zhang, J. Platinum Single Atoms on Tin Oxide Ultrathin Films for Extremely Sensitive Gas Detection. *Mater. Horiz.* **2020**, *7*, 1519–1527. [CrossRef]
89. Li, Z.; Liu, X.; Zhou, M.; Zhang, S.; Cao, S.; Lei, G.; Lou, C.; Zhang, J. Plasma-Induced Oxygen Vacancies Enabled Ultrathin ZnO Films for Highly Sensitive Detection of Triethylamine. *J. Hazard. Mater.* **2021**, *415*, 125757. [CrossRef]
90. Das, S.; Jayaraman, V. $SnO_2$: A Comprehensive Review on Structures and Gas Sensors. *Prog. Mater. Sci.* **2014**, *66*, 112–255. [CrossRef]
91. Masuda, Y. Recent Advances in $SnO_2$ Nanostructure Based Gas Sensors. *Sens. Actuators B Chem.* **2022**, *364*, 131876. [CrossRef]
92. Hiyoto, K.A.M.; Fisher, E.R. Utilizing Plasma Modified $SnO_2$ Paper Gas Sensors to Better Understand Gas-Surface Interactions at Low Temperatures. *J. Vac. Sci. Technol. A* **2020**, *38*, 043202. [CrossRef]
93. Stuckert, E.P.; Miller, C.J.; Fisher, E.R. The Effect of $Ar/O_2$ and $H_2O$ Plasma Treatment of $SnO_2$ Nanoparticles and Nanowires on Carbon Monoxide and Benzene Detection. *ACS Appl. Mater. Interfaces* **2017**, *9*, 15733–15743. [CrossRef] [PubMed]
94. Srivastava, R.; Dwivedi, R.; Srivastava, S.K. Effect of Oxygen and Hydrogen Plasma Treatment on the Room Temperature Sensitivity of $SnO_2$ Gas Sensors. *Microelectron. J.* **1998**, *29*, 833–838. [CrossRef]

95. Chaturvedi, A.; Mishra, V.N.; Dwivedi, R.; Srivastava, S.K. Response of Oxygen Plasma-Treated Thick Film Tin Oxide Sensor Array for LPG, $CCl_4$, CO and $C_3H_7OH$. *Microelectron. J.* **1999**, *30*, 259–264. [CrossRef]
96. Acharyya, S.; Guha, P.K. Enhanced Formaldehyde Sensing Performance Employing Plasma-Treated Hierarchical $SnO_2$ Nanosheets through Oxygen Vacancy Modulation. *Appl. Surf. Sci.* **2024**, *655*, 159640. [CrossRef]
97. Mirzaei, A.; Yousefi, H.R.; Falsafi, F.; Bonyani, M.; Lee, J.-H.; Kim, J.-H.; Kim, H.W.; Kim, S.S. An Overview on How Pd on Resistive-Based Nanomaterial Gas Sensors Can Enhance Response toward Hydrogen Gas. *Int. J. Hydrogen Energy* **2019**, *44*, 20552–20571. [CrossRef]
98. Zhu, M.; Zhang, H.; Zhang, S.; Yao, H.; Shi, X.; Xu, S. Chemoresistive Gas Sensors Based on Noble-Metal-Decorated Metal Oxide Semiconductors for H2 Detection. *Materials* **2025**, *18*, 451. [CrossRef]
99. Hu, K.; Wang, F.; Liu, H.; Li, Y.; Zeng, W. Enhanced Hydrogen Gas Sensing Properties of Pd-Doped $SnO_2$ Nanofibres by Ar Plasma Treatment. *Ceram. Int.* **2020**, *46*, 1609–1614. [CrossRef]
100. Chaturvedi, A.; Mishra, V.N.; Dwivedi, R.; Srivastava, S.K. Selectivity and Sensitivity Studies on Plasma Treated Thick Film Tin Oxide Gas Sensors. *Microelectron. J.* **2000**, *31*, 283–290. [CrossRef]
101. Pan, J.; Ganesan, R.; Shen, H.; Mathur, S. Plasma-Modified $SnO_2$ Nanowires for Enhanced Gas Sensing. *J. Phys. Chem. C* **2010**, *114*, 8245–8250. [CrossRef]
102. Huang, H.; Tan, O.K.; Lee, Y.C.; Tran, T.D.; Tse, M.S.; Yao, X. Semiconductor Gas Sensor Based on Tin Oxide Nanorods Prepared by Plasma-Enhanced Chemical Vapor Deposition with Postplasma Treatment. *Appl. Phys. Lett.* **2005**, *87*, 163123. [CrossRef]
103. Huang, H.; Lee, Y.C.; Chow, C.L.; Tan, O.K.; Tse, M.S.; Guo, J.; White, T. Plasma Treatment of $SnO_2$ Nanocolumn Arrays Deposited by Liquid Injection Plasma-Enhanced Chemical Vapor Deposition for Gas Sensors. *Sens. Actuators B Chem.* **2009**, *138*, 201–206. [CrossRef]
104. Hu, K.; Wang, F.; Shen, Z.; Liu, H.; Zeng, W.; Wang, Y. Ar Plasma Treatment on ZnO–$SnO_2$ Heterojunction Nanofibers and Its Enhancement Mechanism of Hydrogen Gas Sensing. *Ceram. Int.* **2020**, *46*, 21439–21447. [CrossRef]
105. Du, H.; Wang, J.; Sun, Y.; Yao, P.; Li, X.; Yu, N. Investigation of Gas Sensing Properties of $SnO_2/In_2O_3$ Composite Hetero-Nanofibers Treated by Oxygen Plasma. *Sens. Actuators B Chem.* **2015**, *206*, 753–763. [CrossRef]
106. Du, H.-Y.; Wang, J.; Yu, P.; Yu, N.-S.; Sun, Y.-H.; Tian, J.-L. Investigation of Gas Sensing Materials Tin Oxide Nanofibers Treated by Oxygen Plasma. *J. Nanoparticle Res.* **2014**, *16*, 2216. [CrossRef]
107. Jyothilal, H.; Shukla, G.; Walia, S.; Kundu, S.; Angappane, S. Humidity Sensing and Breath Analyzing Applications of $TiO_2$ Slanted Nanorod Arrays. *Sens. Actuators A Phys.* **2020**, *301*, 111758. [CrossRef]
108. Du, B.; Qi, T.; Li, J.; He, Y.; Yang, X. Improving Anti-Humidity Property of $In_2O_3$ Based $NO_2$ Sensor by Fluorocarbon Plasma Treatment. *Sens. Actuators B Chem.* **2021**, *344*, 130268. [CrossRef]
109. Du, H.; Wang, H.; Yao, P.; Wang, J.; Sun, Y. $In_2O_3$ Nanofibers Surface Modified by Low-Temperature RF Plasma and Their Gas Sensing Properties. *Mater. Chem. Phys.* **2018**, *215*, 316–326. [CrossRef]
110. Wu, M.-R.; Li, W.-Z.; Tung, C.-Y.; Huang, C.-Y.; Chiang, Y.-H.; Liu, P.-L.; Horng, R.-H. NO Gas Sensor Based on $ZnGa_2O_4$ Epilayer Grown by Metalorganic Chemical Vapor Deposition. *Sci. Rep.* **2019**, *9*, 7459. [CrossRef]
111. Chang, T.-Y.; Singh, A.K.; Shao, J.-H.; Huang, C.-Y.; Shieh, J.-M.; Wuu, D.-S.; Liu, P.-L.; Horng, R.-H. Performance Improvement of MOCVD Grown $ZnGa_2O_4$ Based NO Gas Sensors Using Plasma Surface Treatment. *Appl. Surf. Sci.* **2023**, *637*, 157929. [CrossRef]
112. Das, M.; Roy, S. Polypyrrole and Associated Hybrid Nanocomposites as Chemiresistive Gas Sensors: A Comprehensive Review. *Mater. Sci. Semicond. Process.* **2021**, *121*, 105332. [CrossRef]
113. Suzuki, T.; Tanner, P.; Thiel, D.V. $O_2$ Plasma Treated Polyimide-Based Humidity Sensors. *Analyst* **2002**, *127*, 1342–1346. [CrossRef]
114. Zhang, Z.; Zhao, L.; Du, H.; Chu, J. Improved NO2 Gas-Sensing Performance of PPy by Hydrogen Plasma Treatment: Experimental Study and DFT Verification. *Sens. Actuators A Phys.* **2023**, *364*, 114848. [CrossRef]
115. Kunzo, P.; Lobotka, P.; Micusik, M.; Kovacova, E. Palladium-Free Hydrogen Sensor Based on Oxygen-Plasma-Treated Polyaniline Thin Film. *Sens. Actuators B Chem.* **2012**, *171–172*, 838–845. [CrossRef]
116. Cao, W.; Nie, J.; Cao, Y.; Gao, C.; Wang, M.; Wang, W.; Lu, X.; Ma, X.; Zhong, P. A Review of How to Improve $Ti_3C_2T_x$ MXene Stability. *Chem. Eng. J.* **2024**, *496*, 154097. [CrossRef]
117. Mirzaei, A.; Kim, J.-Y.; Kim, H.W.; Kim, S.S. Resistive Gas Sensors Based on 2D TMDs and MXenes. *Acc. Chem. Res.* **2024**, *57*, 2395–2413. [CrossRef]
118. Mirzaei, A.; Kim, J.-Y.; Kim, J.H.; Nam, M.-S.; Kim, H.W.; Kim, S.S. Accordion-like $Ti_3C_2T_x$ MXene with High Flexibility for $NH_3$ Sensing in Self-Heating Mode. *Ceram. Int.* **2025**, *51*, 2930–2946. [CrossRef]
119. Wang, Y.; Fu, J.; Xu, J.; Hu, H.; Ho, D. Atomic Plasma Grafting: Precise Control of Functional Groups on $Ti_3C_2T_x$ MXene for Room Temperature Gas Sensors. *ACS Appl. Mater. Interfaces* **2023**, *15*, 12232–12239. [CrossRef]
120. Ansari, H.R.; Mirzaei, A.; Shokrollahi, H.; Kumar, R.; Kim, J.-Y.; Kim, H.W.; Kumar, M.; Kim, S.S. Flexible/Wearable Resistive Gas Sensors Based on 2D Materials. *J. Mater. Chem. C* **2023**, *11*, 6528–6549. [CrossRef]

121. Sharma, R.; Laishram, R.; Gupta, B.K.; Srivastva, R.; Sinha, O.P. A Review on MX2 (M = Mo, W and X = S, Se) Layered Material for Opto-Electronic Devices. *Adv. Nat. Sci. Nanosci. Nanotechnol.* **2022**, *13*, 023001. [CrossRef]
122. Seo, W.S.; Kim, D.K.; Han, J.-H.; Park, K.-B.; Ryu, S.C.; Min, N.K.; Kim, J.H. Functionalization of Molybdenum Disulfide via Plasma Treatment and 3-Mercaptopropionic Acid for Gas Sensors. *Nanomaterials* **2020**, *10*, 1860. [CrossRef] [PubMed]

sensors

MDPI

*Review*

# Conducting Polymers-Based Gas Sensors: Principles, Materials, and Applications

**Rongqing Dong [1,2], Mingna Yang [1,2], Yinxiu Zuo [2,*], Lishan Liang [1,2], Huakun Xing [2,3], Xuemin Duan [2,4] and Shuai Chen [1,2,*]**

[1] Jiangxi Provincial Engineering Research Center for Waterborne Coatings, School of Chemistry and Chemical Engineering, Jiangxi Science & Technology Normal University, Nanchang 330013, China; drongqing@163.com (R.D.); yangmingna1023@163.com (M.Y.); lianglishan09@163.com (L.L.)
[2] Jiangxi Provincial Key Laboratory of Flexible Electronics, Jiangxi Science & Technology Normal University, Nanchang 330013, China; xhk1235@126.com (H.X.); duanxm@jxstnu.edu.cn (X.D.)
[3] Institute of Energy Materials and Nanotechnology, School of Civil Engineering and Architecture, Nanchang Jiaotong Institute, Nanchang 330100, China
[4] School of Chemistry and Chemical Engineering, Jinggangshan University, Ji'an 343009, China
* Correspondence: zuoyinxiu@jxstnu.edu.cn (Y.Z.); shuaichen@jxstnu.edu.cn (S.C.)

**Abstract:** Conducting polymers (CPs) have emerged as promising materials for gas sensors due to their organic nature coupled with unique and versatile optical, electrical, chemical, and electrochemical properties. This review provides a comprehensive overview of the latest developments in conducting polymer-based gas sensors. First, the fundamental gas sensing mechanisms in CPs-based sensors are elucidated, covering diverse transduction modes including electrochemical, chemiresistive, optical, piezoelectric, and field-effect transistor-based sensing. Next, the various types of conducting polymers employed in gas sensors, such as polypyrrole, polyaniline, polythiophene, and their composites are introduced, with emphasis on their synthesis methods, structural characteristics, and gas sensing response properties. Finally, the wide range of applications of these sensors is discussed, spanning industrial process control, environmental monitoring, food safety, biomedical diagnosis, and other fields, as well as existing issues such as long-term stability and humidity interference, and a summary of the biocompatibility and regulatory standards of these conductive polymers is provided. By integrating insights from sensing mechanisms, materials, and applications, this review offers a holistic understanding of CPs-based gas sensors. It also highlights future research directions, including device miniaturization, AI-assisted gas identification, multifunctional integrated sensing systems, wearable and flexible sensor platforms, and enhanced sensitivity, selectivity, and on-site detection capabilities.

**Keywords:** conducting polymer; gas sensor; sensing mechanism; environmental monitoring

## 1. Introduction

Environmental pollution, public health concerns, and industrial safety needs have created an urgent demand for advanced gas-sensing technologies. Rapid industrialization and urbanization have led to significant harmful gas emissions, with pollutants such as sulfur dioxide ($SO_2$) [1], nitrogen oxides ($NO_X$) [2], and volatile organic compounds (VOCs) [3] threatening air quality. Accurate detection of these hazardous gases is essential for evaluating air quality and formulating effective environmental policies to safeguard the atmosphere critical to human health. In industrial settings, real-time monitoring of gas concentrations is equally crucial for safety and process control. It can prevent accidents (e.g., explosions or toxic exposure) and help maintain product quality and production efficiency. For instance, industries such as chemical manufacturing [4] and food processing [5] rely on precise gas detection to ensure safe operations and consistent product quality. In the medical field, analysis of specific components in exhaled breath provides a non-invasive approach to early disease diagnosis and personalized health monitoring [6]. For example, elevated acetone levels in the breath of diabetic patients serve as an important biomarker of their condition and treatment efficacy [7]. These diverse applications underscore that gas detection technology has become indispensable in modern society.

However, while effective for certain purposes, conventional gas detection methods have significant limitations [8]. For instance, gas chromatography-mass spectrometry (GC-MS) offers high sensitivity and resolution for identifying and quantifying gases in complex mixtures, but it requires bulky, expensive equipment and skilled operators. Its complex, time-consuming procedures make GC-MS impractical for rapid on-site or real-time monitoring [9–11]. Similarly, optical spectroscopic methods (such as infrared sensing) enable non-contact and fast detection, yet often suffer from low sensitivity for certain gases and depend on high-cost instrumentation [12]. These drawbacks have driven researchers to explore new sensing materials and techniques to meet the growing demand for gas sensors with higher precision, sensitivity, and selectivity under practical conditions.

To overcome these challenges, conducting polymers (CPs) have emerged as promising gas sensor materials offering unique advantages. CPs are organic polymers with $\pi$-conjugated backbones that endow them with intrinsic electrical conductivity. Moreover, their electrical, optical, and chemical properties can be readily tuned through chemical modification or doping [13], allowing for sensor designs tailored to specific target gases. Metal oxides and metal-organic frameworks (MOFs) have shown excellent adsorption and selective detection performances toward gas sensing, but they often need to work under high-temperature conditions [14,15]. In contrast, the great advantages of CPs in this field are that they can work at room temperature, quickly respond to gases through intermolecular interactions, enable solution processing, have low power consumption, as well as be suitable for flexible integration and portable monitoring [16].

Several CPs materials have been extensively studied in gas sensors, notably polyaniline (PANi) [16], polypyrrole (PPy) [17], polythiophene (PTh) [18], poly(3,4-ethylenedioxythiophene) (PEDOT) [19], and its derivatives PEDOT:PSS [20]. Each of these CPs provides distinct characteristics beneficial for gas detection. For example, PANI features excellent environmental stability and a widely tunable conductivity, which has been leveraged to detect gases such as ammonia and nitrogen dioxide [16]. PPy is easily synthesized and forms uniform conductive films with high electrochemical activity, making it effective for sensing various VOCs [17]. PTh and its derivatives offer unique electronic and optical properties that can be exploited to enhance selectivity toward specific gas molecules [18]. PEDOT is known for its high conductivity and stability [19], while PEDOT:PSS combines the conductivity of PEDOT with the water dispersibility of PSS, enabling solution-processed, flexible sensor devices [20].

Thanks to these advantages, CPs-based gas sensors have demonstrated promising performance across a wide range of applications, including environmental monitoring [21], industrial process control [22], medical diagnostics [23], and food safety assurance [24]. Numerous studies have validated the utility of CPs sensors in such areas; however, despite the considerable progress, several challenges remain. In particular, further improvements are needed in detection sensitivity and selectivity, expansion of the range of detectable gases, and enhancement of long-term stability and reliability. In light of both the achievements and the remaining challenges in this field, this review aims to provide a comprehensive overview of CPs-based gas sensor technology. We discuss the fundamental gas sensing mechanisms and the diverse CPs materials employed, summarize recent advances in various application domains, and finally outline current challenges and future research prospects for CPs-based gas sensors in this rapidly evolving field.

## 2. Conductive Polymer-Based Gas Phase Sensors

### *2.1. Detection Principle*

#### 2.1.1. Electrochemical Sensing

Electrochemical sensing utilizes redox reactions occurring at the electrode–electrolyte interface to detect electrochemically active gases, converting chemical information into measurable electrical signals [25]. Electrochemical gas sensors are predominantly classified into potentiometric, conductometric, and amperometric methods. The potentiometric method operates based on the reaction of the target gas at the working electrode surface, generating a potential difference between the working and reference electrodes, which varies with changes in gas concentration [26]. The conductometric method relies primarily on changes in the conductivity of sensitive materials upon gas interaction to detect the analyte. Among these approaches, the amperometric method has become the most widely utilized due to its high sensitivity and strong quantitative detection capabilities. A typical amperometric gas sensor comprises a working electrode (WE), a counter electrode (CE), and a reference electrode (RE). The WE is the principal site where oxidation or reduction of the target gas occurs, initiating electron transfer and generating a current or potential shift directly correlated to analyte concentration [27]. When detecting oxidizing gases (e.g., $NO_X$, $O_3$), the target molecule accepts electrons at the cathodic interface, generating a measurable cathodic current [28]. Conversely, reducing gases (e.g., $H_2S$, $NH_3$, CO) donate electrons upon oxidation at the anodic interface, producing a characteristic signal. The magnitude of the response follows Nernst's equation ($E = E^0 + \frac{RT}{nF} ln \frac{[O_x]}{[Red]}$) and Faraday's law ($m = \frac{M}{nF} Q = \frac{M}{nF} It$) providing quantitative information on the gas concentration [25,26].

CPs are ideal candidates for electrochemical gas sensors due to their intrinsic redox activity, high electrical conductivity, and tunable surface chemistry [29]. Their $\pi$-conjugated backbones facilitate efficient charge transport, and their functional groups or dopants can be tailored to enhance selectivity toward specific analytes. For example, Serafini et al. [30] developed a wearable electrochemical ammonia gas sensor with core components including PEDOT, electrochemically deposited iridium oxide particles, and a hydrogel membrane. Upon contact with ammonia gas, a dissociation equilibrium within the hydrogel releases $OH^-$ ions, increasing the hydrogel's pH (Figure 1a,b). This pH shift disrupts the redox equilibrium of iridium oxide (IrOx), facilitating electron injection into PEDOT:PSS, consequently reducing its conductivity. Under an applied voltage between two electrodes, a measurable current decrease occurs. Monitoring this current reduction allows for accurate ammonia gas detection. Following ammonia introduction, a rapid current drop illustrates the efficiency of the detection mechanism.

Despite their benefits, electrochemical gas sensors are primarily effective for gases possessing intrinsic redox activity, such as carbon monoxide, hydrogen sulfide, and ammonia. In contrast, inert gases like chlorine or nitrogen require alternative detection approaches or indirect redox mechanisms [31]. Thus, continued optimization of electrode materials, electrolyte compositions, and CPs electrode interfaces remains essential for broadening detection capabilities and enhancing sensor performance.

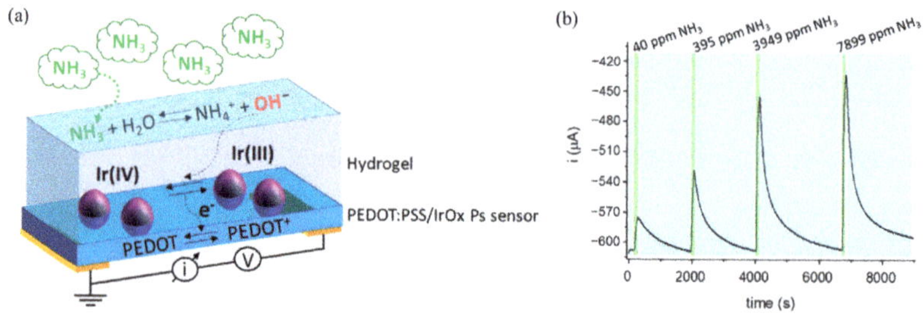


**Figure 1.** (**a**) Schematic of the gas sensor working principle; (**b**) current vs. time response of the $NH_3$ sensor [30].

### 2.1.2. Chemiresistive Sensing

Chemiresistive sensing is based on detecting changes in the electrical resistance of sensing materials upon interaction with target gases [32]. Specifically, gas molecules adsorb onto the sensor's active layer, triggering physical or chemical interactions that alter its electronic properties, such as charge carrier concentration, mobility, or barrier height, thereby producing measurable changes in resistance [33]. CPs are particularly suitable for chemiresistive sensors due to their tunable electrical properties, ease of chemical modification, high sensitivity, and rapid response capability. When a target gas interacts with CPs-based sensing materials, it can function as either an electron donor or acceptor. For example, electron-donating gases (e.g., $NH_3$, acetone) increase electron density in p-type CPs, thereby reducing hole concentration and consequently increasing the polymer's resistance [34]. Conversely, electron-accepting gases (e.g., $NO_2$) withdraw electrons, increasing hole concentration and decreasing resistance.

A representative example is the $SnO_2$/PTh nanocomposite sensor developed by Beniwal et al. [35] for acetone detection. Upon exposure to acetone gas, acetone molecules act as electron donors, transferring electrons to polythiophene (PTh, a typical p-type semiconductor) (Figure 2a). This electron transfer decreases the concentration of holes (majority charge carriers) in PTh, significantly increasing its electrical resistance. Since PTh coats the $SnO_2$ nanoparticles, this resistance change directly influences the composite's overall conductivity, enabling precise acetone detection through resistance measurements. In another study, Chaudhary et al. [36] developed a chemiresistive sensor by combining polythiophene (PTh) with citric acid-functionalized cadmium sulfide quantum dots (CdS QDs) for ammonia sensing. Ammonia molecules donate lone-pair electrons upon adsorption, interacting with the polymer's polarons. This interaction disrupts the charge balance and restricts carrier mobility within the PTh/CdS composite, resulting in a pronounced decrease in electrical resistance [37]. Notably, this sensor demonstrated a rapid response (under 1 s) and recovery (4–8 s), with a clear linear dependence on ammonia concentration, highlighting the exceptional suitability of PTh-based composites for chemiresistive ammonia detection.

Despite significant advancements, challenges remain for chemiresistive sensors, including cross-sensitivity to humidity, baseline drift, and long-term stability [38]. Current research efforts therefore emphasize enhancing selectivity through molecular design and composite engineering and developing strategies to mitigate environmental interference, further advancing CPs-based chemiresistive sensors.

### 2.1.3. Piezoelectric Sensing

Piezoelectric sensing involves converting mechanical deformation induced by gas adsorption into measurable electrical signals, utilizing materials exhibiting the piezoelectric effect [39]. This sensing mechanism relies on changes in mass, viscoelastic properties, or mechanical stress within piezoelectric materials upon interaction with target gases [40]. CPs and their composites are increasingly employed in piezoelectric gas sensors due to their flexibility, ease of modification, and tunable piezoelectric properties. Typically, piezoelectric gas sensors utilize quartz crystal micro-balance (QCM) or polymer-based piezoelectric substrates coated with CPs-sensitive layers. Upon gas adsorption, mass loading or interfacial interactions between gas molecules and CPs coating cause frequency shifts or mechanical stress changes, generating electrical signals proportionate to the gas concentration [41]. The advantage of using CPs coatings includes their customizable affinity toward specific gas molecules and improved mechanical flexibility, crucial for wearable and flexible electronics applications.

Adjaoud et al. [42] developed a flexible piezoelectric sensor utilizing ionic polymer-polymer composites ($IP_2Cs$), incorporating PEDOT:PSS-modified electrodes. Mechanical stimulation induced ion migration within the composite, generating measurable voltage signals due to asymmetric charge distribution (Figure 2b,c). The integration of PEDOT:PSS facilitated improved electrical conductivity and sensitivity, demonstrating significant potential for wearable gas sensing applications. Furthermore, combining CPs with inorganic piezoelectric materials, such as carbon nanotubes (CNTs) or graphene oxide (GO), can further enhance sensing performance. For instance, Pasupuleti et al. [43] prepared a $NO_2$ sensor based on GO-PEDOT:PSS nanocomposites, which showed good stability, enhanced sensitivity, and improved response toward $NO_2$.

Despite promising outcomes, piezoelectric CPs-based gas sensors continue to face challenges, such as sensitivity to humidity and issues related to long-term stability [44]. Current research aims to optimize composite formulations, refine device architectures, and mitigate environmental interferences to facilitate broader practical applications.

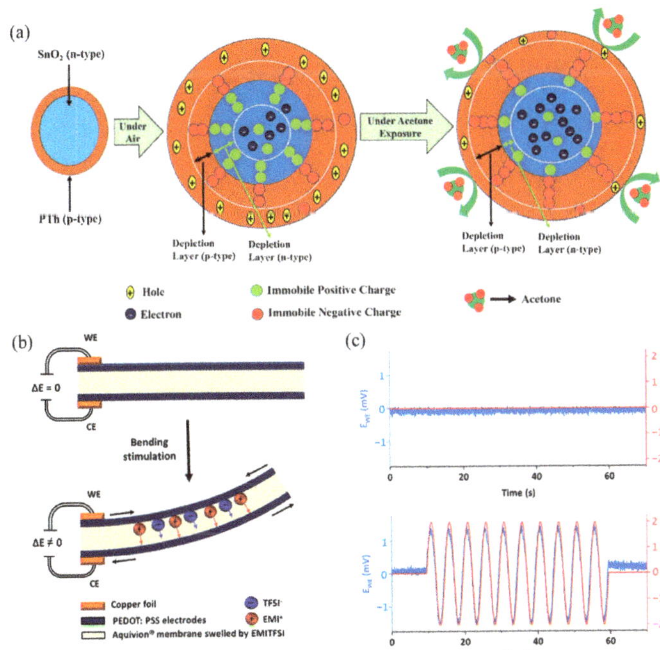


**Figure 2.** (**a**) Schematic of sensing mechanism of the $SnO_2$/PTh nanocomposite toward acetone detection [35]; representation of the piezoionic effect. (**b**) Schematic illustration of the tri-layer before and after mechanical stimulation and (**c**) electrical response (voltage output) of $IP_2Cs$ sensor before and after mechanical stimulation [42].

#### 2.1.4. Mechanical Sensing

Mechanical sensing transforms mechanical stimuli, such as strain, stress, pressure, or deformation, into quantifiable electrical signals by altering conductive networks or electron tunneling paths within conductive materials [45]. CPs have emerged as promising materials for mechanical sensing owing to their inherent flexibility, tunable mechanical and electrical properties, ease of processing, and capability to integrate with various substrates, making them highly suitable for wearable and flexible sensing applications.

CPs-based mechanical sensors typically operate through mechanisms such as changes in tunneling resistance, contact separation, and crack propagation [46]. Under mechanical deformation, the conductive pathways within CPs composites experience structural modifications, altering inter-particle distances and particle alignment, or causing micro-crack formation, significantly impacting electrical conductivity [47]. For instance, Mallya et al. [48] described poly(DTCPA-co-BHTBT)-CB composites used for sensing toluene vapor. The absorption of toluene vapor causes polymer swelling, akin to mechanical deformation effects on CPs composites. This swelling increases the spacing between carbon black particles, disrupting the carbon black network structure and resulting in conductivity changes that enable toluene vapor detection. This exemplifies how CPs-based materials exploit structural changes to induce conductivity variations, achieving sensing capabilities in different scenarios.

Despite their promising characteristics, CPs-based mechanical sensors still encounter challenges regarding sensitivity, stability, mechanical robustness, and resilience to environmental factors [49]. Future research efforts should focus on optimizing composite formulations, engineering hierarchical structures to improve sensor durability, and addressing baseline drift and sensitivity recovery issues for sustained practical applications.

#### 2.1.5. Optical Sensing

Optical sensing detects target gases by monitoring changes in optical properties, such as absorbance, fluorescence intensity, refractive index, and reflectivity, caused by interactions between gas molecules and sensitive materials [50]. In recent years, in addition to traditional fluorescence and absorption sensors, emerging technical directions in this area such as plasmon-enhanced CPs fluorophores [51,52], waveguide-integrated CPs interferometers [53], and cavity-coupled absorption sensors [54] have also been developed. CPs hold particular promise for optical gas sensing due to their intrinsic optical activity, tunable molecular structures, and ease of functionalization, allowing precise modulation of optical responses upon gas exposure.

Typical CP-based optical gas sensors include fluorescence-based [55], absorption-based [56], and refractive index-based [57] sensors. In fluorescence sensing, gas molecules interact with CPs films, resulting in fluorescence quenching or enhancement. Lee et al. [58] developed an optical gas sensor by integrating CMOS-MEMS technology with CP-based fluorescent sensing materials (Figure 3a). This sensor employed fluorescence quenching, where blue LEDs excite the sensing material, producing fluorescence, and exposure to the target gas reduces fluorescence intensity through molecular interactions, resulting in a measurable decrease in photocurrent [59,60]. This technique demonstrates rapid response and high sensitivity, making it suitable for on-site gas detection. Additionally, CP-based absorption optical sensors detect changes in absorption spectra caused by gas adsorption. Liu et al. [61] developed an ethanol gas sensor using polypyrrole (PPy)-modified plastic optical fibers (POFs). Upon ethanol exposure, PPy interacts with ethanol molecules, altering its refractive index and absorption characteristics, subsequently affecting the transmitted optical signal intensity. Such sensors exhibit fast response times and can be easily integrated into compact, portable devices, ideal for practical applications.

Despite significant progress, CP-based optical sensors still face challenges concerning sensitivity limits, humidity interference, and long-term optical stability. Ongoing research focuses on improving the design of sensing materials, optimizing optical signal processing techniques, and developing robust integration strategies to enhance sensor performance for real-world applications [62].

#### 2.1.6. Field-Effect Transistor (FET) Sensing

Field-effect transistor (FET) gas sensors operate by detecting changes in the electrical conductivity of the transistor channel caused by interactions between gas molecules and the active sensing layer [63]. CPs are particularly attractive as channel materials in FET-based gas sensors due to their high electrical conductivity, ease of processing, chemical tunability, and potential for low-cost, flexible, and wearable sensing applications. A typical CPs-based FET gas sensor comprises three electrodes: a source, a drain, and a gate [64]. In operation, gas molecules adsorb onto the CPs channel layer. This adsorption induces charge transfer or polarization effects that alter the charge carrier density and mobility within the polymeric semiconductor channel [65]. These changes modulate the channel current, thereby enabling sensitive and selective detection of target gases. For example, FETs employing polythiophene derivatives or polyaniline as the channel material have demonstrated excellent performance for gas detection. Amer et al. [66] fabricated an ammonia gas sensor based on an organic field-effect transistor (OFET) architecture using polyaniline (PANi) and its derivatives as the active channel materials. When this sensor is exposed to ammonia ($NH_3$) gas, ammonia molecules react with protonated sites in the polymer to form ammonium ions ($NH^{4+}$), releasing electrons from their lone pairs in the process (Figure 3b). For instance, in a PANi:DBSA-doped device, this reaction leads to a decrease in hole density in the polymer, which in turn significantly reduces the transistor's channel current. By monitoring the change in channel current, the ammonia concentration can thus be accurately determined at room temperature. Similarly, PEDOT:PSS, well-known for its excellent conductivity and environmental stability, has also been widely used as an active channel material in FET gas sensors. Owing to its strong interactions with polar analytes such as $NO_2$, $SO_2$, and volatile organic compounds (VOCs), PEDOT:PSS-based FET sensors exhibit high sensitivity, rapid response, and excellent selectivity [67]. Moreover, the solution processability of PEDOT:PSS facilitates the integration of these sensors into flexible and wearable sensing devices.

Despite these promising advances, several challenges remain, particularly related to sensor stability, environmental interference, and baseline drift [68]. Ongoing research efforts are focusing on developing composite CPs materials, implementing surface modifications, and designing advanced device architectures to enhance sensing performance, improve selectivity, and extend the operational stability of CPs-based FET gas sensors.

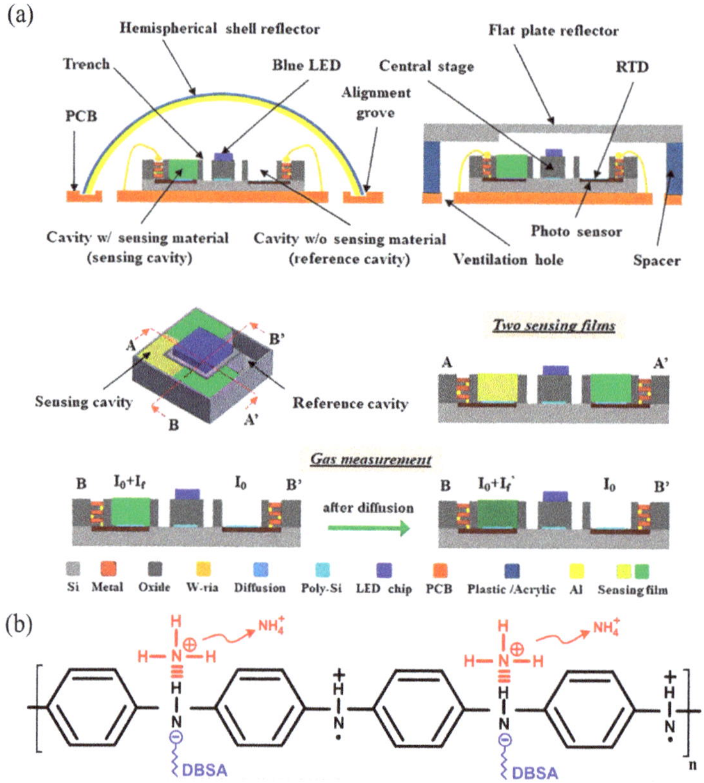


**Figure 3.** (**a**) Schematic of the presented gas sensor design principle and cross-section [58]; (**b**) schematic diagram of the sensing mechanism of PANi:DBSA [66].

## 2.2. Sensitive Material

### 2.2.1. PANi

Polyaniline (PANi) has been extensively studied as a CP for gas sensing owing to its distinctive properties, such as facile synthesis, environmental stability, reversible doping/dedoping behavior, and tunable conductivity [69–71]. PANi can be synthesized by either chemical or electrochemical polymerization [72]. In chemical synthesis, PANi is typically produced via oxidative polymerization of aniline monomers using oxidants like ammonium persulfate. In contrast, electrochemical polymerization enables the direct formation of PANi layers on conductive substrates under controlled potentials, which facilitates sensor fabrication. PANi exhibits notable advantages for gas sensing applications, most prominently a reversible redox behavior that allows efficient electron transfer when interacting with various analyte gases. Yuan et al. [73] demonstrated highly sensitive ammonia sensors by combining protic acid-doped PANi (PA-PANi) with graphene oxide (GO) and reduced graphene oxide (rGO). Their PA-PANi/GO/rGO composite sensor showed a 262.5% greater response at 25 ppm $NH_3$, along with significantly faster response and recovery times, compared to a sensor based on pure PANi, (Figure 4a).

However, PANi also has inherent limitations, including relatively low intrinsic conductivity, poor processability, and limited solubility that constrain its utility in large-scale sensor applications [74]. To address these issues, researchers have extensively explored composite approaches. Bibi et al. [16] enhanced PANi's $H_2S$ sensing performance by incorporating carbon aerogel (CA) in an interdigital electrode structure, which significantly improved sensitivity toward hydrogen sulfide. In this configuration, the porous CA framework provides abundant active sites for gas adsorption, thereby enabling rapid and sensitive detection of ultralow $H_2S$ concentrations. Flexible sensor applications have also benefited from PANi composites. For instance, Wan et al. [75] developed a flexible $NH_3$ sensor by depositing a PANi–carbon nanotube (CNT) composite onto a polyethylene terephthalate (PET) substrate. The resulting sensor exhibited remarkable sensitivity with a detection limit as low as 1 ppm, making it suitable for applications such as breath analysis and food safety monitoring. Similarly, Zhuang et al. [76] combined PANi with multi-walled carbon nanotubes (MWCNTs), achieving a detection limit of 0.3 ppm for $NH_3$ as well as high selectivity and excellent stability under varying humidity and mechanical stress conditions. Furthermore, PANi-based sensors have shown promise in hydrogen detection, which is critical for safety in hydrogen energy applications. Askar et al. [77] investigated various nanostructured PANi materials and found that PANi hollow nanotubes exhibited outstanding hydrogen-sensing capabilities, including a detection limit as low as 1 ppm, a high sensitivity of about 29%, and rapid response and recovery times (15 s and 17 s, respectively) (Figure 4b,c).

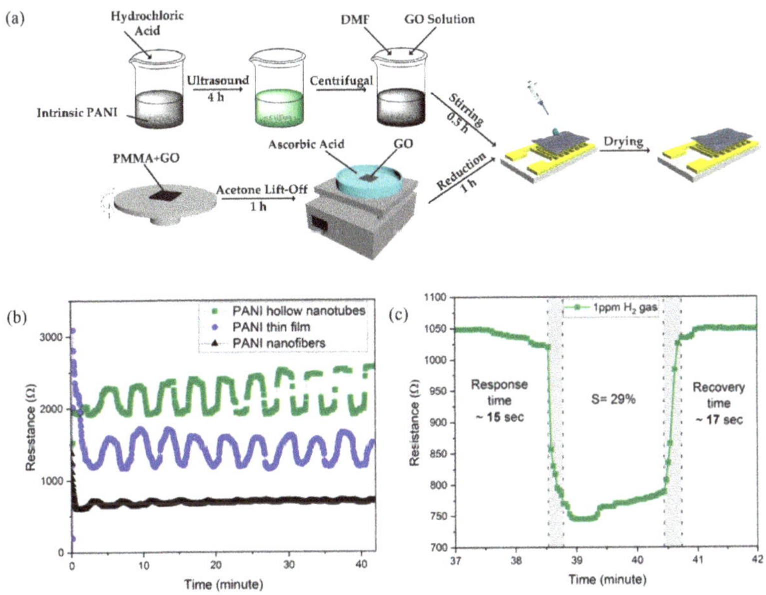


**Figure 4.** (**a**) Schematic of the device fabrication process [73]; (**b**) comparison of responses in different PANi at 1 ppm $H_2$ detection; (**c**) response time of hollow PANi nanotubes sensor at 1 ppm $H_2$ gas [77].

Overall, although PANi-based sensors hold considerable promise for gas sensing, challenges remain in improving their intrinsic conductivity, long-term stability, and environmental resilience. Future research should focus on optimizing composite formulations, refining molecular designs, and developing advanced sensor architectures to overcome these limitations and broaden the scope of practical applications.

### 2.2.2. PPy

Polypyrrole (PPy) is a highly promising CP widely utilized in gas sensing applications due to its intrinsic conductivity, good environmental stability, ease of synthesis, and tunable electrochemical properties [78–80]. PPy can be synthesized via chemical or electrochemical polymerization of pyrrole monomers [81]. Chemical polymerization typically involves pyrrole monomers and oxidants such as ammonium persulfate, while electrochemical polymerization allows precise control of PPy morphology and thickness by adjusting deposition parameters [82].

PPy-based sensors demonstrate notable advantages, including low fabrication costs, straightforward preparation methods, biocompatibility, and suitability for flexible and wearable devices [83]. For instance, Gai et al. [84] synthesized PPy-tetra-β-carboxyl cobalt phthalocyanine tetrasodium salt (PPy-TcCoPc) nanorod composites via one-step in situ polymerization. The synergistic interaction between PPy and TcCoPc significantly enhanced ammonia detection performance, achieving high sensitivity (response of 49.3% at 50 ppm $NH_3$), rapid response (8.1 s), and excellent selectivity, stability, and humidity resistance (Figure 5a). However, pure PPy often exhibits limitations, including relatively low sensitivity, inadequate selectivity, and poor recovery due to its disordered aggregation structure [85]. To address these issues, researchers have combined PPy with various nanomaterials to enhance sensing performance. Santos-Ceballos et al. [86] developed PPy@laser-induced graphene (LIG) nanocomposite sensors through electrochemical polymerization, achieving remarkable ammonia sensitivity with a detection limit of 1 ppm and superior repeatability due to the synergistic interactions between PPy and LIG. Furthermore, PPy-based composites have shown promising results in detecting other hazardous gases such as hydrogen sulfide ($H_2S$) [87]. Al-Sabagh et al. [88] fabricated PPy composites incorporating CuO and $SnO_2$ nanoparticles, demonstrating significant improvements in $H_2S$ detection performance attributed to the nanoparticles' large surface areas and enhanced electrical properties.

Future research directions for PPy-based gas sensors include optimizing composite formulations, improving sensor architectures, and enhancing environmental resilience and stability. Ongoing studies focus on incorporating novel materials with PPy to further refine sensor sensitivity, selectivity, and long-term reliability, promoting broader practical applications across industrial, environmental, and medical fields [89,90].

#### 2.2.3. PTh

Polythiophene (PTh) is a CP known for its excellent electrical conductivity, chemical stability, and easily tunable molecular structure, making it highly suitable for gas sensing applications [91,92]. The backbone of PTh, comprising conjugated thiophene units, facilitates efficient electron transport and strong interactions with gas analytes [93]. These characteristics allow PTh-based sensors to exhibit high sensitivity and selectivity toward specific gases. However, pure PTh sensors commonly face challenges such as limited sensitivity, slow response times, and relatively poor stability under varying environmental conditions [94].

To overcome these issues, researchers have developed composite materials by incorporating various nanomaterials with PTh, significantly enhancing its sensing performance. Bai et al. [95] synthesized flexible gas sensors based on ethylenediamine-modified reduced graphene oxide (RGO) combined with PTh through in situ polymerization (Figure 5b). This composite material exhibited approximately four times higher sensitivity to nitrogen dioxide ($NO_2$) gas compared to pristine PTh sensors, achieving a detection limit as low as 0.52 ppm. Additionally, the flexible nature of the sensor facilitated its integration into wearable devices, demonstrating its potential for practical environmental monitoring applications. Belhousse et al. [18] further illustrated the importance of optimizing PTh layer thickness in sensor design by fabricating sensors using electrochemically polymerized PTh layers on porous silicon (PSi) substrates. They reported that sensors with PTh polymerized over six cycles provided optimal sensitivity, rapid response, and superior stability when detecting carbon dioxide ($CO_2$) and cigarette smoke at room temperature. This highlights the critical role of precise structural control in improving the sensing performance of PTh-based sensors.

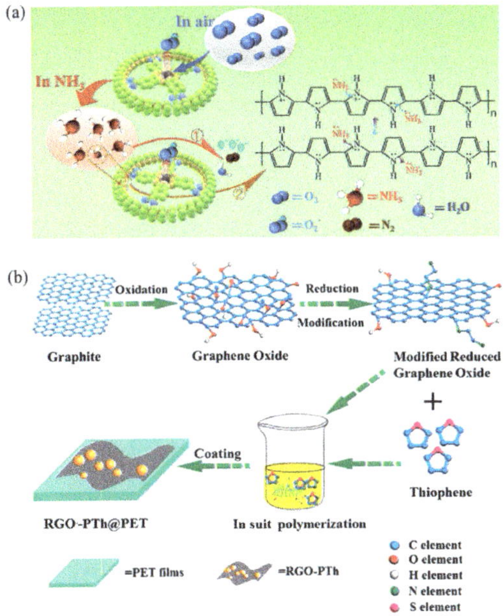


**Figure 5.** (**a**) Response mechanism of the PPy-TcCoPc sensor to $NH_3$ [84]; (**b**) schematic diagram of the preparation process for RGO-PTh hybrid [95].

Future research on PTh-based gas sensors should continue exploring advanced composite formulations and sophisticated fabrication techniques to further enhance sensitivity, selectivity, and stability. Developing flexible, wearable, and cost-effective PTh sensors capable of reliable operation under diverse environmental conditions will significantly broaden their applications in industrial safety, environmental monitoring, and biomedical fields.

#### 2.2.4. PEDOT

Poly(3,4-ethylenedioxythiophene) (PEDOT) is a widely recognized CP known for its excellent electrical conductivity, environmental stability, biocompatibility, and ease of processability, making it highly attractive for gas sensing applications [96]. PEDOT's conjugated molecular structure provides efficient electron transport pathways, enabling rapid electrical signal generation upon interaction with analyte gases [97]. Furthermore, its good chemical stability and resistance to oxidation make PEDOT suitable for sensors operating under various environmental conditions [98]. However, pristine PEDOT-based gas sensors face certain limitations, including susceptibility to humidity-induced interference, moderate sensitivity, and potential baseline drift.

To overcome these issues, researchers have employed composite approaches and structural optimizations. Xiao et al. [99] successfully developed ammonia sensors by creating core-shell structured nanofibers (Figure 6a), incorporating PEDOT as the core sensing material and polyvinylidene fluoride-trifluoroethylene (PVDF-TrFE) as the protective hydrophobic shell. This design effectively reduced humidity interference while maintaining excellent sensitivity to ammonia. Additionally, PEDOT has been effectively combined with nanostructured materials, such as graphene oxide (GO) and metal nanoparticles, to further enhance sensor sensitivity and selectivity. These composite structures significantly improve sensor performance by increasing active surface area, facilitating gas adsorption, and enhancing electron transport properties [100].

Future research directions for PEDOT-based gas sensors include further composite engineering to boost sensitivity and selectivity, developing robust structures to minimize environmental interference, and exploring novel doping strategies to optimize electronic properties [101]. Additionally, fabricating flexible and wearable PEDOT sensors is expected to expand their applicability significantly across environmental monitoring, industrial safety, and medical diagnostics.

#### 2.2.5. PEDOT:PSS

Poly(3,4-ethylenedioxythiophene):poly(styrene sulfonate) (PEDOT:PSS) is a prominent CP widely applied in gas sensors due to its exceptional electrical conductivity, optical transparency, excellent environmental stability, and solution processability [20]. Incorporating the hydrophilic polyelectrolyte poly(styrene sulfonate) (PSS) significantly enhances the water solubility and processability of PEDOT, facilitating fabrication into various sensor architectures through simple methods like spin-coating, spray-coating, or printing techniques [102].

Despite these advantages, pristine PEDOT:PSS sensors still encounter certain limitations, such as moderate sensitivity and susceptibility to environmental humidity interference, which can affect sensor accuracy and stability. Therefore, various strategies have been explored to improve the sensing capabilities of PEDOT:PSS, including compositing with other nanomaterials and employing structural optimizations. For instance, Alves et al. [103] successfully enhanced PEDOT:PSS gas sensor performance by integrating graphene oxide (GO) into the polymer matrix. This PEDOT:PSS/GO composite sensor exhibited significantly improved sensitivity toward methanol, displaying approximately 2.5 times higher response compared to the pure PEDOT:PSS sensor. Additionally, it maintained stable detection performance across a broad humidity range (0–80% RH) and temperatures (21–60 °C), demonstrating practical applicability in complex indoor environments. In another innovative approach, Farea et al. [104] developed PEDOT:PSS/poly(p-methoxyaniline) (PEDOT:PSS/PPA) nanocomposite sensors for carbon monoxide (CO) detection. The composite structure enhanced the selective interaction with CO molecules, significantly improving the sensor's selectivity and sensitivity. The sensor displayed remarkable repeatability, stability, and selectivity, effectively discriminating CO from interfering gases such as acetone and toluene (Figure 6b). However, sensor performance was influenced by humidity, indicating the necessity for additional structural modifications or protective layers to mitigate environmental interferences (Figure 6c).

Future research on PEDOT:PSS-based gas sensors should focus on further enhancing sensor selectivity and stability, particularly by engineering composite structures or developing advanced encapsulation strategies to minimize environmental interference [103]. Expanding application scenarios, such as wearable or flexible sensing devices, is another promising direction that leverages the polymer's intrinsic flexibility and robust mechanical properties, broadening its practical applications in environmental monitoring, healthcare, and industrial safety.

Overall, conductive polymers such as PANi, PPy, PTh, PEDOT and PEDOT:PSS exhibit multiple performances in gas sensing. The sensing mechanisms, target gases, and response times of different materials are significantly different, as shown in Table 1.

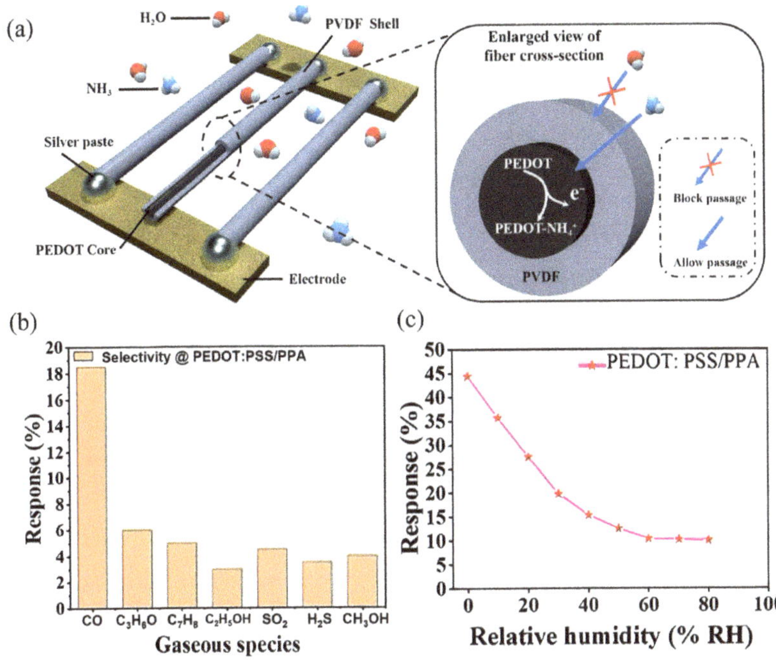


**Figure 6.** (**a**) Schematic diagram of the CSNF sensor structure and its response principle [99]; sensing performance of the PEDOT:PSS/PPA sensor at 100 ppm of CO. (**b**) Selectivity, (**c**) humidity effect on the PEDOT:PSS/PPA sensor [104].

**Table 1.** Comparison of the critical application parameters of different CPs-based sensors for special gas analytes.

| Mechanism | Sensor Materials | Target | LOD | Responce | Ref. |
|---|---|---|---|---|---|
| Chemiresistive | PEDOT:PSS/PPA | CO | 50 ppm | 58 s | [104] |
| Chemiresistive | PANI/$SnO_2$ | $C_6H_6$ | 0.4 ppm | 33 s | [105] |
| Mechanical | poly(DTCPA-*co*-BHTBT)-CB | $C_7H_8$ | 4 ppm | 36.47 s | [48] |
| Chemiresistive | TSP-nAu-PANi | $CHCl_3$ | – | 360 s | [106] |
| Chemiresistive | GO:PEDOT:PSS | $CH_3OH$ | – | 24.47 s | [103] |
| Optical | POF/PPy | $C_2H_5OH$ | 140 ppm | 5 s | [61] |
| Chemiresistive | SWCNT/$C_4$F-PPy | $C_3H_6O$ | 1 ppm | 750 s | [107] |
| Chemiresistive | $SnO_2$/PTh | $C_3H_6O$ | 0.5 ppm | 10 s | [35] |
| Chemiresistive | PANi/AgNWs/Silk | TMA | 1.38 ppm | 90 s | [24] |
| Chemiresistive | PANi | $H_2$ | 1 ppm | 15 s/17 s | [78] |
| Chemiresistive | CA-PANi | $H_2S$ | 1 ppm | 1 s | [16] |
| Chemiresistive | PANi/$MO_X$ | $H_2S$ | <100 ppm | 10 s | [88] |
| FET | Si/PANI:DBSA | $NH_3$ | – | 2 s | [66] |
| Chemiresistive | PEDOT-PVDF | $NH_3$ | 10 ppm | 80 s | [99] |
| Chemiresistive | PA-PANI/GO | $NH_3$ | 25 ppm | – | [73] |
| Chemiresistive | f-MWCNT-PEDOT:PSS | $NH_3$ | <10 ppm | 228 s | [108] |
| Chemiresistive | PPy-TcCoPc | $NH_3$ | 50 ppm | 8 s | [84] |
| Chemiresistive | CdS QDs-PTh | $NH_3$ | 10 ppm | 0.6 s | [36] |
| Electrochemical | PEDOT:PSS/IrOx Ps | $NH_3$ | 8 ppm | 87 ± 9 s | [30] |
| Chemiresistive | PANi | $NH_3$ | 2.5 ppm | 110 s | [22] |
| Chemiresistive | PPy@LIG | $NH_3$ | 1 ppm | 450 s | [86] |

**Table 1.** *Cont.*

| Mechanism | Sensor Materials | Target | LOD | Responce | Ref. |
|---|---|---|---|---|---|
| Chemiresistive | PANi/FMWCNT | $NH_3$ | 1 ppm | 15 s | [77] |
| Chemiresistive | PANi-MWCNTs/PDMS | $NH_3$ | 10 ppb | – | [109] |
| Chemiresistive | RGO-PTh | $NO_2$ | 0.52 ppm | 498 s | [95] |
| Chemiresistive | MWCNTs/PANi | $NH_3$ | 0.3 ppm | 21 s | [76] |
| Piezoelectric | GO:PEDOT:PSS | $NO_2$ | 175 ppb | 35 s | [43] |
| Chemiresistive | PTh | $NO_2$ | 0.25 ppm | 4980 s | [110] |
| Chemiresistive | PANi/BP | $NO_2$ | <2 ppm | 98 s | [111] |
| Chemiresistive | Au-ZnO-PANi | $NO_2$ | <10 ppm | 600 s | [112] |

### 2.3. Main Constituent Materials of the Sensor Device

#### 2.3.1. Substrate

Substrates play a crucial role in CPs-based gas sensors, providing essential physical support and significantly influencing sensor performance, sensitivity, stability, and applicability. Ideal substrates should exhibit compatibility with sensing materials, mechanical flexibility, chemical inertness, appropriate thermal stability, and cost-effectiveness [113]. Common substrates include ceramics, polymers, and flexible materials.

Ceramic substrates are widely utilized due to their excellent chemical stability, high-temperature resistance, and mechanical robustness, making them suitable for harsh environmental conditions or sensors requiring high-temperature processing [114]. Feng et al. [105] employed ceramic substrates in PANi/$SnO_2$ hybrid gas sensors, taking advantage of ceramics' high thermal stability and chemical inertness, effectively ensuring stable performance in detecting ammonia and benzene vapor. However, ceramic substrates possess inherent brittleness, limiting their application in flexible electronics and wearable sensors.

Thus, polymer-based substrates, particularly polyethylene terephthalate (PET), have emerged as promising alternatives. PET substrates offer distinct advantages, including excellent flexibility, durability, and compatibility with diverse processing techniques such as printing and coating methods [115]. Boonthum et al. [108] demonstrated an ammonia gas sensor fabricated from functionalized multi-walled carbon nanotubes (f-MWCNTs) combined with PEDOT:PSS on a flexible PET substrate. This sensor exhibited stable performance even when subjected to bending with various curvature radii (Figure 7a), demonstrating significant potential for flexible and wearable sensor applications. Another emerging substrate material is polyimide (PI), characterized by high thermal stability, superior mechanical properties, and excellent chemical resistance, particularly suitable for sensors involving high-temperature processes or requiring enhanced mechanical strength [116]. Kang et al. [107] fabricated flexible sensors using PI substrates interdigitated with Cu/Ni/Au electrodes. This design ensured strong adhesion between electrodes and substrates, enhancing sensor reliability and overall durability.

Future research in substrate materials should focus on developing multifunctional and composite substrates that further enhance sensor performance, flexibility, and environmental resilience. Advancing substrate technologies will significantly expand the practical applications of CPs-based sensors in environmental monitoring, healthcare, wearable electronics, and industrial safety.

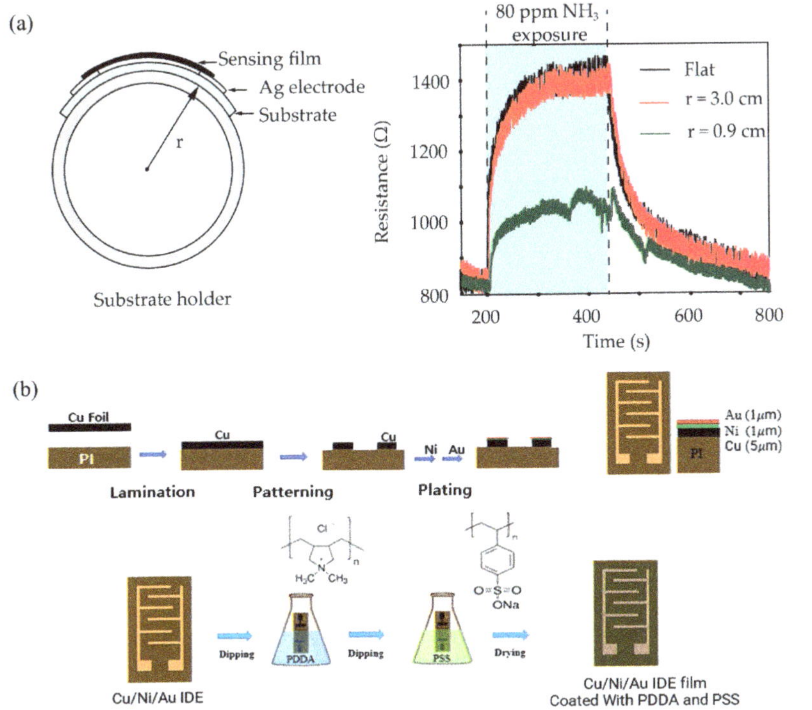


**Figure 7.** (**a**) Schematic diagram of f-MWCNT-PEDOT:PSS gas sensor under bending test and resistance changes in gas sensor measured during flat and bending (r = 0.9 cm, r = 3.0) states [108]; (**b**) schematic showing the preparation of the PI substrates interdigitated with the Cu/Ni/Au electrodes (IDE substrate) [107].

### 2.3.2. Electrode Material

Electrode materials play a critical role in CPs-based gas sensors, directly impacting their sensitivity, selectivity, stability, and overall performance [117]. Ideal electrode materials should possess high electrical conductivity, excellent chemical stability, strong adhesion to the substrate, compatibility with CPs, and resistance to environmental interferences.

Noble metals, such as gold (Au), platinum (Pt), and silver (Ag), are extensively used electrode materials due to their exceptional electrical conductivity, chemical inertness, and stability under various environmental conditions [118]. Park et al. [110] developed gas sensors utilizing platinum (Pt) interdigitated electrodes combined with polythiophene (PTh) for nitrogen dioxide ($NO_2$) detection. The Pt electrodes provided stable and efficient electron transport pathways, ensuring accurate signal acquisition with minimal interference. However, noble metals' high cost limits their large-scale practical application, prompting researchers to seek cost-effective alternatives.

Carbon-based materials, including graphene, carbon nanotubes (CNTs), and laser-induced graphene (LIG), have emerged as promising electrode materials due to their outstanding electrical conductivity, large specific surface area, and excellent mechanical flexibility [119]. Kang et al. [107] designed flexible acetone sensors using single-walled carbon nanotube (SWCNT)-based composite materials with Cu/Ni/Au tri-layer electrodes on polyimide (PI) substrates (Figure 7b). The copper (Cu) provided low-cost conductive paths, nickel (Ni) effectively prevented oxidation and migration, while gold (Au) ensured superior electrical conductivity and chemical stability. This multi-layer electrode configuration significantly enhanced sensor performance and stability.

Transparent conductive oxide electrodes, such as indium tin oxide (ITO), have also gained attention, particularly in optical and transparent sensor applications, owing to their combination of transparency and electrical conductivity. ITO electrodes enable real-time optical and electrical signal monitoring simultaneously, broadening their practical applications in multifunctional sensors [120].

Future research should focus on developing composite electrodes and advanced electrode architectures that enhance sensor sensitivity, stability, and durability [121]. Additionally, efforts should be directed toward identifying alternative cost-effective materials and fabrication methods to facilitate large-scale production and commercialization of CPs-based gas sensors.

#### 2.3.3. Packaging Material

Packaging materials are crucial components of CPs-based gas sensors, providing essential protection and isolation from environmental factors, thereby enhancing sensor stability, reliability, and operational lifespan [122]. Suitable packaging materials should exhibit excellent chemical and thermal stability, mechanical strength, effective moisture resistance, and compatibility with CPs to maintain sensor sensitivity and response consistency [123]. Commonly used packaging materials include epoxy resins, silicone rubbers, polyimide (PI), and biodegradable polymers like polycaprolactone (PCL) [124]. Epoxy resins are widely favored for their strong adhesion, high mechanical robustness, and chemical inertness, effectively protecting sensors from harsh environmental conditions. Silicone rubbers offer superior elasticity, excellent resistance to temperature variations, and chemical stability, making them suitable for flexible sensor applications.

For instance, Safaee et al. [125] utilized polycaprolactone (PCL) as a packaging material combined with single-walled carbon nanotubes (SWCNTs) to fabricate wearable optical microfiber textile sensors for real-time monitoring of hydrogen peroxide in biomedical applications (Figure 8a). PCL was selected for its excellent biocompatibility, chemical stability, and ease of fabrication, enabling secure encapsulation and reliable long-term performance in biological environments [126]. Nevertheless, PCL's relatively slow degradation rate and limited mechanical robustness under high stress conditions present challenges that must be considered in specific applications [127]. Polydimethylsiloxane (PDMS) is another prominent packaging material, particularly beneficial in flexible and wearable sensors due to its outstanding flexibility, biocompatibility, and environmental stability [128]. Lu et al. [129] developed flexible strain sensors using silver nanowires (AgNW), thermoplastic polyurethane (TPU), and PDMS (Figure 8b). PDMS provided critical mechanical flexibility, effectively protecting the sensor under repeated mechanical deformation while maintaining stable sensing performance.

Future research on packaging materials should focus on developing multifunctional composites and advanced packaging techniques, such as encapsulation strategies to minimize environmental interference and enhance durability [130]. Efforts to improve biocompatibility, biodegradability, and mechanical resilience will significantly broaden the practical applications of CPs-based sensors, particularly in wearable devices, healthcare, and environmental monitoring sectors.

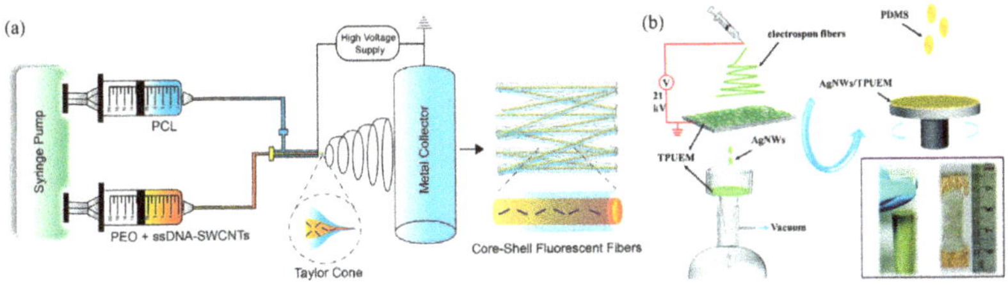


**Figure 8.** (**a**) Core-shell electrospinning setup for the fabrication of the optical microfibrous textiles [125]; (**b**) schematic diagram of ATP strain sensor production [129].

## 3. Application of CPs-Based Gas Sensors

### *3.1. Promising and Broad Applications*

#### 3.1.1. Environmental Monitoring Field

With increasing environmental pollution challenges, accurate and efficient environmental monitoring has become crucial for safeguarding human health and ecological sustainability. CPs-based gas sensors have emerged as powerful tools in environmental monitoring due to their unique properties, including high sensitivity, rapid response, ease of fabrication, and compatibility with flexible and wearable technologies. These characteristics position CPs as ideal candidates for monitoring harmful gases such as nitrogen dioxide ($NO_2$), sulfur dioxide ($SO_2$), ammonia ($NH_3$), volatile organic compounds (VOCs), and greenhouse gases [1–3].

PEDOT:PSS, a widely studied CP, has shown remarkable effectiveness in detecting environmentally harmful gases due to its superior electrical conductivity and chemical stability [131]. For example, Deller et al. [21] developed a voltammetric sensor by integrating PEDOT:PSS with gold nanoparticles (AuNPs) to detect pirimicarb (PMC), an environmental contaminant. This sensor demonstrated high sensitivity and selectivity, with a low detection limit and stable performance, highlighting the applicability of CPs in precise environmental pollutant monitoring.

Polyaniline (PANi) and its composites also exhibit great potential in environmental gas monitoring. Tang et al. [111] designed a room-temperature $NO_2$ sensor by combining PANi with black phosphorus (BP), achieving excellent sensitivity within a concentration range of 2–60 ppm $NO_2$. Similarly, Bonyani et al. [112] fabricated a highly selective $NO_2$ sensor based on Au-modified ZnO-PANi composite nanofibers (Figure 9a), significantly improving sensor performance and reliability.

Despite these advancements, CPs-based environmental sensors still face challenges regarding stability under variable humidity and temperature conditions, long-term operational durability, and cross-sensitivity to interfering gases. Future research directions include developing advanced composite materials, optimizing sensor architectures, and incorporating intelligent data-processing methods to further enhance sensor sensitivity, selectivity, and environmental robustness [132]. Additionally, efforts toward flexible, miniaturized, and wearable sensor systems will enable widespread deployment for continuous, real-time environmental monitoring, substantially improving air-quality management and pollution control strategies.

#### 3.1.2. Industrial Production Field

In industrial production, accurate and real-time monitoring of gas concentrations is essential to ensure product quality, enhance operational safety, and prevent potential accidents caused by hazardous gases. CPs-based gas sensors have gained increasing attention in industrial applications due to their advantages, such as rapid response, high sensitivity, low fabrication cost, and the ability to function effectively at room temperature. Volatile organic compounds (VOCs), including benzene, toluene, and formaldehyde, are common pollutants generated in various industrial processes, posing severe health risks and environmental hazards [133].

CPs gas sensors, owing to their excellent selectivity and sensitivity, have shown significant promise for VOC monitoring in manufacturing environments. For instance, Selvanayakam et al. [106] developed a sensor based on tamarind seed polysaccharide (TSP)-coated gold nanoparticles integrated with polyaniline (PANi) for detecting chloroform vapor in industrial emissions. This composite sensor displayed exceptional sensitivity, rapid response, and robust selectivity, effectively distinguishing chloroform from other interfering gases. Moreover, CPs-based gas sensors are widely employed in detecting flammable gases such as hydrogen, essential for safety monitoring in chemical industries and energy sectors [109]. Dipak et al. [22] demonstrated a hydrogen sensor using PANi nano-ink with a response of about 75% to ammonia (Figure 9b), which can effectively detect very small amounts of ammonia in the environment, providing crucial safety assurance in hydrogen-based energy applications.

Despite these advantages, challenges remain, including the susceptibility of CPs sensors to humidity interference, limited long-term stability under harsh industrial conditions, and potential baseline drift over extended use. Addressing these issues through material engineering, improved encapsulation techniques, and sensor design optimization is essential for advancing practical applications.

Future research should focus on developing advanced composite materials, optimizing sensor design, and integrating intelligent data-processing methods to enhance sensor reliability, stability, and resilience to environmental interference. Additionally, exploring flexible, miniaturized, and low-cost CPs sensors suitable for integration into industrial automation systems will significantly expand their applicability and impact on industrial safety and quality control.

#### 3.1.3. Food Safety Field

Gas sensing plays a crucial role in ensuring food safety, acting as an "invisible guardian" throughout the entire food supply chain—from production to sale—to continuously safeguard food quality. When food begins to spoil, various volatile gases are released. With its high sensitivity, gas-phase sensing technology can rapidly detect these gases, effectively assessing the freshness and safety of food products.

As the most widely consumed meat globally, pork serves as an essential source of animal protein for humans. However, fresh pork is highly susceptible to spoilage during storage. Microorganisms and enzymes can accelerate the decomposition of proteins, fats, and other components, producing biogenic amines such as ammonia, trimethylamine (TMA), and dimethylamine [134–136]. The deterioration of meat quality typically results in noticeable changes in surface texture, color, and odor. Among the numerous indicators used to evaluate pork freshness, TMA is particularly important. Its concentration gradually increases as pork freshness declines, and it emits a pungent rancid odor closely associated with the degree of spoilage. Consequently, accurate TMA detection can effectively and intuitively reflect pork freshness, making it a critical parameter in pork quality assessment.

To address this need, Li et al. [24] synthesized polyaniline (PANi) and silver nanowires (AgNWs) onto silk fibroin fibers (SFF) through an in-situ polymerization method, fabricating a novel, reliable, flexible, and easy-to-use gas sensor. This sensor was employed to detect various gases at 100 μg/L, including TMA, $NH_3$, $H_2S$, $H_2O$, and $C_2H_6O$ (Figure 9c). Compared to other gases, the sensor exhibited a significantly higher response toward TMA, allowing rapid, non-destructive, sensitive, and cost-effective detection of TMA in pork.

Thus, it effectively evaluates pork freshness, preventing consumers from ingesting spoiled meat and ensuring food safety.

Compared with traditional detection methods, gas sensors offer advantages such as simpler operation and shorter detection times. They can provide real-time monitoring of meat freshness throughout food production, processing, transportation, and sales, enabling the timely identification of spoiled products and minimizing economic losses. Consequently, these sensors represent efficient tools for freshness detection in the food industry and contribute significantly to the advancement of food safety technologies.

#### 3.1.4. Medical Diagnostic Field

Accurate, non-invasive, and rapid medical diagnostics significantly enhance the quality of patient care and facilitate early disease detection and monitoring [137]. CPs-based gas sensors offer substantial potential for medical diagnostics due to their high sensitivity, rapid response times, portability, flexibility, and ability to operate at room temperature, making them ideal for analyzing biomarkers in exhaled breath or volatile organic compounds (VOCs) emitted from the human body.

Exhaled breath analysis has become an essential diagnostic approach, providing vital insights into metabolic disorders, infectious diseases, and various physiological conditions. For instance, ammonia in human breath is an important biomarker for the diagnosis of kidney disease. CPs sensors have demonstrated remarkable efficacy in accurately and rapidly detecting trace ammonia levels due to their adjustable conductivity and selectivity [138,139]. Zhu et al. [133] developed a flexible ammonia sensor that uses a combination of polyaniline (PANi) and multi-walled carbon nanotubes (MWCNTs) to achieve excellent sensitivity and accurately detect ammonia in exhaled breath, providing a reliable technical means for non-invasive diagnosis of kidney disease. Moreover, sensors based on CPs have been successfully applied to detect acetone levels in the breath, an important marker of diabetes. Ananda et al. [140] prepared a gas-phase sensor based on PPy composites with ternary oxide $ZnCo_2O_4$ (ZCO) and $MnCo_2O_4$ (MCO) nanoparticles. Both nanocomposites exhibit high sensitivity and selectivity to acetone at room temperature (Figure 9d). The sensor effectively distinguished diabetic breath samples from healthy controls, indicating its considerable potential in non-invasive diabetes monitoring.

Despite these advancements, CPs-based gas sensors still face challenges such as cross-sensitivity to environmental factors, limited long-term stability, and potential signal drift. Future research should focus on enhancing sensor specificity through molecular engineering, integrating advanced composite materials, developing robust sensor designs, and incorporating artificial intelligence for accurate, real-time diagnostics. Further efforts toward creating wearable, low-cost, and flexible diagnostic devices will significantly expand the scope and impact of CPs-based sensors in medical applications, driving their adoption in personalized medicine and continuous health monitoring.

Conductive polymer-based gas sensors are widely used in multiple fields. Materials such as PANi, PPy, PTh, PEDOT, and PEDOT:PSS have different properties. Table 1 summarizes the sensing mechanisms, target gases, and performance data of different materials. Due to diverse application requirements involving multiple dimensions such as sensitivity and stability, and the performance of materials being affected by processes, it is impossible to simply compare which material is the best.

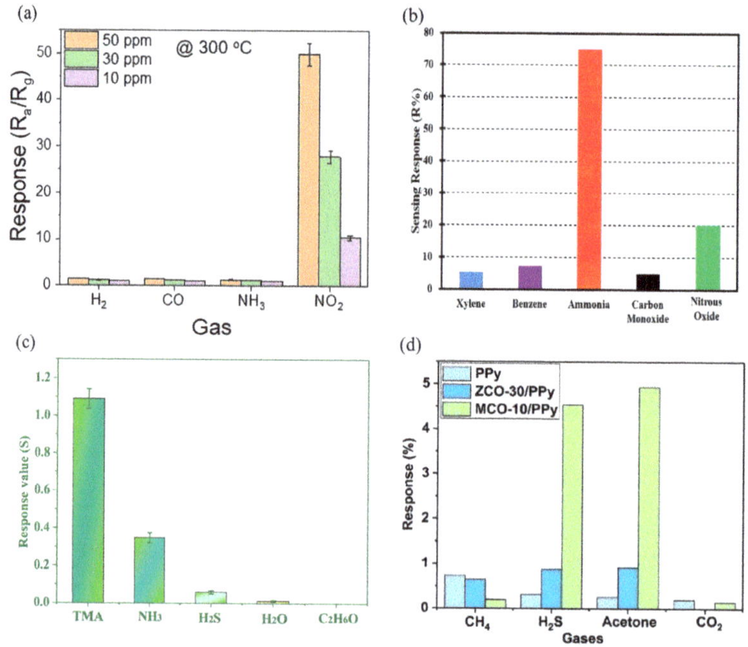


**Figure 9.** (**a**) Selectivity graph of Au-decorated ZnO-PANi (25 wt.%) composite nanofiber gas sensor to interfering gases at 300 °C [112]; (**b**) selectivity of the PANi nano gas sensors [22]; (**c**) sensor response of the PANi/AgNWs/silk composite nanofibers to 100 μg/L of different gases [24]; (**d**) response of PPy, ZCO-30/PPy, and MCO-10/PPy for various gasses along with a comparative response [140].

### *3.2. Critical Issues in Real Applications*

#### 3.2.1. Long-Term Stability and Humidity Drift: Mechanisms and Quantitative Evaluations

Although CPs-based gas sensors can be used to detect multiple gases and exhibit excellent performances, in practical application scenarios, they still face two key challenges: humidity-induced drift and poor long-term stability. Especially when operating in a variable environment for a long time, these problems seriously restrict their reliable working. The root cause of the performance degradation lies in the chemical, electrochemical, and mechanical changes that occur at the interface between the CPs-based sensing layer and the matrix electrode.

In a humid environment, due to the possible cross-reaction between different gases, such sensors may produce similar responsiveness to non-target gases. Therefore, the gas sensing performance may be a mixed response to the target gas and water vapor [141]. On the one hand, water adsorption consumes the adsorption sites of sensing materials, leading to the misjudgment of the concentration of target gases and reducing measurement accuracy. On the other hand, water molecules could interact with polymer chains and disrupt the charge transport path. In addition, the oxidizing or reducing gases may react with the main molecular chain of CPs or their composites, to destroy the conjugated system, reduce the carrier density, and weaken the conductivity and sensing ability of the sensor [142]. For example, in PEDOT:PSS, the PSS shell layer will adsorb water molecules, causing volume expansion, leading to an increase in the distance between PEDOT cores, and thus reducing the carrier transmission efficiency, and further hindering the target gas from reaching the sensing site, prolonging the response time and reducing the sensitivity [143]. By designing a multi-dimensional structure or compounding with inorganic materials, the adsorption sites and charge transfer efficiency of CPs-based sensing materials can be effectively enhanced, thereby improving the responsiveness and selectivity of related sensors. For example, inorganic two-dimensional (2D) nanomaterials such as rGO and MXenes have been introduced to improve their electrical conductivity and enhance their mechanical properties [73,95], using other polymers with good hydrophobicity and thermal stability, such as PDMS and fluorinated polyurethane (PU) to encapsulate these sensors [121,124,125].

Another important challenge for gas sensors is their long-term stability. This not only exists in CPs-based but also in other sensors. This is because during long-term use, sensing active materials may experience performance degradation or even failure due to mechanical external forces, environmental erosion, aging, and other factors. For example, when CPs-based sensors are exposed to air for a long time, due to the obvious dedoping effect, the adsorption sites of CPs may be affected, resulting in a significant decline in sensing performance. At the same time, the presence of oxygen could cause the degradation of CPs, thereby reducing their conductivity [144]. In flexible or wearable devices, mechanical stress cycling may cause delamination at the interface between the CPs films and the substrates, leading to interface failure and functional loss, finally affecting the related sensing performances and shortening the service life of such sensors.

#### 3.2.2. Biocompatibility, Toxicology, and Regulatory Considerations for CP-Based Wearable and Biomedical Sensors

CPs, especially PANi, PPy, PThs, PEDOT, and PEDOT:PSS, have been widely explored for wearable and biomedical sensing platforms. However, translating these materials from laboratory prototypes to clinically viable devices necessitates careful consideration of biocompatibility, cytotoxicity, and regulatory compliance under prolonged skin contact or implantation scenarios, as shown in Table 2.

##### Biocompatibility and Toxicological Assessment

PANi, while conductive and easy to process, may release toxic degradation products (e.g., aniline) if not adequately stabilized. Doping agents and processing residues significantly influence its biocompatibility, Composites or coatings (e.g., with biopolymers like chitosan or silk fibroin) are often employed to improve its safety [145,146].

PPy has demonstrated favorable in vitro and in vivo biocompatibility, so it is generally considered non-cytotoxic, especially when synthesized electrochemically without residual monomers or toxic dopants. Several studies show that PPy-coated scaffolds support cell adhesion and proliferation [147,148].

PThs also have excellent biocompatibility, with few adverse reactions in cell culture and in vivo models. Therefore, they can also promote the adhesion and differentiation of neural stem cells and have significant potential in neural regeneration and bioelectronic devices that require long-term tissue integration [149,150].

PEDOT and PEDOT:PSS, especially when purified or treated to remove excess PSS (e.g., with DMSO or ethylene glycol), exhibit low cytotoxicity and good compatibility with fibroblasts and neuronal cells [151]. As a result, PEDOT:PSS has been used in bioelectronic implants, such as cochlear [152] and neural interfaces [153], supporting its suitability for extended contact with biological tissues.

##### Relevant Standards and Regulatory Framework

ISO 10993 Series [154]: For any device involving skin contact (>30 days) or implantation, ISO 10993 mandates a series of biological evaluations including:

(1) ISO 10993-5 (in vitro cytotoxicity);
(2) ISO 10993-10 (skin irritation and sensitization);
(3) ISO 10993-11 (systemic toxicity);

(4) ISO 10993-6 (implantation effects) if the device is implanted.

FDA Device Classification [155]:

Wearable CPs-based sensors typically fall under Class I or Class II medical devices, depending on the intended use (diagnostic vs. therapeutic) and level of invasiveness. Implantable CPs-based sensors, if developed, would likely be Class III (requiring premarket approval), especially for glucose, neurotransmitter, or gas biomarker detection.

Material Risk Assessment:

For example, PEDOT:PSS was evaluated by the U.S. FDA as part of neural recording systems and passed biocompatibility testing in several device submissions (e.g., for cortical electrodes).

Design Considerations for Clinical Translation

For clinical translation, there are many serious considerations for the material systems design on CPs-based gas sensors, as shown below.

(1) Encapsulation strategies using biocompatible elastomers (e.g., PDMS, TPU) are essential to isolate CPs from direct tissue exposure while maintaining sensing functionality [125];
(2) Incorporation of bioinert and hydrophobic coatings also helps prevent ion leaching and immune response [156];
(3) Long-term implantation trials (animal models >30 days) are necessary to assess chronic inflammation and fibrotic encapsulation [157].

**Table 2.** Some biosafety and regulatory standards on CPs.

| Conducting Polymer | Biocompatibility | Toxicity Concerns | Regulatory Standard | Ref. |
|---|---|---|---|---|
| PANi | Variable; can release toxic aniline, modified forms, or composites safer | Aniline toxicity, residual dopants, and degradation products are concerns | ISO 10993-5, -10; FDA Class I/11 depending on application | [145,146] |
| PPy | Generally good; supports cell adhesion and proliferation | Low toxicity if properly synthesized | ISO 10993-5, -10, -11; preclinical animal studies needed for implants | [147,148] |
| PTh | General, modification or compounding for improvement | The toxicity of degradation products is unknown; residual monomers may be toxic | ISO 10993-5, -10; in some cases, additional in vitro and in vivo testing may be required based on application | [149,150] |
| PEDOT | Excellent; used in neural and cardiac interfaces | Minimal; depends on dopants and processing additives | ISO 10993-1, -5, -6, -10, -11; used in FDA-cleared implants | [151] |
| PEDOT:PSS | Good after PSS removal or treatment (DMSO/EG); low cytotoxicity | Excess PSS may irritate; removal improves safety | ISO 10993-5, -10; reviewed under FDA Class 11 submissions (e.g., neural devices) | [152,153] |

## 4. Conclusions and Prospects

CPs-based gas sensors hold significant promise across numerous fields due to their unique combination of tunable electrical/chemical properties and organic nature, providing versatile solutions for gas detection. This review has presented a comprehensive understanding of this class of sensors by examining their sensing mechanisms, sensitive materials, and device components. CPs gas sensors leverage diverse transduction mechanisms—including electrochemical, chemiresistive, optical, piezoelectric, and field-effect transistor (FET) methods—each offering distinct advantages suited to specific applications. Likewise, a variety of CP materials (such as PANi, PPy, PTh, PEDOT:PSS, and related composites) have been explored, each exhibiting unique gas-responsive characteristics and selectivity toward certain analytes. Key device components (e.g., flexible substrates, electrode configurations, and encapsulation materials) also critically influence sensor performance and durability. By integrating insights from detection principles, material properties, and device design, researchers have greatly advanced CPs-based gas-sensing technology and expanded its practical applicability.

Despite these advances, several challenges and opportunities remain, which define important directions for future research and development. First, improving sensor sensitivity and selectivity continues to be paramount. In complex gas environments, CPs sensors can be vulnerable to interference from other gases, which complicates the accurate identification of target species. Future studies should delve deeper into the interaction mechanisms between CPs materials and gas molecules and optimize polymer structures (e.g., nanoscale morphology and doping) to enhance selective adsorption and reaction with specific gases. Strategies such as incorporating nanostructured additives or forming CPs nanocomposites can create additional reactive sites and more efficient charge-transfer pathways, thereby dramatically boosting sensor response. For instance, integrating CPs with high-surface-area nanomaterials has been shown to improve the detection of gases like $H_2S$ by increasing adsorption sites and electron transport efficiency.

Second, long-term stability and reproducibility are critical issues that need to be addressed. Variations in environmental conditions (temperature, humidity, etc.) often affect CPs sensor baselines and response, leading to drift and inconsistent results over time. To overcome this, it is essential to develop robust sensor designs that minimize environmental susceptibility. For example, through innovative packaging materials and coatings that shield the sensitive layer from ambient fluctuations. Refining fabrication processes to produce uniform and stable polymer films, as well as implementing calibration or compensation techniques, will help ensure consistent performance. Improving the environmental adaptability of CPs sensors (for instance, by integrating humidity/temperature compensation elements) can greatly enhance their reliability in real-world applications, thus improving repeatability and facilitating widespread deployment.

Third, the emergence of intelligent gas-sensing systems is a frontier area poised to elevate CPs-based sensor capabilities. By coupling sensor arrays with advanced algorithms (such as machine learning and artificial intelligence), researchers can create electronic nose systems capable of recognizing complex odor/gas patterns and distinguishing specific target gases within mixtures. Machine learning techniques can analyze the multidimensional data from CPs sensor arrays, filter out interference, and even perform real-time pattern recognition that surpasses the selectivity achievable by materials alone. Recent work has demonstrated that hybrid sensor arrays combined with machine learning can rapidly identify hazardous gases in complex backgrounds, highlighting the power of data-driven approaches for enhancing selectivity and sensitivity. Integrating CPs gas sensors into the Internet of Things (IoT) framework with wireless connectivity and cloud analytics can further enable smart gas monitoring networks that learn and adapt over time. Such intelligent systems represent a promising direction to improve accuracy in complex scenarios (by, for example, self-calibrating for drift or compensating for cross-sensitivity), thereby expanding the practicality of CPs-based sensors in industrial safety, environmental monitoring, and other areas where automated, real-time decision-making is crucial.

Fourth, multi-functional integrated sensing platforms offer another promising avenue for future development. Instead of operating in isolation, CPs-based gas sensors can be combined with other types of sensors (such as temperature, humidity, pressure, or even multiple gas sensors in an array) on a single platform to provide comprehensive environmental data. By monitoring multiple parameters simultaneously, such integrated systems can account for environmental factors and improve overall measurement accuracy and context awareness. For example, real-time readings of ambient temperature and humidity alongside gas concentration allow for automatic compensation of environmental effects on the gas sensor's output, yielding more reliable results. Moreover, multi-gas sensor arrays can be designed to detect a suite of gases at once, enabling a broad-spectrum "electronic nose" capable of profiling complex gas mixtures (e.g., for air quality or breath analysis). The fusion of data from different sensor modalities can thus enhance the selectivity and robustness of the sensing system. In the future, lab-on-a-chip implementations may integrate CPs gas sensors with microfluidic channels, chemical detectors, and electronic circuitry, culminating in portable devices that offer multi-modal sensing and on-site analysis for applications ranging from environmental surveillance to medical diagnostics.

Fifth, exploring new sensor form factors, particularly flexible, wearable, and implantable gas sensors, is an exciting direction to broaden the applicability of CPs-based sensors. Owing to the intrinsic mechanical flexibility of CPs, there is considerable potential to fabricate gas sensors on bendable substrates (plastics, textiles, or even paper), enabling devices that conform to various surfaces or can be worn on the body. In recent studies, researchers have developed wearable CPs gas sensors (for example, a PEDOT-based ammonia sensor on a textile substrate) that can continuously monitor gaseous biomarkers in human sweat or breath. These flexible and wearable sensors open up opportunities for personalized health monitoring (such as real-time breath analysis for medical screening) and on-body environmental exposure tracking for occupational safety. Looking ahead, implantable CPs gas sensors could be envisioned for specialized biomedical applications. For instance, detecting internal gas biomarkers or changes in blood chemistry in vivo, provided that biocompatible materials and safe operation can be ensured. While implantable gas sensors are still largely conceptual, the combination of CPs' biocompatibility (in certain formulations) and their compatibility with soft electronics suggests that future research could yield minimally invasive gas-sensing devices for healthcare. Overall, developing flexible, stretchable, and wearable CPs gas sensors will significantly expand their use cases, allowing integration into everyday objects, clothing, or even the human body, thereby extending gas monitoring capabilities to scenarios that were previously impractical for rigid, conventional sensors.

Sixth, there is a growing interest in sustainable and eco-friendly materials for sensor development, and CPs-based gas sensors stand to benefit from this trend. Unlike sensors based on scarce or non-renewable inorganic materials, CPs sensors can potentially be made with renewable or biodegradable components. Future research may focus on designing biodegradable CPs or composites that maintain excellent sensing performance while being environmentally benign after their service life. For example, incorporating natural polymers or biodegradable matrices with CPs blends could produce sensors that eventually decompose under specific conditions, reducing electronic waste. Such sustainable sensors would be especially valuable in disposable or short-term use applications (like wearable health patches or environmental sensors deployed in large numbers) where device recovery is difficult.

Additionally, green synthesis approaches for CPs (using less toxic reagents or energy-efficient processes) and the recycling or reprocessing of CPs materials are important considerations for making the next generation of gas sensors more sustainable. Embracing eco-design principles in developing CPs gas sensors will help align the field with global sustainability and environmental safety goals, without sacrificing the performance improvements gained in recent years. Beyond these technical innovations, interdisciplinary integration of CPs-based gas sensors with other fields will further expand their impact and unlock novel applications. One such field is catalysis, where integrating gas sensors with catalytic processes could greatly benefit real-time reaction monitoring and catalyst development.

In catalytic reactors or chemical synthesis systems, CPs gas sensors can be used to monitor reaction gases in real-time, providing immediate feedback on reaction progress. For example, a CP sensor placed in a reactor could detect the emergence of a particular gaseous product or the depletion of a reactant, enabling dynamic adjustments to reaction conditions (temperature, feed rate, etc.) to optimize yield and selectivity on the fly. Conversely, incorporating catalytic nanoparticles or enzymes into a CP sensor's sensitive layer can create a hybrid sensor-catalyst material that not only facilitates a specific chemical reaction but simultaneously detects its gaseous products. Such multi-functional composites, possessing both catalytic and sensing capabilities, allow for simultaneous catalysis and monitoring. This dual function is valuable for studying reaction mechanisms; the sensor's readings give insight into intermediate formation and reaction rates, and it can aid in developing smarter catalysts (since the effectiveness of a catalyst under various conditions can be directly observed through the integrated sensor response).

In summary, the convergence of CPs-based gas sensing with catalysis research offers a powerful approach for advancing process control in chemical manufacturing and for innovating new catalytic materials with built-in sensing functionality. Another domain where CPs gas sensors show great promise is biomedical diagnostics. Gas biomarkers are an emerging frontier in medical screening and monitoring. For instance, volatile organic compounds (VOCs) in exhaled breath can serve as indicators for diseases such as diabetes, lung cancer, or infectious diseases. CPs-based gas sensors, with appropriate selectivity tuning, can be tailored to detect specific biomarker gases, potentially enabling non-invasive diagnostic tools. For example, a CP sensor designed to respond to acetone in breath could assist in monitoring diabetes (as breath acetone levels correlate with blood glucose status), and sensors for nitric oxide or other breath VOCs could help in the early detection of respiratory conditions. By integrating these sensors into portable or wearable formats (such as a handheld breath analyzer or a patch that samples skin-emitted gases), patients could perform real-time health monitoring outside of clinical settings.

Furthermore, coupling CPs gas sensors with microfluidic and MEMS technologies can yield miniaturized lab-on-chip diagnostic platforms capable of analyzing exhaled breath or headspace from biological samples with high sensitivity. Such systems could concentrate trace gases and deliver them to the CP sensor array for detection, providing a fast and accurate analysis for personalized medicine. The interdisciplinary collaboration between material scientists, medical researchers, and engineers will be key to optimizing CPs sensor designs for biocompatibility, selectivity to relevant biomarkers, and user-friendly operation, ultimately bringing gas sensing diagnostics into routine healthcare practice. In the energy sector, the integration of CPs-based gas sensors can significantly enhance the safety and efficiency of energy storage and conversion systems. One promising application is in battery technology: lithium-ion batteries and other high-energy devices can release trace gases (such as $CO_2$, CO, or electrolyte vapors) as early indicators of thermal runaway or degradation.

Embedding CPs gas sensors within battery packs or enclosures could allow continuous internal monitoring of such gas evolution, providing early warnings of battery failure or overheating. This real-time sensing could enable proactive measures (like cooling or disconnecting a failing cell) to prevent fires or explosions, thereby greatly improving battery safety and reliability. Similarly, for fuel cells and other fuel-based energy devices, monitoring the concentrations of reactant and product gases ($H_2$, $O_2$, water vapor, etc.) using CPs sensors can help in optimizing performance. Sensors can feed information to control systems to adjust fuel flow or operating conditions for maximum efficiency and detect deviations that might indicate catalyst poisoning or membrane leaks.

Moreover, incorporating CPs gas sensors in environmental control systems of energy facilities (such as hydrogen storage, biogas plants, or carbon capture units) can aid in leak detection and emissions monitoring, contributing to safer and more sustainable energy operations. These examples in energy applications underscore the broad prospects for CPs gas sensors when combined with energy technology development a synergy that can drive innovations in both fields simultaneously. In summary, CPs-based gas sensors are poised to play an increasingly significant role in future sensing technologies and interdisciplinary applications. By addressing the remaining challenges, improving sensitivity/selectivity, enhancing stability, and embracing intelligent algorithms—and by exploiting new opportunities such as multi-modal integration, flexible form factors, sustainable materials, and cross-domain collaborations, the next generation of CPs-based gas sensors will become even more sensitive, selective, robust, and versatile.

Future researchers should continue to focus on these emerging directions and foster innovation at the intersection of materials science, engineering, and application domains. Through sustained research efforts and interdisciplinary collaboration, CPs-based gas-sensing technology will continue to advance, driving breakthroughs in how we detect and utilize gas information in environmental monitoring, industrial process control, healthcare, energy, and beyond. Ultimately, the ongoing innovations will ensure that CPs gas sensors realize their full potential as key components of smart, responsive, and sustainable sensing systems for the modern world.

In addition, artificial intelligence (AI) technology has great potential in the future development of gas sensors on the basis of CPs. In terms of data processing, AI can efficiently analyze the massive and complex data generated by sensors toward real and unpredictable gas analytes, like E-noses for pneumoconiosis screening and diagnosis [158]. For example, deep learning algorithms could deeply mine the response data of sensors under different gas concentrations and environmental conditions and establish accurate gas recognition models, thereby greatly improving the recognition accuracy of sensors for target gases and effectively reducing misjudgments and omissions. In terms of adaptive adjustment, AI may dynamically adjust the working mode and parameter settings of sensors according to real-time environmental parameters and historical data of sensors. For example, when there are large changes in environmental humidity or temperature, the compensation mechanism of the sensor may be automatically optimized through AI algorithms to keep the sensor in the best detection performance at all times and further improve its stability and reliability in complex and changeable environments. Moreover, with the predictive analysis ability of AI, it is possible to predict in advance possible failures or performance declines of sensors so that timely maintenance and replacement can be carried out to reduce the risk of equipment operation and improve the operating efficiency of the entire gas detection system. By deeply integrating AI technology with CPs-based gas sensors, it is expected to promote leapfrog development in this field and create a more intelligent and efficient new era of gas detection.

**Author Contributions:** R.D. conceptualization, software, validation, visualization, writing—original draft; M.Y. and L.L. formal analysis, investigation, writing—review and editing; H.X. and X.D. writing—review and editing; Y.Z. and S.C. conceptualization, supervision, writing—review and editing. All authors have read and agreed to the published version of the manuscript.

**Funding:** The financial support from the National Natural Science Foundation of China (No. 32360620), the Academic Development Project of TongXin Funds (No. 2024161815) is gratefully acknowledged and the Jiangxi Provincial Key Laboratory of Flexible Electronics (No. 20242BCC32010).

**Conflicts of Interest:** The author declares no conflicts of interest.

## References

1. Ding, C.Q.; Ren, Y.; Liu, X.Y.; Zeng, J.J.; Yu, X.P.; Zhou, D.X.; Li, Y.J. Detection and discrimination of sulfur dioxide using a colorimetric sensor array. *RSC Adv.* **2022**, *12*, 25852–25859. [CrossRef]
2. Li, X.; Feng, Y.; Long, J.Y.; Lv, H.F.; Guo, Y.J.; Zu, X.T. MXene-activated graphene oxide enhancing $NO_2$ capture and detection of surface acoustic wave sensors. *Sens. Actuators B Chem.* **2024**, *401*, 135006. [CrossRef]
3. Teixeira, G.D.G.; Esteves, C.; Moro, A.J.; Lima, J.C.; Barbosa, A.J.M.; Roque, A.C.A. An odorant-binding protein based electrical sensor to detect volatile organic compounds. *Sens. Actuators B Chem.* **2024**, *411*, 135726. [CrossRef]
4. Ku, W.; Lee, G.; Lee, J.Y.; Kim, D.H.; Park, K.H.; Lim, J.; Cho, D.; Ha, S.C.; Jung, B.G.; Hwang, H.; et al. Rational design of hybrid sensor arrays combined synergistically with machine learning for rapid response to a hazardous gas leak environment in chemical plants. *J. Hazard. Mater.* **2024**, *466*, 133649. [CrossRef] [PubMed]
5. Ma, M.Z.; Yang, X.T.; Ying, X.G.; Shi, C.; Jia, Z.X.; Jia, B.C. Applications of gas sensing in food quality detection: A review. *Foods* **2023**, *12*, 3966. [CrossRef]
6. Hu, W.W.; Wu, W.W.; Jian, Y.Y.; Haick, H.; Zhang, G.J.; Qian, Y.; Yuan, M.M.; Yao, M.S. Volatolomics in healthcare and its advanced detection technology. *Nano Res.* **2022**, *15*, 8185–8213. [CrossRef]
7. Tao, B.R.; Bai, J.N.; Miao, F.J.; Zhao, M. Passive RFID integrated acetone sensor and ammonia sensor based on $LaFeO_3$/ZnO/rGO composites. *Microchem. J.* **2024**, *207*, 112122. [CrossRef]
8. Zheng, Q.Y.; Yang, M.; Dong, X.; Zhang, X.F.; Cheng, X.L.; Huo, L.H.; Major, Z.; Xu, Y.M. ZnO/PANi nanoflake arrays sensor for ultra-low concentration and rapid detection of $NO_2$ at room temperature. *Rare Met.* **2023**, *42*, 536–544. [CrossRef]
9. Ampong, I.; Zimmerman, K.D.; Nathanielsz, P.W.; Cox, L.A.; Olivier, M. Optimization of imputation strategies for high-resolution gas chromatography-mass spectrometry (HR GC–MS) metabolomics data. *Metabolites* **2022**, *12*, 429. [CrossRef]
10. Alenzy, E.K.; Kandjani, A.E.; Shaibani, M.; Trinchi, A.; Bhargava, S.K.; Ippolito, S.J.; Sabri, Y. Human breath analysis; clinical application and measurement: An overview. *Biosens. Bioelectron.* **2024**, *278*, 117094. [CrossRef]
11. Ziani, I.; Bouakline, H.; El Guerraf, A.; El Bachiri, A.; Fauconnier, M.L.; Sher, F. Integrating AI and advanced spectroscopic techniques for precision food safety and quality control. *Trends Food Sci. Techol.* **2024**, *156*, 104850. [CrossRef]
12. Lou, C.G.; Liu, X.; Wang, Y.; Li, R.K.; Huang, L.; Liu, X.L. Miniature quartz tuning fork-based broad spectral coverage and high detectivity infrared spectroscopy. *Infrared Phys. Technol.* **2022**, *126*, 104322. [CrossRef]
13. Pavel, I.A.; Lakard, S.; Lakard, B. Flexible sensors based on conductive polymers. *Chemosensors* **2022**, *10*, 97. [CrossRef]
14. Liu, Z.Q.; Wang, Y.L.; Li, Y.Y.; Sui, C.M.; Liu, Y.Z.; Liu, Y.Y.; Zhao, Y.D.; Liang, X.S.; Liu, F.M.; Lu, G.Y. Bimetallic MOF derived mesoporous structure of Ru doped $SnO_2$ enable high-sensitivity gas sensors for triethylamine in high humidity. *Sens. Actuators B Chem.* **2024**, *405*, 135275. [CrossRef]
15. Zhao, J.Y.; Li, P.; Zhang, Q.; Ye, Z.L.; Wang, Y.J.; Tian, J.Y.; Xu, Z.; Peng, N.C.; Ren, H.; Zhang, X.H. MOF nanoflowers-based flexible portable NO sensors for human airway inflammation detection. *Chem. Eng. J.* **2024**, *499*, 156184. [CrossRef]
16. Bibi, A.; Rubio, Y.R.M.; Santiago, K.S.; Jia, H.W.; Ahmed, M.M.M.; Lin, Y.F.; Yeh, J.M. $H_2S$-sensing studies using interdigitated electrode with spin-coated carbon aerogel-polyaniline composites. *Polymers* **2021**, *13*, 1457. [CrossRef]
17. Lawaniya, S.D.; Kumar, S.; Yu, Y.; Awasthi, K. Ammonia sensing properties of PPy nanostructures (urchins/flowers) toward low-cost and flexible gas sensors at room temperature. *Sens. Actuators B Chem.* **2023**, *382*, 133566. [CrossRef]
18. Belhousse, S.; Tıghılt, F.Z.; Bennıa, S.; Adjtoutah, S.; Sam, S.; Lasmı, K.; Hamdanı, K. Effect of polythiophene thickness on hybrid sensor sensitivity. *Rev. Adv. Mater. Sci.* **2021**, *60*, 839–845. [CrossRef]
19. Hui, Y.; Bian, C.; Xia, S.H.; Tong, J.H.; Wang, J.F. Synthesis and electrochemical sensing application of poly(3, 4-ethylenedioxythiophene)-based materials: A review. *Anal. Chim. Acta* **2018**, *1022*, 1–19. [CrossRef]
20. Lv, C.H.; Zhou, X.; Chen, C.; Liu, X.H.; Qian, J. Highly sensitive and flexible ammonia sensor based on PEDOT:PSS doped with lewis acid for wireless food monitoring. *Chem. Eng. J.* **2024**, *493*, 152652. [CrossRef]
21. Deller, A.E.; Hryniewicz, B.M.; Pesqueira, C.; Horta, R.P.; Silva, B.J.G.; Weheabby, S.; Al-Hamry, A.; Kanoun, O.; Vidotti, M. PEDOT:PSS/AuNPs based composite as voltammetric sensor for the detection of pirimicarb. *Polymers* **2023**, *15*, 739. [CrossRef] [PubMed]
22. Dipak, P.; Tiwari, D.C.; Samadhiya, A.; Kumar, N.; Biswajit, T.; Singh, P.A.; Tiwari, R.K. Synthesis of polyaniline (printable nanoink) gas sensor for the detection of ammonia gas. *J. Mater. Sci. Mater. Electron.* **2020**, *31*, 22512–22521. [CrossRef]
23. Byeon, J.H.; Kim, J.S.; Kang, H.K.; Kang, S.; Kim, J.Y. Acetone gas sensor based on SWCNT/polypyrrole/phenyllactic acid nanocomposite with high sensitivity and humidity stability. *Biosensors* **2022**, *12*, 354. [CrossRef]

24. Li, Y.H.; Li, Y.X.; Shi, J.Y.; Li, Z.H.; Wang, X.; Hu, X.T.; Gong, Y.Y.; Zou, X.B. A novel gas sensor for detecting pork freshness based on PANi/AgNWs/silk. *Foods* **2022**, *11*, 2372. [CrossRef]
25. Zeng, L.H.; He, M.; Yu, H.H.; Li, D.L. An $H_2S$ sensor based on electrochemistry for chicken coops. *Sensors* **2016**, *16*, 1398. [CrossRef] [PubMed]
26. Wang, L.; Meng, W.W.; He, Z.X.; Meng, W.; Li, Y.H.; Dai, L. Enhanced selective performance of mixed potential ammonia gas sensor by Au nanoparticles decorated $CeVO_4$ sensing electrode. *Sens. Actuators B Chem.* **2018**, *272*, 219–228. [CrossRef]
27. Pereira, J.D.; Monge, J.; Postolache, O. Measurement and applications: Electrochemical sensors and instruments: Main characteristics and applications. *IEEE Instrum. Meas. Mag.* **2024**, *27*, 18–25. [CrossRef]
28. Wu, Z.X.; Rong, L.M.; Yang, J.L.; Wei, Y.M.; Tao, K.; Zhou, Y.B.; Yang, B.R.; Xie, X.; Wu, J. Ion-conductive hydrogel-based stretchable, self-healing, and transparent $NO_2$ sensor with high sensitivity and selectivity at room temperature. *Small* **2021**, *17*, 2104997. [CrossRef]
29. Ghazanfar, E.; Zahoor, H.; Awwad, N.S.; Ibrahium, H.A.; Mir, S.; Ahmed, I. Ternary composites based next-generation supercapacitors electrode material: Emerging trends. *Electrochem. Commun.* **2025**, *20*, 107893. [CrossRef]
30. Serafini, M.; Mariani, F.; Gualandi, I.; Decataldo, F.; Possanzini, L.; Tessarolo, M.; Fraboni, B.; Tonelli, D.; Scavetta, E. A wearable electrochemical gas sensor for ammonia detection. *Sensors* **2021**, *21*, 7905. [CrossRef]
31. Williams, D.E. Electrochemical sensors for environmental gas analysis. *Curr. Opin. Electrochem.* **2020**, *22*, 145–153. [CrossRef]
32. Ghorbani, G.; Taghipour, F. UV-activated chemiresistive gas sensor response curve analysis for the fast measurement of toxic gases. *Sens. Actuators B Chem.* **2024**, *419*, 136396. [CrossRef]
33. Gao, C.S.; Zhang, Y.Y.; Yang, H.R.; Liu, Y.; Liu, Y.F.; Du, J.H.; Ye, H.Y.; Zhang, G.Q. A DFT study of in doped $Tl_2O$: A superior $NO_2$ gas sensor with selective adsorption and distinct optical response. *Appl. Surf. Sci.* **2019**, *494*, 162–169. [CrossRef]
34. Barkade, S.S.; Pinjari, D.V.; Nakate, U.T.; Singh, A.K.; Gogate, P.R.; Naik, J.B.; Sonawane, S.H.; Pandit, A.B. Ultrasound assisted synthesis of polythiophene/$SnO_2$ hybrid nanolatex particles for LPG sensing. *Chem. Eng. Process. Process Intensif.* **2013**, *74*, 115–123. [CrossRef]
35. Beniwal, A. Room temperature operated electrospun nanofibers-based $SnO_2$/PTh sensor for acetone sensing applications. *IEEE Trans. Elect. Dev.* **2021**, *68*, 4084–4089. [CrossRef]
36. Chaudhary, P.L.; Adhyapak, P.V. Cadmium sulfide quantum dots-polythiophene nanocomposite for electrical ammonia sensing. *J. Mater. Sci. Mater. Electron.* **2024**, *35*, 1659. [CrossRef]
37. Sharma, A.; Rawal, I.; Rajpal, A.; Khokhar, A.; Kumar, V.; Goyal, P.K. Highly sensitive and selective room temperature ammonia sensor based on polyaniline thin film: In situ dip-coating polymerization. *J. Mater. Sci. Mater. Electron.* **2022**, *33*, 14071–14085. [CrossRef]
38. Kim, K.J.; Culp, J.T.; Ohodnicki, P.R.; Cvetic, P.C.; Sanguinito, S.; Goodman, A.L.; Kwon, H.T. Alkylamine-integrated metal–organic framework-based waveguide sensors for efficient detection of carbon dioxide from humid gas streams. *ACS Appl. Mater. Interfaces* **2019**, *11*, 33489–33496. [CrossRef]
39. Kim, K.; Kim, J.; Jiang, X.; Kim, T. Static force measurement using piezoelectric sensors. *J. Sens.* **2021**, *2021*, 6664200. [CrossRef]
40. Wang, Z.L. Piezotronic and piezophototronic effects. *J. Phys. Chem. Lett.* **2010**, *1*, 1388–1393. [CrossRef]
41. Bader, N.; Yempally, S.; Al-Ashker, F.; Al-Ejji, M.; Ponnamma, D. Cerium oxide decorated graphene nanolayers filled polyvinylidene fluoride nanofibers as optical piezoelectric sensors. *Macromol. Mater. Eng.* **2024**, *3*, 2400350. [CrossRef]
42. Adjaoud, A.; Nguyen, G.T.; Chikh, L.; Péralta, S.; Trouillet-Fonti, L.; Uguen, N.; Braida, M.D.; Plesse, C. Piezoionic sensors based on formulated PEDOT:PSS and Aquivion for ionic polymer-polymer composites. *Smart Mater. Struct.* **2021**, *30*, 105027. [CrossRef]
43. Pasupuleti, K.S.; Reddeppa, M.; Nam, D.J.; Bak, N.H.; Peta, K.R.; Cho, H.D.; Kim, S.G.; Kim, M.D. Boosting of $NO_2$ gas sensing performances using GO-PEDOT:PSS nanocomposite chemical interface coated on langasite-based surface acoustic wave sensor. *Sens. Actuators B Chem.* **2021**, *344*, 130267. [CrossRef]
44. Yang, L.Y.; Liu, Q.; Li, M.F.; Liu, Y.P.; Li, X.F.; Liu, Q.Z.; Zhu, T.; Lu, Y.; Liu, X.; Wang, D. A high-performance, flexible, and dual-modal humidity-piezoelectric sensor without mutual interference. *Sens. Actuators B Chem.* **2025**, *423*, 136778. [CrossRef]
45. Joshi, V.; Strege, P.R.; Farrugia, G.; Beyder, A. Mechanotransduction in gastrointestinal smooth muscle cells: Role of mechanosensitive ion channels. *J. Appl. Physiol. Gastrointest. Liver Physiol.* **2021**, *320*, G897–G906. [CrossRef]
46. Yi, H.K.; Wang, S.J.; Mei, S.X.; Li, Z. Conductive polymer composites for resistive flexible strain sensors. *Polymer* **2024**, *307*, 127286. [CrossRef]
47. Budiman, B.A.; Adziman, F.; Sambegoro, P.L.; Nurprasetio, I.P.; Ilhamsyah, R.; Aziz, M. The role of interfacial rigidity to crack propagation path in fiber reinforced polymer composite. *Fibers Polym.* **2018**, *19*, 1980–1988. [CrossRef]
48. Mallya, A.N.; Ramamurthy, P.C. Mechanical actuation of conducting polymer in the presence of organic vapor stimulus. *IEEE Sens. J.* **2017**, *17*, 3391–3397. [CrossRef]
49. Wang, D.; Li, X.; Tian, H.; Chen, X.; Nie, B.; Luo, Y.; Shao, J. Flexible strain sensor based on embedded three-dimensional annular cracks with high mechanical robustness and high sensitivity. *Appl. Mater. Today* **2021**, *25*, 101247. [CrossRef]

50. Afsari, A.; Sarraf, M.J. Design of a hydrogen sulfide gas sensor based on a photonic crystal cavity using graphene. *Micro Nanostruct.* **2020**, *138*, 106362. [CrossRef]
51. Liu, W.; Shi, Y.; Yi, Z.; Liu, C.; Wang, F.M.; Li, X.L.; Lv, J.W.; Yang, L.; Chu, P.K. Surface plasmon resonance chemical sensor composed of a microstructured optical fiber for the detection of an ultra-wide refractive index range and gas-liquid pollutants. *Opt. Express* **2021**, *29*, 40734–40747. [CrossRef]
52. Nimbekar, A.A.; Bhatia, P.G.; Deshmukh, R.R. Ammonia sensors manufactured by plasma enhanced grafting of conducting polymers on nylon-6 fabrics. *Synth. Met.* **2021**, *279*, 116840. [CrossRef]
53. Mumtaz, F.; Zhang, B.H.; Subramaniyam, N.; Roman, M.; Holtmann, P.; Hungund, A.P.; Hungund, A.P.; O'Malley, R.; Spudich, T.M.; Davis, M.; et al. Miniature optical fiber Fabry–Perot interferometer based on a Single-Crystal Metal–Organic framework for the detection and quantification of benzene and ethanol at low concentrations in nitrogen gas. *ACS Appl. Mater. Interfaces* **2024**, *16*, 13071–13081. [CrossRef]
54. Rossi, S.; Olsson, O.; Chen, S.Z.; Shanker, R.; Banerjee, D.; Dahlin, A.; Jonsson, M.P. Dynamically tuneable reflective structural coloration with electroactive conducting polymer nanocavities. *Adv. Mater.* **2021**, *33*, 2105004. [CrossRef]
55. Zandi, A.; Amjadi, M.; Hallaj, T. Plasmon-enhanced fluorimetric and colorimetric dual sensor based on fluorescein/Ag nanoprisms for sensitive determination of mancozeb. *Food Chem.* **2022**, *369*, 130967. [CrossRef]
56. Li, M.; Zhou, Q.W.; Wu, Y.Q.; Gao, J.; Zhu, R.; Gao, Q.; Wu, X.J.; Zhang, Y.G. An ultrafast optical sensor for simultaneous detection of $NH_3$ temperature and concentration based on ultraviolet absorption spectral redshift combined with spectral reconstruction. *Sens. Actuators B Chem.* **2025**, *422*, 136631. [CrossRef]
57. Njegovec, M.; Donlagic, D. A fiber-optic gas sensor and method for the measurement of refractive index dispersion in NIR. *Sensors* **2020**, *20*, 3717. [CrossRef] [PubMed]
58. Lee, Y.C.; Liou, C.S.; Chien, T.L.; Tsou, C.; Fang, W. Integration and encapsulation of light-emitting diode and CMOS-MEMS chips for fluorescence quenching gas sensor. *Front. Mech. Eng.* **2022**, *8*, 894060. [CrossRef]
59. Lotfiani, A.; Mohseni, S.M.; Ghanaatshoar, M. High-sensitive optoelectronic SPR biosensor based on fano resonance in the integrated MIM junction and optical layers. *Opt. Commun.* **2020**, *477*, 126323. [CrossRef]
60. Zhao, L.L.; Chen, H.Y.; Tang, Y.; Li, P.P.; Zhu, X.H.; Liu, J.Y.; Liu, M.L.; Zhang, Y.Y.; Yao, S.Z. $Ag_2S$ QDs integration with $MnO_2$ nanosheets for the sensitive detection of Cr (VI) via the redox reaction induced photoelectrochemical variation. *Anal. Chim. Acta* **2023**, *22*, 341471. [CrossRef]
61. Liu, W.Y.; Hu, Y.J.; Hou, Y.L. Ethanol gas sensitivity sensor based on roughened POF taper of modified polypyrrole films. *Sensors* **2020**, *20*, 989. [CrossRef]
62. Wang, H.L.; Chen, D.; Chen, Y.; Liu, J.Y.; Xu, J.X.; Zhu, A.N.; Long, F. Development of novel handheld optical fiber dissolved oxygen sensor and its applications. *Anal. Chim. Acta* **2022**, *1200*, 339587. [CrossRef] [PubMed]
63. Guo, S.S.; Long, W.B.; Hu, Z.X.; Wang, P.; Li, H.Y.; Liu, H. Field effect transistors with semiconductor-sensitized gate for high-sensitivity hydrogen detection. *IEEE Sens. J.* **2023**, *23*, 22651–22659. [CrossRef]
64. Yin, W.J.; Sun, J.; Zhang, Y.; Zhang, Y.; Li, S.S.; Zhu, M.Q.; Hong, H.; Ba, Y.T.; Deng, T. A novel three-dimensional Ag nanoparticles/reduced graphene oxide microtubular field effect transistor sensor for $NO_2$ detections. *Nanotechnology* **2020**, *32*, 025304. [CrossRef]
65. Suh, Y.; Lim, D. Doping-and capacitor-less 1T-DRAM cell using reconfigurable feedback mechanism. *Nanotechnology* **2024**, *36*, 065203. [CrossRef] [PubMed]
66. Amer, K.; Elshaer, A.M.; Anas, M.; Ebrahim, S. Fabrication, characterization, and electrical measurements of gas ammonia sensor based on organic field effect transistor. *J. Mater. Sci. Mater. Electron.* **2019**, *30*, 391–400. [CrossRef]
67. Kumar, A.; Tripathi, D.; Rawat, R.K.; Chauhan, P. Hybrid $MoS_2$/PEDOT:PSS sensor for volatile organic compounds detection at room temperature: Experimental and DFT insights. *ACS Appl. Nano Mater.* **2024**, *7*, 27599–27611. [CrossRef]
68. Yuvaraja, S.; Bhyranalyar, V.N.; Bhat, S.A.; Surya, S.G.; Yelamaggad, C.V.; Salama, K.N. A highly selective electron affinity facilitated $H_2S$ sensor: The marriage of tris (keto-hydrazone) and an organic field-effect transistor. *Mater. Horiz.* **2021**, *8*, 525–537. [CrossRef]
69. Park, C.S.; Lee, C.; Kwon, O.S. Conducting polymer based nanobiosensors. *Polymers* **2016**, *8*, 249. [CrossRef]
70. Ali, S.S.; Pauly, A.; Brunet, J.; Varenne, C.; Ndiaye, A.L. MWCNTs/PMMA/PS composites functionalized PANi: Electrical characterization and sensing performance for ammonia detection in a humid environment. *Sens. Actuators B Chem.* **2020**, *320*, 128364.
71. Alaş, M.Ö.; Güngör, A.; Genç, R.; Erdem, E. Feeling the power: Robust supercapacitors from nanostructured conductive polymers fostered with $Mn^{2+}$ and carbon dots. *Nanoscale* **2019**, *11*, 12804–12816. [CrossRef] [PubMed]
72. Yang, L.Y.; Xu, X.R.; Liu, M.D.; Chen, C.; Cui, J.; Chen, X.; Wu, K.; Sun, D.P. Wearable and flexible bacterial cellulose/polyaniline ammonia sensor based on a synergistic doping strategy. *Sens. Actuators B Chem.* **2021**, *334*, 129647. [CrossRef]
73. Yuan, Y.; Wu, H.Y.; Bu, X.R.; Wu, Q.; Wang, X.M.; Han, C.Y.; Li, X.; Wang, X.L.; Liu, W.H. Improving ammonia detecting performance of polyaniline decorated rGO composite membrane with GO doping. *Materials* **2021**, *14*, 2829. [CrossRef] [PubMed]

74. Beygisangchin, M.; Abdul, R.S.A.; Shafie, S.; Sadrolhosseini, A.R.; Lim, H.N. Preparations, properties, and applications of polyaniline and polyaniline thin films—A review. *Polymers* **2021**, *13*, 2003. [CrossRef]
75. Wan, P.B.; Wen, X.M.; Sun, C.Z.; Chandran, B.K.; Zhang, H.; Sun, X.M.; Chen, X.D. Flexible transparent films based on nanocomposite networks of polyaniline and carbon nanotubes for high-performance gas sensing. *Small* **2015**, *11*, 5409–5415. [CrossRef]
76. Zhuang, Y.; Wang, X.; Lai, P.f.; Li, J.; Chen, L.; Lin, Y.J.; Wang, F. Wireless flexible system for highly sensitive ammonia detection based on polyaniline/carbon nanotubes. *Biosensors* **2024**, *14*, 191. [CrossRef]
77. Askar, P.; Kanzhigitova, D.; Ospanova, A.; Tapkharov, A.; Duisenbekov, S.; Abutalip, M.; Soltabayev, B.; Turlybekuly, A.; Adilov, S.; Nuraje, N. 1 ppm-detectable hydrogen gas sensor based on nanostructured polyaniline. *Sci. Rep.* **2024**, *14*, 26984. [CrossRef] [PubMed]
78. Rahman Khan, M.M.; Rumon, M.M.H. Recent progress on the synthesis, morphological topography, and battery applications of polypyrrole-based nanocomposites. *Polymers* **2024**, *16*, 3277. [CrossRef]
79. Zhuo, H.; Hu, Y.J.; Chen, Z.H.; Zhong, L.X. Cellulose carbon aerogel/PPy composites for high-performance supercapacitor. *Carbohydr. Polym.* **2019**, *215*, 322–329. [CrossRef]
80. Wang, Y.M.; Liu, A.P.; Han, Y.Q.; Li, T.X. Sensors based on conductive polymers and their composites: A review. *Polym. Int.* **2020**, *69*, 7–17. [CrossRef]
81. Cichosz, S.; Masek, A.; Zaborski, M. Polymer-based sensors: A review. *Polym. Test.* **2018**, *67*, 342–348. [CrossRef]
82. Antony, N.; Mohanty, S.; Nayak, S.K. Electrochemical inspection of polypyrrole/chitosan/zinc oxide hybrid composites. *J. Appl. Polym. Sci.* **2020**, *137*, 49561. [CrossRef]
83. Kurian, A.S.; Souri, H.; Mohan, V.B.; Bhattacharyya, D. Highly stretchable strain sensors based on polypyrrole-silicone rubber composites for human motion detection. *Sens. Actuators A Phys.* **2020**, *312*, 112131. [CrossRef]
84. Gai, S.J.; Wang, X.L.; Zhang, R.Z.; Zeng, K.; Miao, S.L.; Wu, Y.Q.; Wang, B. A controllably fabricated polypyrrole nanorods network by doping a tetra-β-carboxylate cobalt phthalocyanine tetrasodium salt for enhanced ammonia sensing at room temperature. *RSC Adv.* **2023**, *13*, 13725–13734. [CrossRef] [PubMed]
85. Dimple; Madan, R.; Kumar, V.; Mohan, D.; Garg, R. Acetone gas sensing behavior of polypyrrole/ZnO nanocomposites synthesized via chemical oxidation method. *J. Mater. Sci. Mater. Electron.* **2024**, *35*, 130. [CrossRef]
86. Santos-Ceballos, J.C.; Salehnia, F.; Romero, A.; Vilanova, X.; Llobet, E. Low cost, flexible, room temperature gas sensor: Polypyrrole-modified laser-induced graphene for ammonia detection. *IEEE Sens. J.* **2024**, *24*, 9366–9374. [CrossRef]
87. Luo, K.H.; Yan, M.; Hung, Y.H.; Kuang, J.Y.; Chang, H.C.; Lai, Y.J.; Yeh, J.M. Polyaniline composites containing eco-friendly biomass carbon from agricultural-waste coconut husk for enhancing gas sensor performance in hydrogen sulfide detection. *Polymers* **2023**, *15*, 4554. [CrossRef]
88. Kabel, K.; Al-Sabagh, A.; Sharara, T.; Badawi, A.M.; Abdel-Rahman, A.; Gado, W. Fabrication of $H_2S$ gas sensor based on PPy/CuO and PPy/$SnO_2$ nanocomposites at room temperature. *Egypt. J. Chem.* **2020**, *63*, 2763–2774. [CrossRef]
89. Xiong, S.; Zhou, J.; Wu, J.H.; Li, H.L.; Zhao, W.; He, C.G.; Liu, Y.; Chen, Y.Q.; Fu, Y.Q.; Duan, H.G. High performance acoustic wave nitrogen dioxide sensor with ultraviolet activated 3D porous architecture of Ag-decorated reduced graphene oxide and polypyrrole aerogel. *ACS Appl. Mater. Interfaces* **2021**, *13*, 42094–42103. [CrossRef]
90. Farea, M.A.; Bhanuse, G.B.; Mohammed, H.Y.; Farea, M.O.; Sallam, M.; Shirsat, S.M.; Tsai, M.L.; Shirsat, M.D. Ultrahigh sensitive and selective room-temperature carbon monoxide gas sensor based on polypyrrole/titanium dioxide nanocomposite. *J. Alloys Compd.* **2022**, *917*, 165397. [CrossRef]
91. Chen, R.; Chen, S.S.; Zhou, Y.L.; Wei, Z.Y.; Wang, H.Y.; Zheng, Y.J.; Li, M.; Sun, K.; Li, Y.F. Unsubstituted polythiophene film deposited via in-situ sequential solution polymerization for chemo-/electrochromism. *Macromolecules* **2020**, *53*, 4247–4254. [CrossRef]
92. Husain, A.; Ahmad, S.; Mohammad, F. Thermally stable and highly sensitive ethene gas sensor based on polythiophene/zirconium oxide nanocomposites. *Mater. Today Commun.* **2019**, *20*, 100574. [CrossRef]
93. Haldar, U.; Mondal, S.; Hazra, S.; Guin, S.; Yeasmin, L.; Chatterjee, D.P.; Nandi, A.K. Tailor made synthesis of water-soluble polythiophene-graft-poly(caprolactone-block-dimethylaminoethyl methacrylate) copolymer and their pH tunable self-assembly and optoelectronic properties. *Eur. Polym. J.* **2022**, *168*, 111124. [CrossRef]
94. Du, Y.N.; Wang, M.Y.; Ye, X.L.; Liu, B.Q.; Han, L.; Jafri, S.H.M.; Liu, W.C.; Zheng, X.X.; Ning, Y.F.; Li, H. Advances in the field of graphene-based composites for energy-storage applications. *Crystals* **2023**, *13*, 912. [CrossRef]
95. Bai, S.L.; Guo, J.; Sun, J.H.; Tang, P.G.; Chen, A.; Luo, R.X.; Li, D.Q. Enhancement of $NO_2$-sensing performance at room temperature by graphene-modified polythiophene. *Ind. Eng. Chem. Res.* **2016**, *55*, 5788–5794. [CrossRef]
96. Brodský, J.; Migliaccio, L.; Sahalianov, I.; Zítka, O.; Neužil, P.; Gablech, I. Advancements in PEDOT-based electrochemical sensors for water quality monitoring: From synthesis to applications. *Trends Anal. Chem.* **2024**, *183*, 118115. [CrossRef]
97. Groenendaal, L.; Jonas, F.; Freitag, D.; Pielartzik, H.; Reynolds, J.R. Poly(3,4-ethylenedioxythiophene) and its derivatives: Past, present, and future. *Adv. Mater.* **2000**, *12*, 481–494. [CrossRef]

98. Jang, C.; Park, J.K.; Yun, G.H.; Choi, H.H.; Lee, H.J.; Yook, J.G. Radio-frequency/microwave gas sensors using conducting polymer. *Materials* **2020**, *13*, 2859. [CrossRef]
99. Xiao, S.H.; Hu, M.J.; Hong, Y.H.; Hu, M.J.; Sun, T.T.; Chen, D.J. Core-shell PEDOT-PVDF nanofiber-based ammonia gas sensor with robust humidity resistance. *Biosensors* **2024**, *14*, 411. [CrossRef]
100. Yang, L.F.; Zhou, Y.; Xu, X.; Shen, Y.Y.; Yan, H.F.; Qin, Z.Y. Interior design of hierarchical micro/nanostructures for enhancing energy storage ability of polyanilines through frozen interfacial polymerization. *Electrochim. Acta* **2021**, *386*, 138448. [CrossRef]
101. Wang, J.; Zhou, Q.X.; Shu, Z.; Wang, M.; Wang, Y.; Yao, Y. Photoelectrochemical sensor for hypochlorous acid detection based on the MWNTs doped PEDOT loaded with BP5 functionalized gold nanoparticles. *Mater. Today Chem.* **2024**, *40*, 102230. [CrossRef]
102. Zhang, X.S.; Yang, W.T.; Zhang, H.N.; Xie, M.Y.; Duan, X.X. PEDOT:PSS: From conductive polymers to sensors. *Nanotechnol. Precis. Eng.* **2021**, *4*, 045004. [CrossRef]
103. Alves, L.S.M.; das Neves, M.F.F.; Benatto, L.; Ramos, M.K.; Eising, M.; de Oliveira, C.K.B.Q.M.; Zarbin, A.J.; Roman, L.S. Influence of nanostructuring sensors based on graphene oxide and PEDOT:PSS for methanol detection. *IEEE Sens. J.* **2022**, *23*, 1845–1853. [CrossRef]
104. Farea, M.O.; Alhadlaq, H.A.; Alaizeri, Z.M.; Ahmed, A.A.; Sallam, M.O.; Ahamed, M. High performance of carbon monoxide gas sensor based on a novel PEDOT:PSS/PPA nanocomposite. *ACS Omega* **2022**, *7*, 22492–22499. [CrossRef] [PubMed]
105. Feng, Q.H.; Zhang, H.H.; Shi, Y.B.; Yu, X.Y.; Lan, G.D. Preparation and gas sensing properties of PANi/$SnO_2$ hybrid material. *Polymers* **2021**, *13*, 1360. [CrossRef]
106. Selvanayakam, S.; Esakkidurai, S.P.; Kalaiyar, S. Conductivity-based gas sensors using tamarindus indica polysaccharide-capped gold nanoparticles for the detection of volatile gases. *ACS Omega* **2024**, *9*, 10640–10649. [CrossRef]
107. Kang, H.K.; Byeon, J.H.; Hwang, H.J.; Jang, Y.H.; Kim, J.Y. Flexible sensor film based on rod-shaped SWCNT-polypyrrole nanocomposite for acetone gas detection. *Polymers* **2023**, *15*, 3416. [CrossRef]
108. Boonthum, D.; Oopathump, C.; Fuengfung, S.; Phunudom, P.; Thaibunnak, A.; Juntong, N.; Rungruang, S.; Pakdee, U. Screen-printing of functionalized MWCNT-PEDOT:PSS based solutions on bendable substrate for ammonia gas sensing. *Micromachines* **2022**, *13*, 462. [CrossRef]
109. Zhu, C.H.; Zhou, T.T.; Xia, H.; Zhang, T. Flexible room-temperature ammonia gas sensors based on PANi-MWCNTs/PDMS film for breathing analysis and food safety. *Nanomaterials* **2023**, *13*, 1158. [CrossRef]
110. Park, C.S.; Kim, D.Y.; Jung, E.Y.; Jang, H.J.; Bae, G.T.; Kim, J.Y.; Shin, B.J.; Lee, H.K.; Tae, H.S. Ultrafast room temperature synthesis of porous polythiophene via atmospheric pressure plasma polymerization technique and its application to $NO_2$ gas sensors. *Polymers* **2021**, *13*, 1783. [CrossRef]
111. Tang, B.L.; Shi, Y.B.; Liu, J.J.; Zheng, C.D.; Zhao, K.; Zhang, J.H.; Feng, Q.H. Low-drift $NO_2$ sensor based on polyaniline/black phosphorus composites at room temperature. *Chemosensors* **2024**, *12*, 181. [CrossRef]
112. Bonyani, M.; Zebarjad, S.M.; Janghorban, K.; Kim, J.Y.; Kim, H.W. Au-decorated polyaniline-ZnO electrospun composite nanofiber gas sensors with enhanced response to $NO_2$ gas. *Chemosensors* **2022**, *10*, 388. [CrossRef]
113. Sun, B.C.; Xu, G.B.; Guan, C.H.; Ji, X.; Yang, Z.H.; Chen, S.R.; Chen, X.; Ma, Y.M.; Yu, Y.Q.; Feng, J.G. Nacre-inspired hierarchical bionic substrate for enhanced thermal and mechanical stability in flexible applications. *Sens. Actuators A Phys.* **2024**, *378*, 115832. [CrossRef]
114. Peng, X.Y.; Zhang, X.Q.; Wang, R.; Chen, Y.K.; Chu, X.M.; Kong, L.; Yan, X.; Kuang, M.X. Printing of carbon nanotube-based temperature and bending sensors for high-temperature-resistant intelligent textiles. *ACS Appl. Electron. Mater.* **2022**, *4*, 1949–1957. [CrossRef]
115. Chen, Y.; Pu, X.; Xu, X.; Shi, M.; Li, H.J.; Wang, D. PET/ZnO@MXene-based flexible fabrics with dual piezoelectric functions of compression and tension. *Sensors* **2022**, *23*, 91. [CrossRef] [PubMed]
116. He, Y.; Guo, J.P.; Bi, C.C.; Hao, Z.F.; Li, C.P.; Wang, Q.; Han, X.G. Sandwich-like polyimide nanofiber membrane of PEO-based solid-state electrolytes to promote mechanical properties and security for lithium metal batteries. *Int. J. Hydrogen Energy* **2025**, *109*, 1266–1273. [CrossRef]
117. Chen, X.J.; Gao, X.X.; Nomoto, A.; Shi, K.; Lin, M.Y.; Hu, H.J.; Gu, Y.; Zhu, Y.Z.; Wu, Z.H.; Chen, X.; et al. Fabric-substrated capacitive biopotential sensors enhanced by dielectric nanoparticles. *Nano Res.* **2021**, *14*, 3248–3252. [CrossRef]
118. Sawada, H.; Borisenko, K.B.; Ohnishi, I.; Jimbo, Y.; Okunishi, E.; Kirkland, A.I. STEM and elemental analysis by EDS and EELS for two-dimensional atomic structure containing Au and Cu. *Microsc. Microanal.* **2019**, *25*, 1776–1777. [CrossRef]
119. Yu, X.T.; Hu, X.Q.; Ma, X.; Chen, G.Y.; Ling, Y.Q.; Jin, T.; Chen, Y.F. Carbon nanotubes for optimizing response stability and speed of acrylate hybrid copolymers humidity sensor. *Polym. Compos.* **2024**, *45*, 10516–10526. [CrossRef]
120. Choi, C.; Schlenker, E.; Ha, H.; Cheong, J.Y.; Hwang, B. Versatile applications of silver nanowire-based electrodes and their impacts. *Micromachines* **2023**, *14*, 562. [CrossRef]
121. Li, L.B.; Tang, X.N.; Wu, B.; Huang, B.Y.; Yuan, K.; Chen, Y.W. Advanced architectures of air electrodes in zinc–air batteries and hydrogen fuel cells. *Adv. Mater.* **2024**, *36*, 2308326. [CrossRef]

122. Tay, Y.S.; Yang, L.; Zhang, H.; Kor, H.B.; Zhang, L.; Liu, H.; Gill, V.; Lambourne, A.; Li, K.H.; Chen, Z.; et al. Ruggedized sensor packaging with advanced die attach and encapsulation material for harsh environment. *Microelectron. Reliab.* **2023**, *150*, 115115. [CrossRef]
123. Maharshi, V.; Ahmad, I.; Agarwal, A.; Mitra, B. Wafer level hermetic bonding and packaging using recrystallized parylene. *J. Micromech. Microeng.* **2022**, *33*, 014004. [CrossRef]
124. Duarah, R.; Aleksic, I.; Milivojevic, D.; Rameshkumar, S.; Nikodinovic-Runic, J.; Padamati, R.B. Development of nystatin-based antifungal, biodegradable polymer composite materials for food packaging via melt processing approach. *J. Appl. Polym. Sci.* **2023**, *140*, e54663. [CrossRef]
125. Safaee, M.M.; Gravely, M.; Roxbury, D. A wearable optical microfibrous biomaterial with encapsulated nanosensors enables wireless monitoring of oxidative stress. *Adv. Funct. Mater.* **2021**, *31*, 2006254. [CrossRef]
126. Da Silva, A.C.; Córdoba de Torresi, S.I. Advances in conducting, biodegradable and biocompatible copolymers for biomedical applications. *Front. Mater.* **2019**, *6*, 98. [CrossRef]
127. Balaburov, E.; Kamaraj, M.; Doyle, S.E.; Ahmadi, Z.; Di Bella, C.; Nisbet, D.R.; Moulton, S.E.; Caballero Aguilar, L.M. Noninvasive biosensing 3D scaffold to monitor degradation: The potential of fluorescent PCL and PLGA for tissue engineering. *J. Appl. Polym. Sci.* **2024**, *141*, e54759. [CrossRef]
128. Cai, X.X.; Wang, X.Q.; Bian, F.P.; Li, J.Y.; Zhou, R.X.; Hu, J.W.; Lin, S.D. Flexible sensor based on conformable, sticky, transparent elastomers for electronic skin. *Chem. Eng. J.* **2024**, *498*, 154934. [CrossRef]
129. Lu, L.J.; Wei, X.D.; Zhang, Y.; Zheng, G.Q.; Dai, K.; Liu, C.T.; Shen, C.Y. A flexible and self-formed sandwich structure strain sensor based on AgNW decorated electrospun fibrous mats with excellent sensing capability and good oxidation inhibition properties. *J. Mater. Chem. C* **2017**, *5*, 7035–7042. [CrossRef]
130. Seok, S. Polymer-based biocompatible packaging for implantable devices: Packaging method, materials, and reliability simulation. *Micromachines* **2021**, *12*, 1020. [CrossRef]
131. Kim, T.G.; Ha, S.R.; Choi, H.; Uh, K.; Kundapur, U.; Park, S.; Lee, C.W.; Lee, S.H.; Kim, J.; Kim, J.M. Polymerizable supramolecular approach to highly conductive PEDOT:PSS patterns. *ACS Appl. Mater. Interfaces* **2017**, *9*, 19231–19237. [CrossRef] [PubMed]
132. Fu, L.; You, S.X.; Li, G.J.; Li, X.X.; Fan, Z.C. Application of semiconductor metal oxide in chemiresistive methane gas sensor: Recent developments and future perspectives. *Molecules* **2023**, *28*, 6710. [CrossRef] [PubMed]
133. Lin, T.T.; Lv, X.; Hu, Z.N.; Xu, A.S.; Feng, C.H. Semiconductor metal oxides as chemoresistive sensors for detecting volatile organic compounds. *Sensors* **2019**, *19*, 233. [CrossRef] [PubMed]
134. Mir, N.A.; Rafiq, A.; Kumar, F.; Singh, V.; Shukla, V. Determinants of broiler chicken meat quality and factors affecting them: A review. *J. Food Sci. Technol.* **2017**, *54*, 2997–3009. [CrossRef]
135. Ruedt, C.; Gibis, M.; Weiss, J. Quantification of surface iridescence in meat products by digital image analysis. *Meat Sci.* **2020**, *163*, 108064. [CrossRef]
136. Andre, R.S.; Facure, M.H.; Mercante, L.A.; Correa, D.S. Electronic nose based on hybrid free-standing nanofibrous mats for meat spoilage monitoring. *Sens. Actuators B Chem.* **2022**, *353*, 131114. [CrossRef]
137. Tripathi, K.M.; Kim, T.; Losic, D.; Tung, T.T. Recent advances in engineered graphene and composites for detection of volatile organic compounds (VOCs) and non-invasive diseases diagnosis. *Carbon* **2016**, *110*, 97–129. [CrossRef]
138. Zhou, T.T.; Zhang, P.; Yu, Z.Z.; Tao, M.; Zhou, D.L.; Yang, B.; Zhang, T. Light-driven, ultra-sensitive and multifunctional ammonia wireless sensing system by plasmonic-functionalized $Nb_2CT_X$ MXenes toward smart agriculture. *Nano Energy* **2023**, *108*, 108216. [CrossRef]
139. Cox, L.A., Jr. Re-assessing human mortality risks attributed to PM2.5-mediated effects of agricultural ammonia. *Environ. Res.* **2023**, *223*, 115311. [CrossRef]
140. Ananda, S.R.; Kumari, L.; MV, M. Studies on room-temperature acetone sensing properties of $ZnCo_2O_4$/PPy and $MnCo_2O_4$/PPy nanocomposites for diabetes diagnosis. *Appl. Phys. A* **2022**, *128*, 669. [CrossRef]
141. Wang, B.; Jian, Y.Y.; Jiang, X.; Liu, T.Q.; Yang, D.Y.; Zhao, Z.H.; Liang, Y.P.; Feng, H.R.; Yao, M.S.; Xie, P.Y.; et al. Controlling response of polyaniline toward humidity by self-assembly fatty acids. *ECS J. Solid State Sci. Technol.* **2022**, *11*, 037001. [CrossRef]
142. Ibanez, J.G.; Rincon, M.E.; Gutierrez-Granados, S.; Chahma, M.; Jaramillo-Quintero, O.A.; Frontana-Uribe, B.A. Conducting polymers in the fields of energy, environmental remediation, and chemical-chiral sensors. *Chem. Rev.* **2018**, *118*, 4731–4816. [CrossRef] [PubMed]
143. Sezen-Edmonds, M.; Yeh, Y.W.; Yao, N.; Loo, Y.L. Humidity and strain rate determine the extent of phase shift in the piezoresistive response of PEDOT: PSS. *ACS Appl. Mater. Interfaces* **2019**, *11*, 16888–16895. [CrossRef]
144. Liu, X.H.; Zheng, W.; Kumar, R.; Kumar, M.; Zhang, J. Conducting polymer-based nanostructures for gas sensors. *Coord. Chem. Rev.* **2022**, *462*, 214517. [CrossRef]
145. Youssef, A.M.; Hasanin, M.S.; Abd El-Aziz, M.E.; Turky, G.M. Conducting chitosan/hydroxylethyl cellulose/polyaniline bionanocomposites hydrogel based on graphene oxide doped with Ag-NPs. *Int. J. Biol. Macromol.* **2020**, *167*, 1435–1444. [CrossRef] [PubMed]

146. Tran, N.P.; Okahisa, Y.; Okubayashi, S. Chitosan aerogel containing mechanically fibrillated fibroin and its model test for wound dressing. *Polym. Eng. Sci.* **2025**, *65*, 1340–1349. [CrossRef]
147. Humpolíček, P.; Kašpárková, V.; Pacherník, J.; Stejskal, J.; Bober, P.; Capáková, Z.; Radaszkiewicz, K.A.; Junkar, I.; Lehocký, M. The biocompatibility of polyaniline and polypyrrole: A comparative study of their cytotoxicity, embryotoxicity and impurity profile. *Biomater. Adv.* **2018**, *91*, 303–310. [CrossRef]
148. Li, Y.M.; Zhao, R.; Li, X.; Wang, C.Y.; Bao, H.W.; Wang, S.D.; Fang, J.; Huang, J.Q.; Wang, C. Blood-compatible polyaniline coated electrospun polyurethane fiber scaffolds for enhanced adhesion and proliferation of human umbilical vein endothelial cells. *Fibers Polym.* **2019**, *20*, 250–260. [CrossRef]
149. Sarvari, R.; Massoumi, B.; Zareh, A.; Beygi-Khosrowshahi, Y.; Agbolaghi, S. Porous conductive and biocompatible scaffolds on the basis of polycaprolactone and polythiophene for scaffolding. *Polym. Bull.* **2020**, *77*, 1829–1846. [CrossRef]
150. Molina, B.G.; Cianga, L.; Bendrea, A.D.; Cianga, I.; Alemán, C.; Armelin, E. An amphiphilic, heterografted polythiophene copolymer containing biocompatible/biodegradable side chains for use as an (electro) active surface in biomedical applications. *Polym. Chem.* **2019**, *10*, 5010–5022. [CrossRef]
151. Zhang, Z.; Tian, G.; Duan, X.; Chen, H.L.; Kim Richie, D.H. Nanostructured PEDOT coatings for electrode–neuron integration. *ACS Appl. Bio Mater.* **2021**, *4*, 5556–5565. [CrossRef] [PubMed]
152. Tan, F.; Li, X.R.; Li, X.; Xu, M.X.; Shahzad, K.A.; Hou, L. GelMA/PEDOT:PSS composite conductive hydrogel-based generation and protection of cochlear hair cells through multiple signaling pathways. *Biomolecules* **2024**, *14*, 95. [CrossRef]
153. Gou, S.C.; Li, P.X.; Yang, S.; Bi, G.Q.; Du, Z.H. High-performance MXene/PEDOT-PSS microscale fiber electrodes for neural recording and stimulation. *Adv. Funct. Mater.* **2025**, 2424236. [CrossRef]
154. Thangaraju, P.; Varthya, S.B. ISO 10993: Biological evaluation of medical devices. In *Medical Device Guidelines and Regulations Handbook*; Springer International Publishing: Cham, Switzerland, 2022; pp. 163–187.
155. Aronson, J.K.; Heneghan, C.; Ferner, R.E. Medical devices: Definition, classification, and regulatory implications. *Drug Saf.* **2020**, *43*, 83–93. [CrossRef] [PubMed]
156. Qu, F.D.; Zhang, S.D.; Huang, C.Z.; Guo, X.Y.; Zhu, Y.; Thomas, T.; Guo, H.C.; Attfield, J.P.; Yang, M.H. Surface functionalized sensors for humidity-independent gas detection. *Angew. Chem. Int. Ed.* **2021**, *133*, 6635–6640. [CrossRef]
157. Cai, Y.; Liang, B.; Chen, S.D.; Zhu, Q.; Tu, T.T.; Wu, K.; Cao, Q.P.; Fang, L.; Liang, X.; Ye, X.S. One-step modification of nano-polyaniline/glucose oxidase on double-side printed flexible electrode for continuous glucose monitoring: Characterization, cytotoxicity evaluation ametal oxidemetal oxidemetal oxidend in vivo experiment. *Biosens. Bioelectron.* **2020**, *165*, 112408. [CrossRef]
158. Zhang, Y.Q.; Xuan, W.F.; Chen, S.; Yang, M.N.; Xing, H.K. The screening and diagnosis technologies toward pneumoconiosis: From imaging analysis to e-noses. *Chemosensors* **2025**, *13*, 102. [CrossRef]

MDPI AG
Grosspeteranlage 5
4052 Basel
Switzerland
Tel.: +41 61 683 77 34

*Sensors* Editorial Office
E-mail: sensors@mdpi.com
www.mdpi.com/journal/sensors

www.ingramcontent.com/pod-product-compliance
Lightning Source LLC
Chambersburg PA
CBHW061807210926
51850CB00006B/631